GOD & THE UNIVERSE
LIFE & DEATH

A SEARCH FOR MEANING

GOD & THE UNIVERSE

LIFE & DEATH

A SEARCH FOR MEANING

What is the meaning of the existence of an incredibly complex universe of radiant energy which metamorphoses into biological organisms housing mind and self-awareness, able to construct exquisite abstractions--and wonder about the meaning of everything?

Paul A. Barron

Publication Division
Suburban Printing Services, Inc.

SECOND PRINTING

Publication and Printing: Publication and First Printing, September 1996, Alexandria, Virginia. Second Printing, with corrections-revisions, April 1997, by :

Publication Division--Suburban Printing Service Inc.
5380A Eisenhower Ave., Alexandria, Va. 22304

For Further Information:

Publication Division--Suburban Printing Service Inc.
5380A Eisenhower Ave., Alexandria, Va. 22304
Tel: (703) 370-5800
Fax: (703) 751-8637

Library of Congress Catalog Card Number 96-086265

ISBN 0-9654266-0-2 (pbk)

ISBN 0-9654266-1-0 (hdbk)

ONE-PAGE BOOK DESCRIPTION

God & the Universe, Life & Death

A Search for Meaning

The Search for Meaning: Human life searches for meaning, wonders about God, about the universe, and the significance of human life, the meaning of death, and asks *the root question in all of metaphysics*: Why is there *something* rather than nothing? Does God exist? Did the universe result from act of God or from natural forces? Or are the two the same?

Is the universe eternal or will it some day end? Does the universe have purpose and what is our role in it ? Is God transcendent--outside of space and time? Is God immanent, reachable within us?

God and Cosmology: With the overriding theme of Meaning, the book addresses meaning as it relates to: cosmology, the origin and nature of the universe, the Big Bang, relativity theory, quantum uncertainty and quantum reality, unification & superstring theory, parallel universes, antimatter. It treats of the shifting relationships between science and theology, of religiosity, of God as essential to meaning, differing perceptions of God, definitions and proofs of God, including 'scientific' proofs.

Meaning of Life: The book addresses the imperative of man's search for meaning, the role of religion, the felicitous yet perverse nature of the universe. In the process, it provides a wealth of philosophical insights on meaning as relating to God, the universe, life, and death. It discusses the emergence of life, consciousness, and the remarkable quantum phenomenon we call the human mind. It also explores the *anthropic principle* involving the constants of physics which, by remarkable coincidence, are precisely those needed for the emergence of life and mind.

Meaning in Death: The book discusses meaning in and of death; death and the nature of the universe; consciousness and death; the nature of mind as a survivable phenomenon; afterlife and immortality; death and views of salvation; what kind of afterlife; Near Death Experiences; parapsychic aspects of death; death as relating to time & quantum theory; grief and coping with death.

Free Will and Meaning: Because so much of what is meaningful is traceable to whether there is true free will or choice, down through the years, science and philosophy have prioritized this problem in the pursuit of meaning. The book explores whether all that happens is predetermined, as St. Augustine declared, or governed by random chance; whether we have true free will; and what physicists have to say about the physics of free will as in relativity theory and quantum theory.

DEDICATED TO:

Wife Ruth,
For her patience, encouragement
and assurance that the imperfect world
has an explanation.

Mother,
Who knew God.

TABLE OF CONTENTS

> Theology is but the science of mind applied to God....Truth is everlasting but our ideas of truth are not. Theology is but our ideas of truth classified and arranged....Theologies are well in their place but repentance and love must come before all other experiences. --H. W. Beecher

PREFACE & FOREWORD

The Human Quest for Meaning: The human quest for meaning is a major driving force in human life. Critical to finding meaningfulness is some coherent and underlying framework of knowledge about the universe and God, about life and death.

This work seeks to provide a compendium of challenging classical and contemporary thought as a foundation for finding meaning, for identifying redeeming aspects of our existence, and comprehending the scheme of things from which may be judged one's place and purpose in the world, and in the society in which we find ourselves.

The Search for Rationality: We wonder about existence in general, and our personal existence in particular, uncertain or ambivalent about its rational nature. Access to essential knowledge from which to grasp our reality is critical.

This base of knowledge must address fundamentals such as: the origin and nature of the universe, the emergence of life and mind, the nature of consciousness and self-awareness, the source of morality and abstractions, the subtle influences of quantum theory, and the existence of something which we call 'God.'

Source of General Edification and Instruction: With its intellectual, scientific, philosophical-theological content, this book is addressed first to the public-at-large, to all who search for meaning. Apart from its important function as a reference source for concepts and sources of meaning, it should be especially useful for devotees of philosophy-theology and cosmology.

Some Questions--The Role of Mind: What is the meaning of an incredibly complex universe of radiant energy which metamorphoses into biological organisms housing mind and self-awareness, able to construct exquisite abstractions--and wonder about the meaning of everything?

Is our existence and sequence of experienced events predestined, or just happenstance, the product of random chance? And what is the role of the human mind in constructing reality? Are we, as some scientists suggest, part of some computer-generated *virtual* reality? Why is the human mind somehow tuned to the physical and quantum structures of the universe. Is everything traceable, as some physicists suggest, to the mathematical underpinning of the universe? And why does the human mind resonate with this mathematical substrate?

Increased Blurring of Science and Theology: Questions about the meaning of the universe and existence of human life, in the past century, were addressed primarily by philosophy-theology. Increasingly, though, it is addressed by science as the separation between theology and science has become blurred. Scientific explanations are emerging for phenomena earlier considered matters of philosophy or theology. Of particular relevance is the emergence of 'quantum-genesis' theories, suggesting explanations of what was *before* the creation of the universe, before the Big Bang--a subject once the domain of theology.

Meaning and Death: Critical to seeking the meaning of life is understanding and directly confronting the phenomenon of death, the *reality framework* within which death occurs, and how to anticipate and cope with its devastating effects. Deep in the human psyche is a need for some justification for our hopes and beliefs. We need understanding, to undergird our persistence, to rationalize and accept the present, and to build confidence in projecting into the future.

Origin of the Universe--Stranded without Meaning? The meaning of the universe and life is naturally influenced by one's view of the *origin* of the universe. We have a tendency to view 'meaning' in the context of the existence of a personally-involved Creator who provides meaning via the implications of a Divine Creation, and through biblical revelation.

But if, as 'quantum cosmology' suggests, the universe was self-generated--the chance product of primordial pre-existing quantumuncertainty--then, as astronomer and Congregational minister Hugh Ross declares, in his *The Creator and the Cosmos*, humankind may be on a trip to nowhere, stranded in the abyss of meaninglessness.

Sensibilities About God: As philosophers have observed, theists and non-theists, believers and skeptics, philosophize about God's existence and nature, often arriving at opposite conclusions. Reviewer reactions to this book varied from suggesting a pro-personal God bias on my part, to displaying an absence of religious conviction. Also challenged was the suggestion of metaphoric equivalence of Big Bang or quantum creation to a Divine Creation.

My intention was simply to explore the world's wisdom without a preconceived direction. For those who have unwavering faith, nothing here will dissuade; it may reinforce conviction. However, it may well influence those for whom doubt lingers, perhaps toward belief. For those who are convinced *nontheists*, be careful. The

"Concluding Thoughts" Chapter gives expression to some of my own personal impressions, not so much about God's existence as about the meaning of the term.

Punctuation Protocol: Where quoted material is italicized, quotation marks that might be placed around the italicized words are generally omitted, except where dictated by some special purpose. Where the Chicago Manual of Style might suggest or permit double quotation marks, single quotation marks may be used here to reduce 'busyness' appearance, except where needed to signify the cited author's precise word usage. Some hyphenation is functional rather than dictionary-dictated, to optimize word and letter spacing. Where appropriate, certain words are not hyphenated, such as spacetime, superforce, supergravity.

Table of Contents: Since an Index is not included in this printing, the Table of Contents has been expanded to show subject sub-categories set out in certain chapters. Some index-type information is also contained in the "Acknowledgements."

Captions: To hasten reading and comprehension, 'speed-reading' and reading enhancement, advocates suggest first scoping the book by examining the table of contents, the foreword and introduction, and then skimming the chapters for key thoughts. In this book, subjects and ideas have been carefully captioned throughout the book to promote easy scanning, perusal and retention.

Acknowledgments: The Acknowledgments-Bibliography section pays "Special Tribute" to the many authors whose thoughts have influenced the ideas expressed here. For those who find useful ideas expressed here, those books are recommended as must reading.

Paul A. Barron
Boca Raton, Florida, 1997

> Two books in which the actions of God are recorded, the Bible--theology, and the 'Book of Nature' --Natural Theology. These are two windows from which we can view the world, two human enterprises, sharing many common features.
>
> Freeman Dyson

PART I---INTRODUCTION & PROLOGUE
CHAPTER 1
INTRODUCTION--MEANING AND PHILOSOPHY

DESCRIPTION OF 'PARTS'

For convenience, the material is divided into sixteen 'Parts' as follows:

Part I, Introduction and Prologue, provides insight into the philosophic under-pinnings of the search for meaning and goes on to describe the scope and subject matter of the book with the overriding theme of *meaning,* as it relates to God, the universe, life and death.

Part II, Meaning and Science, discusses meaning in the context both of science literacy and intellectual literacy.

Part III, IV, V and VI deal with cosmology, quantum, and unification theories, the mystique of relativity theory, the mystery of time, the Big Bang, the breakdown of the forces, black holes, pulsars, gravity waves, superstrings, and parallel universes, antimatter, and the creation theories. They provide classical thoughts on *how cosmology is an integral aspect of meaning.*

It reviews some Stephen Hawking thoughts on cosmology and gravity effects, Big Bang physics, including recent confirmations of Big Bang assumptions. It also reviews alternative creation theories, especially Hawking's quantum cosmology in which quantum mechanics fuzzes out a distinct time boundary or beginning of the universe. Also discussed is whether the universe is open or closed and the end-of-the-universe scenarios.

Part VII, The Universe As Mathematics, discusses mathematical underpinning of the universe, and mathematics topics such as the implications of mind being able to tune in on the universe's mathematical resonances.

Part VIII, Cosmic Purpose, explores purpose and meaning in cosmic terms and the efficacy of the anthropic principle addressing the extraordinary coincidences in the constants of physics being precisely those necessary to promote the emergence of life, mind, and consciousness.

Part IX, God and Science, Part X, Defining God, and XI, Proving the Existence of God, reviews the science-theology border and tensions, the emergence of religiosity in societies, God as necessary for meaning, changing notions of God, human life's efforts to define God, finite aspects of God-- limitations on omnipotence, and so-called arguments for and proofs of God, including so-called scientific proofs invoking the anthropic principle and Omega Point theory, God as requiring no proofs, and arguments against God's existence.

Part XII, Life, provides classical philosophical thoughts on life. It discusses the nature and origins of life; what distinguishes life and non-life; consciousness, artificial intelligence, the human brain; mind over matter--mind's influence over illness; sanctity of life--right to life--right to death; where life comes from; genes--the blueprint for reproduction; human life's destiny in the cosmos and mind as destined to alter the universe.

Part XIII, Search for Meaning in Life, is devoted to identifying and discussing meaning in life. It provides random classical thoughts on meaning in life, discusses the search for meaning in life in terms of: religion and meaning, the multiplicity of meanings, other thoughts on meaning, the endless search for meaning, why bad things happen to good people, loneliness and meaning, and risking one's life for another.

Part XIV, Death and Meaning, provides random classical thoughts on death and discusses: finding meaning in death; death and the nature of the universe; consciousness and death; death--afterlife and immortality; death and salvation, what kind of afterlife; near-death experiences, parapsychic aspects of death; time, quantum theory, and death; and grief and acceptance of death.

Part XV, Predetermination and Free Will, provides random classical thoughts on the subject of predetermination and the existence of free will. It discusses the views of philosophers and scientists on the problem of whether our universe permits us true free will. It also touches on notions of mathematician Frank Tipler and physicist Roger Penrose on free will in terms of the physics of relativity and quantum theory, as expressed in his *The Physics of Immortality.*

Part XVI, Concluding Thoughts, contains a recap of the theme of the book--and some final thoughts about the future, God, philosophy, and physics, and emerging concepts affecting our perceptions of God. There is also an "Epilogue" with the rationale for prayer to the mystical and unprovable God that defies human understanding, especially where faith is elusive.

MEANING AND PHILOSOPHY

The Search for Meaning as the Practice of Philosophy: In *Basic Teachings of the Great Philosophers*, S.E. Frost notes that everyone, regardless of vocation or station in life, being human and endowed with a highly-developed brain, must think--and in so doing is following the "pathway to philosophy." Even denial of interest in philosophy is expressing philosophy. This is so because 'philosophy' is the amalgam of all components of meaning which the world presents to us.

Fundamental Philosophic Questions in the Search for Meaning: But what are the fundamental philosophic problems confronting us which the great philosophers have identified through the ages? There are varying responses, of course, but the most cited are Immanuel Kant's three: (1) Does God Exist? (2) Is there life after death? and (3) Do we have free will? The reason suggested for Kant's limiting these to three is that others are either *subsumed* in these three, or seen as of secondary significance.

But various historians of philosophy have listed other various groupings of philosophic questions. In *Basic Teachings of the Great Philosophers*, S.E. Frost summarizes philosophic concepts and expands the Kant's three into the following ten question-categories (paraphrased):

> (1) What is the nature of the universe? (2) What is human life's place in the universe? (3) What is good and what is evil? (4) What is the nature of God? (5)Are all occurrences fated, that is, laid out in time and so predestined, and the related question, Do we have free will? (6) Do we have a soul and is it immortal? (7) What is the best form of state and what is the source of the state's authority? (8) What is education and its purpose? (9) Is mind another form of matter or separate, and is its fate tied to matter? (10) What is the source of our ideas and the laws of thinking?

Questions Addressed: In one way or another, all these have been addressed in this book. Another commonly-cited question is whether language is adequate to communicate knowledge and human thought. Still another question relates to the extraordinary manner in which the underlying *structure of the universe can be expressed mathematically*--whether the universe *is* mathematics, as asserted by P.W. Atkins in *Creation Revisited.* What is the meaning of existence--life and death--if, as some scientists suggest, we are part of some computer-generated *virtual* reality?

All Knowledge Empirical? All Metaphysics Unauthoritative? Still another philosophic enigma is addressed by the 'logical positivists,' like Ludwig Wittgenstein and the 'Vienna Circle' philosophic group, suggesting that all 'knowledge' must be empirical, derived from sense experience, otherwise it is groundless. Consequently, in this view, metaphysical notions--theology, philosophy, ethics--are described as 'nonsense,' used in the sense of something less than authoritative.

By definition, of course, the concept of metaphysics as nonsense also *undermines all philosophy*--including these assertions of nonsense. It should be pointed out, however, that physicists like Paul Davies and Frank Tipler demonstrate that much of what earlier was *metaphysical*, has been found to be founded *in physics*, including such things as the nature of space and time, dimensionality, and even the origin of the universe, invoking 'quantum cosmology.'

We Stand on the Shoulders of the Great Philosophers: Getting back to Frost, he says that the history of philosophy is the story of how different philosophers have created different patterns, proposed different solutions to the "puzzle that is human experience." We are, he says, at the apex of centuries of struggle with the problems of human kind, *standing on the shoulders of the great philosophers* of the past, who say to us, in effect: This is what the universe of human experience has meant to us. Start your own thinking from here.

The Universal Quest for Meaning-- 'Real Philosophy': In *Real Philosophy, An Anthology of the Universal Search for Meaning*, philosophy professors Jacob Needleman and David Applebaum undertake a survey on the "universal search for meaning," which they describe as "real philosophy," and "the love of wisdom."

Their "real philosophy" is not limited to Western thought, but includes 'readings,' excerpts from the spiritual traditions of the Eastern spiritual wisdom, all contributing to what they describe as the goal of philosophy, which is nicely summed up in a single word--*guidance*.

The Existential Imperative: Cannot Live Without Meaning: The underlying fundamental drive "within the human psyche," they conclude, has its source not so much from biological or social impulse as from *inborn thirst for meaning*. They note famous death-camp survivor and psychiatrist Viktor Frankl's conclusion, in *Man's Search for Meaning*, that the deepest driving force within every human being is the *existential imperative, the need for meaning;* that no one can live without meaning--that which distinguishes the human species from other life of which we are aware, and *defines our humanity*.

Material and Science Progress Not Sufficient for Meaning: Needleman and Applebaum observe the insufficiency of material and scientific progress alone to provide an understanding of the meaning of our lives. They recall how down through the years philosophers have struggled with questions of human identity, purpose and ethics. They recognize the trend of society turning to other sources for meaning, such as religion, psychiatry, political ideology, and mystical teachings, emerging in older religions and spawning new religions.

Foundations of All Human Activity Being Reevaluated: The authors see this "real philosophy" as a "separate force" in human life that combines an inner search (spiritual) and an outer discovery process (science), covering a middle-ground between intimate self-questioning and ordinary interaction with others.

They see a trend in the physical sciences which sometimes blends Eastern thought on the unity of all things with the 'new physics.' Included are new views of health and illness which address the whole person, invoking 'holistic' medicine. So also are there new concepts of consciousness and the quantum. And so, they suggest, the *intellectual foundation of all human activity is being reevaluated.*

I simply believe that some part of the human Self or Soul is not subject to the laws of space and time. --Carl Jung

CHAPTER 2
PREVIEWING THE BOOK

Meaning--God, the Universe, Life and Death: This book explores a number of fundamental questions in metaphysics and physics, addressing our preoccupation with *meaning*, both in the work-day world of daily life and experience and in the broader context of purpose and meaning in cosmic terms.

The chapter provides insight into the considerable scope and subject matter of the book, with the overriding theme of 'meaning' as it relates to what we are accustomed to include in our notions of existence: the universe, life, death, and for many, the most important, 'God.' Necessarily, it is steeped in philosophy, theology, quantum and classical physics, cosmology, both conventional and quantum.

The Search for Meaning: Human life searches for meaning, wonders about God, about the universe, and the reason for life, especially human life, and, upon reflection, asks *the root question in all of metaphysics:* Why is there *something* rather than *nothing*?

Did the universe result from act of God or from natural forces? Or are the two the same? Is the universe eternal or will it some day end? Does the universe have 'purpose' or is it random chance--and what is our role in either case? Does God 'exist' and is his existence *transcendent*--apart from the universe and outside of space and time? Is God *immanent*, reachable within us? Is our universe all there is or are there other universes and how might they affect us?

Are We Alone in the Universe? The possible existence of other life in the universe has been the subject of philosophic speculation for centuries. Given the universe's age and likely existence of planets similar to Earth, the scientific assumptions favor multiple emergence of intelligent beings throughout the universe's vastness . Programs like the NASA SETI or search for extraterrestrial intelligence program have sought to home in on likely wavelength transmissions.

Early in this century it was thought that Mars, in astronomical terms, next door, so-to-speak, had *canals*, giving rise to exciting speculations about an existing or once-existing Martian civilization. After the 1976 Viking Orbiter data disclosed no canals on Mars, thoughts of other advanced life in our own solar system faded, but perhaps there is possible microbial life on moons of the larger outer planets of Jupiter and Saturn. The 1996 finding of a lunar south pole ice field will facilitate moon living.

Life on Mars? In *Monuments of Mars*, Richard Hoagland, space science and exploration consultant, speaks of the startling *Mars Face*, part of the Martian geology photographed in 1976 by the Viking Orbiter satellite, in the Cydonia region, a collection of pyramid and sphinx-like geological formations. It contains a conglomeration of lines and angles similar to those of the structures in Egypt, and Inca and Mayan structures in Central America.

Martian Life-Suggesting Rock on Earth: Publicized in August, 1996, a small rock found in Antarctica in 1984 is theorized to have been hurled into space from an ancient asteroid strike on Mars to fall to earth some 13,000 years ago. The rock's mineral & organic residues, and structures implying bacteria, suggest the rock as formed some 3.6 billion years ago when Mars was conducive to organisms.

Earlier, thoughts of life elsewhere in the cosmos took backseat, despite the searches for extraterrestrial messages such as the NASA (SETI) Search for Extraterrestrial Intelligence project. Although that program was discontinued in 1993, there are similar American and other national programs, such as the privately-funded Harvard-Smithsonian radio telescope Beta (Billion channel Extraterrestrial Assay) project. NASA has an 'Origins' program to search for life-supporting planets.

Finding Other Planets--Sharing the Universe: In *Are We Alone,* and in a February, 1996 Time article, physicist Paul Davies highlights the proposition that humankind is on the threshold of a truly revolutionary revision of our most fundamental concepts on science, religion, and *the entire structure of our basic belief systems.*

That revolution is signaled by recent confirmation of what, for decades, had previously been merely assumed, namely, *the existence of planets around other stars.* This has renewed earlier speculations that somewhere in the vastness of the larger cosmos, other forms of intelligence *must* share our universe. The stars involved are 47 Ursae Majoris in the Big Dipper, 200 trillion miles from Earth, and star 70 Virginis, in the Virgo constellation. Both *permit water*, a prerequisite for life.

Though the size and composition of these particular planets are unlikely to support human life, some atmospheric form of biological life is possible and these planets may have smaller, more life-prone Earth-like moon-planets. The point is that confirmation of life elsewhere in the universe will forever alter our anthropocentric model of Earth and humankind's special status-- and our philosophy and theology.

The Competing Paradigms of Decay and Progress: Davies speaks of the "rival paradigms" of the two parallel processes that characterize the universe. At the same time that the universe proceeds inexorably toward ultimate chaos through the thermodynamic process of the decay of high order, a counter drive works to build pockets of high order and complexity, emerging spontaneously through processes of self-organization, starting with galaxies, stars and planets.

The Drive toward Complexity and the Spread of Intelligence: This countervailing drive toward pockets of complexity goes on to organic materials, then life and mind, and now the myriad of complexities introduced by human intelligence and human life's exponentially developing technology. This is dramatically illustrated by human life's development of artificial intelligence which, as physicists Freeman Dyson and Frank Tipler suggest, provides a framework for the spread of intelligence throughout the Milky Way galaxy and eventually the entire universe.

Directionality, Mind, and the *Omega Point God:* Of profound significance, perhaps, is to recognize that this mind-generating complexity-building process evidences *directionality*, a proclivity somehow implicit in the design of the cosmos, reflecting an evolutionary trend toward life, mind and consciousness, and implying that *mind* underlies the universe. This spread of intelligence or mind goes on until, perhaps, as Frank Tipler suggests in *The Physics of Immortality*, it reaches the *Omega Point*, which he defines as representing God of the future who, because of the relativistic and quantum coexistence of all time, also *exists now*.

Life--A Natural Process or Divine Intervention? But getting back to extraterrestrial life, the possible, if not likely, existence of other highly intelligent civilizations, of course, seriously undermines the anthroprocentric notion of our divine uniqueness that makes mankind the central focus of God's creation. This is especially so if, as speculated, we turn out to be one of the more primitive of the life forms which have evolved. Still unclear, though, is whether this 'directionality' toward life and mind is a purely natural phenomenon or involves some kind of external *intervention*--or whether the two are the same.

Pessimism and Optimism about Our Presence in the Universe: Davies cites Bertrand Russell's famous depressing pessimism about the futility of "all the noon-day brightness of human genius," given a universe destined to end in ultimate chaos, and Nobel-prize-winning biologist Jacques Monod's equally depressing reference to human life as "alone in the unfeeling immensity of the universe" from which it has "emerged only by chance."

But, he asks, what if there is "systematic progress alongside decay?" For those seeking a "deeper meaning or purpose" behind our presence here, he observes, the presence of other life forms in the universe would impart "a spectacular boost, implying that we live in a universe that is in some sense getting better and better rather than worse and worse."

God and Cosmology: The book addresses meaning as it relates to cosmology, the origin and nature of the universe, including such subjects as the Big Bang, relativity theory, quantum theory and quantum reality, the universe as a computer, parallel universes, and superstring & unification theories. It also deals with meaning in terms of shifting relationships between science and theology, of religiosity, of God as essential to meaning, and of changing approaches to the meaning of the term 'God.'

Significance of New Proofs of the Big Bang: In the Creator and the Cosmos, astronomer-minister Hugh Ross argues, not unpersuasively, that the existence of God is demonstrated by the new supporting evidence for the Big Bang, like the NASA COBE satellite findings of primordial ripples in spacetime; by the extraordinary anthropic coincidences of the life-promoting constants of physics; and by certain compatible or corroborative statements in the Bible.

Prevailing View of the Nature of the Universe? If there is a prevailing scientific view of the universe it probably favors a quantum-chance-created universe, devoid of purpose, although allowing for human initiative in constructing purpose and meaning.

Some physicists like Paul Davies, see a purposeful and designed universe, based on the anthropic principle involving the remarkable life-supporting coincidences in the amalgam of the various providential strengths of the physical constants and other aspects of the laws of physics. Davies also speaks of the infinite regression of causation resting in logical necessity--*which some may call God.*

Others, like Nobel-Laureate Steven Weinberg and Oxford chemistry professor P.W. Atkins, see a "pointless" or "purposeless" universe, the product of laws of physics and their mathematical foundations, although physicists like Steven Hawking provocatively ask what "breathes fire into the equations" to bring mathematics to a physical reality.

Other Profound Consequences of the Big Bang: But verification of the Big Bang has other profound implications, not the least of which is that all present diversity, divergence, disharmony, conflict, irreconcilabilities, both human and throughout nature, derive from an original cosmic perfect unity. All of our remarkable diversity and discord derives from a single exponentially small *quantum* particle representing the perfect symmetry of unified physics reflected in the superforce and the *relativistic singularity* from which our present spacetime and dimensionality unfolded.

Human Life Finding New Meanings: In *Cosmic Dawn*, Johns Hopkins astrophysicist Eric Chaisson points to the fact of *human life finding new meanings*, a "relevance to cosmic evolution, an underlying motive for universal change." Through the life and death of billions of stars, new forms of matter generate life, *a kind of cosmic reincarnation* making us more than happenstance products of the unfolding process.

He declares that we are "agents of the universe commissioned by it to probe itself." And together with our galactic neighbors, if they exist, he says, we may eventually gain control of the resources of the universe, redesigning it to suit our purposes and, in effect, *ensuring for our civilization a sense of immortality.*

Meaning of Life and Religion: This book addresses man's imperative search for meaning, the role of religion, the felicitous life-propensity of the universe, and provides a wealth of philosophical insights on meaning as relating to God, the universe, life, and death.

It discusses the emergence of life, consciousness, and the remarkable quantum phenomenon we call the human mind. It also explores the *anthropic principle* involving the constants of physics which, by extraordinary coincidence, are precisely those necessary to assure the emergence of life and mind at some time in the universe's unfolding process.

Death--Most Troubling Aspect of Existence: The book goes on to discuss meaning associated with the phenomenon of death, including: death and the nature of the universe; consciousness and death; the nature of mind as a survivable quantum phenomenon; afterlife and immortality; death and views of salvation; what kind of afterlife; near-death experiences; parapsychic aspects of death; death as relating to time & quantum theory; and particularly, grief and coping with death.

Free Will and Meaning: Because so much of what is meaningful is traceable to whether there is genuine free will and true choice, science and philosophy have highlighted this problem in the treatment of meaning. The book explores the physics of whether all that happens is predetermined, leaving us without true choices, or whether we have true free will, and what physicists have to say about the physics of free will in terms of relativity and quantum theories.

Exploring physics leaves us with a nagging dilemma. While quantum theory implies an underlying *universe of chance*, the physics of time, in terms of theories both of quantum mechanics and classical relativity, signify the coexistence of all time--suggestive of *predestination*.

Exploring the God-Physics Interface--Mind and Mathematics: These materials draw on the work of many physicists, philosophers and theologians who have addressed the relationships between science and theology. In *The Mind of God*, for example, physicist Paul Davies explores the interface between science and theology, discusses quantum cosmology and the anthropic principle and what may have influenced our physics. In the process, he explores ultimate causation, logical necessity, and the mysterious manner in which the human mind resonates with the physical structure of the universe and its mathematical substrate.

Is Theology or Physics the Broader Framework? The exploration of meaning, involving as it does, both science and theology, raises the question of which is the broader frame of reference, science or theology. Although impulse suggests theology as broader, recent developments in physics, especially 'quantum cosmology' have complicated the question--and not to everyone's liking.

At the End, Still Mystery: Paul Davies observes that his work was an effort to "trace the logic of scientific rationality" to answer "the mystery of existence" and *where the chain of reasoning might end*, where each explanation of the universe is found contingent on another. A closed logical scheme explaining everything will fail, he says, and the search inevitably will generate either "an infinite regress," or a mysterious, "closed loop" explanation. At the end, he declares, *mystery will always prevail.*

A Deeper Level of Explanation--God? But there must be a *deeper level of explanation,* he feels. Whether one wishes to call that deeper level, 'God,' he suggests, is concerned with definition and taste. He has come to realize that mind-- conscious awareness of the world, is fundamental, not just a "meaningless quirk of nature."

Humans Wonder about the Universe, God, and the Existence of Life: Human beings wonder about the universe, God, and life's very existence. Did the universe result from act of God or from natural forces--or are the two the same? Is the universe eternal or will it some day end--and how? Down through the ages human life has looked up at the night sky and wondered about the universe, about life, about God. Where did the universe come from--and why are we here?

Preoccupation with Existence: Is there 'universal mind,' as some suggest, and if so, what is its relationship to God. If God created the universe, does He continue to be 'personally involved' as theists suggest, or is the universe now self-governing, as deists propose? And what of human existence--how does one find meaning in that existence? This preoccupation with *existence* has been the object of philosophic thought from the earliest times.

In his *Perfect Symmetry* and *The Cosmic Code*, physicist Heinz Pagels explores the nature of the universe and describes its history in terms of the symmetry-breaking process from an initial state of unified forces. Like physicist Stephen Hawking in *A Brief History Of Time*, Pagels wonders why *something* exists at all rather than *nothing,* the ultimate perfect symmetry.

Existential Concerns--The Root Question of All Metaphysics: From human life's reflections on the subject of 'existence' have come the ideas and notions commonly embraced in the school of philosophy known as 'existentialism.' Pagels notes that existentialist philosopher Martin Heidegger views this *preoccupation with existence as representing the root question in all of metaphysics*: Why is there any being at all rather than nothing, a question raised alike by philosophy and science, leading, of course, to questions about the existence and role of God and the meaning of nature.

What Is Our Place in the Universe? In *A Brief History of Time*, Stephen Hawking asks: Where did the universe come from, and where is it going? How and why did it begin? Did the universe have a beginning, and if so, what happened *before* then? He cites St. Augustine in *The City of God* and Immanuel Kant, in the *Critique of Pure Reason*, as dwelling on such questions.

In his Conclusions, Hawking notes that "We want to make sense of what we see around us and to ask, What is the nature of the universe? What is our place in it and where did it and we come from?" These questions go to the heart of science and theology and are best considered in the context of known scientific and theological concepts.

More Questions: What, if anything, existed 'before' the universe--or could there be a 'before' if time was a part of the creation? And how could the primordial Big Bang radiation transform to bring into existence sentient beings able to wonder about God and the universe, purpose and meaning, predestination and free will, and a legion of exquisite abstractions.

Meaning of and in the Universe--Science and Theology: Any discussion of meaning of the existence of the universe and of its extraordinary design, which permits the appearance of consciousness, sentient beings capable of reflecting on abstractions, must address questions which are at the interface between science and theology. To be noted are mellowing contemporary attitudes, reciprocating accommodation, and emerging recognition of the kinship between these once more stridently adversarial fields.

Organizations have emerged dealing specifically with science and theology, such as 'Reasons to Believe,' which regularly publishes articles on new developments in science thought to reinforce the concept of God as the Creator. Its President, Congregational minister-astronomer-cosmologist Hugh Ross, is author of a number of books on the science and theology of creation.

Compactified Dimensions? The Fourth Quarter of 1995 issue of Reasons to Believe, for example, explores "A Super Theory of Superstrings Reveals A Super Creator." The article describes the work of physicists Ed Witten and Nathan Seiberg, reducing an entire field of mathematics to a simplified statement. Involving 'string theory,' it hypothesizes the split of the starting universe of ten dimensions into six 'compactified' dimensions and four dimensions which expanded into our observable three space and one time dimension.

The New Age Culture and Its Challenge: In "The Plain Truth" publication of the Worldwide Church of God for May/June 1996, Keith Stump in his *Sharing Your Faith with a New Ager*, addresses the remarkable growth of "New Age"culture as a philosophy and quasi-religion of sorts, outside of and competing with the conventional Christian religious framework. He goes on to provide a "Crash Course on the New Age," more fully discussed in the chapter on Religion and Meaning.

Do a Better Job of Reconciling Science and Spirituality: He sees this New Age culture as a challenge to conventionality and suggests that if Christianity is to meet the challenge, "churches will have to do a better job of answering questions about God's existence, about the nature of the mind and consciousness, the concept of the spirit, the reality of space and time, the reconciliation of science and spirituality, the compatibility of psychology with Christianity."

Fragility of Life and Threats to Human Life's Existence: In past centuries, plagues have decimated populations, giving rise to fears of the end of human life. Unchecked, diseases, like AIDS or other potential emergent diseases, are suggested as having human-life-destruction potential. Human life's own depredations--setting Kuwait's Oil fields afire, releasing large-scale atomic radiation, destroying the ecological balance through uncontrolled carbon dioxide and methane emissions--may signal the ultimate decline and disappearance of human and perhaps all life.

An Asteroid to End Life on Earth? Periodically, natural threats to earth's habitability have appeared, such as the 1989 FC asteroid, a half-mile diameter 'cosmic jaywalker,' which recently made one of its annual passes at the earth, a mere half-million miles away. It is expected that, drawn by gravity, the asteroid will one day crash either into earth, Mars, or our moon, with the force of 20,000 one-megaton hydrogen bombs, enough to destroy life as we know it. If it destroys our civilization, the human race, what then would have been the purpose or meaning of our existence?

End of the Dinosaur Period: Sixty-six million years ago, during the dinosaur period, a six-mile-wide comet or asteroid, striking at 40,000 mph, may have set in motion disturbances in the planet's climate and ecology which wiped out two-thirds of Earth's species living at that time, including the dinosaurs.

Examination of sedimentary deposits, such as in Montana, suggest the occurrence of asteroid or comet strikes on this planet about every twenty-six million years, resulting in massive dust clouds which alter the ecosystem to wipe out flourishing life species. It is thought that the normal orbit of these bodies in the solar system is disrupted by some recurring event which results in a major gravitational shift toward the planets. Indian physicists suggest another explanation for this life-extinction, massive volcanic eruptions caused by the earth passing through a cloud of "dark matter."

Current Concerns over Vitality of the Planet: There is a current and deep concern over global warming which may be in process through the greenhouse effect caused by carbon-dioxide, methane and other noxious emissions, altering earth's atmosphere so as to resemble that of non-life-supporting Venus, being studied through the NASA Magellan satellite.

Human life-produced chemical emissions threaten the atmosphere's ozone layer which protects human kind from deadly ultraviolet radiation. But human life's ingenuity, if properly marshaled, might find deployable technological solutions to global warming, such as by diminishing solar radiation through suitable screening materials in space--as by using the 'mass drivers' suggested by Princeton University scientist G.K. O'Neill.

Exploring the Path of Cosmic and Biological Development: For reasons we will explore, the laws of physics are such as virtually to assure the presence of intelligent life forms in the universe at large. We know that there is at least one such intelligent life form, on planet Earth, which is driven to explore the universe, to find out how and even why the universe exists and to probe into what may be our role and destiny.

But this life potential, brought to fruition on this particular planet at this time in the unfolding universe, is not guaranteed to persist. Moreover, the life which exists now is fragile and perhaps quite temporary. So it is appropriate that we seek some understanding, and perhaps deeper appreciation, of the science involved in the formation of astrophysical structures and emergence of life.

A Revolution in Human Thought: Many are only dimly aware of the revolution in human thought about the nature of our universe, its origins and mysterious quantum characteristics that require a sweeping revamping of earlier notions of reality derived from classical physics. This is important because of the shift in views of the universe from a predestined, cause-and-effect universe to one governed, at the subatomic level, by quantum uncertainty.

The 'new physics,' revolving around quantum mechanics, implies that a nonmaterial universe of information and mind underlies what we have been accustomed to regard as physical structure, something mysteriously different from the scientific-materialism labels previously applied. Consciousness, self-awareness, and subjectivity, as nonmaterial adjuncts to physical brain structure, have new quantum and metaphysical implications.

Blending Theological and Scientific Explanations for Creation: The Big Bang theory of the start of the universe from a *singularity,* representing a unity of space, time, matter and dimensionality, implied a beginning of time and gave some support for a universe-initiating act of God. But the recent emergence of 'quantum cosmology,' factoring in quantum theory, through the work of physicists like Stephen Hawking and Roger Penrose, contemplates a pre-big-bang 'unstable' quantum state permitting 'natural,' uncertainty-state self-generation of quantum-particle universes, supposedly without supernatural intervention. The subject inevitably generates scientific speculation and theological concern.

The God-Physics Interface: These materials draw on the work of many physicists, philosophers and theologians who have addressed the relationships of science and theology, such as Paul Davies, Professor of mathematical physics, at the University of Adelaid, Australia. In his *The Mind of God* Davies discusses questions about the 'timeless,' 'immutable,' 'perfect' God who creates a 'contingent,' changing and time-related, imperfect, universe.

Davies also explores quantum cosmology, the anthropic principle and what may have influenced our physics, especially so-called 'initial conditions' of the starting universe. Stephen Hawking, for instance, notes the high order 'initial conditions' of this universe as having been extraordinarily unlikely. Far more likely, he feels, would have been chaotic models, raising the question of why our universe was gifted with high order favorable to the formation of life and consciousness.

The material also explores 'reasoning' and 'logic,' and the mysterious manner in which the human mind resonates with the physical structure of the universe and its mathematical substrate. Also discussed are such subjects as mathematician Frank Tipler's treatment, in *The Physics of Immortality*, of his "Omega Point God Theory," and our universe as a possibly computer-simulated quantum reality.

Is Theology or Physics the Broader Framework? The exploration of meaning, involving as it does, both science and theology, raises the question of which is the more fundamental--the broader frame of reference--science or theology? Although intuition suggests theology as broader, recent developments in physics, especially 'quantum cosmology' have complicated the question--and not to everyone's liking. There are, of course, a variety of views on this, as eloquent as they are diverse.

As we shall see, physicists like Paul Davies and mathematician Frank Tipler suggest science as the proper framework. Theologian-scientists like John Polkinghorne and Russell Stannard, and astronomer-minister Hugh Ross, question this conclusion. It is interesting to observe that while theological materials increasingly discuss astronomy, cosmology, and theoretical physics, physicists increasingly discuss God and metaphysics. Once incompatibly-divergent views may be finding some harmony.

Inhibitions and Sensibilities over Science and Theology: Scientists cite Jacques Monod's often-quoted admonition in *Chance and Necessity,* against scientific intrusions into the subject of values or purpose. But many physicists, like Freeman Dyson, Heinz Pagels, Paul Davies, Frank Tipler, Stephen Hawking and others, whose ideas are explored here, find little constraint in Monod's famous *caveat*.

In his *Infinite In All Directions*, Princeton physicist Freeman Dyson, though mindful of Monod's proscription, nevertheless deals at length with metaphysics, purpose, the relationship of Divine and human mind. In a more flagrant challenge, physicist Paul Davies unabashedly devotes entire books, like his *God and the New Physics* and *The Mind of God*, to exploring God-physics relationships.

Discussion of Different Approaches: Physicist Paul Davies, whose book *The Mind of God*, is one of those discussed here, reviews the 'natural' sources of creation, with emphasis on the regression of creation-causation to 'logical necessity,' the 'logic of mathematics,' and 'ultimate good,' all of which one may call 'God,' though ostensibly part of some 'natural' rather than supernatural process.

Also discussed, is theoretical physicist Stephen Hawking's *A Brief History of Time*, in which Hawking offers a quantum cosmology involving creation by quantum self-generation and asks some now-almost legendary questions about physics and 'the Mind of God.' Cited also, is Oxford chemistry professor Peter W. Atkins' *Creation Revisited*, which describes everything that exists as having its source in mathematics, "without outside intervention."

To assure some balance in outlook, also reviewed is astronomer-minister Hugh Ross's *The Creator And The Cosmos,* which rejects quantum cosmology's universe self-generation, in favor of the relativistic Big Bang and the anthropic principle as near mathematical proofs of God as the source of creation and therefore the existence of objective meaning as revealed in the Bible.

Stephen Hawking: Incorporated in this book, along with some of Davies' thoughts, are provocative thoughts of other renowned scientific figures such as mathematician Frank Tipler in his *The Physics of Immortality*, dealing with the emerging God in his Omega Point Theory. It also includes reference to ideas of Oxford chemistry professor P. W. Atkins in his *Creation Revisited,* proposing that mathematics fully accounts for the universe, without, as he puts it, "outside intervention."

Since the world of physics has focussed so much attention on Cambridge University theoretical physicist Stephen Hawking's quantum cosmology and unification theory, outlined in his *A Brief History of Time*, some of his thoughts are explored.

Astronomer-Minister Hugh Ross : Noted here is the work of astronomer-cosmologist and minister Hugh Ross, who, in his *The Creator and the Cosmos*, tracks the Big Bang and recent developments in astrophysics to passages in the New Testament. He argues that the Big Bang, in the context of the anthropic principle, confirms the truth of the God of the Bible and therefore of a universe designed for a purpose--which gives it meaning, revealed in the Bible.

In the quantum cosmology of physicists like Stephen Hawking, a primordial quantum void triggers eruption of a highly-energized spacetime universe-particle. While relativistic physics implies a singularity of zero time and zero radius or volume, quantum uncertainty effects, he suggests, would "smear out" time and space boundaries, leaving no true Creation--with the consequent theological negatives about the role of God. Astronomer-minister and cosmologist Hugh Ross views Hawking's quantum cosmology explaining the start of the universe as essentially speculative metaphysics. There are, of course, many points of view on this central question of primal cause.

What's at the End of the Explanatory Chain? Superlaw? God? Questions about the origin of the universe, comments physicist Paul Davies, in *The Mind of God*, lead to a process of persistent inquiry in pursuit of the fundamental cause. Like Hawking, Davies asks whether the causal chain can stop somewhere, perhaps with God or with some superlaw. Davies continues. If God is the final cause, who or what made God? If the laws of physics are the ultimate source, who or what created them? And, as physicists like Stephen Hawking and Heinz Pagels wondered, why should there be something rather than nothing?

Ultimate Questions As Beyond Science--Beyond Human Understanding: Sooner or later, muses Davies, we have to accept something as given, whether this is God, or logic, or laws, or some other *underpinning* for existence. Are the fundamental questions on existence unanswerable? Instinctively, he believes that we are unlikely to get to the source despite relentless pursuit of rational inquiry.

Must Be a Deeper Level of Explanation--Is This God? But, he insists, there must be a "deeper level" of resolution"which one may wish to call "God" He has come to the point of view that "mind, conscious awareness of the world," is not an insignificant caprice of nature, but "an absolutely fundamental facet of reality."

Meaning without God? Finally, the Concluding Thoughts Chapter revisits the proposition that the search for meaning may end before it starts, unless we deal with the concept the existence and nature of God. The chapter expresses some of the author's reactions to this challenge.

> In spite of the spectacular success of modern science, it would be foolish to suppose that the fundamental questions concerningthe existence of God, the purpose of the universe or the role of mankind in the natural and supernatural scheme has been answered by these advances....[But] the very conceptual framework in which the religious questions are posed can be altered by scientific advances. -- Paul Davies

PART II---MEANING AND SCIENCE

CHAPTER 3
THE EXISTENCE OF MEANING

Human Life Wonders about God, Life, Death, and Meaning: In *A Brief History of Time*, British theoretical physicist Stephen Hawking muses that "We find ourselves in a bewildering world. We want to make sense of what we see around us and to ask: What is the nature of the universe? What is our place in it and where did it and we come from? Why is everything the way it is?"

From time immemorial, human life has wondered about the nature of the universe, about God, and the meaning of life. Why does life begin, struggle to survive and achieve, only to end just when a measure of wisdom and success is achieved or in prospect? How does one rationalize such cynical, compassionless reality ? Is there any such thing as objective meaning, and who or what determines that? Is meaning *universal*, or is it subjective and personal to each of us?

Questioning the Existence of Meaning: Most of us believe that the universe, and therefore life, has some underlying purpose and meaning, and consciously or otherwise, seek that meaning. But there are some, challenging thinkers, who deny the existence of objective meaning in life, and for that matter, question that the universe's existence has anything to do with purpose or meaning.

The great Russian novelist-philosopher Leo Tolstoy asserted that "The meaningless absurdity of life is the only incontestable knowledge accessible to man." Albert Camus, the French existentialist novelist and playwright, observed that "We turn toward God only to obtain the impossible."

British philosopher and logician Bertrand Russell, in *A Free Man Worship*, observed that man is the product of random events; that "the whole temple of Man's achievement must inevitably be buried beneath the debris of a universe in ruins"; that the "soul's habitation" can be safely built "only on the firm foundation of unyielding despair."

We Emerge from Nothingness in Search of Meaning: Each of us emerges into this life, unconsulted, from whence we do not know, to become a new *existent* on a planet called Earth, orbiting a star we call the Sun, in a galaxy we call the Milky Way. Each of us is a separate *awareness* of the universe, a new organism we call human 'life,' assigned some kind of undefined 'life-charter,' the *meaning* of which is not only elusive and vague but changes with time.

Compared with other life forms on this planet, as members of the human species, we come with deluxe equipment, a host of characteristics: individual subjectivity and self-awareness; an intuitive sense of a relationship to a higher 'being' we call God; a sense of conscience and duty; an inbred sense of the preciousness and fragility of life; and, for most at least, an intuitive desire or compulsion to *remain here,* despite our known, time-constrained destinies.

Inborn, Imperious Need for Meaning: French Nobel Prize-winning biologist author Jacques Monod is known for his views on human life's presence as the result of chance--genetic mutation. Monod cites human life's driven search for meaning as an "imperious need...which is inborn...inscribed somewhere in the genetic code." Physicist Heinz Pagels, in *The Cosmic Code*, says that *the universe is a message written in code, a cosmic code--which it is the duty of science to decipher.*

Life, the Odyssey in Search of Meaning: We come here equipped with the natural curiosity of childhood. As we mature, we find ourselves with a *need for reassurance* that we are not some kind of random happening, an accident of nature, and that our existence has some special significance or meaning. For many at least, life becomes an odyssey--a *lifelong search for meaning.*

Failure in this quest for meaning results in the sense of futility that diminishes the body and tugs at the soul. It leads, sometimes, even to a voluntary *surrender* of the life charter--suicide-- trading-in the chips for whatever, if anything, may follow. This can happen when one concludes that either there is *no* meaning, or a meaning which is inconsequential, or encumbered with baggage too burdensome or painful to accept.

How Do You Get from Radiant Energy to Life? If you have lost, or are about to lose, a loved one, you will have desperate need to understand more about the implications and meaning of death in our universe. No single event in our lives generates more questions about the meaning of existence and life, of the universe and purpose, than does the life-cessation, the death or imminent passing of a loved one--or the contemplation of one's own death.

But there is no realistic way to understand these things that does not involve knowing something about the universe, its origin and nature, whether it is can be said to have purpose. And how does one explain the incredible metamorphosis from the radiant energy of the Big Bang to the existence of sentient beings that wonder about the meaning and purpose of the universe and life?

Getting from Life to Abstractions: How do you get from physical human life to the concept of the *abstractions* so critical to meaning? Physicists trace the growth of order from quantum beginning to evolving abstractions, values and basic moral concepts derived from the underlying information system of the universe which influences the cosmic evolutionary process.

Philosopher, theologian, and mathematician Gottfried Wilhelm von Leibniz, author of *Theodicy,* like Immanuel Kant, argued that continued life after death was assured by the existence of certain necessarily-existing abstractions. In his view, Justice, for instance, was a built-in characteristic of the universe. Since we don't get equal treatment in this life and are subjected to disproportionate inequities, unhappiness, suffering and injustice here, there must, of necessity, be life continuance elsewhere--to square things. Godliness is fairness and fairness requires it; it is a necessary and mandatory characteristic of the universe-- or so he argues.

Life, Subjectivity, Meaning and Quantum Reality: Philosopher-cosmologist Immanuel Kant reasoned away the Thomas Aquinas 'proofs of God' but then reasoned God back into existence via the existence of basic *a priori* concepts, moral values and duty, implicit in his theory of innate moral mandates and the *categorical imperative.* These values, it is thought by some, are *wrought into* matter, into our very existence, by our very composition, part of the quantum information system of the universe.

Where Do Abstractions Come from? In considering whether the universe is purposeful, reference is often made to the subject of the abstractions which influence if not govern life to make it meaningful. Why do such abstractions exist and how do they originate?

Physicist Richard Feynmann describe how abstractions are the product of increasing order and complexity associated with the growth of mind. Cosmic evolution starts with a quantum eruption of a mini-universe superforce of unified physics.

And from the universe's initial negative gravitational energy comes our positive energy, radiation, then matter, stars, and planets. Then comes the oceans, the atmosphere, organic matter, and finally, life forms and human beings. Pockets of increasing order and complexity bring consciousness, self awareness, and finally sophisticated abstractions: love, hope, beauty, compassion, forgiveness, reverence, and all those intangible concepts which give life its meaning.

Two Contexts of Meaning: The *Encyclopedia of Philosophy* speaks of two contexts of meaning. The *first* sense is related to divine purpose, whether a supernatural intelligence created us 'to serve some end,' a reference to existence and *purpose of life in general*. The *second* context relates to finding meaning in aspects of everyday living, conduct, and goals.

Noted, are the views of scientists like astronomer Fred Hoyle that we really have "scarcely a clue" as to our significance. Of course there are philosophically-inclined scientists, such as physicists Freeman Dyson and Paul Davies, who suggest evidence of design, implying purpose and the existence of a *guiding intelligence.*

A 'Guiding Intelligence': Some physicists are more comfortable in describing this 'guiding intelligence' without commitment to whether this represents a *natural* or a *supernatural* phenomenon. And there are some, of course, who wonder whether the two are perhaps the same.

In considering whether the 'guiding intelligence' may be a reference to or synonym for 'God,' some scientists, following Monod's admonition against science getting into values, express themselves in subtle ambiguities of one sort or another, using terms like 'cosmic mind,' 'cosmic intelligence,' or even more indirect expressions like 'directionality' or they refer to natural patterns of increase in organization or complexity.

Demonstrating God's Existence? In *The Symbiotic Universe,* a fascinating analysis of *the anthropic principle*, physicist George Greenstein comes *razor-edge close* to demonstrating what some refer to as *mathematical proof* of the existence of God. And because he realizes how uneasy his science peers are about such matters, he proceeds, painstakingly, to extricate himself from the God implications of his all-too persuasive arguments. He comes close to proving the existence of God, or so it appears, and then, with heightened detachment, disavows any intention of addressing God's existence.

Origin of the Universe As Critical to Existence of Meaning: The meaning of the universe, and of life, is naturally influenced by one's view of the origin of the universe. We view 'meaning' in the context of the existence of a personally-involved Creator who provides meaning through the act of creation and biblical revelation.

But if, as quantum cosmology suggests, the universe was quantum-uncertainty-self-generated, astronomer-minister Hugh Ross, in his *The Creator and the Cosmos*, declares that humankind is left stranded without objective meaning and at the whim of constructed meanings according to our subjective fantasy.

Do We Construct Our Own Universe? These materials discuss meaning, values, purpose, sources of meaning. Distinguished writers point out that each person's perceptions are influenced by genetics, social, environmental factors and a particular society. In *Masks of the Universe* Edward Harrison describes the subjective universes of different societies, in which *each person constructs his own universe* within the broader framework of that society.

This book examines: the search for meaning as an imperative and meaning in various contexts such as religion and meaning, the meaning of meaning, ideas like imperfect world, bad things happening to good people, materialism and transcendentalism, determinism and free will, and meaning in death.

Lives of great men all remind us
We can make our lives sublime,
And, departing, leave behind us
Footprints on the sands of time.
--H.W. Longfellow

CHAPTER 4
MEANING--SCIENCE AND INTELLECTUAL LITERACY

Background--Some Thoughts of Physicists, Theologians and Philosophers: These materials draw on the work of physicists, theologians, philosophers who have explored the significance of new developments in cosmology, physics, and the interface between science and theology. This background helps to provide a base of enabling scientific and philosophic literacy needed in our time to explore meaning at both cosmic and human levels.

A number of physicists have seen fit to give expression to their thoughts on the God-physics frontier, where lurk the foundations of the meaning of the universe. In this context, thoughts of Paul Davies, Professor of mathematical physics, at the University of Adelaid, Australia, and Princeton physicist Freeman Dyson are explored, together with similar or divergent views of others.

Scientific Literacy for Exploring Meaning: In *Science Matters--Achieving Scientific Literacy,* Robert Hazen and James Trefil, address the need for background knowledge for scientific literacy, citing the fact that only seven percent of adult Americans qualify as scientifically literate.

And what is science literacy? They suggest it is knowledge of "a mix of facts, vocabulary, concepts, history and philosophy." Examples of this needed knowledge include: genetic engineering, ozone holes, DNA, microelectronics, superconductor vs. semiconductor, particles--basic physics and chemistry--an atom vs. a molecule, energy concepts, some nuclear physics, cosmology--origin and development of the universe, relativity, quantum theory, and origin of life.

Professional Scientists May Also Be Narrowly Focussed: In the discussion they note that professional scientists may possess just as narrow a focus as any other professional class and one should be mindful of this when a science Nobel laureate "speaks *ex cathedra* on issues outside his or her own field of specialization."

Science Not the Only Route to Understanding the World: But, they acknowledge, science is not the only route to an understanding of our world. Religion and philosophy help us grasp the 'meaning of life' through reasoning processes and without experimentation or mathematics, while art, music, and literature provide us with a "kind of aesthetic non-quantitative knowledge." But, they suggest, science does complement this other knowledge, offering insights on different aspects of the universe.

Science--A Way of Learning about the Cosmos and Our Place in It: In their *Epilogue--The Role of Science*, Hazen and Trefil observe that science is a way of learning about the cosmos and our place it. It is through science that we discover physical laws governing matter, energy, forces, and motion applying to "every life-form and every world."

But, they note, human beings are only one "small, recent step" in the life's four-billion year evolution on earth, and in the process, countless species have appeared and disappeared. We are but one species among millions struggling for survival. It would be folly, they insist, to ignore the realities of our critical role in the planet's ecology.

Search for Meaning of Our Unique Role Through Philosophy and Religion: But, observe Hazen and Trefil, we cannot deny that human beings play a unique role because we are the only species in the planet's history to have learned to employ resources and mold the environment, to probe our world through science, to "appreciate our world through the arts" as well as through philosophy and religion, and to search for the meaning of our unique role.

Human beings have a driving curiosity, and it is science that permits us to indulge that drive toward understanding the physical universe. Science, say Hazen and Trefil, is a "great human adventure, with formidable challenges and priceless rewards, unimagined opportunities and unparalleled responsibilities."

It is science, they suggest, that permits us to explore "backward in time," "outward through space," and to discover "unity in the workings of the cosmos." And, equipped with that knowledge, we can fight disease and influence our environment.

Echoing Einstein's concern over the consequences for mankind of his energy-matter equivalency equation, $E = MC^2$, Hazen and Trefil conclude their book with the *portentous proposition*: That science gives us "the means to predict the consequences of our actions and perhaps, with wisdom, to save us from ourselves."

Meaning and the Mystery of Language: A corollary to scientific literacy is 'intellectual literacy' and the role of language and communication in the search for meaning. It is said that *the search for meaning is the essence of philosophy.* To discuss meaning-- philosophic ideas and concepts, it is crucial to consider how mental ideas, perceptions, are expressed and communicated. We normally think of language as able to give full expression to our thoughts and our experience, our needs, hopes, fears, plans, our views on religion, politics, and aspects of every-day life in the world.

Language Shortcomings: Occasionally, however, we are frustrated by our inability to verbalize our thoughts. And at that point we recognize the shortcomings of words to communicate all our ideas and feelings. Abstractions are sometimes complex, difficult to understand, and even more difficult to preserve in the transfer between minds--to convey by language.

Language is dynamic, evolving with new needs for description of emerging concepts and technology, like quantum uncertainty, virtual reality, cyberspace and the INTERNET. Is the failure of words, fully or accurately to communicate ideas, a failure in language itself or does it reflect deeper or inherent fluency limitations?

Effort to Apply the Logic of Mathematics to Language: In *Real Philosophy*, authors Jacob Needleman and David Applebaum observe that such questions are obviously important since the pursuit of meaning, whether of life, or the universe, or God, all depend on how thoughts are interpreted, described, and communicated.

The drive for precision clarity in communication has in the past resulted in marginally successful efforts to *apply the logic of mathematics to language,* as reflected, for example, in the work of mathematics-oriented philosophers like Bertrand Russell, Alfred North Whitehead, and Ludwig Wittgenstein.

Interestingly, these philosophic-linguistic analyses have also led to revision of earlier ideas on the efficacy of philosophy in the field of science, and its more vague role in the field of metaphysics. To communicate some of these ideas the authors include excerpts from the work of several philosophers.

Wittgenstein and the Use of Language: In *Tractatus, Logico-Philosophicus*, and *Philosophical Investigations*, mathematician-philosopher Ludwig Wittgenstein, protégé of Bertrand Russell at Cambridge, speaks of two levels of use of language. The first is the physical world of science in which meaning can be effectively formulated and language-described because the propositions can be sense-verified as the source of knowledge, a concept associated with the philosophical school of 'logical positivism.'

The second is the more esoteric field of metaphysics, in which propositions, concepts, or beliefs, are viewed as neither tautological nor empirical. Their truth is unverifiable and speculative and therefore seen as possessing limited authenticity. Metaphysical ideas cannot be sensed but only mentally conceived. Here the function of philosophy is considered limited to the clarification of language.

Symbolic Logic Not Usable for Esoteric Philosophic Problems: Thus philosophic problems may be addressed through the analysis of meaning employing questions about language but cannot be solved by use of procedures of symbolic logic. Ordinary language is the medium of resolution, as Bertrand Russell, George Moore, and Ludwig Wittgenstein thought. Philosophy unrelated to science, in this view, possesses no factual content and is considered essentially a process for the clarification of logical confusion.

A Caution--Philosophy Does Play a Role in Extending Physical Knowledge: A word of caution seems appropriate on this 'limited' role of philosophy. Einstein's description of the process by which he arrived at his relativity theories suggests that extending our factual knowledge of the universe through pure philosophic speculation does indeed pay off in physical discovery. Physicists like Steven Weinberg and Heinz Pagels note that esoteric concepts of *symmetry* and *beauty* in mathematical and physical theory do indeed play a significant role in theory development and evaluation, as in the case of breakthroughs in unification theory.

Of equal recent significance and interest, however, is the fact that much of what in the past was viewed as metaphysics is now emerging as physics. Examples are: the free will question in the light of relativistic and quantum physics; the hypothesized quantum state existing before the Big Bang; and the mathematical exponentials of the constants of physics associated with the anthropic principle.

> Your right [neocortex] brain is your...intuitive brain....You might call your right brain your *mystic*, since its wisdom comes from some hidden source, *perhaps a cosmic source*. --Richard Bergland

PART III--COSMOLOGY--BACKGROUND

CHAPTER 5
COSMOLOGY--INTRODUCTION

Cosmology--Study of the Universe: Cosmology is generally defined as the study of the origin, evolution, structure, and fate of the universe. Is the universe eternal or did it have a beginning in time? If there was a beginning, who or what is responsible for its existence? Will the universe some day come to an end? How and why does the physical universe generate mind, consciousness and life? How can abstract laws of physics generate physical reality? Was the universe created by God or do we live in a 'causeless,' self-initiating universe triggered by the pre-big-bang primordial quantum void? Are there other universes and, if so, what might they be like?

Such questions are of particular interest to astronomers, physicists, and, of course, to theologists.The meaning of existence, and the universe and of life, necessarily involves cosmology and inevitably and immediately raises the question of the existence of God. A God-created universe implies *purpose*, from which all meaning may be constructed. Deriving meaning from a self-generated or an eternal universe, with no beginning, and no *Creator*, is something else.

Cosmology--Theological and Scientific Implications: Why do we study cosmology? Not only astronomers and physicists, but theologians and philosophers study cosmology, because, in a sense, it is the study of *existence*. Each of the disciplines has its respective areas of interest, scientists for science, and philosophers and theologians for its relationship to concepts of *existence* and the existence and nature of *God,* and for insights into purpose and meaning of and in the universe. In Romans, St. Paul observed that we can learn things about God from examination of his works.

Associating Meaning and Justice with Existence of God: We are used to associating purpose, morality and justice with the existence of God, looked upon as the controlling influence on human behavior. The famous Russian 19th century philosophic novelist Fyodor Dostoyevsky, author of acclaimed *Crime and Punishment*, was concerned with our earthly suffering, the meaning of life, and the nature and justice of God.

In his classic, *The Brothers Karamazov*, he writes of the lives and relationship of four brothers to a depraved and vindictive father, and offers his oft-cited observation that

If God does not exist, then everything is permitted. Most people probably share this concern for God's existence as critical to meaning and the existence of duty and morality.

Contemporary Cosmology and Einstein's Relativity Theories: Although seriously studied by early societies such as the Egyptians, Babylonians, Greeks, and Chinese, 'modern' cosmology is recognized as beginning early in the 20th century with Einstein's publication of his special theory of relativity. This theory demonstrated that space and time were associated, a *continuum,* described as 'space-time' or 'spacetime' and interrelated to electromagnetic transmission speed--the speed of light.

Hawking's Picture of the Universe: In the field of recent cosmology, no one is better known, or more cited, than British theoretical physicist and cosmologist, Stephen Hawking. So a brief review of some of Hawking's thoughts on the beginning and nature of the universe, expressed in *A Brief History of Time*, will be discussed.

Some Questions about the Universe: What do we know about the universe, and how do we know it? asks Hawking. Where did the universe come from, and where is it going? Did the universe have a beginning, and if so, what happened 'before' then, he asks. What is the nature of time? Will it ever come to an end? He notes that recent breakthroughs in physics suggest answers.

Newton--Motion, Gravity, and 'First Cause': Hawking cites the thoughts of Aristotle, Ptolemy, Copernicus, Keeler, and Galileo on early models of the universe. Then came Isaac Newton's monumental *Naturalis Principia Mathematica,* which Hawking describes as probably the most significant single work in the physical sciences. Newton advanced his theory of how bodies move in space and time, and the 'law of universal gravitation' in which bodies are attracted toward each other by a force proportionately to their mass and proximity.

According to early cosmologies and Western Biblical tradition, he points out, the universe started at a finite and not very distant time in the past and it was felt necessary, he notes, to have a "first cause" to explain the existence of the universe.

The Expanding Universe--The Big Bang: Hawking cites St. Augustine's conclusion, in *The City of God*, that the universe could not have been around all that long, perhaps since about 5,000 B.C. according to biblical family-history sequence. Hawking notes that in his *Critique of Pure Reason*, philosopher Immanuel Kant explored the question whether the universe had a beginning in time and limit in space, arguing that there were equally valid reasons for a 'timeless' and a 'created' universe. But, says

Hawking, it was American astronomer Edwin Hubble that made the discovery that the distant galaxies were *receding*, and therefore the universe must be expanding. The expansion implied, of course, that the universe must have had a starting point or beginning, and this, of course, contributed to the theory of the Big Bang.

While most religious groups welcomed the Big Bang Theory as at least confirming the Biblical Genesis concept of a Creation event, some adherents of biblical *inerrancy* find difficulty in accepting it even as an allegorical confirmation of biblical truth, particularly regarding the age of the universe and the evolutionary versus instantaneous appearance of human life. But more on this later.

Causation: One could imagine, says Hawking, that God created the universe at the instant of the big bang, or afterwards so as to give the impression of a Big Bang. But, he notes, it would be meaningless to suppose that it was created *before* the Big Bang. As we shall see, the question of 'causation,' before the existence of the very phenomenon that permits causation, namely time, becomes a critical issue in the comments of physicists like Paul Davies in his *God and The New Physics* and *The Mind of God.*

The Question of Universe's Origin--Before and after the Creation: The approach taken by most scientists, Hawking says, is to divide the problem of a theory that describes the universe into *two parts*. One, is the question of the laws governing change with time. The second, is the question of the 'initial state' of the universe. Some feel, he notes, that science should be concerned with only the first part--that the second, the initial conditions--what existed before the universe, is metaphysics or theology.

Being omnipotent, he says, God could have started the universe off *any way he wanted.* But, he notes, it appears that God chose to make it evolve in a "very regular way" and according to certain laws. So he feels that it is not unreasonable to suppose that there are also *laws that determine the initial state*.

Two Theories Involved--Relativity Theory and Quantum Theory: The universe, Hawking points out, is described by two basic partial and inconsistent theories, the general theory of relativity, describing the effects of the force of gravity and the large-scale structure of the universe, and the theory of quantum mechanics, dealing with the subatomic or extremely small-scale phenomena. The major theme of his book, he says, is to search for a "new theory that will incorporate them both--a *quantum theory of gravity.*" And, he notes that we already know a good bit about the predictions of a quantum theory of gravity.

Must *Combine* the Two Theories: If the universe is not "arbitrary," he says, you must ultimately combine the partial theories into a "complete unified theory" that will explain *everything in the universe*. But , he notes, this assumes our rationality and freedom to observe the universe and draw logical deductions.

If there really is a complete unified theory, he adds, it would also presumably determine our actions and so the theory itself would determine the outcome of our search. We might expect that the reasoning abilities that natural selection has given us would be valid also in our search for a complete unified theory, though the discovery of complete unified theory may not assure the survival of our species.

Our Goal: A Complete Description of the Universe: He comments that since the dawn of civilization people have not been content to see events as "unconnected" and "inexplicable." We still yearn to know why we are here and where we came from and, he declares, our goal is "nothing less than a complete description of the universe we live in."

Creation Theories Require Review of *Relativity* Theory and *Quantum* Theory: Hawking's 'creation' theory depends on having some understanding both of relativity and quantum theories and so he includes chapters to touch on these theories in order to explain his "quantum cosmology." Accordingly, before discussing 'Creation Theories,' in Part VI, our Part IV addresses Cosmology--Relativity theory and Part V reviews quantum theory.

> Einstein's general relativity: The single greatest human creative thought achievement in the history of science.--Edward Teller

CHAPTER 6
COSMOLOGY AND GOD

Cosmology and God: In chapter 5, Cosmology--Introduction, we discussed cosmology, referring to the relationship of the Big Bang to the Genesis Creation scenario. Differences between the two are readily evident. While some religious organizations like the Vatican accept the some-fifteen-billion-year-old Big Bang event as confirmation of the Genesis allegory, certain religious conservatives, holding to biblical inerrancy, challenge the Big Bang 'cosmic' time account.This disagreement is further discussed in the chapter on Religion, Denominations, and Meaning.

Troubling Questions Remain About God and the Nature of the Universe: We noted the theories about 'self-generation' of the universe versus divine Creation. There is more to this subject of great interest which has been addressed by physicists like Stephen Hawking on 'quantum cosmology' and Paul Davies who writes scholastically and comfortably about God involvements.

Davies Features Questions About the Creation and Cause and Effect: In *the Mind of God*, physicist Paul Davies asks: Can the universe create itself? Was there a creation event? Was a supernatural act required? Did God cause the Big Bang? Was there a Creation Event? Davies discusses 'cause and effect' pointing out that it is natural to try to explain the universe through a succession of causal events. Would this chain of cause-and-effect ever end, he asks?

That there is an ultimate 'something' that must have started it off, he notes, is deeply-ingrained in Western culture, not open to scientific inquiry but resting in the *supernatural*. He refers to this as the 'cosmological argument' for the existence of God.

The Self-starting Universe of Quantum Uncertainty: This approach was hard to challenge, he notes, until recently when new theories of physics surfaced, involving the quantum theory uncertainty principle. As we have seen, the exponentially small starting-point beginning universe was implied in the Big Bang, derived from Einstein's relativity equations, and confirmed by astronomer Hubble's finding of an expanding universe, and the finding of the cosmic background fossil radiation left from the Big Bang. This in turn led to ideas of a causeless, self-initiating universe, triggered by the 'quantum uncertainty' inherent in a pre-big-bang primordial quantum void.

Cyclical Time in Ancient Cultures vs. Judeo-Christian Linear Progression: Davies speaks of the cyclic views of time in ancient cultures, the repeating cycles of Eastern religions, so-called cyclic cosmology, such as involved in the Hindu Brahman life cycle similar to Babylonian, Egyptian, and Mayan cosmologies. The Greek philosophic cyclic nature of time, says Davies, was inherited by the Arabs.

But much of the present world view of European cultures can be traced to the 'monumental clash' between pagan Greek philosophy and the Judeo-Christian tradition that God created the universe at some specific moment in the past, and that subsequent events form an "unfolding unidirectional historical progression-- the Fall, the Covenant, the Incarnation and Resurrection, the Second Coming."

In their anxiety to adhere to *linear,* rather than *cyclic,* time, he says, the early Church Fathers denounced this cyclic world view. Thomas Aquinas, while acknowledging Aristotle's philosophical arguments that the universe must always have existed, appealed for belief in a cosmic origin on biblical grounds, notes Davies.

The Separate Creator: A key feature of the Judaic-Christian creation doctrine, observes Davies, is that the Creator is entirely separate from and independent of his creation. So God's existence does not automatically ensure the existence of the universe. In this view, the universe came into existence at a definite instant in time as an act of deliberate supernatural creation by a *pre-existing* being.

Genesis and the Big Bang: Davies speaks of the Genesis description of the creation as borrowed from the then contemporary Middle East mythology, suggesting that it is "long on poetry and short on factual details." It is interesting, in this context, to note the varying treatment given the scientific relevancy of the creation scenario as laid out in Genesis.

A precise literal comparison of the Genesis creation sequence and the Big Bang sequence, based on the breakdown of original symmetry and phased separation of the forces, leaves much of Genesis in question, including the question of whether Genesis can be said to support creation *ex nihilo.* But much of Christianity, like the Vatican, is reasonably comfortable with the Big Bang theory since it is a scientific reaffirmation of the central critical fact of a creation event, questioned by earlier scientific theory of an eternal universe.

Like Paul Davies in *The Mind of God,* many if not most physicists avoid comparisons between the Genesis and physics explanations of the creation, or if enticed into comment, point to various significant variations in substance and sequence between the two accounts, in essence, questioning technical reconciliation between the two, other than the fact of a creation event.

But Hugh Ross, an astronomer-astrophysicist and a devout minister, in his *The Creation and the Cosmos,* not only finds no difficulty in harmonizing Genesis with Big Bang theory but finds a remarkable correlation between New Testament passages addressing the pre-existing God, and the latest findings in astrophysics. He explores biblical phenomena like Jesus walking on water and three persons in one, which seem impossible in a world of three space and one partial time dimension but explainable in dimensions which exist beyond those we experience.

However, he vigorously questions scientific theories about what existed *before* the Big Bang, what produced the Big Bang, and he speaks of 'quantum cosmology' and 'quantum genesis,' as essentially speculative metaphysics.

Did God Cause the Big Bang? This subject brings Davies to review his difficulty with the picture of God as "a sort of pyrotechnic engineer... igniting the big bang and then sitting back to watch the show." He observes that, as discussed earlier, "a supernatural creation cannot be a causative act in time, for the coming into being of time is part of what we are trying to explain."

He cites British physicist Russell Stannard, who draws the analogy between God and the author of a book which exists in its entirety. Although we humans read it "in time sequence from the beginning to the end," God's creativity has to be seen as "permeating equally all space and all time; his role as Creator and Sustainer merge."

Pitfalls in Invoking God: But apart from the problem of time, he notes, there are other "pitfalls in invoking God as an explanation for the big bang." These he discusses as the book progresses. He quotes an imaginary but stimulating dialog between a "theist" and an "atheist" in which the theist, for one thing, notes that the atheist acknowledges that despite theories, the reason for the Big Bang lies outside the scope of science.

The atheist observes that by invoking God to explain the Big Bang, the theist is introducing further complicating features. Also there is no way to test theist's God hypothesis, thus invoking science philosopher Karl Popper's proposition that a scientific fact must be falsifiable--experimentally verifiable.

Is the Universe a Closed System Eliminating the Need for God? Davies goes on to discuss Hawking's universe where quantum effects fuzz out any distinct time boundary and therefore there is no distinct creation point--or need of a Creator. Does this mean that the universe created itself? asks Davies. His reply is that the universe of spacetime and matter is "internally consistent and self-contained." Its existence does not require anything outside of it; specifically, *no prime mover is needed.*

Davies asks, Does this eliminate need for God? Can the universe be regarded as a closed system containing the reason for its existence? The laws of physics, he responds, permit the universe to "take care of itself, including its own creation." But, he acknowledges, this fails to explain where do the laws allowing for self-creation come from, something he later addresses.

The Doctrine of *Ex Nihilo*: Davies asks whether these new quantum cosmology developments square with the doctrine of creation *ex nihilo*? God bringing the universe into existence cannot, he reminds us, be regarded as a temporal act because time came into being with the creation. In modern cosmology [meaning quantum cosmology], he says, one should no longer think of spacetime as "coming into existence," he says, but that spacetime simply *is,* there being no "initial event" and so, he observes, all moments "have a similar relation to the Creator."

He acknowledges that this image of God is far removed from the twentieth-century Christian God and Creation act; that it is closer to the 17th century philosopher Spinoza's God where the physical universe takes on aspects of God's existence, such as being 'eternal' and 'necessary.'

So Where Are We in the Causal Chain? Earlier Davies had discussed 'cause and effect,' the effort to explain the universe through a succession of causal events, leading to the question whether this chain of cause and effect ever ends. That there is an ultimate 'something,' that must have started it off, he noted earlier, was the belief deeply ingrained in Western culture and this 'something' rested in the supernatural, the cosmological argument for the existence of God as the primal cause.

But as we noted, new theories of physics, revolving around quantum cosmology, introduced the idea of the self-starting universe randomly erupting through quantum uncertainty. This primordial state was devoid of space, time, forces or matter but, as physicist Heinz Pagels described it, having all the potentialities of the laws of physics.

So in the causal chain we are at the point where the Big Bang is a quantum-uncertainty eruption event traceable to the laws of physics. But where do the laws of physics come from? In his following chapter on the "Laws of Nature," Davies explores the question of the nature of the laws of physics--and possible sources.

But in summary, then, the universe is, as Davies hypothesizes, self-created by the laws of physics, physics represents mathematics, mathematics represents logic. So where does logic come from? Perhaps logic exists of *necessity* and Davies notes, not even God can violate logic.

In *Creation Revisited*, Oxford British chemistry professor P.W. Atkins asserts that creation is self-initiated *without outside intervention,* because underlying everything, including the primordial 'nothing,' is *mathematics, a universe-self-generating mechanism.* In Atkins view, nothing else is needed to "breathe fire into the equations," to use Stephen Hawking's famous phrase. So Atkins' view is that *mathematics does it all!* Atkins recognizes, of course, that mathematics represents logic and so underlying their differing terminology, Atkins and Davies are conceptually close.

Creation of the Universe Traceable to Logical Necessity? Davies concludes by referring to the path from physical phenomena, through the laws of physics, to mathematics and ultimately logic, and in doing so, opens up the "beguiling prospect that the world can be understood through the application of logical reasoning alone." Could it be, he asks, that much if not most of the physical universe might be as it is as the result of *logical necessity*? This, he suggests, calls for inquiry into the nature of the laws of physics, the "laws of nature."

God as an Amorphous Expression for Mathematics? Atkins' treatment, in *Creation Revisited,* of the remarkable role of mathematics in accounting for the underlying structure of the universe, suggests the need to explore the implication that the 'creator,' in some sense may, perhaps, be perceived as an amorphous expression for the laws of physics with their mathematical underpinning.

We noted P.W. Atkins' provocative proposition of mathematics as the source of the creation, without "outside intervention." If God is ultimately impersonal mathematics and the laws of physics, is God then a 'natural' rather than 'supernatural' God? And if God is, in a sense, impersonal mathematics, where do personhood and highly personal abstractions come from?

The Natural God? The central idea of God as representing primarily the natural rather than supernatural processes--essentially physics and the derivative sciences, with their roots in mathematics--has been a recurring theme for some years now. This idea was reflected, for example, in Sir James Jeans' observation that "God is a mathematician."

But, if the question posed is: "Who or what created the universe," and if the answer proposed is: 'physics-mathematics,' then some, theologians included, may respond that this answer begs the question since there is left unanswered the question: What is the source of these concepts and how do such abstract mathematical concepts generate physical structure for a material universe?

This is Stephen Hawking's now famous and oft-quoted question in *A Brief History of Time:* "What is it that breathes fire into the equations and makes a universe for them [the laws] to govern?"

Is the Unified Theory so Compelling as to Generate the Universe? But Hawking asks another question, suggesting a possible answer. "Is the ultimate unified theory so compelling," he asks, "that it brings about its own existence?" In *Creation Revisited,* Atkins, in effect, says yes to Hawking's last question.

Hawking, on the other hand, concludes: "Although science may solve the problem of how the universe began, it [science] cannot answer the question: Why does the universe bother to exist" [His way of saying that science cannot answer the question of what breathes fire into the equations]. And, Hawking declares, "I don't know the answer to that."

Atkins Goes All the Way--The Universe Is Mathematics! But in *Creation Revisited,* British scientist P.W. Atkins is willing to dispense with intermediaries like unification or logic as compelling creation and asserts that creation is self-initiated *without outside intervention,* because underlying everything, including the primordial "nothing," is *mathematics, a universe-self-generating mechanism.* In Atkins view, nothing else is needed to "breathe fire into the equations," *mathematics does it all!*

Ultimately a Theological Question? But this really leaves us with two questions: First, what introduces mathematics and the laws of physics into existence? and, second, *how* do mathematical abstractions self-generate physical existence? In *Creation Revisited,* Atkins says, in essence, that both these effects are self-initiated because 'nothing,' it turns out, is a *numerical expandable entity.*

Some scientists agree; others, like Hawking, disclaim knowledge of the answer. Some are not opposed to these questions being classified as theological--at least for now.

Einstein and God: Einstein is credited with having laid the groundwork for contemporary cosmology and literally redefined the universe with his relativity theories. His frequent references to God led many theologists and religious figures to want to know more of his views of God and the creation. Danish quantum physicist Niehls Bohr is said to have cautioned Einstein to stop telling God what to do.

Einstein's Views on God? Einstein's views on God are sometimes described as enigmatic. But apparently he mostly wavered between deism (God set up the universe as a self-operating mechanism without further divine involvement) and pantheism (God is the universe and everything, including humanity, is part of God).

To the dismay of the many religious leaders with whom he had dialogue, he had difficulty with the concept of a personal watchful theistic God who rewarded, protected, and punished as earned, given the realities, horrors and absurdities of everyday life and God's complicity in a predetermined, deficient world .

But nevertheless, his belief in God was so strong that one obstacle to his accepting quantum uncertainty was his conception that God would not use random chance to play out the development of the universe--that God would not 'play dice' with the universe. Einstein was equally perplexed that uncertain quantum systems are, according to Bohr's Copenhagen quantum interpretation, affected and resolved by *observation.*

COSMOLOGICAL CHAUVINISMS?

Scientific Discoveries as Proof of God? In his *The Creator and the Cosmos*, astronomer, cosmologist, and Congregational Minister Hugh Ross notes the deep philosophical and theological implications of this study of cosmology. He refers to certain recent scientific discoveries as proof of the existence of God and of the truth of Biblical statements. Some physicists, like Paul Davies, don't go quite so far but do suggest that the exponentially fine-tuned constants of physics suggest *purpose.*

Meaning and a Quantum-Uncertainty-Generated Universe: A God-designed universe certainly implies meaning and purpose, while one born of quantum uncertainty can have 'no objective meaning,' insists astronomer-minister Hugh Ross in *The Creator and the Cosmos*. Where chance quantum effects spontaneously self-generate the universe, says Ross, "morality and religion" in such a chance universe, while "temporarily useful" are *ultimately irrelevant*, he declares.

Exploring Cosmology For Life's Meaning: If the universe is created by God, observes Ross, then there is "ultimate reality" beyond the universe, represented by the Creator, the source of life, which, he declares, establishes its meaning and purpose. God's personality, he notes, "defines personality" and God's character "defines morality." So the study of cosmology, the origin and development of the universe, is the exploration of "the basis for any meaning and purpose to life."

Cosmological Chauvinisms: Ross is concerned about *cosmological chauvinism,* theological, philosophic, and scientific, each of which competes for recognition as authentic, when what is needed is the *interdisciplinary* rather than the exclusionary approach, since each alone distorts meaning.

Fundamental Questions about the Existence of Meaning: In this view, examples of scientific chauvinism may be found in P.W. Atkins' *Creation Revisited* concept of the universe's chance inception as "purposeless" and in Steven Weinberg's *The First Three Minutes* and the later *Dreams of a Final Theory*, characterizing the universe as "pointless," although not denying that we are free to construct meaning. Weinberg discerned *no cosmic imperatives* that mandated value postulates.

What About God? In his later *Dreams of A Final Theory*, in his chapter called *What About God?*, Steven Weinberg notes that if there were anything we could discover in nature that would give us some special insight into the handiwork of God, it "would have to be the final laws of nature." But Weinberg does not envision finding "ultimate principles" that do not require explanation in "deeper principles." And, reluctantly, he concludes that we are unlikely to find any "sign of the workings of an interested God in a final theory."

Meaning and a Divine Creation : As astronomer-minister Hugh Ross points out in his *The Creator And The Cosmos*, the meaning *of* the universe and meaning *in* life will naturally be influenced by one's view of the origin and nature of the universe. Western thought, of course, is accustomed to pursuing meaning of the universe in the context of the existence of a theistic, personally-involved God who purposely created, and continues to watch over, the universe, suitably designed for the human race, as chronicled in the Biblical Genesis sequence.

CHAPTER 7
PHYSICS, UNIFICATION THEORY & GOD

Nobel Laureate in Physics and Challenger of Intellectual Thought: Steven Weinberg, Nobel Laureate for his contribution to identifying unification of the electromagnetic and weak nuclear forces, is a member of the Royal Society of London, the U.S. National Academy of Sciences, as well as recipient of many honorary degrees.

Recognized as one of the world's leading theoretical physicists, Weinberg is also noted for his intellectual candor and willingness to challenge institutional thinkers on matters involving philosophy, especially the philosophy of physics. His *Dreams of a Final Theory* introduces the reader to some of his stimulating if not iconoclastic thinking about the universe, unification theory and God. While religious scientists may not fully agree with all his observations relating to physics and God, his views are all the more noteworthy since he rarely indulges in such discussion

In his *The First Three Minutes*, Weinberg had given expression to certain thoughts which seemed to flow from the nature of the universe, its laws and underlying quantum randomness, from the functioning of the fields and forces, and from the unexplainable tragic events in life and the likely destined end of life in the universe.

A 'Final Theory' As Part of Our Intellectual History: In the Preface of his *Dreams of a Final Theory,* Weinberg notes his intention to lay out for the nonphysicist the issues raised by the idea of a final theory as "part of the intellectual history of our time" and, in the process, weave in concepts of modern physics. He acknowledges the contributions of physicists, theologians and editors to his book.

The Central Role of Matter Replaced by Principles of Symmetry: In his Prologue, he notes that through the amalgamation of relativity with quantum mechanics a new view of the universe has emerged, in which the key role of matter has been replaced by symmetry principles, some of which, he says, are hidden by the present state of the universe.

And, he says, on these principles has been built a theory of unification which is well-accepted at least as to unification of electromagnetism and the weak force into the *electroweak* force. Although in a somewhat arbitrary fashion, it also brings in the strong nuclear interactions.

Occasional Glimpses of a Final Theory: He observes that the Congressional and professional debate over the justification for the Texas Super-Conducting Super-Collider compelled him to try explaining the objectives of elementary particle studies. Although the validity of our present theories are acknowledged to be limited, occasionally, he suggests, we catch flashes of a final theory of *unlimited validity.*

Role of Avoiding Infinities in the Road to a Final Theory: When we find universal truths about nature, he notes, we attempt to explain them as deducible "from deeper truths." Chemistry works the way it does, he observes, because of the physics of the electron and electromagnetic fields. We may be led mathematically to part or all of the final theory, he is reminded, through some method of avoiding the 'infinities' which emerge in unification mathematics.

And, he adds, the new 'string' theory may be the mechanism for avoiding the infinities which surface in the effort to reconcile relativity and quantum mechanics. He goes on to observe that the *aesthetics* of certain physics theories may serve as a useful guide in scientific research.

Many See an Endless Regression of Deeper Principles: Many take it for granted, says Weinberg, that we will uncover an endless chain of deeper principles and find it difficult to envision arriving at *final* physical principles that need no rationalization in still "deeper principles." He cites science philosopher Karl Popper to the effect that every explanation may be further explained by more universal theory; that there can be no explanation which does not need further explanation.

But Infinite Regression Can't Be Based on Not Having Final Theory: Although this view of an endless process might prove correct, he says, it can't be based simply on the fact that no one has yet found a final theory. Some searches *do* end, he observes. He notes in passing, Stephen Hawking's suggestion some time ago that the supergravity theories earlier in vogue might approach a final theory, something which did not come to pass. But, he points out, Hawking never claimed that a final theory had already been achieved. If history is any guide, Weinberg says, it seems to him to suggest that there *is* a final theory--one still to be identified.

If No Fundamental Law Exists, Then 'Metalaws'? More radical is the suggestion that at the end we *may find that there is no law at all*. Physicist John Wheeler, he points out, has occasionally suggested that there is *no* fundamental law and that all the laws are imposed on nature by the way that we make observations. But, it seems to Weinberg that this view merely postpones the problem of the final laws.

Wheeler's "world without laws," he says, still needs *metalaws* to tell us how our observations impose regularities on nature. And among these metalaws is quantum mechnics itself. He expects that all efforts to do without fundamental laws will simply result in the introduction of meta-laws that describe how what we now call laws came about.

More Ominous Possibility--Final Theory As beyond Our Intelligence: A more likely and more ominous possibility is that there *is* a final theory but one which we will never learn because it is beyond the potential of human intelligence. Another more imminent concern is that resources may not be made available for the effort, noting the difficulty in funding the Super Collider project, which, as it turned out, was terminated on national priority considerations.

Final Law As a "Discontinuity in Human Intellectual History": His own guess is that there *is* a final theory, and we *are* capable of discovering it. We may, he notes, even be able to find a candidate for such a final theory among today's string theories. The discovery of the final laws of nature, Weinberg declares, will mark a *discontinuity in human intellectual history*, the most acute since the onset of modern science.

Why Should the Final Theory Describe Anything Like Our World? But why, he asks, should the final theory describe anything like our world? The possible explanation might be traceable to the *principle of fecundity,* that in some sense, all logically possible universes exist with their own sets of fundamental laws.

The principle hypothesizes that there are different universes, possibly subject to different laws. If this principle is true, he notes, then our own quantum-mechanical world exists, but so do many other universes with altogether different principles, not just a matter of the so-called constants of nature varying from one part of the universe to another.

The Significance of Other Universes: But, if these other universes are totally unknowable, he notes, the statement of their existence would be without consequence except to preclude questions concerning their *nonexistence*. In *A Brief History of Time*, Stephen Hawking made the same observation about the significance of other universes which we cannot know, citing philosopher Wittgenstein. The problem, says Weinberg, is trying to be logical about a question that is not subject to logical treatment.

The Principle of Fecundity Supports Anthropic Reasoning: Weinberg notes that this "principle of fecundity" (multiple universes) would provide another support for use of 'anthropic reasoning' to help explain the final laws of our universe. There would be many conceivable kinds of universes with laws inhospitable to life, but any scientist who asks why the universe is the way it is would have to be living in a life-supporting universe. While both Weinberg and Hawking question the usefulness of the anthropic principle they nevertheless feel drawn to it in certain contexts.

Could There Be Only One Logically Isolated Theory? An extreme possibility, he notes, is that there is only one "logically isolated theory" with no undetermined constants that is consistent with the existence of intelligent beings who can speculate on a final theory. If this could be demonstrated, he remarks, then we would be near a "satisfactory explanation of why the world is the way it is."

He concludes this topic with the nostalgic observation that with the finding of a final theory we may regret that nature has become less the subject of "wonder and mystery." If this occurs, he visualizes scientists of the future envying today's physicists, still on the voyage to discover the final laws.

What About God? In his chapter, *What About God?* Weinberg quotes the Bible, "The heavens declare the glory of God; and the firmament showeth his handiwork." But since King David's day, he observes, the sun and stars have lost their 'special status.' We now understand that stars are glowing gas spheres, held intact by gravity, and prevented from collapse by thermonuclear heat pressure emanating from nuclear reactions in the stars' cores. So, he notes, the stars signify nothing more about the glory of God than do the "stones on the ground around us."

Weinberg notes that if there were anything we could discover in nature that would give us *some special insight into the handiwork of God*, it must be the final laws of nature. And knowing these laws, we would have in our possession the "book of rules that governs stars and stones and everything else." So, he acknowledges, it was understandable that Stephen Hawking should discuss the laws of nature in the context of "the mind of God."

Physicists Use of the Term 'God': Weinberg notes theologian Paul Tillich's observation that among scientists only physicists seem capable of using the word 'God' without embarrassment. Whatever one's religion, or lack of it, he notes, it is an *irresistible metaphor* to relate the final laws of nature to "the mind of God," as Stephen Hawking did.

Elastic Views on Term 'God': Some people, he notes, have views of God that are so elastic that it is inevitable that they will *find God* wherever they look for Him. The word 'God,' he notes, can be given any meaning we like. But if the word is to be of any use, he suggests, it should be taken to mean "an interested God, a creator and lawgiver who has established not only the laws of nature and the universe but also standards of good and evil, some personality that is concerned with our actions, something, in short, that is appropriate for us to worship." This functional or pragmatic definition of God might well be added to the various definitions of God discussed in the chapter on Efforts to Define God.

Using God As an Abstraction: He is reminded that scientists and others sometimes use the word "God" to mean something "so abstract and unengaged" that such a God is hardly distinguishable from "the laws of nature." Weinberg cites Einstein as once saying that he believed in "Spinoza's God who reveals Himself in the orderly harmony of what exists, not in a God who concerns himself with fates and actions of human beings."

God As 'Order' or 'Harmony': What difference does it make to anyone, asks Weinberg, if we use the word 'God' in place of 'order' or 'harmony,' except perhaps to avoid the accusation of being an atheist, having no God. Though we are free to use the term 'God' in that way, it seems to him that it makes the concept of God *not so much wrong* as *unimportant*.

Will We Find an Interested God in the Final Laws of Nature? Weinberg asks whether we will find a personal interested God in the final laws of nature. There seems something almost absurd in asking this question, he responds. Not only because we do not yet know the final laws, but more because it is difficult enough to imagine being in the possession of ultimate principles that do not need any explanation in terms of deeper principles.

Premature as the question may be, however, it is hardly possible, he says, not to wonder whether we will find any answer to our profoundest questions, "any sign of the workings of an interested God, in a final theory." He thinks that we will not.

Demystification of the Heavens and Life--Evolution: Weinberg goes on to discuss the *demystification* of the heavens by discoveries such as those of Copernicus, Galileo, and Hubble, and the demystification of life by persons like Darwin and Wallace, and by the life-related discoveries in chemistry and biology. He is reminded of the opposition to theories of evolution and natural selection, and cites University of California Law School Professor Phillip Johnson who argues that there is no "incontrovertible experimental evidence" that evolution is not guided by some divine plan. *Naturalistic* evolution according to Johnson, does not provide a very good explanation for the origin of species.

Weinberg thinks Johnson goes wrong here because Johnson has no feeling for the problems that any scientific theory alway has in accounting for what we observe. It seems to Weinberg a "profoundly important discovery" that we can get very far in explaining the world without invoking divine intervention--in biology as well as in the physical sciences.

On Conflict Between Science and Religion: One often hears, he is reminded, that there is no conflict between science and religion. But, he says, the meaning of religion is defined by what religious people really believe, and it is his sense that the majority of religious people would be surprised to learn that religion had nothing to do with *factual reality*.

Though there are still countless things in nature that we cannot explain, he acknowledges, we think we know the principles that govern the way they work. For those who see no conflict between science and religion, he observes, religion's withdrawal "from the ground occupied by science is nearly complete."

No Special Status for Life? Judging from this historical experience, it is his guess that, though we will find beauty in the final laws of nature, we will not find any special status for life or intelligence. And *a fortiori*, that we will find "no value or morality" standards, or hint of "a God who cares about such things." Although we may find these things elsewhere, he suggests that it will not be found in the laws of nature.

Grappling with the Theodicy: The "God of birds and trees," he feels, would have to be also the "God of birth defects and cancer." Religious people, he reminds us, have grappled for millennia with the theodicy; the problem posed by the existence of suffering in a world supposedly ruled by a good God. They have suggested "ingenious solutions" in terms of various supposed divine plans.

He will not argue with these solutions, or supply one of his own. Remembrance of the Holocaust, he says, leaves him unsympathetic to attempts to justify the ways of God to humans. If there is a God that has special plans for humankind, then, He has "taken very great pains to hide His concerns for us." It woud seem "impolite, if not impious," he ventures, to " bother such a God with our prayers."

Need Evidence of Some Special Role for Intelligent Life: Weinberg observes that if we found *some special role for intelligent life* in the final laws, we might then conclude that the creator who established these laws was "in some way specially interested in us."

Wheeler's Observer Universe: Of course, the quantum theory observer requirement suggests a special human role. Weinberg notes physicist John Wheeler's impression that according to the standard Copenhagen interpretation of quantum mechanics, a physical system cannot be said to have any definite values for correlated quantities, like position or energy or momentum, until these quantities are measured by some observer's apparatus. So for Wheeler, intelligent life is required to give meaning to quantum mechanics.

But, he says, other physicists, like himself, prefer looking at quantum mechanics in terms of a wave function that can describe "laboratories and observers," as well as atoms and molecules, as governed by laws not materially dependent on observers.

The Values of the Physical Constants As Favoring Life: He notes that some scientists make much of the fact that some of the fundamenal constants have values that seem remarkably well-suited to producing intelligent life in the universe. He suggests that it is not yet clear whether there is significance to this observation. But even if there is, he says, it does not necessarily imply "divine purpose." As we have discovered more fundamental physical principles, he says, they seem to have less to do with us.

John Polkinghorne--Science and Theology Finding a Home: Weinberg reminds us that people would not in any event have expected to learn about God from scientific discoveries. He cites theologist-physicist John Polkinghorne as arguing eloquently for a theology "placed within an area of human discourse where science also finds a home," that would be based on religious experience such as revelation, just as science is based on experiment and observation.

Different Religious Experience Versus Consensus on Scientific Discoveries: Those who think that they have had religious experiences of their own have to judge for themselves the "quality" of that experience, he observes. But the great majority of the world's religious are relying on revelations supposedly experienced by others. While this might not seem so different from the theoretical physicist relying on the experiments of others, there is, he declares, a very important distinction.

In contrast, the statements about God, or anything else derived from religious revelation, points in radically different directions, he suggests. Unlike the consensus on scientific discoveries, after thousands of years of theological analysis, we are, he feels, no closer to a common understanding of the lessons of religious revelation.

Steven Weinberg's 'Pointless' Universe: Steven Weinberg got into real trouble with his reference to a *pointless* universe. In his best seller *The First Three Minutes,* he had declared that given the quantum mechanical random chance creation scenario, the likely ignominious end of the universe, and the tragedy and pain and suffering inflicted on life in the universe, that *The more the universe seems comprehensible, the more it also seems pointless.*

Given his acclaim as one of the great theoreticians in physics, a Nobel Laureate, his statement had immediate and dramatic responses, some with acqiescence, but more with disapproval, and some with even dismay. Physicist Heinz Pagels, for example, while agreeing that the self-generating universe does not suggest purpose, expressed the view that there is still room for purpose and that science must continue its pursuit of truth.

The Poll on Weinberg's 'Pointless' Universe: Weinberg cites the results of Alan Lightman and Roberta Brawer's interviews with twenty-seven cosmologists and physicists on what they thought of that remark. With certain qualifications, he notes, ten agreed with him and thirteen did not.

Of the thirteen disagreeing, however, three did so, he comments, because they did not see why anyone would expect the universe to have a point. And Harvard astronomer Margaret Geller asked, Why should it have a point? What point? It's just a physical state. And Princeton astrophysicist Jim Peebles remarked, "I'm willing to believe that we are flotsam and jetsam." Another Princeton astrophysicist, Edwin Turner, agrees with him but suspected that Weinberg had intended the remark merely to annoy the reader.

Weinberg's favorite response was that of his colleague, astronomer Gerard de Vaucouleurs who said that he thought the remark was "nostalgic." Indeed it was, Weinberg acknowledged, nostalgic "for a world in which the heavens declared the glory of God."

God As a Necessity for Meaning: With respect to Weinberg's reference to the "pointless" universe, many persons, especially religious persons, see a universe without God as meaningless. In *The Creator And the Cosmos*, astronomer-minister Hugh Ross indulges in a clear and cogent expression of these sentiments.

He declares that if the universe is 'accidental' (quantun-uncertainty created), then the universe has "no objective meaning," and so it follows, he observes, that life, "has no meaning." In that situation, everything is determined by a "mechanical chain of events." Morality and religion, though "temporarily useful," he says, are "ultimately irrelevant."

The Origin of the Universe As the Basis for Meaning: But if the universe is "created," then there must be "reality beyond the confines of the universe" and it is the Creator that is that ultimate reality. Thus the Creator is not only the the source of life, but "establishes its meaning and purpose." And he adds, it is the "Creator's personality that defines personality" and "the creator's character that defines morality." So to study the origin and development of the univese, he observes, is to investigate "the basis for any meaning and purpose to life." Thus, he concludes, "cosmology has deep theological and philosophical ramifications."

But Mainstream Physics Goes Only As Far As the Anthropic Principle: But the urgent necessity of Ross's theistic God to assure meaning of the universe and of life is probably at variance from mainstream physicists. Some, like Paul Davies in *The Mind of God* and George Greenstein in *The Symbiotic Universe*, are willing to give some credence to versions of the anthropic principle as possibly suggesting purpose.

The 'strong' version of that principle holds that there is evidence of 'design' or purpose in the remarkable coincidences reflected in the exponential precision in the constants of physics which assure life in the universe. Physicists Davies, Greenberg, Rothman, Kistiakowski, and others, appear inclined in this direction.

Different Perspectives on the Anthropic Principle: But some scientists, like Stephen Hawking, Heinz Pagels, Steven Weinberg, and P. W. Atkins, do not see a universe which evidences purpose, as derived from the 'strong' anthropic principle which emphasizes the incredible coincidence of precision fine-tuning of the constants of physics to allow for life and consciousness. These physicists simply endorse the 'weak' anthropic principle which, in effect, means little more than the fact that we are here necessarily means that the constants of physics are such as to allow for life.

An Infinity of Universes to Assure Life? Physicists are inclined to the view that the quantum-uncertainty creation of the universe, and in particular, the "Many Worlds" quantum theory interpretation, runs against the idea of a universe which is 'purposeful' in human terms. But other physicists argue that quantum uncertainty is God's special device to assure free will, an interesting universe, and one in which everything is not predetermined, as in 'cause-and-effect' classical physics.

Physicist Fred Alan Wolf is a devotee of the 'many worlds' interpretation of quantum uncertainty which posits the existence of an infinite number of universes to satisfy "all possible quantum states." In his *Parallel Universes*, Wolf goes so far as to suggest that the creation of infinite universes is God's mechanism to assure life *somewhere* in that infinity.

A 'Final Theory' and Hawking's 'The Mind of God': We noted earlier Weinberg's observation that if there were anything we could discover in nature that would give us some special insight into the handiwork of God, it would have to be the final laws of nature which would represent the "book of rules that governs stars and stones and everything else." So, he understands why Stephen Hawking would refer to the laws of nature as "the mind of God."

The Irresistible Metaphor: But it becomes clear that Weinberg does not share Hawking's optimism about any final theory as having significance relating to the mind of God. Whatever one's religion or lack of it, he observes, it is an "irresistibe metaphor" to speak of the final laws of nature in terms of the mind of God. But as other physicists like Paul Davies have observed, Weinberg envisages no "ultimate principles" that do not require explanation in some further regression, or some "deeper principles," as he puts it.

Conflict between Science and Religion? Weinberg believes that science will find no "hint of God" or "standards of value or morality," although we may find these "elsewhere." He is unaware of any scientist who finds scientific evidence for a divine being, although astrophysicist Hugh Ross, as we discuss, does look upon scientific discoveries as doing just that.

It would be wonderful, he observes, to find in the laws of nature a plan prepared by a concerned creator in which human beings played some special role. "I find sadness," he acknowledges, "in doubting that we will." And nothing is accomplished, he feels, by identifying the laws of nature with "some sort of remote God," as Einstein did.

God and Occam's Razor: Interestingly, though, many physicists, conscious of the potential for infinite regression of explanations, do prefer to keep God available as the ultimate explanation of everything. After all, the time-honored principle of Occam's razor favors the simplest among competing options. Weinberg recognizes, however, that many physicists do maintain a "nominal affiliation" with their family faith but, he notes that few pay much attention to their religions's theology.

Here he is, in effect, denying that the anthropic principle which addresses the fact of the incredible coincidences in the laws of physics which aggregate to permit life, signifies any discernible special meaning or special purpose, as physicists like Paul Davies, John Gribbin, Martin Rees, and George Greenstein seemingly imply.

Criticizes Religious Liberals: Although he expresses concerns over religious conservative views, he is openly critical of "religious liberal" belief that people can believe in "different mutually exclusive things" *without any of them being wrong.* He points to the evolving views of scientists who adjust their conclusions again and again about matters that can be subjected to objective experiment and verification. So, he asks, how can we rely on the claims of religious tradition or sacred writings as somehow imbued with certain knowledge about metaphysical-unverifiable matters which are "beyond human experience"?

Why Not Tilt Our Beliefs in Favor of Consoling Religion? He notes that just as we had to learn in growing up to resist the temptation of wishful thinking about ordinary things like lotteries, so also "our species has had to learn in growing up that *we are not playing a starring role in any sort of grand cosmic drama.*

But, he asks, if we do not find spiritual comfort in the laws of nature, why should we not find spiritual comfort elsewhere--"in spiritual authority or in an independent leap of faith?" If we can adjust our beliefs, for example, to minimize the pain of death of a loved one, then why not do so? He replies, cryptically, that he sees "no scientific or logical reason not to seek consolation by adjustment of beliefs--only a moral one, a point of honor."

The Honor of Resisting Temptation: He acknowledges that science is unlikely to provide the consolations offered by religion in facing death. But, he muses, the "honor of resisting this temptation" [to tilt to the consoling], is only "a poor substitute for the consolations of religion," but this resistance is "not entirely without satisfactions of its own."

We have noted Weinberg's point of view that steers clear of finding signs of divine purpose in the laws of physics, probably representing the predominant view among physicists. But there are others in the field who see something that Weinberg does not.

Hugh Ross As the Astronomer and Biblical Inerrancy Advocate: Besides his career work in astronomy and cosmology, astronomer-minister Hugh Ross, author of *The Creator and the Cosmos,* is founder of "Reasons To Believe," a nondenominational religion-related organization which adheres to the doctrinal statements of the National Association of Evangelicals, as well as the International Council on Biblical Inerrancy. He lectures frequently on subjects which address the physics and theology of cosmology.

In the last chapter, "The Point," of *The Creator and the Cosmos*, he recounts his lecture experience involving audiences of scientists where emphasis is placed on science discovery as it tends to confirm New Testament passages.

Recent Discoveries As Proof of the Big Bang and of God: Ross's book emphasizes the recent scientific developments which, he feels, demonstrate the validity of Einstein's general relativity theory giving rise to the Big Bang Theory which, in turn, demonstrates the existence of God.

NASA Cosmic Background Radiation Measurements: He cites the recent NASA COBE Explorer satellite measurements of the cosmic background radiation left from the Big Bang, showing minute temperature variations in the early radiation, taken as the basis both for galaxy formation and as supporting the theory that there was a Big Bang, that the Big Bang expansion process was initiated by an exponential 'inflation,' and the temperature variations initiate the gravitational processes leading to galactic formation.

Proof of God: Ross expresses the view that incontrovertible proof of God is provided by this discovery, reinforced as it is by other recent corroborative radiation measurements, together with the remarkable life-supporting constants of physics, involving the *anthropic principle*.

He further cites the June 12, 1995 announcement of Johns Hopkins University astronomers that a NASA space shuttle mission with the Astro 2 Observatory Ultraviolet Telescope sent back views into very distant intergalactic clouds, showing spectral lines for helium in amounts exactly corresponding to the 24 percent of the starting universe's initially-created hydrogen would have been fused into helium by the Big Bang fusion cauldron within the first three minutes from the Big Bang.

He also cites passages from the New Testament involving seeming miracles, which, he feels, are now explainable by the dimensionality of space and time.

Disappointed over Non-Committal Responses: Ross gives expression to his disappointment over the uncertain or noncommittal responses to what he views as powerful arguments for scientific confirmation of Bible statements. The "beauty" of the scientific evidence God has made available through the discovery process, he remarks, should satisfy the doubts of two large segments of society: (1) those whose barriers to personal faith are intellectual, and (2) those whose barriers come from "personal pain or stubborn rebellion" which, he notes, surface in the form of "intellectual objections."

Drawing Near to God: It is awesome, declares Ross, to behold the character of the Creator in what He has made--in the elegant architecture of the universe, a galaxy, the sun, the earth, the moon, and in a human being. Some people, he notes, are struck by the wisdom, power and care of the Creator, while others see merely an amazing coincidence--or perhaps even the work of extraterrestrials.

Not everyone seems to see God in all this, laments Ross. Drawing near to God, calling on Him "in truth," observes Ross, begins with humbly acknowledging who we are--His creation, and who He is-- "the Divine Maker and Provider of all things, including a way across the gulf that divides us from Him." The Bible tells us, he says, that "if we draw near to Him He will draw near to us."

Scientific Evidence Proportionate to Resistance to the Truth: Ross concludes with the proposition that we are now being flooded with scientific evidence of his existence because God measures out evidence in proportion to the resistance to His truth. And, Ross laments, the Western world, having the most wealth, discretionary time, education , and technology, responds that "we humans deserve all the credit," that "humanity is deity."

Humanity As Deity: By humans deserving all the credit, Ross is referring, seemingly, to the quantum theory concept that man's current observations of the universe may have retroactively collapsed the universe's initial quantum uncertainty state into our present reality.

His reference to "humanity is deity" is seemingly directed at physicist Frank Tipler's Omega Point Theory that human intelligence spreads throughout the universe until the collective intelligence evolves into the Omega Point God which, in quantum and relativity theory, then occupies all of time to give us our present God. Ross ends with an appeal that the ample evidence of God be used to "build our own faith and the faith of others."

> For me it suffices to wonder at these secrets and to attempt humbly to grasp with my mind a mere image of the lofty structure of all that there is.--Albert Einstein

PART IV--COSMOLOGY-- RELATIVITY THEORY

CHAPTER 8 A RELATIVITY

SOME RANDOM THOUGHTS ON RELATIVITY

Einstein's general relativity: The single greatest human creative thought achieve-ment in the history of science.--Edward Teller

Relativity: The greatest Feat Of Human Thinking.--Max Born

Two great revolutions gave birth to the new physics: quantum theory and the theory of relativity. The latter almost exclusively due to the work of Einstein, is a theory of space, time and motion. Its consequences are as equally baffling and profound as the quantum theory. --Paul Davies

Between 1904 and 1917, or between his 25th birthday and his 38th, [Einstein] had rebuilt the universe in his head. --Nigel Calder

Special relativity: The Law of all creation. --Nigel Calder

Light from distant stars at great speed for long periods of time across vast areas of space, travel no distance whatever in spacetime. --Edward Harrison

Space by itself and time by itself are doomed to fade away into mere shadows, and only a kind of union of the two will preserve an independent reality. --Hermann Minkowski.

RELATIVITY THEORY

General Relativity Equations Define the Origin and Fate of the Universe: We noted earlier that modern cosmology dates back to Einstein's relativity theories, especially the 1916 theory of general relativity, applying relativity to bodies in motion, introducing gravity as equivalent to acceleration, and gravity as determining the geometry of space-time. *It is the equations of general relativity theory that define the origin, expansion, and destiny of the universe.* As an important aspect of cosmology, a brief review of relativity theory will be helpful.

Cosmology--Relativity Theory and the Big Bang: At one time, cosmology, the study of the universe, was dominated by theology, the account of God's Creation. But major reorientation came with the advent of Einstein's then shocking theory of general relativity in 1916. Einstein's general relativity equations had profound implications for physics and theology, for they demonstrated mathematically that the universe was not eternal, as many had thought, but had a *beginning.*

In *The Fingerprint of God*, astronomer-minister Hugh Ross, after summarizing how general relativity mathematics demonstrates the Big Bang, notes that "Somehow, from beyond itself, the universe came to be....a limited time ago ...[with] monumental implications [the existence of God as Creator].

The theory predicts, in part: (1) that the wavelength light frequency should shift as it comes within the gravity influence of a massive object--the so-called electromagnetic wavelength spectrum red-shift; (2) the existence of *gravitational waves;* (3) the existence of bizarre gravitational incongruities, 'black holes,' containing *singularities* where time stops, gravitation is infinite, and the laws of conventional physics break down; and (4) the Big Bang start of the universe.

Special Relativity: *Special relativity theory's* underlying concept is that: the basic components of the universe--time, space, mass, and motion, previously thought to have been 'absolutes,' are interdependent and relative, to each other and to a universal constant, the speed of light--referring to the velocity of electromagnetic (EM) transmissions, EM being one of the 'fundamental forces' that give us our dimensionality and, in essence, make up the universe.

Implied in this 'interdependency' and 'relativeness,' are the seeds of unification theory--that space, time , electromagnetism, and the fundamental forces, derive from some earlier unified state, as Einstein theorized.

Reference to the speed of light as a *universal constant* means that the speed of light (or any other electromagnetic or gravitational transmission velocity, such as that of radio waves, X-rays, gamma rays, ultraviolet rays, infrared rays, or microwave rays) in the vacuum of space is the same in all inertial frames--for all freely-moving observers.

Unlike classical mechanics, where 'separate' speeds are added or subtracted, forward speed of a spaceship does not increase the speed of light directed forward from the spaceship--or decrease the speed of light directed from the back of the forward-speeding spaceship. Also, the speed of light is the same for all observers. There is no absolute reference system permitting the measurement of absolute motions in space.

This reveals that space and time are part of the *single continuum* of *space-time,* and interrelated with the force of electromagnetism. Equally revealing is the proposition that mass or matter is a form of *energy*, rest energy, described by Einstein's famous equation declaring the equivalence of energy and mass or matter in its relationship to the speed of light: $E = MC^2$ --Energy equals mass times the speed of light squared.

General Relativity: *General relativity* extends special relativity principles to accelerating bodies. It defines gravity not as a force of attraction operating in flat spacetime but as *geometry*, the effect of matter (mass) to curve surrounding space-time so as to dictate motion and paths of affected bodies. As physicist Paul Davies observes in *Superforce,* the pull across space is now 'spacewarp.'

Newton's gravity attraction theory was that gravity was a force like others, in effect acting in flat spacetime. But general relativity asserts that spacetime is curved by mass, so bodies (and light) are impelled to move in *geodesics*, orbits curved in three-dimensional space, but representing *straight paths in four-dimensional spacetime*.

General relativity also asserts that the effects of gravity and acceleration are the same, equivalent. General relativity also asserts, as did Newton, that gravitational mass (weight) and inertial mass (push resistance) of different materials of the same mass are equal. This means that different compositions of mass have the same gravitational acceleration, at least in empty space where there is no atmospheric resistance.

Changing Concepts of How Gravity Works: In *Was Einstein Right?* Clifford M. Will points out that at first, gravity was thought of as causing an "action" at a distance, a concept later replaced by the thought of gravity as creating a field of force, like electromagnetism. But Einstein came up with a startlingly different alternative, that *mass produces the gravitational effect of warping surrounding spacetime; gravity is, in effect, geometry.*

Thus Einstein's "leap," says Will, was the principle of *equivalence of gravity and spacetime curvature*--that gravity was, in effect, *behavior* caused by geometry. So the gravitational force one feels, when not in free fall, is the effect of motion in curved space-time; *gravity is curved spacetime geometry*. The mathematics of relativity led Einstein to seek the mathematical underpinnings of a universe in which all forces and fields, space and time were interrelated so as to suggest a primordial unification state.

General Relativity Quantifies Space-time Curvature: Will notes that although Einstein spoke of "equivalence," curved spacetime and general relativity, without really distinguishing them, that general relativity is now thought of as an entity separate from equivalence and spacetime curvature.

The equivalence principle, Will notes, says only that spacetime is curved. But it doesn't tell by *how much*. It is general relativity, says Will, that provides a set of mathematical 'field equations' *permitting calculation of the amount of spacetime curvature generated by any given matter.*

In effect, general relativity provides the mathematics to *quantify the amount of spacetime curvature* produced by a given mass. So, says Will, the *equations* specify the amount of curvature, and the *equivalence principle* specifies how matter responds to gravity--the motion of bodies, like the Earth in gravity-curved spacetime---along geodesics.

Gravity as Geometry: So gravity represents acceleration, the effect of mass to influence the geometry of spacetime, to generate the motion of bodies along paths curved in the fourth dimension of time.

Gravity also has the effect of bending light or other electromagnetic transmissions via the relativity-related warping of surrounding spacetime by mass. As Heinz Pagels puts it in *Perfect Symmetry*, under relativity theory, *gravity was reduced to geometry; and matter, space and time became unified.* The gravity effect between any two bodies is said to derive from the exchange of massless 'messenger,' *graviton particles*.

According to general relativity, the wavelength or frequency of light should shift as it comes within the gravitational influence of a massive object--so-called gravitational red-shift. Two other predictions of general relativity are the existence of *gravitational waves* and the existence of *black holes,* the ultimate gravitational effect in which *matter is exponentially concentrated and time stops*.

Einstein Had Rebuilt the Universe in His Head: Relativity theories, which provide the mathematical framework for the Big Bang and remarkable insights on the nature of the universe, were largely the work of a single human mind, that of Albert Einstein, at the time, an obscure Swiss patent examiner.

In announcing his relativity theories Einstein had created an *astounding new threshold both for physics and cosmology.* The world would never be the same. In his commemorative *Einstein's Universe*, Nigel Calder observes that between 1904 and 1917 Einstein, by his relativity theories, *had rebuilt the universe in his head,* accomplishing the most revolutionary scientific breakthroughs ever conceived by the mind of human life.

Nuclear physicist Edward Teller described Einstein's work as *the single greatest human creative thought achievement in the history of science.* The great quantum physicist Max Born described Einstein's work on relativity as "the greatest feat of human thinking." Einstein's 1979 birth centennial was marked world-wide by academic and scientific symposia, conferences, books, and other commemorative activities.

Elevating Cosmology and Redefining the Universe: In presenting his theories of special and general relativity, Einstein had elevated the study of physics and cosmology to new levels, providing new insights into age-old cosmic mysteries. Space, time, energy, matter, and motion, had been perceived as essentially independent absolutes. And, as Newton had described it, gravity had been viewed as the fundamental force of 'attraction.'

But by his special relativity theory, Einstein mathematically demonstrated that space, time, mass, and motion are not absolutes but inextricably related, interdependent, the single constant being the speed of light. And by his simple but portentous equation $E = MC^2$ (Energy equals inertial mass times the speed of light squared), Einstein had declared that mass was energy whose motion was arrested, 'rest energy.'

The Entire Universe Is Just Energy: So that simple special relativity energy-matter equivalency equation told the world that all of our supposedly physical universe of such incredibly diverse forms as radiation, matter, galaxies, planets, light, air, diamonds, automobiles--and people, is reducible to a single component--energy.

But he went further, formularizing the constancy of the speed of light, identifying gravity as a form of acceleration, focussing on the nature of time as a dimension of the universe, and redefining the nature of space and time as a single continuum. In declaring matter as rest or locked-up energy, and as convertible to motion energy, he had, of course, formularized the atomic and hydrogen bomb. The world was now a far more perilous abode. His famous letter to President Roosevelt on this subject was motivated by the concern that which nation first achieved the atomic bomb would determine the war's outcome.

Expanding Universe, Black Holes, Creation and End of the Universe: As time passed, it was realized that the application of Einstein's general relativity equations *implied* some astounding conclusions: that the *space* in the universe is ***expand-ing***, later confirmed by astronomer Edwin Hubble's finding that the galaxies were receding.

Tracing the expanding universe back to its smaller states signified that the universe must have originated in the explosion of an infinitely massive, hot, cosmic mini-universe speck, exponentially smaller than a proton, representing the beginning of space, time, energy-matter, and all the fundamental forces and fields. The relativity equations also mandated that the ultimately reduced state would represent *black holes*, where the classical laws of physics break down. Time, space, and matter would no longer be definable by classical physics.

Relativity--Demonstrating Cosmic Interrelationships: Building on the work of such pioneers as Galileo, Kepler, Newton, Mach, Gauss, Reimann, Fitzgerald and Lorentz, Clifford, Poincaré, Minkowski, and others, Einstein synthesized their work, broadening their concepts to show the interdependency and relativity of space, time, mass, motion, to a constant, the speed of light, setting the stage for the scientific world's drive for the mechanics and mathematics of *unification theory.*

Newton's Gravity Redefined As Geometry: In general relativity, Newton's 'gravitational attraction' became the geometric curving effect of mass or matter on surrounding space and time; and space and time became the single entity of a *four--dimensional spacetime continuum*. Gravitation, in Einstein's general relativity theory, became acceleration, not in the reference frame of flat three-dimensional space of Euclidian geometry but in three dimensional space that is curved in the fourth [time] dimension, explained by non-Euclidian (Reimannian) curved-surface geometry.

As astronomer, science writer Timothy Ferris observes in his monumental work, *Coming Of Age In The Milky Way*, within the framework of the four-dimensional spacetime continuum, "gravitation *is* acceleration, the acceleration of objects as they glide along 'world lines,' paths of least action traced over the slopes of a three dimensional space that is curved in the fourth [time] dimension."

The Accelerating Spaceship: Some of these interrelationships are illustrated by measurements relating to an accelerating spaceship departing from Earth, taken by a theoretical observer located elsewhere (in a frame of reference independent of the accelerating spaceship's frame of reference). The Earth, or another spaceship, for example, would be such an elsewhere or independent frame of reference for measurements of events on the accelerating spacecraft.

M*easured by this theoretical independent observer*, as the spaceship's speed or velocity increases, several so-called *relativistic interdependent events* are occurring: The *time* on the spacecraft's clock, as measured by the observer, is slowing, the *length* of the spaceship, in the direction of its motion, is shrinking, and the space-ship's inertial mass is increasing. For a spaceship reaching the speed of light, time would stop, mass would be infinite, and the *spaceship would cease to have length* in the direction of its motion!

Contraction Formula for Constancy of the Speed of Light: Again, as measured by the observer, these three relativistic changes resulting from the increase in the spaceship's velocity, namely, the slowing of time, shrinkage in length in the direction of motion, and increase in mass, *vary* in such proportion as will preserve the constancy or *invariance* of the speed of light. The proportionality formula for preserving this constancy is variously referred to as: the Lorentz contraction formula, the Fitzgerald-Lorentz contraction formula, or the Einstein-Lorentz contraction formula.

The Irish physicist Fitzgerald and Dutch physicist Lorentz first looked upon these as *actual* physical changes. But Einstein perceived the formula as a mathematical measurement device to reflect *relativistic* rather than physical effects.

Size and Distance in Space, and Duration in Time, Are Relative: Relativity says that there is neither *absolute* size or distance in space, or *absolute time* or duration. The size, or distance occupied in space, of one object can only be determined relative to its interdependents of time, motion, and mass.

Subjective Judgments of Size: We are quite accustomed to the idea of relative size because that's the way we deal with it in practical terms--by comparing the size of one thing to that of another. One judges his own size either by comparing it with that of someone else, or with something taller or shorter, or by using units of a measuring device like a yardstick or ruler.

Thus a person is tall or short only as compared with another who is correspondingly shorter or taller, or as compared with some uniform standard. In relativity theory, distance or size is also relative, becomes longer or shorter interdependently with time, velocity and mass.

We're all familiar with the idea of relativeness of time. What is a long or short time? What is the length of a year? Our planet's year represents one orbit of our planet around our star, the sun; our day, one rotation on our planet's axis; and one hour, as one circuit of our clock's hour hand, representing one twenty-fourth of our planet's turn on its axis.

But the axis-turns and orbits of other planets--the shorter Venus orbit or the longer Mars orbit, or still longer Jupiter orbit--represent their 'days' and 'years.' This brings to mind the problem of the length of the first 'day' in the Genesis six-day creation sequence.

We all experience the relativity of *subjective impressions* of time. The 'watched pot' never boils; expected guests take forever to arrive; pleasurable events seem to end quickly; painful events seem endless. Impressions of time duration are affected by one's age. A child's time passes slowly; an adult's, more quickly, unless, of course, waiting for the dentist, or hungry and in a long dinner line.

At Rest and in Motion: Regardless of the motion of an object generating light, or the motion of an observer, the velocity of light relative to an observer is, as noted earlier, invariant or constant while, on the other hand, time, size, distance in space, mass, and motion are all relative to the observer's frame of reference.

The motion of two moving satellites in space is related to which satellite is selected as the frame of reference. Viewed from the frame of reference of either satellite considered as being at rest, the other may be considered the one in motion.

From earth's frame of reference, of course, *both* of two moving satellites are considered in motion. For that matter, from the frame of reference of either satellite, which may be considered at rest, both the other satellite *and the Earth* will be perceived as in motion. As physicists point out, there is the exception that Earth's motion may be considered as absolute in reference to the universal cosmic back-ground radiation. But in general relativity, even an object at rest in three space dimensions is in motion along its world line in a four-dimensional spacetime continuum curved by the presence of mass.

Light and the Sum of Velocities in Classic Physics: Before Einstein's relativistic physics, the speed of light was thought to vary according to the motion of the observer. If you flashed light from the front of a moving train, you would, it was thought, 'sum up' the two velocities to yield a combined speed of the train and light.

But special relativity recognizes a special characteristic of electromagnetism, an absolute velocity for light which remains invariant or constant regardless of the motion either of the observer or of the source of the light. So, as discussed earlier, the *sum of velocities*' rule of classical mechanics does not apply to electromagnetic transmissions such as light.

> Light from distant stars at great speed for long periods of time across vast areas of space, travel no distance whatever in spacetime. --Edward Harrison

CHAPTER 8 B
THE MYSTERY OF TIME

RANDOM THOUGHTS ON TIME

The distinction between past, present, and future is only an illusion, however persistent. --Albert Einstein

We physicists work with time every day, but don't ask me what it is. It's just too difficult to think about.--Richard Feynman

Space by itself and time by itself are doomed to fade away into mere shadows, and only a kind of union of the two will preserve an independent reality. --Hermann Minkowski.

All reality ...is a one-way process in time; unitary; continuous, irreversible; self-transforming; and generating variety and novelty during its transformation. --Julian Huxley

Time goes, you say? Ah no! Alas, Time stays, we go. --Austin Dobson, The Paradox of Time

Time is the moving image of eternity. Time and heaven came into being at the same time. --Plato

The leading idea which is present in all our researches, and which accompanies every fresh observation...is Time!--Time!--Time! --George Scrope

Spinoza--Measure, time and number are nothing but modes of thought or rather of imagination. --Spinoza

As if you could kill time without injuring eternity. --Thoreau

Time is the life of the soul. --Longfellow

Time is the measure of motion. --Aristotle

It is manifest that time and eternity are not the same. --Aquinas

God alone is--according to an eternity...not measured by time...one who by one single *now* fills the *ever*.... --Montaigne

Time is so fleeting that if we do not remember God in our youth, age may find us incapable of thinking about him. --Hans Christian Anderson

But time and chance happeneth to them all. --Ecclesiastes

My days are swifter than a weaver's shuttle. --Job

He who has seen present things has seen all, both everything which has taken place from all eternity and everything which will be for time without end; for all things are of one kin and of one form. --Marcus Aurelius

The past does not exist except in our memory, nor the future, except in our expectations of it. The most a clock gives us is the time an instant ago--David Allen

What we take for the history of nature is only the very incomplete history of an instant. --Denis Diderot

So no object or phenomenon can be properly understood by considering its ext-ension in space alone. All things have a shape in time as well as in space.... Time is perceived as the way of measuring the progressive change that is building the universe even while we live and take part in it.... We must compare the whole of time to the act of creation. --Louise. B. Young

Time is nature's way of keeping everything from happening at once--George Zebrowsky

ST. AUGUSTINE AND TIME

An Intuitive Metaphysical Insight: No discussion of time, be it philosophical, theological or scientific, gets far without exploring St. Augustine's remarkable work, one of the most profound and intuitive metaphysical insights of early times; insights which mysteriously anticipate the advent of Einstein's relativity theories.

"What, then, is time?" asks St. Augustine. "If no one asks me, I know what it is. If I wish to explain what it is to him who asks me, I do not know." And so, in the fifth century, St. Augustine expresses the perplexity in understanding the meaning of time that persists to this day. A teacher of rhetoric, St. Augustine sought to define the nature and characteristics of time.

St. Augustine's View of the Direction of Time and of Creation: St. Augustine was one of the first to speak of time as having linear direction. Anticipating the general views of contemporary science, he also spoke of the six days of creation as being allegorical descriptions of relatively lengthy periods, and our creation as being from primordial matter from which the species evolved in the general order indicated in Genesis.

Time As Beginning with the Creation of the Universe: In his *Confessions*, St. Augustine dwelt at length on the matter of time. As noted earlier, he is often cited by physicists as anticipating the Big Bang theory in asserting that time began with the creation of the universe. This idea of time beginning at creation has led to many questions, not the least of which is the question of what was God doing with the 'nothingness' prior to creation and the existence of time. How could God have 'planned' the creation, a temporal activity, before time came into being?

But St Augustine Goes on: "If before heaven and earth were made there was no time, then what is meant by the question 'What was God doing then?' If there was not any time, there was not any then." He added that to say there was a time when time was not, is as absurd as saying this world was when it was not. It is remarkable that, fifteen hundred years ago, a person with such intuitive insights was discussing such ideas in a manner so compatible with sophisticated contemporary science.

Where Are the Future and the Past? St. Augustine asked, "Where are the future and the past?" Wherever they may be, he says, they are there not as the future or past but as the present, for, he notes, "the future is not yet and the past is no more." The present of things past, he said, is memory, the present of the present is sight, and the present of the future is expectation.

David Allen of the National Institute of Standards and Technology adopts this fifteen-century-old philosophic view when he observes that "The past does not exist except in our memory, nor the future, except in our expectations of it. The most a clock gives us is the time of an instant ago."

"I Do Not Even Know What I Do Not Know!" We speak of the one-way direction of time as toward the future. But St. Augustine spoke of time as from the future, passing by way of the present, into the past. Some would agree that while the thermodynamic and electromagnetic arrows of time points to the future, from the past through the present, human perception of the flow may, as St. Augustine thought, be from the future to the present and into the past.

This was St. Augustine's perception as he said, “In other words, it passes from that which does not yet exist, by way of that which lacks extension, into that which is no longer.” He goes on: “...I do not know what time is...or perhaps I do not know how to express what I know.” Expressing the eternal human dilemma, he plaintively declares: “Alas for me, I do not even know what I do not know!”

GOD BEFORE TIME

The Paradox of Time: Paul Davies, in *God And The New Physics*, discusses the idea of God ‘planning’ the universe from nothing. As St. Augustine had noted, God created time when He created the spacetime universe, a proposition consistent with Einstein's general relativity theory. But planning, even for God, says Davies, is a temporal activity, something performed in time, doing something now for a future effect. Without time there is no distinction between past, present or future.

So how could God ‘plan,’ an act in time, before the existence of time, which allows temporal activity like planning? Theologian Paul Tillich addresses this by his view of God as existence itself, not constrained by its laws of physics.

A God Who Transcends Time? Physicist Paul Davies points out the problem of a God who transcends time; that most of the activities and qualities of God--planning, answering prayer, sitting in judgment--only make sense in the context of time. If He transcends time--in his omniscience, knowing the future, why, asks Davies, is He concerned about the fight against evil or human progress, since He knows all outcomes.

The very act of creation, says Davies, is an act in time. A caused creation involves time in the form of cause and effect. It is, by definition, temporal and "cannot be applied to a state in which time does not exist." Says Davies, The naive image of God existing before the universe is clearly absurd if time did not exist--if there was no before.

Resort to a Quantum Theory for ‘Retroactive Causation’: But Davies goes on to discuss phenomena in the laws of physics which might permit the creation prior to the existence of time, to avoid conflict with the Genesis account. So to reconcile the absence of time with God's providing for the universe to come into being, he invokes quantum theory. Davies describes a quantum form of causation, one not requiring present causation for future effect. This brings him to *retroactive causation*, faster-than-light phenomena, a reversed time causality involved in quantum theory.

He quotes physicist John Wheeler's reference, in *Genesis and Observership*, to the quantum phenomenon in which the actions of the observer in the future, "defines what happens in the past--even in a past so remote that life did not then exist." Davies also goes on to discuss theologian Tillich's God as apart from or 'beyond' time, the condition existing 'before' the commencement of time, before the geometric dimensionality of our spacetime universe.

STARTING EARLY--WHAT IS TIME?

Early Philosophic Notions: The subject of time has always been a source of wonder. Earlier societies developed many philosophic and theological notions having to do with time. Plato declared that time is the moving image of eternity; that time and heaven came into being at the same time, a thought also expressed by St. Augustine. Time is the measure of motion, said Aristotle, a concept strangely consistent with Einstein's theory of special relativity.

In his *Confessions*, St. Augustine asserted that time began with God's creation of the universe, a remarkable insight which, although once questioned, as we have noted, is consistent with the Big Bang theory derived from Einstein's general relativity theory.

Time--The Measure of Life--The Preoccupation of Poets and Philosophers: Although everyday conversation doesn't focus much on the laws of physics, time is commonly a subject of endless speculation, dwelt on by poets, philosophers and theologians. They wondered whether time was something physical. Spinoza said that: measure, time and number are nothing but modes of thought or rather of imagination, a concept strangely consistent with modern quantum theory in which physicists suggest our perception of time as a quantum mental construct, a virtual reality created by the human mind as the quantum observer.

From the beginning, subjective ideas of time, so-called psychological time, have been highlighted, even in biblical literature. Job observed: "My days are swifter than a weaver's shuttle." " But time and chance happeneth to them all," says Ecclesiastes.

Time is part of one's constant awareness, daily routine, and existential concerns. Any thoughts about life, death, change, history, planning the future, reminiscing about the past, all invoke concepts of time.

Time, Life and Death, History, and Divinity: Because time is naturally associated with existence--with the cessation of life, death-- the subject of time is powerfully arresting. To this day, figures of Father Time and the Grim Reaper come readily to mind on New Year's Eve. Memorial eulogies turn inevitably to the role of time. It is time which permits the history of human life, pride in the past, hope for the future, and fundamental philosophic ideas.

Quantum Theory, Consciousness, Death, and Time: Elsewhere we discuss the nature of the mind, of consciousness, and whether it exists in time. Physicist Paul Davies notes that the mind and consciousness are nonmaterial entities that exist somewhere. He asks where they exist and *whether they exist in time*. He concludes that our very perception of time, require consciousness to exist in time.

Can Consciousness Survive Death, the Cessation of Its Organic Framework? Quantum physics, Davies notes, leads us to a *different conclusion* on this question than does classical physics and biology. It appears that concepts of holism, quantum theory, spiritualism and religious transcendentalism all allow for survival in time through physical death-induced disembodiment of the consciousness.

This holds out the hope, Davies notes, of *making scientific sense of immortality*, focussing more light on notions of "pre-existence" and "post-existence," and in exploring the "temporal dilemma [that] runs through all discussions of immortality."

What Is Time? But what is time? Physicist Richard Feynman noted that "We physicists work with time every day, but don't ask me what it is. It's just too difficult to think about." Some would say that time is simply what a clock measures, a succession of events. But the subject of time is not quite that simple. Although there are many aspects of time that still baffle science, there are now many things about time that we do know and can describe.

Time Viewed Differently: Time has been viewed differently in different cultures and at different times. In some religions, time, particularly human time on Earth, is viewed as running in cycles in which people die and are reborn. Some ancient Greek philosophers perceived time as an illusion, an interesting thought in the light of Einstein's coexistence of all time and references to past, present and future as illusion.

Newton believed in a time and space that were absolute and unchanging, a view superseded by Einstein's relativity theories which see time and space as a continuum, time as elastic, and interdependent with space, motion, and gravitational influences.

Time and Early Physics: For much of history, as we have noted, the nature of time was a mystery and the subject primarily of theological and philosophical thought. Galileo is credited with first introducing physical aspects of time with his mathematical formula for relating time to distance in falling objects.

The distance of fall, said Galileo, was proportional to the time of fall, squared: one foot the first second, four feet the next second, etc. Then came Newton's laws of motion, involving time. In Newton's second law, for example, F = MA, force equals mass times acceleration, where acceleration is the rate at which velocity changes in time.

Newton's View of Time As Independent and Flowing Evenly: But Newton viewed time as a phenomenon "flowing equably without relation to anything external." Newton's description of time as having an even flow, and independent of anything else, while popular in philosophy, had to give way to Einstein's special relativity theory in which time is not independent, does not flow at a fixed rate, and, in a sense, since it is dimensional, may not really even be said to 'flow.'

Time and Special Relativity: Special relativity theory, displacing Newton's absolute time, says that time is not independent and does not flow evenly. Time is interdependent with space, mass, and motion, the speed of light being the constant in their interrelationships. There is no single 'now.' Time varies between observers in different motion. In general relativity, time and distance vary according to the velocities of accelerating or decelerating observers.

Gravity As Acceleration in Space Curved in the Dimension of Time: Replacing Newton's physics of gravity as an attraction force, Einstein's general relativity theory holds that gravity causes objects to accelerate, implying the equivalence of gravity and acceleration. So gravitation, rather than being Newton's 'force of attraction,' is acceleration--the effect which mass has to induce motion in the surrounding three dimensional space that is curved in the fourth or time dimension, as explained by the non-Euclidian geometry of curved space.

As astronomer and science writer Timothy Ferris observes in his monumental work, *Coming Of Age In The Milky Way*, within the framework of the four-dimensional spacetime continuum, "gravitation is acceleration, the acceleration of objects as they glide along 'world lines,' paths of least action traced over the slopes of a three dimensional space that is curved in the fourth [time] dimension."

Time Emerges from the Quantum: It turns out, according to contemporary physics, as expressed for example in the writings of physicists like Stephen Hawking and Paul

Davies dealing with unification theory, that all the physical components of our geometry and structure, including space, time, the fundamental forces and fields, were constituent parts of unified physics. And they emerged from a primordial mega-universe quantum state, devoid of separated space, time, forces or fields, but filled with the potential of physics to create these components through a separation process.

The quantum-genesis model of creation has the universe starting as a random fluctuation in the primordial mega-universe quantum void which generates a mini-universe point or bubble of space and time, infused with a superforce of unified physics described as a state of supersymmetry. This 'point' undergoes an exponential expansion which subsides into the Big Bang model in which the unified fundamental forces and fields separate in a symmetry-breaking sequence giving us our present geometry and structure.

In Quantum Theory, the Future Affects the Present: In this early stage of the universe, time, space, energy, and matter, the forces and fields are all one. It is this picture that characterizes time as a quantum effect. The introduction of quantum mechanics in the study of time has produced some strange theories on the direction of time.

As we shall see, new concepts have evolved in which the future affects the present, in seeming contradiction to ideas about the arrows of time pointing to the future from the past via the present. Recent tests may show, says physicist Fred Allen Wolf in *Taking The Quantum Leap*, that the future may actually communicate with the present in the human nervous system.

TIME AND RELATIVITY THEORY

Relativity Theory Says That Time Is Interdependent: As noted earlier, rather than Newton's absolute time, special relativity theory says that time, space, mass, motion are interdependent, the only absolute or constant being the speed of light. Time and distance vary according to the differing velocities of observers and each reference body has its own particular time. There is no meaning in a statement of the time of an event, a now, unless we know the reference body to which the time refers.

Time Is Elastic: Relativity established the elasticity of time. Time stretches or shrinks related to the speed of whatever measures it and related to the amount of mass which happens to be in the neighborhood of the measurer. As an object speeds up, to an independent observer its length shrinks in the direction of motion, its mass gets heavier, and its time expands (slows).

Gravity is a distortion of spacetime caused by the presence of mass. Gravity is the force exerted by mass to curve both space and time in its vicinity. As gravity increases, time slows, so time runs faster in lower gravity space. When gravity increase reaches a critical point, as in a black hole, *time stops altogether*, as discussed in this and the chapter on Relativity.

Time and the Invariance of the Speed of Light: Einstein's proposition of the relativeness of 'now,' the so-called 'relativity of simultaneity,' shows that every reference body has its own particular time--that a particular time can have relevance only to the observer's location (his local time).

As discussed earlier, formulas of the so-called Lorentz transformation, worked out by Dutch physicist Lorentz, modify measurements of time and distance in accordance with the velocity of each system of reference so as to preserve the velocity of light as an absolute, described as invariant or constant. Thus the Lorentz equations require that the dimensions of an object contract along the direction of its motion as it accelerates, by just the amount necessary to assure that the speed of light remains constant.

Constancy of Speed of Light and Classical Addition of Velocities: Einstein spent a year in trying to resolve this problem of the contradiction between the idea of the speed of light as being invariant, an absolute or constant, and the classical mechanics theory that objects in motion should be added or subtracted to reflect their relative velocities.

In classical physics, the speed of a bullet fired forward from a moving vehicle is added to the vehicle's speed to determine the bullets's velocity. Why, then, does light, projected from a speeding spaceship in the direction of travel, move only at the speed of light rather than a speed reflecting both velocities?

Einstein and the Nature of Time: In developing his theory of special relativity, Einstein had spent a year after the Michelson-Morley experiments which had failed to prove the existence of an 'ether,' in trying to resolve the problem of the contradiction between the invariance of the velocity of light and the mechanics theory of added velocities. Einstein says, he finally struck upon the answer.

The Speed of Light Relates to the Nature of Time: Meeting his friend and helpmate Michele Besso, he announced to Besso that the solution to the invariance of the speed of light had to do with the nature of time. "An analysis of the concept of time," was the answer, he declared, contrary to earlier conceptions of time as an independent absolute, as Newton and others had asserted.

Time, he said, cannot be absolutely defined. There is "an inseparable relation between time and signal velocity." There is no absolute time or distance (length). *Time and distance are essentially inseparable parts of a single mathematical reality*. Thus, the question of whether two events are simultaneous cannot be answered in absolutes but is dependent on frames of reference, and the greater the distance between events the greater the problem in determining simultaneity.

As Martin Gardner observes, in *The Relativity Explosion*, it's not just having difficulty in determining what the truth is about the time of an event, *there is no actual truth.* One's local time, one's 'now,' has relevance only to one's spot in the universe, said Einstein, something "beautifully unexpected," declared physicist Edward Teller.

Time and Einstein's Special Relativity: Einstein's special relativity proposition of the relativity of simultaneity (no absolute 'now') shows that every reference body has its own particular time--that each observer has his own time scale. The fallacy of applying the sum of velocities to the speed of light lies in the hypothesis that the time of an event can be determined without reference to the motion of the system of reference. Light travels rectilinearly [in straight lines] in all directions at a constant speed. Einstein described it as the theory of "invariance."

Gravitational Effects on Time: We recall that gravity affects the geometry of three dimensional space by curving it in the fourth or time dimension, as explained by the non-Euclidian geometry of curved space. This geometric effect has its counterpart effect on time to produce time dilation. Just as gravity curves adjoining space, gravity slows the passage of time, as measured by an observer at some distance from the gravitational field that induces the dilation.

The ultimate time dilation is produced by the gravitational force within the 'event horizon' of a black hole. From an outside observer's reference point, the passage of time in that region is suspended altogether. Of course, for observers *within* this region of the black hole, time would appear unchanged. So this represents the ultimate in different times between observers.

Elasticity of Time Illustrated by Gamma Ray Time Dilation: Speaking of time dilation, astronomers Bohdan Paczynski at Princeton and Tsvi Piran of Hebrew University and Harvard had predicted that time elasticity, that is, the time-stretching or dilation effects of relativity would be verified from cosmic gamma ray bursts in which the time duration of bursts from great cosmic distances would be stretched or dilated as the radiation travels across the vastness of the universe of expanding space triggered by the Big Bang.

The NASA Compton Gamma Ray Observatory, launched in April 1991, detected gamma ray bursts from all directions in the universe from objects apparently not producing other visible or invisible wavelength radiation. Physicists theorize that they may be coming from cataclysmic encounters between black holes and neutron stars. At an American Astronomical Society meeting in January 1994, Dr Jay Norris of Goddard Space Flight Center reported that the dim bursts were typically twice as long as the bright bursts, an illustration of relativity theory's time-dilation effects.

'Now'-- the Point between the Future and the Past: What does 'now' mean? It is the point between the future and the past. The Minkowski diagrams imply that the 'now' is infinitely short, it is gone in the very second of its contemplation. Is time therefore simply a continuity or succession of points? It is worth noting the interesting parallel of the concept of the thinking process. The philosopher David Hume defined the mind, our thinking processes as simply a "bundle of impressions," a succession of ideas, the film frames of the movie of our lives.

Lord Bertrand Russell, the mathematician-philosopher-iconoclast, about the beginning of the twentieth century, used the movies to analogize life as a *discontinuous process*, despite its appearance of continuity.

In his *God and the New Physics*, Paul Davies discussing relativity notes that past, present and future are equally real, all are simply there. Objects move about through space in time. One person's past is another person's present, and another's future. Our perception of time as passing, time-flux, is closely associated with perceptions of self and he muses that the "secret of mind," as we discuss in the chapters on Quantum

Reality and Life, will be solved only when we "understand the secret of time," citing Hofstadter's "whirling vortex of self-reference" that produces our awareness of self, our consciousness."

Gravity As Geometry: Gravity is a distortion of spacetime geometry, and both space and time are curved or warped by the gravitational effects of mass. Gravity slows time down. Time runs faster in lower gravity space. When gravity increase reaches a critical point, as in the ultimate gravity environment of black holes, time stops altogether.

A Black Hole--A Rapid Route to Eternity: Increasing gravity will force a star's implosion into a black hole and finally into a 'singularity' in that black hole. Physicist Paul Davies describes a black hole as representing a rapid route to eternity, *the end of time*. A spacefarer inside a black hole will be "beyond the end of time...unable to return to the outside universe again, because the outside universe will have happened."

The Conversion of Space into Time: Davies explains the mutual distortions of space and time as a "conversion of space (which shrinks) into time (which stretches)." A second of time, he notes, corresponds to an "awful lot of space--about 186,000 miles." The perception of what is happening, however, depends entirely on the frame of reference or point of view from one's own position. While an independent observer in one frame of reference notes the slowing of time on the clock of an accelerating spaceship, astronauts aboard that accelerating spaceship observe no such slowdown of time.

Thus, according to relativity theory, there is no so-called absolute time or present moment, just the time measurement of each observer. But more than that, time cannot be separated from space, it is a single continuum, spacetime. The 'now' on earth is different from the now on another planet. Paul Davies says that for a distant quasar 'now' could refer to "any interval over billions of years." One person's past is another's present or future.

Gravity As Acceleration in Space Curved in the Dimension of Time: Gravity, according to Einstein's general relativity theory, causes objects to accelerate, implying the equivalence of gravity and acceleration. So gravitation, rather than being Newton's 'force of attraction,' is acceleration: the effect which mass has to induce motion in the surrounding three dimensional space that is curved in the fourth or time dimension, as explained by the non-Euclidian geometry of curved space.

As astronomer and science writer Timothy Ferris observes in his monumental work, *Coming Of Age In The Milky Way*, within the framework of the four-dimensional spacetime continuum, "gravitation is acceleration, the acceleration of objects as they glide along 'world lines,' paths of least action traced over the slopes of a three dimensional space that is curved in the fourth [time] dimension."

THE TWIN PARADOX

The Twins Paradox--Twins, at Different Speeds, Age Differently: In *The Mathematical Theory of Relativity*, Einstein pointed to a remarkable characteristic of special relativity, that of two "living organisms," the length of time of one, spent in flight at the speed of light, would be "a mere instant," while generations would have passed for the corresponding organism remaining in the "original position."

This phenomenon, known as the 'twin paradox' is entertainingly described in such publications as *Nova--It's About Time*, and in Sagan's *Cosmos*. If one twin set off in a spacecraft to travel the galaxy and accelerated at one earth gravity, to within one percent of the speed of light, returning to Earth after twelve years, he would find his twin an old man, fifty earth-years having passed. If he had spent 56 years traveling the galaxy, he would find that, upon his return, Earth (if still there) would be four million years older!

In Masks of the Universe, Edward Harrison, speaking of this 'clock paradox,' refers to both twin ages as being measured by the same criteria, their respective "clocks, grey hairs and heartbeats." He adds another dimension of definition, through the ensuing discussion between the twins.

The stay-behind twin, observing the youth of his itinerant twin, comments that things are unfair, that he was cheated, that the traveling twin is 'years younger.' The traveling twin replies that the stay-behind is genuinely older, he has lived his age, slept more nights, eaten more meals and read more books; that their age difference is real--the result of different velocities in the geometry of spacetime.

And so the traveling twin is saying to the stay-behind twin, you are older because you had a "longer" life, spent more time of "your time." I'm younger because I have "spent" or had, less of "my time." The relevancy of elapsed time, based on each observer's frame of reference, is at play here. In one sense, time seems to have been conveniently contracted for one twin and, by comparison at least, "stretched" for the other. But such is the geometry of spacetime. When things, including people, have different velocities and their world lines are not parallel, they do not 'live' the same time.

In *Atoms of Silence*, physicist Hubert Reeves gives the illustration of a person taking his dog out walking. The dog runs all around his walking master, the dog's tail wagging faster than the running dog. At the end of the 'walk,' the dog has aged less than his master, and the tail, less than the dog! If you run, perhaps you are running for your life by slowing down your time!

Spacetime Distance Traveled by Light Always Zero: In *Masks of the Universe*, Harrison points to Minkowski's great discovery that the distance between points of a triangle--the length of their lines, is different in spacetime. The Pythagorean theorem (square of the hypotenuse of a right triangle equals the square of the other two sides) is different in spacetime. There is a different rule for measuring the distance between two points in spacetime. In observing two events, your time interval squared, minus your space interval squared, he says, equals my time interval squared minus my space interval squared--even if we live in different space and time worlds.

So consider the question, what is the distance in spacetime between two events connected by a light ray? He answers that *the spacetime distance traveled by light rays is always zero*. Thus, he observes, light from distant stars at great speed for long periods of time across vast areas of space, "travel no distance whatever in spacetime."

Speed and the Hubble Effect: Hubert Reeves describes a space ship traveling at ninety percent of the speed of light. The galactic universe in front of the spaceship appears blue and, to the rear, red, reflecting the color perception of blue in the approach and red in the departure. As the spaceship slows, the universe ahead appears red again, reflecting the Hubble effect of the expanding universe of galaxies moving away from each other from the still-functioning Big Bang.

Intrinsic Speed of the Earth: He hastens to point out the small apparent differences in color attributed to the 'intrinsic' speed of the earth (less than one percent of the speed of light) of thirty kilometers per second, around the sun, the sun's 200 kilometers per second speed about the axis of the Milky Way galaxy, and the several hundred kilometers per second speed relative to our galactic cluster or supercluster.

SEEING THE PAST

Startling Features of Time: Hawking points out that when we look out at the universe, we see it as it was in the past. So far as we know, we can't actually see objects or events in the future. But gazing through telescopes turned on the nearest star, Proxima Centauri, we see it as it was four light years ago, the time it has taken for light to travel to Earth--four light years.

So when we look at our sister galaxy, Andromeda, we actually view the past. We see the Andromeda Galaxy as it appeared millions of years ago. Though unlikely, it might no longer exist--or exist in that configuration. An Andromeda civilization now looking at Earth would find no Earth human species to contact because they would be looking a the Milky Way Galaxy as it was millions of years ago.

The same is true of our own star, the sun. When we look at the sun we see it as it existed some eight minutes ago, the time it takes for the sun's electromagnetic transmissions to reach Earth. If the sun ceased to exist (trouble!), the telltale cessation of electromagnetic and gravitational waves would not manifest itself to us for the same eight minutes.

Now May Be in Some One Else's Past or Future: Relativity theory tells us that our present moment, our 'now,' may be in the past or future of distant moving observers; that each observer's clock has its own time that may not agree with the clocks of other observers. As Stephen Hawking puts it, only the speed of light appears the same for all observers, no matter how they are moving. In Newtonian physics, says Hawking, different observers would agree on the time light has taken (absolute time) for a journey but not agree on how far the light has traveled.

In Relativity, however, observers must agree on how fast light travels but disagree on the distance the light has traveled and therefore also disagree on the time (time is relative) it has taken. *Relativity puts an end to absolute time* and each observer has his own measure of time. The time of a moving spaceship 'speeds up' and 'slows down' as the spaceship is observed to accelerate or decelerate. Time, as observed in this manner, even stops altogether--in the vicinity of a Black Hole.

COSMIC TIME

General Relativity Allows for Absolute Cosmic Time: We have seen how relativity theory demonstrates that time is not independent or absolute, as Newton thought, but interdependent with space or distance, and mass and motion; that time will vary between observers whose motion differ.

So how, then, do intelligent creatures elsewhere in the Galaxy, or the universe at large, reckon the time representing the age of the universe? Well, say the physicists, the time which has elapsed since the beginning, the Big Bang, does have a 'precise meaning' and general relativity allows an absolute and independent cosmic time valid for all observers in the expanding universe.

How Do We Determine the Time Elapsed Since the Big Bang? Determining the age of the universe, the time that has elapsed since the beginning, requires knowledge of the mass density of the universe. If known, this figure, notes Morris, would not only give us the age of the universe but also tell us whether we live in an open or closed universe. The answer revolves around the amount and type of 'curvature,' that is, whether the curvature is 'negative' or 'positive.'

The Curvature of Space and Time: Gravity, the force exerted by matter, represents the curvature of space and time caused by the presence of matter. Consequently, the amount of mass or matter in the universe is the measure of gravitational effect to produce curvature. So, via gravity, the overall mass determines the average curvature of space and time in the universe.

Curvature may be 'negative' or 'positive.' If positive, the universe is 'closed,' its expansion will eventually cease and contraction will follow into the Big Crunch. This is a 'finite' universe with an end to time. If the curvature is negative, it is 'open;' the universe will expand forever and is said to extend, as Morris says, "infinitely in every direction." This means also that time will have no end, although it will cease to have meaning when thermal equilibrium or ultimate chaos is reached, since nothing further can happen.

The Critical Density the Key to the Age of the Universe: As noted earlier, if the matter in the universe exceeds the critical density, the omega point, the universe is closed and it is said that time will end. The critical density is 5×10^{-27} kilograms per cubic meter or, as Morris notes, approximately three hydrogen atoms per cubic yard.

Up to now, only some ten percent of the required mass is accounted for by detectible matter, although as discussed earlier, the so-called dark matter may account for most or all of the so-called 'missing mass.'

General Relativity Allows for a Cosmic Time: So the question of time elapsed since the Big Bang, and whether the universe is open or closed, can be determined by reference to the mass density of the universe. These questions are answered also if the rate at which the expansion is slowing, deceleration, could be determined. So, says Morris, general relativity is "paradoxically, less relativistic than the special theory...[and] does allow one to define a cosmic time that can be applied to the universe as a whole."

So What is 'Cosmic Time?' While there are a number of explanations of cosmic time, in general it is related to expansion of the universe. Morris notes that it is a statistical concept and can be defined as "the time that would be measured by an observer who was moving along with the average expansion of the universe." While individual stars and galaxies have different velocities , on average, movement will "coincide with that of the universe."

Reeves' Cosmic Time: Physicist Hubert Reeves, in *Atoms of Silence*, describes the history of the universe and asks what is hidden behind the complex reality we call time--what is cosmic time in the history of the universe? Since Einstein, there is no longer the single, absolute time as conceived by Galileo and Newton, but, as he points out, an infinite number of times related to geometric characteristics, the speed of whatever measures it and the amount of mass in the neighborhood of the measurer.

Introduced, is the concept of a 'spacetime' complex in which space and time, part of a single geometric concept, is influenced by speed as measured by an observer. Increase speed of the observed object and its time, as observed, slows. Matter or mass also slows time--and distorts space. You wish to slow down time? Take a high-speed jet airplane ride--or a spaceship trip.

Cosmic Time--For All Red Spectrum Observers: Having set this definitional stage, Reeves says that cosmic time is thus the time of observers for whom the deep sky is uniformly red, denoting the expanding universe. While all times are valid, he says, cosmic time has the advantage of being the same for earth, the sun, and the stars--most matter, the time used to measure "the age and size of the universe."

Entropy Direction of Disorder Versus Increasing Order: The Belgian Nobel-Prize-winning physical chemist, Ilya Prigogine, developed mathematical models of increasing order in the universe, in contrast to the second law of thermodynamics, that change in the universe is in the direction of decreasing order or increasing disorder or entropy.

In his lecture, "Only An Illusion," at the Nehru University in New Delhi, in 1982, Prigogine uses this quotation of the illusion of the irreversibility of time, to range over physics and philosophy, providing a wealth of philosophic commentary on the significance of irreversibility and immutability on the one hand, and countervailing processes of self- organization and increasing order, on the other.

THE BEGINNING OF TIME

Does Time Have a Beginning? Is time an independent objective reality? Does time have a beginning or is it eternal? If it first came into existence with a 'creation,' as St Augustine declared and Relativity-Big-Bang theory implies, how does one describe any *preexisting* condition? The very question suggests a state of things for which we do not yet even have terminology.

The Beginning of Time--Science and Theology: The question of the beginning of time, once purely theological is now as scientific a question as theological. Before relativity theory, which provided the mathematical framework for a dynamic rather than static universe, and introduced the now-accepted fourth-dimensional character of time, the commonly-held impression was of time as something eternal, mysterious, and existing independent of space.

Time '0' and the 'Planck Wall': In the Big Bang sequence, time is considered to start not at Time = 0 but with the so-called 'Planck Wall,' at 10^{-43} second, the earliest point where relativistic physics can be said to function. Classical physics is said to break down at the Planck wall, time having no meaning earlier than this point. According to superstring and supersymmetry theory, at this point, gravity and the other forces, all of physics in effect, are thought to have been united.

Planck Dimensions: The Planck length, says John Gribbin in *In Search of The Big Bang*, is 2×10^{-33}, and the Planck mass is 2×10^{-5} gram. A particle of these dimensions, he notes, would be a *black hole*. The Planck measurements, based on the work of the physicist Max Planck, represent the smallest length, time, and mass, that have any meaning.

The Planck time, is defined, incidentally, as the time it takes light to travel across the Planck length, which happens to be some 10^{-20} times the distance 'across' a proton. Planck measurements define the time scale and distance scale "where the quantum effects of gravity dominate the structure of spacetime."

Time Has No Meaning before the Beginning of the Universe: In *A Brief History of Time,* British theoretical physicist Stephen Hawking asks "Did the universe have a beginning, and if so, what happened *before* then? What is the nature of time? Will it ever come to an end?" Hawking observes that the concept of time has no meaning before the beginning of the universe.

He refers to the religious cosmologies--Jewish/Christian/Muslim--that have the universe starting "at a finite, and not very distant, time in the past." But most people, he notes, believed in an eternal universe, until 1929 when astronomer Edwin Hubble discovered that the galaxies were moving rapidly away, the universe is expanding and that there must have been a starting point, ten to twenty billion years ago, when everything was at the same place and when the universe was *infinitesimally small and infinitely dense*.

Receding Galaxies Brings the Beginning into Science: This discovery, he says, *brought the question of the beginning of the universe into the realm of science.* Hawking says that time had a beginning at the big bang, in the sense that earlier times simply would not be defined. He observes that "One can imagine that God created the universe at the instant of the big bang...but it would be meaningless to suppose that it was created before the big bang." Although an expanding universe, does not preclude a creator, he muses, it places limits on "when he might have carried out his job!"

Relativity, Galactic Redshift & Cosmic Background Radiation: Hubble's receding galaxies was not the only proof of a starting point in the universe. Three events gave credibility to the Big Bang theory which posits a beginning of meaningful time: Einstein's general relativity theory, astronomer Hubble's redshift-findings of the receding galaxies, and discovery of the cosmic Big-Bang fossil background radiation.

All confirmed the occurrence of a cataclysmic universe-expansion-causing event emanating from an incredibly hot, infinitesimally small, infinitely dense (a condition of infinite space-time curvature) starting point--the Big Bang. These starting-point conditions imply, of course, a beginning of time, as the Bible implies and as Saint Augustine had concluded.

Quantum Theory and the Subatomic Starting Point: But the subatomic starting point implied by gravity aspects of relativity, required the factoring in of theory dealing with the subatomic--quantum theory. And so there has more recently evolved a *quantum-gravity* theory, the 'no time boundary' theory of Stephen Hawking, described in his *A Brief History of Time.* It employs 'imaginary' rather than 'real' time, although Hawking says that in a sense, this so-called 'imaginary' time may be *real* (and real time imaginary).

Before quantum theory was introduced to address cosmological questions beyond the scope of Einstein's general relativity theory, the mathematics of general relativity theory required a starting point for the universe, a creation event--the *Big Bang,* out of which all space, time and energy-matter emerged.

No Boundaries and the Implications for the Role of God: But by introducing quantum theory, says Hawking, a new possibility arises other than from general relativity. It is that "space and time might form a [self-contained] finite, *four-dimensional space without singularities or* boundaries, like the earth surface earth but with more dimensions," and even accounting for the observed arrow of time.

When he speaks of the absence of singularities or boundaries, he is saying, in effect, that there is a primordial, pre-existing prevailing condition, an eternal mega-universe; that while some quantum event set in motion the emergence of our branch of the mega-universe, that, technically speaking, there is *no real creation*, no beginning or end of the universe, or of time. He goes on to observe that this possibility, *has profound implications for the role of God as Creator*.

The Universe Before the Beginning? Earlier scientific descriptions of the standard Big-Bang model of an expanding universe traced events going back in time to the so-called Planck wall, the first 10^{-43} second, the point at which relativity physics breaks down.

'Before' that, if one can use this term, one might presume that 'physics' was sort of winding up to kick things off. But how, one might ask, can the universe start at some period of time after time = 0? One asks what was going on before the Planck wall of 10^{-43} second? How can there be something before the beginning?

In *A Brief History Of Time*, Hawking speaks of *two kinds of time*, real and imaginary; that in the presence of concentrations of mass and energy, the two kinds of time are "at right angles to each other"; that time is smeared out so that what appears to be an abrupt Big Bang is a holistic pattern arising like a hot water steam bubble.

Quantum Gravity: So what applies where relativity ceases to govern? Well, as we discuss in the chapters on cosmology, this is the point at which physicists say that a quantum theory of gravity is required. At these dimensions, exponentially smaller than a proton, only quantum physics can provide a foundation for answers. The essence of this approach is that all we see in the universe, our matter, energy, dimensionality, the fields and forces, space and time, were all quantum effects which evolved to our present state.

Random Quantum Fluctuation and Unified Physics: Quantum theory suggests a primordial quantum void in which a random fluctuation generates an incredibly small mini-universe point or bubble of spacetime governed by a superforce of unified physics, a state of supersymmetry.

In a quantum leap, the point goes through a momentary exponential inflation, creating all matter and energy in the process and the unified forces, first gravity, then the other forces and fields, broke off in a symmetry-breaking sequence from which we derive our three space dimensions and one time dimension.

The mathematical reconciliation of these events calls for a geometric state of unification with ten or 26 dimensions all but our four dimensions of which remained "compactified" as minute invisible energy forms at the time that three-dimensional space, and time, broke off to give us our four-dimensional structure.

MEASURING TIME

Measuring Time: Encyclopedias note the origins of three familiar units of time: the year, the month, and the day as rooted in the natural cycles observed by ancient peoples, the seven-day week as possibly originated in the Middle East from the ancient Sumerian reckoning using gradations of 60: days divided into 24 hours, hours into 60 minutes, and minutes into 60 seconds.

Earth Rotation and Time Zones: Here, elapsing of time is based on natural phenomena such as rotation of the Earth in which the stars and the sun disappear, reappear, and return to their earlier positions. Compton's Encyclopedia notes that Time related to star motion is 'sidereal' time. The time for a star to seemingly make one circuit is a 'sidereal' day; for the sun to make one rotation (four minutes longer), a 'solar' day--which is longer or shorter depending on distance from the sun. Our clocks are set to the mean solar time, the annual average solar day.

By international agreement, the world is arbitrarily divided into 24 time zones of about 15 degrees longitude. The longitudinal line through Greenwich, England, was adopted as the zero meridian-- 'Greenwich mean time'-- but now largely replaced by 'Universal time,' based on atomic clocks using cesium-atom vibrations, although still more accurate atomic clocks are in the works. Some countries advance regional standard time one hour in the spring and then back in the fall, called 'daylight savings' time, to get the economies and benefits from longer light evenings.

Nobel Prize for Breakthrough on Time Measurement: Quantum mechanics describes atomic absorption and discharge of energy in jumps or 'quantum leaps.' Recently awarded, was the Nobel prize in physics for discoveries which resulted in breakthroughs in the precision of time measurement based on these quantum leaps of atomic particles from one energy state to another. The result is atomic clock time measurement, with incredibly small variation, useful in measurements such as of relativity-testing phenomena, continental drift, and tracking of objects.

The October 12, 1989 Nobel prize for physics was awarded to Norman Ramsy of Harvard, Hans Dehmelt of the University of Washington, Seattle, and Wolfgang Paul of the University of Bonn, for providing two ways to measure atomic energy oscillations, by directing atomic beams through electric fields, or capturing a single atom in radio waves for longer-term observation, days or years. The new measurements offer time measurement far more accurate than the present variations of one second per thousand years.

A decaying atom, one going from one energy state to a lower state, emits or radiates the excess energy as electromagnetic transmission of particles which may take the form of photons, light quanta or particles. These quantum leaps result in countable waves of electromagnetic radiation. Currently, the standard definition of a second of time is based on the 9,192,631,770 oscillations between energy states of the cesium atom.

THE NATURE OF TIME--A VIRTUAL REALITY?

Time Is Relative, Not Absolute: Newton's absolute time disappeared with Einstein's special relativity theory finding that the speed of light is invariant, the same for all observers, that time is elastic and dilated by speed. Clocks run at different rates according to the relative motion of observers. For example, a speeding spaceship's clock will be observed to run slower than an Earth clock. An observer on the spaceship sees the Earthbound clock running slower. According to Einstein's general relativity theory, time is also slowed by gravitation so that an Earth clock runs slower than a nonmoving clock in space.

Time, Relativity, and Thermodynamics: Since Einstein's relativity theory, time cannot be explored without considering two laws of physics, Relativity theory and the Second Law of Thermodynamics. Time in special relativity represents a dimension of the universe in which space and time are interwoven, part of a space-time continuum. Einstein's mathematics teacher, Hermann Minkowski, observed that "Space by itself and time by itself are doomed to fade away into mere shadows, and only a kind of union of the two will preserve an independent reality."

As science writer Louise B. Young observes in *The Unfinished Universe*, "All things have a shape in time as well as in space.... Time is perceived as the way of measuring the progressive change that is building the universe even while we live and take part in it.... We must compare the whole of time to the act of creation."

In a sense, as physicists note with melancholy, the universe pays with its life for the privilege of time. The universe starts enormously hot, a high energy state like a wound-up clock, unwinding thermodynamically, spending energy building pockets of increasing order, like galaxies, planets and organic life, as it unwinds.

The universe, if 'open,' may be expanding ultimately toward chaos--where nothing can happen--and time becomes meaningless because there can be no further change. Or, if the universe is 'closed,' has sufficient mass, the universe may someday cease expansion and start contracting, perhaps to its original starting point, at which time comes to a stop, as it does in a black hole.

The Irreversible, Unidirectional Character of Time: The unidirectionality and irreversibility of time is noted by Julian Huxley: All reality, he says is a one-way process in time; unitary; continuous, irreversible; self-transforming; and generating variety and novelty during its transformation. Louise Young devotes her *The Unfinished Universe* to exploring "this dynamic interpretation of the temporal process" as having "deep implications for philosophy and science."

It is, says Young, "the logical culmination of the positive and unidirectional view of time---which started with Judeo-Christian thought and led, through the Darwinian perception of biological evolution to the broader world view which recognizes the important role of time in the whole cosmic process."

Do the Past and Future Exist? St. Augustine suggested that the past exists only as memory, the future as anticipation. But Einstein, in his consolation letter to the sister of his dear departed friend, insisted that one of the inferences drawn from relativity theory is that distinctions between the past, the present, and the future are illusion--another way of saying that all time coexists.

The Nature of Reality: Other physicists have drawn on Einstein's mysterious comment in discussions of the nature of reality derived from relativity and quantum theory. For example, as we discuss later, in his treatment of the meaning of "distance" in physics, George Zebrowski speaks of this "grand monism...where ordinary intuitions about space and time just don't apply...which suggests that everything is in the same place and does happen all at once."

Dimensionality and the Arrow of Time: The 'dimensionality' of time is derived from Einstein's theory of general relativity in which time, in physics, represents the fourth dimension of the universe, and time and space are, in mathematician Minkowski's description, part of a single continuum, spacetime. The 'direction of time' in the world of large objects like planets and people, in which 'the past and future look different,' is toward the future, that is, from the past to the present and into the future.

This 'arrow of time' is derived from the Second Law of Thermodynamics which deals with the universe's stock of 'order,' representing the potential for change or the capacity for performing work. As this supply of order diminishes with time, the universe winds down toward its 'ultimate state.'

Time and the End of the Universe: What will be this ultimate state of the universe? As we have noted, the universe will follow one of two paths. In the first, the now expanding universe, if 'closed' will contract, under the influence of sufficient gravitational mass, perhaps to a black-hole singularity state such as that from which the Big Bang is said to have occurred. In the second path, where the universe is 'open' (insufficient mass for gravity to stop its expansion), the universe will continue to expand and unwind into an ultimate state of disorder, entropy or chaos, where temporal events can no longer happen and time will have no meaning.

Some Fundamental Questions about Time: We explore here certain aspects of both the physics and metaphysics of time, especially what relativity theory and quantum theory tell us. When did time begin, and if it ends, when will that be? What is meant by time being *elastic*, and how can we 'slow' the passage of time, according to relativity theory? What is meant by references to time and space as being interchangeable? What has the speed of light to do with traveling back in time?

What is meant by the irreversibility of time in the macro universe of galaxies and people, and what are the 'arrows of time?' What is meant when it is said that the spacetime distance traveled by light rays is always zero? What is the relationship between God and time? How do we describe God's existence 'before' the beginning of time?

What are Space and Time? As baffling as space and time seemed when physicists first explored their implications in the context of relativity theory in the early part of this century, their meaning seemed even more uncertain in the light of the emerging physics of quantum theory dealing with subatomic physics.

This strange theory suggested: that reality itself is *the product of observer participation*; that while space and time may have (unknowable) physical properties apart from our perceptions, in our quantum-observer reality, space and time consist of what the thinking mechanisms of our minds construct them to represent, a concept earlier proposed by philosophers such as the 18th century philosopher and cosmologist Immanuel Kant.

A Virtual Reality? Earlier in this century, physicists gave expression to their speculations that quantum theory supports the strange concept that our own subjective observations may dictate objective reality. George Zebrowski, in his fascinating *The Enigma of Distance*, Omni Magazine for November 1992, notes that ours may be a "virtual reality...a mental construct, built up within minds to the point where we experience it as a literal reality of three dimensions and time....*Its true nature may be quite different from what we experience....*"

Perhaps, he muses, these interpretations of space, distance, dimensionality, and time are, as Einstein expressed it, "free creations of the human imagination, means devised for easier comprehension of our sense experience."

Space and Time as Fields: Physicist Paul Davies, in *Superforce*, notes the belief of most physicists that the spacetime concept ceases--cannot be carried back in the superforce unified era. He notes the view that perhaps space and time are to be treated as *fields* which also froze out of the "pregeometric elements," so that the forces, fields and space and time were indistinguishable parts of the mix. In the superforce era, spacetime would not have yet "jelled into a coherent form."

The Forces As Additional Dimensions: In unification theory models, the dimensions representing our geometry of three spatial and one time dimension are complemented by additional 'invisible' space dimensions which manifest themselves as forces. So in this theory, as physicist Paul Davies observes, both forces and energy-matter are related "manifestations of space and time," geometry put in place when the universe was incredibly less than one second old.

The Interdependence of Cosmology and Quantum Forces: In the unification or superforce period, the universe, says Paul Davies, would have been merely a "collection of ultra-simple components--the raw materials from which God fashioned all of space, time and matter."

Davies goes on to observe the growing scientific perception that the infinitely large, described by cosmology, and the infinitely small, described by quantum effects, coalesce to provide a unified description of the cosmos. It is, he says, a description in which "the ultra-microscopic structure of matter is intimately connected to the global structure of the universe, each influencing the development of the other."

Quantum Genesis and Pre-universe Unity: Quantum genesis and unification theories, considered in other chapters, suggest what may have existed before the Big Bang, and describe the symmetry-breaking process through which we derived space, time, matter, the forces and fields, and present dimensionality.

The theory postulates that out of the primordial mega-universe void, empty of space, time, and matter, erupted a cosmic particle, exponentially smaller than a proton. In this particle was unified everything common to our present universe of 'broken symmetry,' including space, time, matter, dimensionality, fields and forces. This cosmic particle represented a single unified state representing and governed, it is thought, by the 'master law' of all physics.

Shatter a Unity and Inflate It--to Get Space and Time: In his discussion of 'distance,' Zebrowski speaks of this "grand monism...where ordinary intuitions about space and time just don't apply...which suggests that *everything is in the same place and does happen all at once*." Perhaps, he notes, "space...is what you get when you shatter a unity and then inflate it so that space is what exists between the pieces that still seem to 'know' each other."

Two Great Scientific Revolutions in Physics: Physicist Paul Davies, in *God and the New Physics*, notes that two great revolutions gave birth to the new physics, quantum theory and the theory of relativity. Almost exclusively due to the work of Einstein, relativity, he says, is a theory of space, time, and motion. He notes that the consequences of relativity are as baffling and profound as the quantum theory, challenging "many cherished notions" about the universe. Never is this more so, he notes, than in the theory's treatment of time, a subject of intense and longstanding concern in all the world's great religions.

Time, Individuality, Memory and the Soul: Concerns over death often lead to questions of why God, in the creation, had to impose time, and its mortality consequences. But without time there is no experience, memory, cause and effect, or planning; nothing can 'happen.' There would be no hope for the future, forgiveness for the past, prayer made or answered. We can't even conceive of a universe without time. Without time there is no personal identity, notes physicist Paul Davies, no soul or experience. All these things are wrapped up in time.

Mathematician- philosopher Gottfried Leibniz argued that God had no choice in going with some of the world's limitations; that God created the most perfect world possible. To some of his contemporaries, of course, this acknowledgment of lim-itation on God's omnipotence was heresy. Einstein often wondered whether God had real choices in configuring the laws of physics. Time and its inevitable effects, such as death, are intrinsic to, and inescapable from, our quantum-relativity universe.

THE ARROWS OF TIME

Defining the Direction of Time: We see references in theology, philosophy, and general literature to the "arrow of time." Scientists refer to more than one "arrow of time." Physicist Paul Davies cites the book, *Time's Arrows*, Scientific Attitudes Toward Time, in which physicist Richard Morris explains "the essential puzzles and paradoxes of time's enigmatic arrows."

The so-called 'arrows of time' are essentially descriptive labels for defining the 'direction of time' as reflected in the behavior of certain physical phenomena like thermodynamics, electromagnetism, and the expansion of the universe.

Speaking of Paul Davies, while most physicists are reluctant to bring God into discussions of physics, Paul Davies, in *God and the New Physics*, and in *The Mind of God*, plunges right ahead to discuss God in relation to certain aspects of physics, such as those involved in questions about God-- and time. As we have seen, one such question relates to how time could have existed 'before' the advent of the universe in which, according to Big Bang theory, time first came into being.

What Are the Arrows of Time? Physicists sometimes speak of *five arrows of time*, four physical and a metaphysical fifth arrow. Most discussions of the 'arrow of time,' involve the thermodynamic arrow of time, reflecting concepts of physics dealing with 'order' and disorder or 'entropy.'

Here, the arrow of time flows into the future, a state of decreasing order in which certain pockets of order, like the things we build, may increase. But, under the laws of thermodynamics, these increases are at the expense of decreasing order in the universe at large.

The second law of thermodynamics says that 'the entropy of an isolated system never decreases' (or order in an isolated system never increases), and an increase in order (decrease in entropy) must be paid for by disorder elsewhere in the universe at large.

The 'Four Physical Arrows' Involve the Laws of Physics: These include the thermodynamic arrow, the universe-expansion arrow, the K-meson arrow, and the electromagnetic arrow. A 'fifth' arrow deals with metaphysics, the physchological arrow, the human, subjective perception of the direction and duration of time.

We should note at the outset that the term 'flow' or 'passage' of time is sometimes questioned as failing to take into account relativity theory's view that, like the dimensions of space, all time is laid out at the creation as a dimension of the universe, part of the spacetime continuum. So that it is we who do the flowing or passing in the dimension of time, along our time lines. As Austin Dobson said, in *The Paradox of Time*: "Time goes, you say? Ah no! Alas, Time stays, we go."

The Thermodynamic Arrow: In the nineteenth century, the Austrian physicist Ludwig Boltzman, pioneer in thermodynamics, thought of the primordial universe as an eternal state of chaos or disorder. In such a thermal equilibrium state, incidentally, time does not exist, that is, it has no meaning because time is a geometric dimension of a spacetime universe which is wound up, that is, in a state of order, when there is heat to produce change and meaningful events.

Time and the Second Thermo Law: It is the operation of the second law of thermodynamics, by which our universe winds down from high order (low entropy) to a state of low order (high entropy) that produces time. So time, says Paul Davies, is that part of the spacetime geometry which represents the "potential for change."

Physicists note that it is this second law of thermodynamics, mandating increasing disorder and the ultimate end of the universe as total chaos, that also mandates the irreversibility of the arrow of time. This 'second law' is reflected in the process by which the universe winds down from a higher to a lower state of order, order, here, represents the 'potential for performing work.'

Stated in terms of disorder or entropy, as the universe or any system winds down, entropy increases until the stock of order is depleted and the system reaches 'thermal equilibrium,' a state of total chaos where nothing further happens, the process of 'change' ceases, in effect, making temporal events impossible.

When we make ice from water by refrigeration, or produce a car by use of energy and materials, or for that matter, introduce biological materials, like plants or people, these represent pockets of increased order or decreased entropy, sometimes referred to as 'negative entropy' or, simply, 'negentropy.' Since this increased order is produced at the expense of greater disorder elsewhere, there is no violation of the second law of thermodynamics which mandates decreasing order or increasing entropy in the universe at large.

The Second Law of Thermodynamics Creates the Arrow of Time: The amount of order, potential for work or change in a system, physicist Morris suggests, can be described as non-equilibrium or dis-equilibrium, representing difference in energy levels which causes change. Physicist Sir Arthur Eddington says that it is the second law which creates the arrow of time telling us that "the past and future look different."

Paul Davies says that the law imposes a "time asymmetry" on the world that distinguishes past from future. And violating the second law, that order may only decrease, says Davies, would be "tantamount to reversing the direction of time." So for most practical purposes the arrow or direction of time is set by the second law of thermodynamics.

Where Did the Entropy Gap or Disequilibrium Come from? Where did the high order or low entropy, the 'disequilibrium,'come from? One theory says that God created the high order to provide the physical mechanisms for the development of life and mind. Another speaks of the energy of the universe as equaling zero, the positive energy being derived from the negative gravitational field from which the expansion was driven. Physicists, like Paul Davies, suggest that the source of the disequilibrium or high order, which fixes the thermodynamic arrow of time, was the exponential inflation of the mini-universe point of spacetime generated by the original quantum fluctuation.

Steven Hawking and the Arrow of Time: British theoretical physicist Steven Hawking generated a world-wide interest in the physics of time through his popular *A Brief History of Time*. He notes that to unify gravity and quantum mechanics one has to introduce *imaginary time* which is indistinguishable from directions in space. In space you can go north or south, and in imaginary time you can go backward or forward in time.

The explanation usually given for no memory of the future, he observes, is that it is forbidden by the second law of thermodynamics, in which "in any closed system, disorder or entropy always increases with time...one example of ... the arrow of time [and] ...a form of Murphy's law: Things always tend to go wrong!"

Where Does the Difference between Past and Future Come from? But in "real time" there is a big difference between these directions in time. He asks, "where does this difference between the past and the future come from? Why do we remember the past and not the future?" He notes that the "laws of science" (quantum theory) do not distinguish between the past and the future.

Hawking's Three Arrows of Time: He goes on to refer to "at least" three different arrows of time: the thermodynamic arrow, in which disorder or entropy increases; the psychological arrow, our subjective perception of the passage of time; and the cosmological arrow, the direction of time in which the universe is expanding.

Why Do the Arrows of Time Point in the Same Direction? He argues that the finite but no boundary condition of the universe, together with the weak anthropic principle can explain why all three arrows point in the same direction; that the psychological arrow is determined by the thermodynamic arrow; and that the thermodynamic and cosmological arrows will not always point in the same direction; but that it is only when they do point in the same direction that conditions will permit intelligent life to ask these questions.

Using Quantum Gravity to Explain the Thermodynamic Arrow: He goes on to explain that one has to use a quantum theory of gravity to understand how the universe began, the exponential inflation, the density fluctuations caused by the uncertainty principle, leading to galaxies--and people--a state of lumpiness and dis-order as time passed. It is this sequence, he notes, that would explain the thermo-dynamic arrow of time.

Would People Live Their Lives Backward in a Collapsing Universe? Would the arrow of time reverse in a contracting, collapsing universe? Would order decrease in a collapsing universe as it increased in the expanding universe? Would people live their lives backward? At first, Hawking thought this might be the case. But other models showed that the no-boundary condition of the quantum-gravity model im-plied that disorder would continue to increase during the collapse and so the thermodynamic arrow would weaken but not reverse.

Recollapse Occurs Too Late for Intelligent Life: Hawking goes on to point out that the no boundary proposal implies that the universe is expanding at very close to the critical rate at which it would just avoid recollapse, and so collapse will be at a remote time when matter (protons and neutrons) will have decayed into light and radiation. The universe will then be close to total disorder and consequently with no strong thermodynamic arrow which is required for life. So intelligent life, he says, could not exist in the ultimate stages of expansion or of a contracting phase.

God and the Second Law of Thermodynamics: In *God And The New Physics*, physicist Paul Davies provides a fascinating description of some of the possible manipulations that might be used to postpone the 'heat death' end of the universe where matter is dissolving. He discusses whether the ancient superintelligence which is said to exist in the universe, a sort of 'natural' rather than supernatural God, would be able to prevent the end of the universe through various manipulations of physics.

But in the end, he says, a natural God just can't circumvent the second law of thermodynamics. The ordered universe is unwinding inexorably. "Only a supernatural God could truly wind it up again," says Davies.

We are reminded that it is this second law of thermodynamics, mandating increasing disorder and the ultimate end of the universe, that also mandates the irreversibility of time. As Davies puts it, the law imposes a "time asymmetry" on the world that distinguishes past from future. So violating the second law, says Davis, would be "tantamount to reversing the direction of time."

Quantum Particles May Reverse the Arrow, Go Back in Time: Physicists point out that while the thermodynamic direction of time applies to molecular aggregates, objects in the macroscopic universe of galaxies, planets, kitchen appliances and people, at the quantum or subatomic level of protons and electrons, this arrow of time, dictated by the second law of thermodynamics, does not apply to individual particles. So, the arrow of time does not apply to subatomic particles in the same way as it does to macroscopic objects. Particles need not always go forward in time, as macroscopic objects like people do. *Particles may go back in time*.

It is interesting to note, however, that the direction of time for large molecular systems, like people, isn't irreversible because of physical impossibility but because of quantum statistical unlikelihood. So there is an infinitesimal chance (like once in the history of several universes) that one might reverse the arrow to go back rather than forward in time.

Quantum Waves Moving Backward from the Future: When concepts of the direction of time are subjected to reexamination in the context of quantum mechanics, strange, almost surrealistic effects are produced. In the transactional quantum interpretation, one of the newer interpretations of quantum theory, quantum waves of probability move 'backward through time from the future.'

In *Taking the Quantum Leap*, physicist Fred Allen Wolf notes that "a physical system can only actually appear to an observer if the quantum wave representing it propagates both from the present into the future and backward through time from the future to the present." What is even more startling is his citation of this principle as operating in human physiology.

The Future Communicating with the Present Through the Human Mind: In tests, human subjects underwent a conscious experience of physical sensation prior to the actual cerebral neuronal brain activity needed to register such an earlier physical sensation. Neurophysiologists were led by these tests, says Wolf, to a "delay-and-antedating hypothesis."

So, asks Wolf, *where does the awareness of a sensation come from if the brain hasn't yet registered the awareness?* The answer, says Wolf, may be a surprising new discovery, that the future actually communicates with the present in the human nervous system.

The Positron--An Electron Moving Back in Time? Based on an idea from phys-icist John Wheeler, Nobel-prize physicist Richard Feynman theorized that anti-particles, like antielectrons, known as positrons, may reverse their direction of time. Thus, Feynman suggested, using Wheeler's notion, *a positron might actually be an electron moving backward in time*.

In this view, the *positive charge of the positron might be an illusion*; the so-called positron-electron annihilation might represent the same particle reversing time direction. The electron might "zig-zag back and forth," the negative-charge particle going forward and taking on the appearance of a positive-charge particle moving back in time.

Physicists like to cite John Wheeler's conversation with Richard Feynman in which Wheeler is said to have whimsically declared that he knew why all electrons have the same charge and mass. They are, he said, all *the same electron*, moving-back-and-forth-in-time! Although this possibility is intriguing, of course, physicists remain ambivalent about the theory.

The Expansion-of-the-Universe-Arrow: The second arrow of time is associated with the expansion of the universe in which the density of the universe is decreasing with the expansion, establishing an arrow of time, which, like thermodynamics, moves to the future. Although some suggest a mathematical interrelationship of the two arrows because order is decreasing with lower temperature and expanding space, Morris suggests that such interrelationship has not been established.

Gravitational Behavior Suggests the Existence of Dark Matter: With the expanding universe predicted by general relativity and Big Bang theory, matter in the past was hotter and denser based on past regression. Whether there is enough matter in the universe for gravity to retard expansion of the universe and start a contraction, is yet undetermined. To date, insufficient mass has been identified to class the universe as 'closed.'

But the gravitational behavior of the galaxies suggests more mass than is reflected in the illuminated or visible matter. Gravitational behavior of the galaxies implies the existence of galactic gravity-generating, unilluminated galaxy-halo-matter, and similar invisible matter in the great galactic voids.

This unilluminated matter, referred to as dark matter, suggests the possibility at least, of the existence of sufficient matter and density (mass) to account for most or all of the so-called 'missing mass' necessary to reach the 'critical mass' required for a contraction, a flat or closed universe. Cosmologists designate this mass by the Greek letter, Omega, and call it the omega point. If the mass is less than one, the universe will expand infinitely, if more than one, the universe will contract to end in the Big Crunch.

Does the Second Law Apply to Gravitational Fields? In *Time's Arrows*, Richard Morris points out that little work has been done on computing the entropy of gravitational fields, or for that matter, determining whether entropy is characteristic of gravity. So, he notes, we do not know if the second law applies to gravitational fields as it does to matter to which the thermodynamic arrow is related. Developments in gravitational entropy since Morris' book and prior to this writing, have been inconclusive.

The Third Arrow, K-Meson Decay, Allows Decay That Is Not Reversible: The third arrow, K-meson decay, is an oddity. Nuclear reactions, notes Morris, do not demonstrate an arrow of time. But there is one apparent exception, the Kahn or neutral

K meson, which participates in "reactions which are not symmetrical in time," the only particle with such characteristics. The unstable K meson decays in less than a millionth of a second after its creation.

While a "three-particle decay" of the K-meson is "symmetrical in time," permitting an inverse process, a "two-particle" decay is not time-reversible. Well, the two-particle decay, although rare, has been found, thus exhibiting irreversibility, a "time-asymmetry" permitting the K meson to "distinguish between two directions of time." The reason for this? Nobody knows, says Morris.

The Electromagnetic Arrow: Electromagnetic waves, like light, X-rays, radio, ultraviolet and infrared waves, all seem to go forward in time, travel in the direction of the future rather than the past. Light from the sun, the planets, and the galaxies originated in the past. Light from the sun, for example, reaches Earth after about eight minutes. This electromagnetic (EM) arrow is not surprising, says Morris, because otherwise we could send EM messages into the past "in violation of the principle of causality."

A Mystery--EM Equations Allow Waves Traveling to the Past: But, says Morris, in *Time's Arrows*, there is a baffling problem with the EM arrow since the laws of electrodynamics are "symmetrical in time," the equations not distinguishing between past and future. In theory, then, a light bulb could absorb rather than radiate light waves. Two possibilities are suggested. The first is that, as a matter of quantum uncertainty, this is possible but highly "improbable."

The second possibility, says Morris, is the theory advanced by physicists John Wheeler and Richard Feynman, that EM radiation travels both into the future and past. We just aren't aware of this radiation into the past, in this theory, because of radiation emission and absorption processes in the universe, in which time-reversed re-emissions into the past from "absorbers in the future" are canceling out current radiations into the past and intensifying current radiation into the future.

Radiation to the Future As Being from the Present and the Future: In this theory, current radiation into the future would be absorbed by an absorber in the future, which re-emits the radiation as time-reversed waves into the future absorber's past. These re-emissions from the future interfere with present radiation into the past via "destructive interference" in which the crests of one wave coincide with the troughs of the other to cancel each other out and produce darkness. Thus, time-reversed waves from the future absorber's re-emissions into the past, destructively interfere with our current waves into the past.

Radiation to the Past As Replaced by Re-emissions from the Future: But, in this theory, these time-reversed waves from the future also interact with present radiation into the future, to augment or intensify the present's radiation into the future to make up for the amount of the present's radiation into the past. So, says Morris, what we see is "nothing but an ordinary forward-traveling wave." We do not see the re-emission radiation from the future because it appears as ordinary radiation going forward in time.

Morris notes that this Wheeler-Feynman theory of the electromagnetic arrow going back and forward in time suggests a connection between the EM arrow (electrodynamics) and cosmology. The reason is that the theory requires a universe that will some day contract, in other words, one with sufficient mass to rule out a forever-expanding or 'open' universe.

The Psychological Arrow of Time: What is the relationship between the physical arrows of time going toward the future and the psychological arrow of time--which happens also to point to the future? The psychological arrow is so mysterious, says Morris, that *some philosophers have questioned the very existence of time*. Subjectively, we are conscious of a 'passage of time, a 'now' which continually moves toward the future. The laws of physics, however, deal only with time as a dimension, with no need of the concept of a series of 'nows.' As a dimension, time is simply a line that extends indefinitely in both directions.

The Time As a Line from the Big Bang to the Big Crunch: In one direction the time line extends back to the Big Bang. In the other direction it extends in the direction of the future, with two possibilities. One is the 'closed' universe in which the universe contracts in the Big Crunch to diminish back into the singularity from which it emerged. The other is the 'open' universe in which the universe expands indefinitely, to the point where time ceases to have meaning because the universe reaches the equilibrium state of total entropy or chaos. This is a state of chaos, characterized by thermal equilibrium in which nothing further can happen, no change can occur, and time and space cease to have meaning.

Time and Subjectivity--No Need of Subjective Now: We noted earlier that while there is the 'subjective' consciousness of a 'flow' or 'passage' of time, a 'now' which continually moves toward the future, the laws of physics deal only with time as a dimension. There is 'no need' of the concept of a series of subjective 'now,' time being simply a line that extends indefinitely in both directions.

Philosophy Insists on Subjectivity: Nevertheless, philosophy treats time and its arrow as a subjective phenomenon. The perception of the rate at which time 'passes' depends on one's moods, age, preoccupations, the time of year, and many other influencing factors. The 'passage' of time is slow when one has nothing to do, fast when one is engaged in pleasurable pursuits--playing baseball or watching a good show, slow when one awaits the dentist, and so on.

But Morris is not prepared to accept mental attitudes toward time as altogether subjective. One sees an object like a baseball as being 'in motion,' not just as occupying a series of positions--of 'now.' The same with music where one not only hears a note in each 'now' but is aware of a musical theme composed of an aggregate of these notes or 'now.' He cites the example of a baseball struck by the bat in a baseball game as an objective fact, one observed by everyone. Of course, each of the 'everyone' can have an identical subjective impression.

EINSTEIN, DEATH, AND THE ILLUSION OF TIME

The Illusion of Time: "...the distinction between past, present, and future is only an illusion, however persistent." These were the words of Albert Einstein to Michelle Basso's sister, upon hearing of the death of his dear friend and compatriot; words presumably designed to diminish the hurt of her brother's passing.

What did he mean by this mystical statement? What is the relationship between time and death? How is it in the nature of time that Einstein might use it as a message of comfort to the bereaved? What are the implications for life and death if all time is "there," if distinctions between past, present and future are an illusion?

Distinction between Past Present and Future As Illusion: "Michelle has preceded me a little in leaving this strange world. This is not important. For us who are convinced physicists, the distinction between past, present, and future is only an illusion, however persistent."

These were the words of Albert Einstein, contained in a letter to the sister of his old friend and valued helpmate Michelle Basso, at Besso's death. In his later years Besso had written Einstein again and again pursuing the meaning of time; what is time and what is meant by irreversibility of time, he had persisted.

"Time Cannot Be Absolutely Defined": Earlier, on the occasion of meeting his friend Michelle Besso, on whom he had been trying out his theories, he announced to Basso that "an analysis of the concept of time" was the "answer," he declared. "Time cannot be absolutely defined, he said, and there is "an inseparable relation between time and signal velocity."

Fascinated with Einstein's new approach to time, Besso sought further clarifications from Einstein. What were the implications of this theory of time on life? Einstein's incremental responses to Besso's queries further intrigued Besso, as they have so many others.

Suggested Deeper Meaning in Einstein's Grasp of Time and Reality: Einstein had commented that Besso's preceding him "a little in this strange world...is not important," that "the distinction between past, present, and future is only an illusion, however persistent." Einstein was a very serious, contemplative person who sought meaning in our universe. These were not idle words.

Philosophic or Theological Implications beyond Pure Physics? The implications from this time-oriented letter of comfort from Einstein to Besso's sister, for many, at least, are that Einstein had grasped not only the 'light' and velocity invariance concepts of special relativity but that his discovery of certain aspects of the nature of spacetime somehow went beyond pure physics--into metaphysics.

He seemed to have grasped some deeper meaning in our reality; something intuitively reassuring; something about time that appeared to Einstein to render even the concept of death in our universe less significant or less menacing, that this thought should lend comfort to his friend Besso's sister.

Similar Reassurances in Quantum Theory: Certainly his thoughts would bear on the central theme of existential philosophy, to say nothing of their theological implications. It is interesting to note, as we do in the chapter on Quantum Reality, that similar 'reassurances,' something 'less menacing,' also surface in quantum theories, suggesting, according to many physicists, the existence of a nonmaterial interconnected universe of information-mind, of which our consciousness is part, unaffected by death of the physical body.

Why Did His Discovery Somehow Give Him More Serenity? But getting back to Einstein and time. He was prepared to accept his friend's death with more serenity than might have been expected but for his conceptions of the meaning of time in the universe. Sadly, his dialogue with Besso came to an end before a more analytical explanation of his beliefs unfolded. So his deeper thoughts elude us. But we can and do speculate on what he meant by these references in connection with the chapters on death.

CONSCIOUSNESS, DEATH, AND TIME

Quantum Spacetime: We have explored, in several chapters, certain aspects of physics, especially quantum theory. We have also dealt with our physical existence in our spacetime universe of four dimensions, and the possible existence of other dimensions, and even of parallel or adjoining universes which may or may not share our physical characteristics and with which we may or may not be able to communicate.

Having noted that mind and consciousness are nonmaterial entities that exist *somewhere*, Paul Davies, in *The Mind of God*, discusses where mind or consciousness might be located, and whether it exists in time (our spacetime dimensionality).

Consciousness As Not Existing in Our Spacetime: Davies discusses the idea of spacetime being divisible into abstract quantum subunits of the universe, extending beyond our presently conceived spacetime, leaving an "elsewhere" of quantum spacetime bits, and of mind as possibly imbedded in a higher-dimensional space. So, he notes, *consciousness may not be in the space of our universe at all.*

Consciousness and Time: But what about consciousness and time? He asks whether consciousness is in our time dimensionality. You recall our earlier discussions about whether God exists 'in time,' and the problems of how He may be thought of as performing temporal functions, such as planning, if He does not exist in time. Well, Davies feels that both human mental processes that are time-dependent, and our very perception of time, require our consciousness to exist in time.

Survivability of the Consciousness: There are different opinions on whether the individual human consciousness and self-awareness can survive the dissolution of its organic framework--death. Both reductionism and scientific materialism suggest that consciousness is dependent on its originating organic format. The concepts of holism, quantum theory, spiritualism and religious transcendentalism, on the other hand, allow for disembodiment of the consciousness, an independent existence and a re-integration of individual consciousness with the cosmic consciousness.

Survivability of Mind As a Quantum Effect: Physicist Paul Davies, as we've noted, speaks of the survivability of the consciousness since it is not a physical part of our spacetime universe. Certain versions of quantum theory, as we discuss in the chapter on Quantum Reality, suggest that the consciousness is part of an underlying cosmic mental fabric, consisting of nonphysical information and software and so it has potential for surviving death of the body with which it is associated. This is another way of saying that quantum aspects of time may lead to different conclusions about life survivability than classical physics.

Making Scientific Sense of Immortality: Davies goes on to discuss a number of principles and concepts relating to individuality and says that some of these ideas "hold out the hope that we can make scientific sense of immortality, for they emphasize the essential ingredient of mind as information." He cites the analogy that Beethoven's Fifth Symphony does not cease to exist when an orchestra finishes playing it. So, he concludes, "the mind may endure by transfer of the information elsewhere."

While research efforts leave little room for the traditional idea of the "soul," he observes, "they leave open the possibility of survival of the personality." It is interesting to note here the oft-repeated religious proposition that our physical life is given us to refine and hone the spirit given us at birth through the wealth of mellowing and enlarging experience that our physical lives produce.

Time: Preexistence and Postexistence: The question of where our consciousness comes from sometimes becomes the question of where did our 'existence' come from and is that existence eternal? If we had some kind of 'pre-existence' in some inchoate form, as Plato asserts, then this will influence our view of its continued eternal character, the post-death continuation.

It was St. Augustine's seemingly prescient view that all time exists, having been laid out at the creation, strangely enough, a notion supported by Einstein's relativity theory of the geometry of the spacetime continuum. In this context of physical theory, at least, it might be asserted that we exist from the beginning and throughout eternity.

Immortality and the 'Temporal Dilemma': In his discussion as to whether the soul, consciousness, transcends time, Paul Davies has difficulties with "timeless souls." He cites the Catholic Encyclopedia: "The notion that God has a supply of souls that are not any body's in particular until he infuses them into human embryos is entirely unwarranted by the evidence....The soul is created by God at the time it is infused into matter."

He notes that this same "temporal dilemma runs through all discussions of immortality." He notes that Jesus spoke of "life everlasting," implying unending time but he refers to the difficulties of reconciling this with God's laws of physics, suggesting that "there may be no 'everlasting' anyway."

TIME IS JUST THERE-- A DETERMINISTIC UNIVERSE?

Chronons and the Minkowski Diagram: A Nova, Public Broadcasting Service program, *Its About Time*, interestingly discusses the movie analogy of the passage of time as a steady stream or a division into bits called 'chronons'; the Minkowski hourglass diagram of the spacetime continuum; and a discussion of whether our future is already laid out on the fixed 'world line,' waiting for us to encounter it. A brief summary of its primary aspects may be helpful.

The Minkowski Diagram: The Russian mathematician Minkowski, as we noted, was a teacher of Einstein and the prime contributor to Einstein's grasp of the relationship of the dimension of Time to the dimensions of space, the spacetime continuum. Mathematicians use a Minkowski diagram to discuss space and time: a grid of the three dimensions of space: up and down, left and right, forward and back. It is popular to use the analogy of a movie with multiple frames of film to discuss increments in the passage of time.

In the movie analogy, the single frame of film represents an event, the three dimensions caught at a point in time. Take depth and up and down dimensions and we have the dimensions of space, fixed on the stopped frame of film. Draw into this frame a direction of time from bottom to top, and a point which lives its life limited to this single dimension.

The World Line--The Hourglass: By joining the frames of film, the quanta or instants of time, so to speak, you can see the track it is leaving, a line the mathematicians call the 'world line,' the point's history, its beginning, activity, and its end. The vertical line is the direction of time, the horizontal line, space. The radiating lines are called light lines, based on the speed of light. Their crossing point is the present, the 'now' line.

It resembles an hourglass. Above the 'now' is a region called the 'absolute future,' the hourglass sands yet to fall. Below the 'now' point is the 'absolute past,' the sands already fallen. On both sides there's a region of space and time, called the 'absolute elsewhere,' about which, "as yet we know nothing."

Time Doesn't Pass, Its Just There: "Alas, time stays, we go," said Austin Dobson, in "The Paradox of Time." According to views based on relativity, time doesn't 'pass,' it is just there. Your consciousness of the present, moves along your world line, experiencing aspects of your life as you reach them.

The principle says that the portion of your world line that is beyond the now point is already there and you simply 'happen' on them when you reach those points. Similarly, the portion of the world line before the now point is still there. In this concept, one might say that things don't 'happen' to you, you happen to these events by encountering them.

Arthur Eddington, originator of the "the arrow of time," in *The Nature of the Physical World*, points out that this 'one-way' characteristic of time is "recognized by consciousness....equally insisted upon by our reasoning faculty," but does not appear in physics except in the context of a "number of individuals," where the arrow represents "progressive increase" of randomness.

What Happens to Free Will? One implication of this view of the nature of time, as a dimension in geometry, is that one does not chose his fate, he encounters it. Are his 'free-will' choices as he moves along really free will? Is one predestined to make those choices and to what extent may they be said to be free? As we discuss in the chapter on Predeterminism and Free Will, the argument is made that this exercise of

will is genuinely elective because it is unpredictable. The discussion there addresses the question of whether deterministic time in relativity is in conflict with quantum uncertainty.

Thus, one is predestined to have made certain elective choices. Of course, this perspective of time being laid out is based upon the relativity model of the universe. What is the effect of adding quantum theory, dealing with uncertainty and randomness, to relativity theory contemplating a universe laid out at the creation? This is discussed in the chapter on *The Problem of Free Will.*

SPEED OF LIGHT AND TIME TRAVEL TO THE PAST

'Tachyons' and the 'Other Side' of the Speed of Light: In the chapter on Relativity, we examine Einstein's special relativity speed-of-light limitation on the speed of communication and travel because of its pessimistic implications for communications between intelligent civilizations. The limitation bears also, of course, on the geometry of time, the slowing down of time with increase of speed or mass, as well as the phenomenon of the irreversibility of time in the thermodynamic and electromagnetic arrows of time which governs the macro universe of everyday life.

Well, theoretical particles dubbed 'tachyon' are hypothesized which operate on the 'other side' of the speed of light limitation; particles which must travel faster than the speed of light.

Travel to the Past? Elsewhere we speak of these as-yet undetected tachyon. In easy-to-understand terms, Michael Talbot writes, in *Quantum Reality*, and Nick Herbert, in *Faster Than Light*, that faster-than-light transmissions are not precluded by Einstein's special relativity theory, citing the hypothesized faster-than-light particles--tachyons.

We have noted that while time may not be irreversible in the quantum world, it is irreversible in our macro universe of stars and people. But implied in the existence of faster-than-the-speed-of-light tachyon is the theoretical possibility of time travel to re-experience the past. Is time, then, not irreversible?

The Quantum Cosmic Connection: Because of the geometry of mass, speed of light, and time relationships, we also discuss instantaneous communication between particles, the so-called 'quantum connection' which is hypothesized to connect and permit simultaneous 'knowledge' between distant particles, as well as para-psychological phenomena of 'instantaneous' communications between widely separated human beings, a phenomenon at variance from the speed-of-light rule for elec-tromagnetic transmissions.

In our chapter on Quantum Reality we discuss the quantum cosmic connection, the concept that underlying our macro physical world of people and physical objects lies a nonmaterial fabric of mind and information in which everything in the universe is interconnected, part of the whole, permitting instantaneous communication between any parts of this whole, regardless of its location in the physical universe.

TIME AND PARALLEL UNIVERSES

Availability Now of the Past and the Future: Relativity and quantum theories offer new insights into the nature of time, possible time travel, and the availability of both the past and the future, in the present.

Relativity, Quantum Theory and Time: The chapter on Relativity speaks of time as a geometric dimension of the universe so that all time--past, present, and future--coexist as a single continuum. Time is there for us to encounter as we move along the world line. Physicist Fred Allen Wolf, noted quantum physicist and author of such books as *Taking The Quantum Leap*, observes, in *Parallel Universes, The Search For Other Worlds*, that we "march on in time....through an imaginary dimension of space called time, moving at the speed of light."

But quantum theory introduces even more startling concepts of the nature of time and what we perceive as a concrete physical universe, so-called 'physical reality.' This physical reality, according to the Copenhagen quantum theory interpretation, involves subjective observer-dependent quantum uncertainty states, in which not only is the universe not so physical as we thought, but, according to the 'many worlds' quantum interpretation, quantum events in this universe interconnect, perhaps in a higher dimension of space, with 'parallel universes.'

Parallel Universes and Time: In his *Parallel Universes, The Search For Other Worlds*, physicist Wolf notes how science fact is "stranger than science fiction." Our universe, he says, may be part of a complex of parallel universes, "occupying the same space as our own in some ghostly manner...perhaps only atomic dimensions away, but in a higher dimension of space"-- what physicists call superspace.

Parallel Universes Required by Absence of Big Bang Observers: We have noted that according to the Copenhagen interpretation of quantum uncertainty, the universe we experience is one which results from the presence of observers; that everything in and about the universe was resolved by the acts of observation that resolved its quantum state from alternative possibilities.

Well, at the time of the Big Bang, quantum theory says that because there were then no observers, the universe was in some "superposition of quantum possibilities" and so, says physicist Wolf, there must have been parallel universes then because quantum uncertainty rules require the occurrence of all possible scenarios.

But, asks Wolf, what was the universe's initial radius, who measured it, and when was it measured? Answers physicist Wolf, "It is our observations *now* that are determining the past," including the Big Bang!

New Notions of Time and the Human Mind: And, it is suggested, these concepts are producing new notions of time and the functioning of the human mind. He goes on to discuss our decision-making process, suggesting that the mathematics of quantum physics demonstrates that *the future as well as the past enter our present time*, that time is not fixed, time travel is possible, such as through spacetime warps created by black-holes; that if parallel universes exist, as he believes, that time travel must be theoretically possible.

Time Machines and Death: Do time machines exist now? According to Einstein's relativity they exist within black holes and *possibly within electrons*, says Wolf, so that our bodies, through electronic feedback, become time machines. Quantum theory, he says, "implies the existence of both the past and the future, simultan-eously" with our own time. The implications of these ideas, of course, impinge on our notions of the meaning of death.

TILLICH AND TIME & SPACE

Tillich Addresses the Arrow of Time: Scientific notions of time and space have not gone unnoticed in theology. Some theologians have sought to consider theological implications. Time is often depicted as the inexorable force pitted against the struggle of life for survival. Theologian Paul Tillich ventured into physics to provide a scientific backdrop for certain theological principles.

In his *Theology of Culture*, theologian and teacher Paul Tillich, speaking of the religious dimension in human cultural activities, discusses the irreversibility of time. He characterizes this as the *Struggle Between Time And Space*. Time and space, he says, are the main structures of existence, by which he means "being finite or being in time and space...the powers of universal existence, including human existence."

The Struggle between Time and Space: Addressing both theology and physics, he proceeds to take up the age-old struggle of human life against the enemy of time, noting the struggle against the inexorable process implied in the arrow of time. Human soul, human history, and "nature itself," Tillich says, are significantly determined by this "struggle between time and space." He speaks of time as the "fourth dimension of space," and of turning around the physical process so that time, "deprived of its most essential quality--to have a direction...is under the full control of space."

Time Deprived of Its Power by Circular Motion of Continuous Repetition: Although it represents time's victory that both time's direction and the life process from birth to death cannot be reversed, nevertheless, the life process can be *repeated.* And so, says Tillich, the direction of time is 'deprived of its power by the circular motion of continuous repetition.' Each new life repeats the law of birth and death, he notes. While this may diminish our significance as individuals, and be of little solace to us as individuals, it forces us to think beyond ourselves, to recognize the greater unity of which we are a small part.

Time's Final Victory: Human life Acts toward Something beyond His Death: But the final victory of time against space is possible, for Human life is "able to act toward some-thing beyond his death," he can have history and can "transcend even the tragic death of families and nations, thus breaking through the circle of repetition toward something new."

Tillich goes on to discuss the "turning point" in the struggle between space and time as represented in the "prophetic message," and the birth of prophetism out of paganism. While the pagan "gods" pass with their peoples, God lives on. "Time is fulfilled in history and history is fulfilled in the universal Kingdom of God."

Tillich walked onto the stage of history at a time in this century when the scientific and theological worlds were trying to deal with the newly-introduced concepts of Einstein's relativity. One wonders how he would have juxtaposed quantum theory and theology.

Paul Tillich's Existentialism And Time: Paul Tillich, discussing "time" in his Theology of Culture, also devotes much effort to existentialism, and the religious dimension in human life's cultural activities. The analysis of existential finitude inevitably "culminates in the analysis of Time," says Tillich, also noting that "the insight that existence is distinguished from essence by its temporal character is as old as the philosophy of Existence."

He discusses the distinction between existential or immediately-experienced Time from dialectical timelessness on the one hand, and from the "infinite quantitative measured Time of the objective world," on the other hand. This "qualitative" Time of "Personal experience" is, he says, the general theme of existential philosophy.

Existentialism and Time--Schelling and Kierkegaard: Schelling, says Tillich, distinguishes three qualitatively different kinds of Time: pre-temporal, temporal and post-temporal, assuming a beginning and end. Kierkegaard seeks escape from measured and objective Time through his Augenblick, "the pregnant moment in which Eternity touches Time and demands a personal decision." Kierkegaard seeks avoidance of the objectivity of the Past through his idea of Gleichzeitkeit, which treats all history as "contemporaneous with the pregnant moment," claiming that "a repetition of the Past is a Present possibility."

Nietzsche: The Weight Of Eternity Assigned To Every Moment: Nietzsche, says Tillich, escapes from infinite quantitative Time via his doctrine of "eternal recurrence" which assigns to every moment the "weight of eternity." In Bergson's struggle against quantitative, objective time, time is "subjected to space."

Heidegger: Temporality Means Care--Kant: Time As 'self Affection': But the most "radical" of the distinctions between existential, immediately experienced time, and quantitative, objective Time, says Tillich, is Heidegger's, influenced perhaps by his impending death: "Temporality is the genuine meaning of care," and Care is finite Existence. Heidegger's analysis of Kant is that time is defined by "self affection," grasping for one's personal existence. So *temporality is existentialism.*

Time, Existentialism, and Meaning: It is said that human preoccupation with time is traceable to concern over our continued existence, the central theme of existentialism. The study of the entire field of philosophy is thought by many to be rooted in the awareness of our finitude, our limited duration, our concern for existential continuation--avoidance of the nothingness which might follow. The search for meaning in and of the universe is often viewed as a function of identifying the purpose of life which, in turn, is inherently tied to the passage of time and the corresponding reduction in the duration of purpose-fulfillment opportunity.

Discussions of time and existentialism frequently lead to examining the thoughts of Martin Heidegger, despite his unfortunate dalliance with German National Socialism and its elevation of will over reason. This fall from grace in the world of philosophy, discussed in Victor Farias' *Heidegger and Nazism*, is attributed to his having been in the wrong place at the wrong time, caught in the unfortunate grip of the contemporary German patriotic preoccupation of his time and place and his inability, later, unequivocally to recant.

Heidegger and Time: Although Heidegger did not consider himself an existententialist, his writings classify him as such. Influenced by Soren Kierkegaard who has been viewed as a pathfinder in the existential school of philosophic thought, Heidegger, in turn, had considerable influence on French existentialism, including Jacques Lacan and Jean-Paul Sartre In this school, precedence must be given to the nature of Being, one's own subjective existence in time and space.

In this context, each individual is a sentient, independent consciousness, responsible for choices based on subjective perceptions of value and truth, and having a deep anxiety over the inevitability of discontinuance and ultimately the relegation to 'nothingness.' And so time becomes the measure of our existence and its potential, and root of our concerns.

Involvement: Heidegger observes that our concerns are often obscured by the loss of selfhood in the cares of daily life. He rejects the distinctions between consciousness and the external world, favoring the basic mode of being in the world, as "involvement." Action and knowledge are inextricably related, being human is being open, "being in the world," and relating to something or someone other than himself.

For many, the very meaning of life is the sum total of those involvements. It is interesting to note that Einstein sensed this deep, concern and gave expression to similar concerns over our habitually too-limited involvement with the rest of humanity--and of our awareness of unity in the cosmos. And so again we come back to criticality of time.

Being in the World: Heidegger's Dasein or "Being in the world," has been described as characterized by a constant state of care, anxiety, solicitude, and "being toward death" which pervades being, and openness to the existential possibilities for being, especially being with and for others--the possibilities for "involvement" and the awareness of the potential for death.

So the idea of transition from an 'existent' to a 'non-existent,' inevitably translates into the importance and meaning of time, almost an equivalency. As theologian Paul Tillich puts it, "For the whole of Existential philosophy the analysis of finitude culminates in the analysis of Time." As the subject of deep human concern, time, a succession of discrete moments, is inextricably wound up with human existence. In this inseparability, according to Heidegger, past, present and future are equally real.

Heidegger and Physics: It is worth considering that Heidegger lived during the period of the development of the theories of relativity, the new concepts of space-time geometry and the evolution of quantum mechanics. The quantum theories of time as being a succession of frames, particles of time, the hypothesized chronons, were likely known to him as was the quantum constant of Max Planck and the particle uncertainty concepts of Werner Heisenberg and Erwin Schrödinger.

It is interesting to observe the relationship between these scientific concepts, especially Einstein's quantum treatment of time and Heidegger's strikingly parallel philosophic treatment of time. Einstein's mysterious reference to distinctions between past, present and future as being illusory and Heidegger's equally enigmatic reference to the past and future as being as "real" as the present, may not be mere coincidence.

Human life's Preoccupation with Existential Concerns: This preoccupation with the fact of existence has pervaded philosophy down through the ages, even before Plato and Aristotle. As physicist Heinz Pagels points out in *Perfect Symmetry*, the existentialist, Martin Heidegger, views this preoccupation as representing the root question in all of metaphysics of why is there any Being at all rather than Nothing.

Pagels asks, "who needs the complexity of the world in the face of the simplicity of absolute nothingness....I feel anxiety as I imagine the unfathomable, silent abyss of nothingness which might have been....That odd feeling provoked by our recognition of the reality of existence rejuvenates our wonder and curiosity about the universe...And that wonder is the beginning of science. But perhaps Pagels has, in a way, more clearly identified the essentiality of existential anxiety; that it provides the basic drive for relentless pursuit of knowledge of the nature of the universe, to comprehend the reason for our presence, and to envision our ultimate destiny.

> Time is nature’s way of keeping everything from happening at once.
> George Zebrowsky

CHAPTER 9
CONFIRMATION OF THE BIG BANG

General Relativity Mathematics Confirmed--Galactic Recession: The general relativity theory equations, suggesting an expanding or contracting universe, was confirmed early in the century, by astronomer Edwin Hubble's finding, from analysis of the downward shift toward the long wavelength red end of the light spectrum, that the distant galaxies were *receding.* This was a confirmation of Einstein's equations that the universe's space was elastic, it was indeed *expanding* from the Big Bang. Cosmology had suddenly become a primary focus of physics, astrophysics, and astronomy.

Big Bang 'Cosmic Background Radiation' Confirmed: After the first few minutes, the expansion and cooling of the universe lowered electromagnetic radiation frequency to permit the formation of matter elements, like hydrogen and helium. Some fifteen billion years later, the Big Bang *cosmic background radiation*, would have cooled to a present temperature of a few degrees above absolute zero.

The existence of this residual radiation from the Big Bang, pouring in from all directions in the universe, was confirmed in 1965 by Bell Laboratory radio astronomers Arno Penzias and Robert Wilson--for which they received the Nobel prize. But this still left the question of how the great galactic structures had formed, given early radiation uniformity

NASA Finding of Temperature Ripples in the Fabric of Spacetime: The scientific community was stunned in April of 1992 when, at the meeting of the American Physical society, NASA announced the results of the Cosmic Background Explorer (COBE) satellite computer-assisted measurements of faint microwave echoes of primordial temperature ripples in the fabric of primordial spacetime, ripples triggered, perhaps, within the *first trillionth of a second* of the universe.

Map of the Earliest Cosmic Structures--Confirming the Big Bang: These ripples, temperature variations of 31 millionths of a degree, are considered to represent 'hills and valleys' in the primordial fabric of space-time, a map of the *earliest cosmic structures* or shapes in the radiation of the early universe, which has taken some fifteen billion light years to reach us. Considered as virtually 'viewing the birth of the universe,' physicist Stephen Hawking termed this finding *the discovery of the century, if not of all time*.

These findings were thought to confirm the hypothesized initial exponential inflation of the universe, and the follow-on Big Bang sequence. It was also viewed as providing information on 'the missing link' between today's galactic structures and the hypothesized smooth starting universe. In effect, it was thought to explain how the inflation-caused uniformly-smooth starting universe was transformed into today's complex structures of galaxies and voids.

Findings of Big Bang Helium: Additional proof of the Big Bang also came in June of 1995 when Johns Hopkins University announced its NASA space shuttle Astro 2 Observatory findings of distant primordial intergalactic clouds containing spectral lines evidencing primordial heavier-than-hydrogen elements of helium, and of trace quantities of deuterium, lithium, boron and beryllium.

These primordial elements, whose creation predates the formation of the galaxies and the stars, were computed to have been fused from hydrogen within the first four minutes after the Big Bang, and to *exist in the very amounts consistent with the electrodynamics calculations* of 24 percent of the universe's hydrogen as converted to helium and trace elements during those first minutes of the universe.

Support for the Big Bang: Apart from Einstein's big-bang equations, astronomer Edwin Hubble's finding of the *receding galaxies,* evidencing the expansion of space, and the Penzias and Wilson finding of the low, smooth, universal *primordial cosmic background radiation,* convinced most physicists of the validity of the Big Bang theory. The still-unresolved question was of the origin of the galaxies, in view of the original radiation uniformity.

As we have noted, most cosmologists agree that the Big Bang represents scientifically proven fact, despite recent Hubble Telescope astronomy data which raises questions about the correct numbers for the Hubble constant and age of the universe. These findings, suggest that the universe may be *younger than some of the stars*, at least according to accepted stellar physics. So questions are still raised about the nature of the Big Bang.

Further Evaluation of the COBE Findings: In the 1995 Epilogue to his *The Light At The End Of The Universe,* Michael Lemonick, wondered about the state of cosmology theory after the reevaluations following the spectacular 1992 NASA COBE announcements described above. In retrospect, he notes, after the initial excitement of the temperature variations, those findings were not sufficient to quite "vindicate" any prior theory, nor, for that matter, very much "change" anything.

The initial enthusiasm over COBE fluctuation findings, reflected in the Hawking characterization as "the most important discovery of the century, if not all time," and George Smoot's famous quote, "like seeing God," had pretty much subsided. Those who had earlier been excited about 'cold dark matter,' the related assumption of a starting inflation event, and an Omega-one universe (in which the density for the universe is precisely sufficient to slow expansion but never to initiate a collapse) were still enthusiastic, but essentially unaffected by COBE.

Following COBE, he says, those who believed in inflation, still do, and that those who believed that Omega is likely *one*, for the most part, continue to do so. Dark matter, though, may now be perceived as more likely a mixture of hot and cold, the hot being small-mass neutrinos (which would explain large-scale galaxy formations), the cold still being the conventional CDM--cold dark matter (which handles the small-scale formations).

Asked about his reference to the COBE findings as "the greatest discovery of the century if not of all time," Stephen Hawking answered that his enthusiasm was based on the findings being consistent with the initial universe inflation theory, which, of course, was viewed as solving so many problems.

One problem now is the recent new Hubble Space Telescope findings suggesting the universe as perhaps only eight to twelve billion years old. This is startling since that is younger than the age of some of the estimated fourteen billion-year-old stars--and, of course, younger than the earlier Big Bang fifteen-to-twenty billion year universe age estimate.

Lemaître's Cosmic Egg: The brilliant, obscure Belgian cosmologist, mathema-tician and priest, Georges Lemaître, applying Einstein's general relativity equations and Hubble's finding of increasing galactic separation, then chronicled a beautiful creation scenario. In that picture, the present universe was expanding from a *creation point,* in which an incredibly small, dense and hot particle--the *cosmic egg*, underwent a cataclysmic explosion which continues to power today's still-expanding universe.

If, according to Einstein's relativity equations, the universe is now expanding, reasoned Lemaître, then it must have been progressively smaller in the past. But the backward-in-time size-regression could not be infinite, and so the process must halt at some minimum state, thought to consist of an infinitely dense, hot, and small state, now designated a *singularity.*

Lemaître dubbed this ultimately diminished-state the *cosmic egg*, and its cataclys-mic explosion, the "big noise," later whimsically dubbed the "Big Bang" by physicist Fred Hoyle-- who, at the time, actually advocated the steady-state eternal universe.

The Genesis Creation Event? The discoveries of uniform cosmic background radiation, thought to be fossil remains of the Big Bang, and Hubble's finding that the universe was still expanding, were looked upon by most physicists as cinching the Big Bang theory. So the steady-state theory of an eternal universe gave way to general acceptance of a *creation* event. Many interpreted this as science reinvigorating the Genesis creation account.

Some Religious and Scientific Views of Cosmic Evolution Differ: The theory left open the question of *what caused* the Big Bang or what, if anything, may have been an 'earlier' state of things. In the face of these increasing indicators of the nature of the start of the universe, the Catholic Church and other denominations have shown receptivity to accepting this view of the origin of the universe. Thus the six-day Genesis creation sequence is viewed as essentially allegorical, suitable for the human species in its earlier state of unsophistication where there was no practical use for a more precise physics handbook.

Strict biblical constructionists, and so-called 'creationists,' on the other hand, insist on the literal or so-called 'inerrant' view of Genesis in which everything, the head-ens, the planets, the earth and the human race, were all miraculously created, some five-thousand or so years ago, within the actual earth-revolving, six-day period, as the Genesis sequence depicts. This is further discussed in the Chapter on Religion, Denominations and Meaning.

Two great revolutions gave birth to the new physics: quantum theory and the theory of relativity . The latter, almost exclusively due to the work of Einstein, is a theory of space, time and motion. Its consequences are as equally baffling and profound as the quantum theory. --Paul Davies

CHAPTER 10
THE PLANCK WALL--
UNIFIED FORCES BREAK DOWN

The 'Planck Wall' and the Big Bang Sequence: We noted earlier the pre-Big Bang events, in which the universe starts as a quantum fluctuation in the eternal mega-universe quantum nothingness, generating a point of unified, space, time, forces, energy-matter which then underwent a sequence of symmetry breaks into our present force and geometry state.

The general relativity theory equations which predict the Big Bang, as we noted, also predict an exponentially dense, small and hot starting point called a *singularity*. This starting point represents the beginning of our universe--*the beginning of time*. This period or point is sometimes referred to as the *Planck wall* or the *Planck era*.

It is the time at 10^{-43} second that is considered the 'beginning of time' and the birth of the universe. Anything before that is viewed as meaningless, *smeared out* by quantum uncertainty. According to supersymmetry theory, this is the point at which gravity and the other forces--all of physics in effect--are thought to have been unified in a single superforce.

Gravity Splits Off: But it is also at this 10^{-43} second birth point of the universe, the Planck wall, that the first separation from unification occurred--the split-off of gravity. Before the split, then, gravity and the related geometry concepts of relativity, were thought to be part of the quantum singleness, the superforce, when all physics may be said to have been in a unified quantum state.

The Grand Unification Theory Period: The period from 10^{-43} to10^{-34} microsecond, is the so-called GUT or 'grand unification theory' period when gravity had split off but the *other forces were still united*. Following the Big Bang creation sequence starting point of 10^{-43} second, the next landmark, says John Gribbin, in his *In Search Of The Big Bang*, is 10^{-36} second when the GUT or grand unification period of the all-but-gravity unified forces comes to an end by the split-off of the strong nuclear force, at a 10^{26} K temperature.

So the period of 10^{-43} to 10^{-35} second is called the GUT era, the period governed by the GUT or Grand Unified Theory during which fundamental forces, the strong, the electroweak (the combined weak nuclear and electromagnetic forces) are unified.

At 10^{-9} second, at a temperature of 10^{15} K, the unified *electroweak* force splits off to separate the electromagnetic force from the weak nuclear force. At this point, the four forces are now separated, and the universe is a particle *soup,* sometimes described as consisting of free quarks, leptons, and photons.

Particle Interactions: Physicist Heinz Pagels, in *Perfect Symmetry*, describes this particle soup as consisting of interacting particles, primarily, quarks, electrons, positrons, photons, neutrinos and antineutrinos. The interaction continually created and destroyed electrons and positrons, annihilating into photons, and photons colliding to create electrons and positrons.

As the soup cools, the interaction in the radiation soup, says Pagels, included a *small contamination of protons and neutrons,* from which, says Pagels, came all the galaxies, stars, and the planets, including Earth, everything we see or detect in the universe of today. In the further cooling process the quarks combine to form protons and neutrons, the Yang Mills fields condense into gluons that bond quarks into protons and neutrons.

The Break in Thermal Equilibrium: At fourteen seconds after time = 0, the temperature, says Pagels, is one billion degrees, cool enough to take the electrons and positrons out of thermal equilibrium with photons and neutrinos so that positrons annihilated would not be recreated, leaving quarks, electrons, neutrinos and photons.

We jump then to three minutes after the Big Bang, with temperatures cooled enough to slow particle movement to permit the proton and neutron *contamination* to combine in nuclei--the hydrogen isotope deuterium, and helium.

Planck Terms: The Big Bang sequence, considered to start with the so-called 'Planck Wall' or 'Planck era,' as noted, is the time at 10^{-43} second considered the point in time when the universe was born and before which time would have no meaning. According to unification and supersymmetry theory, at this time point, as noted earlier, gravity and the other forces, all of physics in effect, are thought to have been unified.

The "Planck *length,*" says John Gribbin *in In Search Of The Big Bang*, is 2 X 10^{-33}, and the "Planck *mass"* is 2 X 10^{-5} gram. He says that a particle of these dimensions *would be a black hole*. The Planck measurements, based on the work of the physicist Max Planck, represent *the smallest length, time, and mass, that have any meaning.*

The 'Planck *time*,' is defined, incidentally, as the time it takes light to travel across the Planck length, which happens to be some 10^{-20} times the distance 'across' a proton. Planck measurements, as noted, are viewed as the smallest measurements that have any meaning. They define the time scale and distance scale "where the quantum effects of gravity dominate the structure of spacetime."

The distance scale of Planck units is that at which gravity equals the other forces, at temperatures over 10^{32} K. So, he says, the time from Time = 0 to the first 10^{-43} second of the universe is *dominated by quantum gravity* and since we have no accepted theory of quantum gravity, the moment of creation is "unknowable."

> Any statement which attempted to express a consensus of scientists about religious and philosophical questions would miss the main point. There is no consensus among us.--Freeman Dyson

CHAPTER 11
BLACK HOLES, PULSARS, GRAVITY WAVES, ANTIMATTER

Relativity and Black Holes: As noted earlier, one prediction of general relativity was the theoretical existence of *black holes.* These represent the ultimate distortional effect of the force of gravity on mass and time. Increasing gravity will overcome other resisting forces and totally curve surrounding space and time, to bring time to a stop and mass to an infinitely compressed point. This gravity compressed point is of zero radius and called a *singularity,* a term thought to have been originated by physicist John Wheeler back in the 1960's.

The Black Hole's Event Horizon: The boundary or 'event horizon' of the black hole is the area surrounding the black hole within which the gravitational field and consequent spacetime warp forces light rays or matter along a path so bent that they cannot escape their *Schwarzschild* 'confinement radius' named for its mathematical author. In the absence of light, which cannot escape a black hole to signal its existence, the existence of a black hole must be extrapolated from calculated gravitational effects on nearby detectible phenomena outside the confinement radius.

Rips in the Fabric of Spacetime: In *Steven Hawking's Universe*, John Boslough quotes some of physicist Stephen Hawking's thoughts on relativity-predicted black holes. Black holes, says Hawking may be viewed as "rips in the fabric of space and time" created by unimaginable gravitational forces. Black holes are thought to be at the center of some, and perhaps all galaxies, and of quasars--so-called quasi-stellar objects at the center of certain distant galaxies which are the universe's most powerful stellar devices, one-trillionth the size of our galaxy but generating a thousand times its energy.

Space and Time Come to an End: Within black holes, says Hawking, "space and time as we normally perceive them, come to an end." Hawking thinks that general relativity effects would have created millions of tiny black holes in the Milky Way and other galaxies, formed by the incredible pressures at the start of the Big Bang.

Stars as Fusion Reactors: Stars are fusion reactors in which hydrogen protons are fused to form an isotope of hydrogen--deuterium, and, in turn, helium, carbon, oxygen and the heavier elements. When these fusion fuels are spent--fusion-nucleosynthesis, the process which produces the gravity-resisting pressures ceases and the implosion force of gravity then takes over to force the star's collapse.

Neutron Stars or Black Holes: If the star's mass is greater than about 1.4 solar masses, the 'exclusion principle' resistance is overcome to collapse the star to an incredibly compact neutron state, a *neutron star*. If the collapsing star's mass exceeds about 3.6 solar masses, the implosion or collapse continues to the black-hole state. The black hole represents the final stage in the death of a high-mass star before it reaches the final or *singularity* state of infinite density and spacetime curvature.

Singularities: It is the black hole, notes Boslough, that masks the singularity from the rest of the universe, *creating a break with the ordinary spacetime around it.* The boundary or 'event horizon' of the black hole is the area surrounding the black hole within which the gravitational field and consequent spacetime warp forces light rays or matter along a path so bent that they cannot escape their confinement. The confinement radius is known, as noted earlier, as the *Schwarzschild radius* named for the German astronomer Karl Schwarzschild who first mathematically constructed this effect.

Physicists Roger Penrose and Stephen Hawking showed that according to general relativity there *must* be a "singularity of infinite density and space-time curvature within a black hole....rather like the big bang at the beginning of time, only it would be an end of time for a collapsing body ...."

The Cosmic Censorship Hypothesis: In *A Brief History of Time,* Hawking speaks of the weak and strong *cosmic censorship hypothesis*--that singularities occur only in places like black holes where "they are decently hidden from outside view by an event horizon." He calls this the *weak hypothesis*, which protects observers *outside* a black hole from "the consequences of the breakdown of predictability that occurs at the singularity but does nothing for the poor unfortunate astronaut who falls into the hole."

Hawking goes on to say that, in some solutions of relativity equations, an astronaut drawn into a black hole might see a naked singularity, avoid hitting it, "fall through a wormhole and come out in another region of the universe."

Travel into the Past Via a Black Hole? He goes on to say that in the "strong cosmic censorship hypothesis," in a "realistic solution of the equations of general relativity," the singularities would always lie either *entirely in the future* (like collapsed-star singularities), or *entirely in the past* (like the Big Bang). It is greatly to be hoped, he volunteers, that some version of the censorship hypothesis holds because close to naked singularities *it may be possible to travel into the past.*

Black Holes, Quantum Gravity, and Unification: The mathematical expression of the physical effects at the event horizon of, and within, black holes are thought to provide the conceptual framework for the interaction of the effects of relativity, quantum theory, and unification theory, representing all the vital interactions of forces, fields, space and time--everything that explains the structure and functioning of the universe. But more on this later.

Some Impacts and Demonstrations of Relativity Theory: Perhaps the greatest initial impact of general relativity was its prediction that since the universe must be expanding, the universe must have had a *creation event,* a beginning--the *Big Bang* from an incredible, exponentially-small state which came to be described as a *singularity*, a state not fully explainable in terms of relativity theory but requiring the invocation of *quantum theory or mechanics*, the science of the subatomic.

This, of course, led to the generation of concepts of *quantum gravity* and also gave impetus to the search for *unification theory.* Relativity theory and so-called *relativistic effects* have had a powerful influence on physics and especially on astronomy, giving rise to the search for and discovery of: *black holes* and rotating black holes, and the nature of rapidly rotating neutron stars called *pulsars,* and other phenomena, such as 'gravitational lensing,' and time dilation such as evidenced by distant cosmic gamma radiation behavior.

Other relativistic objects are the super-illuminated, beacon-type, super-dense objects at the edge of the universe, so-called "quasi-stellar" X-ray sources, and so named *quasars,* a likely source of black holes. These incredibly powerful objects, moving away from us at large velocities, demonstrating their great distances from us in the expanding universe, are thought to be at the center of the nucleus of galaxies.

Quasars: In *Was Einstein Right, Putting General Relativity to the Test,* physicist Clifford M. Wills states that a popular model for a typical quasar involves a super-massive black hole, weighing perhaps *100 million solar masses*, at the center of the galactic nucleus. He notes that the black hole is ferociously consuming stars and gas, likely one solar mass of material per year. As the material approaches the hole, he says, friction from collisions with other material heats it to an enormous intensity making it visible on Earth.

Relativity and the Black Hole in Cygnus X1: Wills goes on to explain how, applying relativity theory, the Uhuru orbiting X-ray satellite observations of an X-ray generating binary star system called Cygnus X1, permitted the identification of a compact-mass object orbiting a star, as a black hole. The object could not be a white dwarf star because its mass exceeded the 1.4 solar Chandrasekhar mass for a white dwarf star.

And neither could it be a neutron star whose relativistic mass cannot exceed certain limits, so this left the black hole as the only object both massive enough to influence the companion's orbit and compact enough to permit the short term X-ray fluctuations. So general relativity is used to infer the presence of black holes.

Confirmation of the Colossal Black Hole in Galaxy M 87: Black holes have been thought to be at the center of pinwheel galaxies like the Milky Way. In May of 1994, NASA and Baltimore's Space Telescope Science Institute at Johns Hopkins announced persuasive confirmation of the Hubble Telescope's sighting of a supermassive black hole at the center of Galaxy M 87 in constellation Virgo, some 50 million light years from Earth.

The corrected Hubble scope disclosed what was expected, a spiraling disk or ‘pancake’ cloud of hot gas swirling toward the carnivorous galaxy black hole which is swallowing the highly compressed gas that is heated to about $18{,}000^{0}$ F.

The 60-light-year diameter gas cloud disk is rotating at an estimated 1.2 miles per hour. Its rotation speed and distance from the gravitational source demonstrate a mass and density equal to between two to three billion suns, *concentrated in an area with a solar-system diameter*. The celestial mechanics of this description depicts a black hole.

With such a voracious appetite, the M87 galaxy black hole may, some day, consume the entire galaxy! Our own Milky Way galaxy might also be so consumed--after awhile-- if as speculated, it too has such a black hole. One needs remember that these gargantuan gravitational collapses are occurring in a universe still expanding from the Big Bang and, in a sense, pictorialize the processes of what may some day become the Big Crunch of the collapsing universe, assuming sufficient collective mass.

Gravitational Lense Effects: Another general relativity effect is represented by the discovery of the *gravitational lense* phenomenon, the gravitational effect of galaxies to warp the surrounding spacetime, bending light in its path to Earth, just as the sun's mass bends light from sources in the universe behind the sun in its path toward Earth. Associated Press articles in January of 1994 refer to an international team of astronomers detecting instances of stars suddenly becoming brighter in the outer fringes of certain galaxies attributable to MACHOS. These are Jupiter-sized so-called ‘massive compact halo objects’ which are dark and invisible to earth telescopes.

They are detected, says Christopher W. Stubbs of UCAL, Santa Barbara, because they cause stars to brighten, when they orbit between Earth and those stars, through the relativity effects known as *gravitational lensing*--which represents the galactic gravitational warping of space-time. The light is bent as it strikes the MACHO, producing the brightening effect.

Gravitational Waves: Einstein had also predicted, as noted earlier, the existence of *gravitational waves*, ripples in the curvature of space traveling at the speed of light, resulting from interacting bodies such as a supernova explosion, or the collision of stellar objects. Changes in force fields produced by such interactions would not be felt instantaneously but would travel out in waves at a speed equal to the velocity of light.

Stephen Hawking points out that the movement of the earth around the sun creates gravity waves reflecting the loss of *gravitational energy,* a loss which *will eventually bring the earth and sun into collision*, after awhile--about 1,000 million, million, million years, he says.

Physicist Clifford Wills, in *Was Einstein Right*, points out the enigma which existed earlier of whether gravity waves were real, physically measurable waves of gravitational energy, as Einstein had predicted, or just a *mathematical* description of the result of certain gravitational events. Also, assuming the technology for detecting such waves, how would one demonstrate the actual release of gravitational energy by interactions of massive bodies without waiting for a chance cataclysmic event like a supernova explosion in our galaxy?

Binary Pulsar Measurements; Proof of Relativity and Gravity Waves: Physicists Joseph Taylor and Russell Hulse received the Nobel Prize in October 1993 for their relativistic discoveries, using the Arecibo Radio Telescope in Puerto Rico, from observations of a binary star system, one star being a *pulsar*, a rapidly rotating neutron star.

Their observations and elaborate calculations resulted in one of the most complete and sophisticated confirmations and validations of the principles of general and special relativity. One was confirmation that the binary pulsar star system was, consistent with Einstein's predictions, generating gravity waves.

Interestingly, Clifford Wills had elaborately chronicled Taylor and Hulse's exhaustive efforts in his *Was Einstein Right, Putting Relativity to the Test*. The emission of gravity wave energy was evidenced, as he notes, *by relativity-predicted increase in the orbital speed of the two stars and a decrease in their orbital separation.* The actual gravity waves were not, of course, actually detected visibly, but evidenced as described.

What is more, Wills notes, improvements in data analysis by these two physicists permitted them actually to measure the effects of gravitational red shift and time dilation, and thereby *measure the mass of the pulsar and of the companion separately.*

Speaking of pulsars and gravitational effects, a February 26, 1994 Associated Press report quoted astronomer Alexander Wolszczan and colleagues as stating that they had found evidence of at least two planet-sized objects in irregular orbits around a very dense, energy-pulsing star (a pulsar), about 1300 light years from earth. Evidence of the planets is the irregularities in the pulses caused, it is thought, by the gravitational pull of the orbiting planets on the pulsar and on each other.

Elasticity of Time Illustrated by Gamma-Ray Time Dilation: Speaking of time dilation, astronomers Bohdan Paczynski at Princeton and Tsvi Piran of Hebrew University and Harvard had predicted that 'time elasticity,' that is, the time-stretching or time-dilation effects of relativity would be verified from cosmic gamma ray bursts in which the time duration of bursts from great cosmic distances would be stretched or dilated as the radiation travels across the vastness of the universe of expanding space triggered by the Big Bang.

The NASA Compton Gamma Ray Observatory, launched in April 1991, detected gamma ray bursts from all directions in the universe from objects apparently not producing other visible or invisible wavelength radiation. Physicists theorize that they may be coming from cataclysmic encounters between black holes and neutron stars.

At an American Astronomical Society meeting in January 1994, Dr Jay Norris of Goddard Space Flight Center reported that the dim bursts were typically twice as long as the bright bursts, an illustration of relativity theory's time-dilation effects.

To understand the controlling mechanisms that give us our universe one must understand the interrelationship of relativity with quantum theory which will be discussed in chapter 13.

Matter and *Antimatter*--Where's the Other Half of the Universe? According to quantum theory, one-half of the universe should be made of *antimatter.* When positive matter particles erupt from the energized vacuum of so-called empty space, which they randomly do, the corresponding negative charge proton particles also emerges, each moving in opposite directions.

The Big Bang represents just such an emergence of matter from the energized primordial void. We know that our sun, planets, our Milky Way galaxy, our galactic cluster, and even our 100-light-year diameter supercluster of galaxies, are all positive matter structures. We also know that galaxies have collided in the past without instantaneous self-annihilation into gamma radiation. So we're faced with the ques-tion of *where is the equal amount of antimatter* that would have emerged in the Big Bang? Are some galactic superclusters made of antimatter?

So Far No Antimatter Found: Physicists have been driven to use antimatter detectors to identify evidence of telltale antimatter cosmic rays from possible distant antimatter galactic sources. Since antimatter protons could be from antihydrogen, they could also be the result of cosmic ray collisions. But heavier negative charge nuclei like those from anticarbon could only have come from the Big Bang or from antimatter galaxies, since the only other source would be fusion in antimatter stars. *So far, no antimatter protons have been found.*

It is hypothesized that when the initial particles and antiparticles condensed from the void at 10^{-34} second after time zero and self-annihilated, only one in a billion matter rather than antimatter particles was left, and this residual supposedly represents the universe's composition. But this residual excess, characterized as a symmetry or CP violation, implies that the laws of physics are out of balance in favor of matter, giving matter and antimatter different decay formulas. So most of the initial Big Bang matter self-annihilated. *But for this symmetry violation, we wouldn't be here.*

The Search for Antimatter: In *The Anti ± Matter Mission,* April 1966 Discover magazine, Gary Taubes discusses this problem and notes that *antimatter galaxies could* make up half the universe and he takes up this discussion, citing the work of Nobel-prize-winning physicist Samuel C.C. Ting in promoting the NASA Alpha Magnetic Spectrometer. That project would search for antimatter particles, for antimatter particles whose source might have been dark matter (supposedly 80% of the universe's matter), and even for particles which might be *anti-dark matter particles.*

Science fiction buffs have long written of use of antimatter to power space ships and to destroy enemy ships and evil planets and, of course, we must keep our distance from visiting antimatter extraterrestrial visitors.

Antimatter Chemistry Potential: In the New Federalist of June 24, 1996, *The Gates to the Universe of Antimatter, Part II,* Charles B. Stevens reports on the May Workshop on Antimatter in Italy dealing with needed facilities for dark matter research on antimatter atoms and antimatter chemistry and spectroscopy. This involves possibly startling developments.

Antimatter, for instance, may actually *rise* in a gravitational field instead of falling. Beaming the negative-charge antiproton, with 2000 times the negative-charge electron's mass, at helium, for example, may produce radically new types of atoms and subsequent molecules, with vast new potential for exotic atomic and molecular chemistry.

> The work that Roger Penrose and I did between 1965 and 1970 showed that, accordingto general relativity, there must be a singularity of infinite density and space-time curvature within a black hole. This is rather like the big bang at the beginning of time, onlly it would be an end of time for the collapsing body.... At this singularity the laws of science and our ability to predict the future would break down.--Stephen Hawking

CHAPTER 12
OPEN OR CLOSED UNIVERSE--
END OF THE UNIVERSE?

OPEN OR CLOSED UNIVERSE?

Most Matter in Intergalactic Space, The Missing Cosmic Mass: The 1995 NASA Astro Observatory 2 findings, incidentally, also support the conclusion that most matter in the universe must exist in intergalactic space rather than within the visible galaxies. One of the greatest mysteries in astronomy relates to the cosmic *missing mass.* Only some ten percent of the mass required for universe expansion and contraction dynamics, the so-called *critical mass,* is thought to be accounted for. The remaining ninety percent, called the 'missing mass' because it is not visible or directly detectible in galactic structures, is referred to as 'dark matter.'

Only Ten Percent of the Cosmic Critical Mass Traceable: In January of 1996, the American Astronomical Society announced that at least fifty percent of the calculated cosmic mass needed to account for gravitational galactic dynamics of the Milky Way was in the form of newly-detected gravitationally-identified objects in the Milky Way halo. This estimate was based on the mass of likely halo objects, such as visually undetectable white-dwarf stars, a type of burned-out star which will some day be the fate of our own star, the sun.

The Critical Density--Eternal Expansion or Big Crunch Contraction? One of the questions raised by relativity mathematics of a universe expanding from the Big Bang is whether the universe is (1) 'open,' that is, having insufficient gravitational mass to arrest expansion, (2) 'closed,' having sufficient mass to stop expansion and start contraction toward the 'Big Crunch,' or (3) whether it has exactly the 'critical' mass, eventually reaching a state in which it will neither contract or expand.

Present Mass Measurements Suggest an Open Universe: 'Visible' matter, the illuminated galactic structures and other electromagnetically-detectible matter, is known to represent *less than ten or perhaps even five percent* of the needed 'critical' mass. Various forms of so-called 'dark matter,' like invisible galactic halo matter, possibly neutrinos with mass, and other more exotic candidates, are hypothesized as existing, which doubtless account for some of the 'missing' mass. But this limited mass implies that the universe is more likely open.

Measuring the Rate of Deceleration of Expansion--Hubble Constant: But another determinant of the 'open-closed' question involves the decrease in the universe's *rate of expansion,* reflected in the equation $v = Hr$, v representing velocity, r the distance between objects, and H, the *Hubble constant,* the velocity-distance proportionality value.

The Hubble constant, presently measured by comparative brightness of distant galaxies, declines more rapidly in a closed universe. But since galactic brightness changes over time, this criterion is less than totally reliable and the rate of deceleration requires further measurement.

New Deceleration Measurement Criterion? Ohio State astronomer Barbara Ryden proposes to use the measurement of shapes of gigantic distant space objects to determine the decline in the Hubble Constant. Using ideas of Princeton cosmologist Bohdan Paczyński and Lawrence Livermore National Laboratory cosmologist Charles Alcock, light waves from receding galaxies are stretched toward the red end of the light spectrum.

So differences in redshifts between the nearer and further extremities should be greater than that of near objects. But needed for these measurements are really gigantic cosmic objects. *Sheets of galaxies,* spread over spherical voids hundreds of millions of light years in diameter, are large enough to show such deceleration distortion. But University of Washington cosmologist Craig Hogan thinks that data from NASA satellites planned for launch in 2001 and 2005 will actually expand forever or eventually contract in the big crunch.

Use of the Hubble Time: Also usable to measure age of the universe is to determine the 'Hubble time,' the time elapsed since the Big Bang, supposedly the time required to account for the distance between galaxies, assuming that the present rate applies during this period. If the earlier expansion rate was higher, the expansion time would be less than the Hubble time.

Relativity mathematics shows that a closed universe expands faster than an open universe. A universe with an age less than two-thirds Hubble time, will be closed; otherwise it is open. Two thirds of the Hubble time is called the critical age of the universe. Before the recent Hubble observation findings suggesting a younger age of eight to twelve billion years, the age estimates were often rounded to fifteen billion years, within the range of ten to twenty billion years.

Is the Universe Older Than Some of the Stars? As noted, one of the methods used to determine whether the universe is open or closed is measuring the time since the Big Bang, the age of the universe, assuming that the present rate of expansion holds throughout this age, the so-called 'Hubble time' required for a galaxy to reach present distance from our galaxy since the original Big Bang event.

Disturbing to earlier notions of the physics of Big Bang theory are conclusions, from recent Hubble Space Telescope findings, of the likely age of the universe as between *eight and twelve billion years.'* This happens to be *less than the age of some stars*, at least according to accepted physics of stellar thermonuclear dynamics, something quite disconcerting to those satisfied with the existing Big Bang configuration.

Eternal Life in an Open or Closed Universe: In his *The Physics of Immortality*, mathematician Frank Tipler describes a universe in which intelligence spreads through the universe to become the *Omega Point God* who, at the end of time, resurrects all earlier life, thus assuring us a form of immortality--after awhile.

Tippler hypothesizes a *closed* universe on his *assumption that the laws of physics must be such as to preserve life*, signifying a closed universe, since, in his hypothesis, an open universe would die a heat death, ending all. He provides fascinating supporting mathematical hypotheses for this, though later Hubble observations may imply a Hubble constant at variance from his assumptions.

In *Infinite in All Directions*, on the other hand, because of the huge 'missing mass,' Freeman Dyson *assumes* an open universe in which intelligence also spreads to pervade the universe, and ultimately to control its physics and thus assure eternal life in some form, despite the 'heat death' normally associated with an open universe.

The 'Dark Side' of the Universe: A number of physicists have addressed this question of the nature of the so-called 'missing' or 'dark' matter. In *The Dark Side of the Universe*, physicist James Trefil explains that the matter in our familiar world of illuminated matter--people, planets, stars, and galaxies--is but a small fraction of the total mass of the universe. This is inferred from: galactic halo effects, rotation gravitational behavior, galactic clustering implying dark matter in the voids between galaxies, and certain mathematical models of the starting universe.

The Critical Mass for Contraction or Continued Expansion of the Universe: Trefil tracks universe formation, from Big Bang to present, exploring the problem of how galaxies could form so early, given the early uniformity of the universe, and discussing whether the cosmic mass is less, more, or equal to, the *critical* mass which determines whether the universe is open, closed, or precisely between.

Alternative Forms of Dark Matter: He explores the various alternative forms this matter may take, including: massive neutrinos (neutrinos with some mass), black holes, unilluminated stellar and planetary bodies, 'shadow' universe gravitational effects, 'axions,' supersymmetric particles, 'wimps,' and cosmic strings.

END OF THE UNIVERSE?

Fate Of The Universe: While many physicists assume a closed universe--one in which the gravitational mass will force contraction to a 'Big Crunch'-- Trefil suggests that at the time of the writing, extrapolations from existing evidence suggest that the expansion rate and mass are *balanced.* The expansion would slow forever, matter finally reverting back to radiation, all life presumably ceasing, although as noted earlier, physicist Freeman Dyson, in *Infinite in All Directions*, hypothesizes life devising some continuation method.

At a Loss for Words in This Gloomy Scenario of the Dying Universe: Confronted with this end of the universe scenario, Trefil observes, "both scientists and poets" are at a loss for words. He cites the gloomy observation of Nobel laureate Steven Weinberg's in *The First Three Minutes*, that "The more the universe seems comprehensible, the more it also seems pointless."

The Ending Universe--No Memories of 'Existence': Trefil asks, "If after billions of years in the future, there is to be no life, no intelligence, no memory of the struggles of humanity," *what point will there have been for a totally-extinguished existence?* After wrestling with this question, Trefil offers his personal resolution of how he would handle his own "everyday decisions."

He Will Know Whether He Was the Best He Could Be: What he finally realized, he notes, was that while "in a quadrillion years the universe will be a cold, expanding sea of radiation," with no one to "remember," the point is tha*t he* will know tomorrow what *he* has done--whether he was the best he could be. And, he declares, "in the end, my friends, that is all that matters."

He is at least implying the ultimate subjectivity of meaning. But not all would agree that the disappearance of universes, with no one to remember the trials and tribulations experienced by their civilizations through the ages, was 'pointless' or without meaning, even if that meaning signified the ultimate futility of history.

Cosmology As Earlier the Subject for Theology: Down through the centuries, cosmology was the preserve of theology and religion, the origin of the universe viewed as metaphysics, a subject for philosophic and theological speculation. As we have noted, science became relevant early in the twentieth century when Einstein's general relativity theory launched the Big Bang theory of the start of the universe.

The Physics of Quantum Cosmology: A New Branch of Physics? In *The Left Hand of Creation, the Origin and Evolution of the Expanding Universe,* the authors, British astronomy professor John D. Barrow and American Professor of astronomy and physics Joseph Silk, note that in the 1980's, cosmology entered a new and exciting phase. It became a branch of physics in which the study of the very largest structures of the universe--galactic and supergalactic, and the study of the very smallest--elementary particles, coalesced to become the physics of *quantum cosmology.*

They cite the 1992 NASA COBE satellite findings in the cosmic background radiation, left from the Big Bang, of primordial temperature ripples in the fabric of spacetime existing at the start of the Big Bang. These fluctuations, they note, represent the "elusive link between present day structures and events close to the beginning of the universe."

Barrow and Silk go on to describe the creation sequence, and the convergence of elementary particle physics with astrophysics, relativity theory, Hubble space telescope findings, negative energy, vacuum quantum physics, spacetime curvature, stellar nucleosynthesis, unification and symmetry concepts, primevil chaos, elasticity of time, universe expansion, and omega or balance-point density of the universe.

Their Conclusions: In their chapter, 'Conclusions and Conundrums,' Barrow and Silk note that "The aim of modern cosmology is a complete reconstruction of the past history of the universe," pointing out that theories of how the universe unfolded are never "proved" but serve only as "guides" to later theories.

Unresolved Conundrum--What Preceded the Big Bang? Barrow and Silk note that there is one obvious, unresolved conundrum relating to the Big Bang, namely, what *preceded* the Big-Bang event? If the Big Bang involved a space-time 'singularity' [a state of infinite density, temperature, and zero volume or radius], then it delineated an absolute 'edge' to the entire universe of space and time, and the answer to our question of what existed 'before,' they note, is simply 'nothing,'except that it is a very special 'nothing.'

Actually, quantum cosmology suggests that the pre-big-bang 'nothing' state is a mega-universe chaotic quantum uncertainty state of potentiality, sometimes described by physicists as 'pregeometry,' or like Fred Alan Wolf, as 'superspace,' though 'superspace'would have no actual separated space, time, matter, or geometry, as we know them.

Going Back 'before' the Singularity Involves Quantum Theory : 'Before' the singularity, there existed neither space, time, matter, or motion, things common to our universe, say Barrow and Silk. Only by assuming an extension to the known laws of physics, can the universe be traced backwards *beyond the singularity*. If the big-bang singularity could be replaced by "something more moderate," they suggest, such as some quantum-gravity manifestation, then this "program of retrodiction" might be possible.

Anarchy As the Only Law of Nature? But it is possible, they suggest, that "complete anarchy" may be the only real law of nature. It has been suggested that the appearance of symmetry in nature is an illusion, that the rules governing which symmetries exist may be "purely random."

Preliminary investigations, they say, suggest that even if the choice is random among all the allowable ways nature could behave, orderly physics can still result, with the "appearances" of symmetry. This may be, they note, because our observations of the world are at energies far below the Planck temperature of 10^{32} degrees Kelvin.

Natural Selection May Take Place--Quantum Unpredictability May Apply: They suggest the possibility that some natural selection process may take place in which, as the temperature of the universe falls, fewer of the entire range of possibilities have significant effect on elementary particle behavior. As we trace our way back in time, approaching the Planck energy condition at the beginning, they suggest, it is likely that chaos and unpredictability reign.

And if this is so, they observe, not only physicists but physics itself owe their existence to our "low-energy-world." So there are some real uncertainties, not only in our knowledge of the laws of nature and their uniformity, but even in the very concept of what we mean by laws of nature.

The Appeal to Cosmology for Evidence of a Creator: Religious thinkers have often appealed to cosmology for evidence of a Creator. Barrow and Silk suggest that we have seen that, in cosmological theories, *the role of Creator is, for all practical purposes, assumed by the naked big bang singularity*. Anything, they believe, can emerge from such a singularity unless, they point out, there exists some rules, unknown to us, that control its behavior.

Does a Naked Singularity Have Any Other Characteristics of a Creator? Does a naked singularity have any other characteristics of a Creator? Fortunately not, respond Barrow and Silk, for human life could, in principle, one day *create its own local naked singularity* by concentrating enough matter into a confined region of space to create a black hole within which a naked singularity may appear. They suggest that the "theological status" of singularities must be "dramatically lowered" if they can be generated by human life.

The act of creation, as evidenced by matter forming a naked singularity, may still require a Creator. Naked singularities of themselves, "do not qualify as deities." They also point out that if human life cold initiate a naked singularity, in the process, it would also have had to develop the means to *destroy space and time*. Fortunately, they note, there is little likelihood of "politicians" funding singularities.

The Quest for Knowledge More Important Than the Knowledge? Barrow and Silk conclude with the observation that the human *quest* for knowing the structure of the universe outweighs the value of finding it because it "motivates the creative power of the human imagination."

About 50 years ago, they note, a group of cosmologists were quizzed on what single question they would submit to 'an infallible oracle' who could answer only 'yes' or 'no.' They credit priest-physicist Georges Lemaître's response as wisest: "I would ask the Oracle not to answer in order that a subsequent generation would not be deprived of the pleasure of searching for and finding the solution."

> [Upon the advent of non-Euclidian geometry]: I acknowledge...that I have a Euclidian earthly mind and how can I solve problems not of this world? ...I advise you never to think about it...especially about God, whether he exists or not. All such questions are utterly inappropriate for a mind created with an idea of only three dimensions. And so I accept God...His purpose...utterly beyond our ken. --Fyodor Dostoevsky

PART V--QUANTUM THEORY, SUPERSTRINGS, AND PARALLEL UNIVERSES

Introduction: Because of the frequent references to terms involving quantum theory, such as quantum reality, quantum cosmology, quantum gravity, quantum effects, quantum uncertainty, this Part addresses certain aspects of quantum theory associated with (1) quantum uncertainty, (2) subatomic particles--the fields and forces, (3) Superstrings--String Theory, (4) the nature of 'reality' in the context of quantum theory, and (5) parallel universes.

CHAPTER 13
THE UNCERTAINTY PRINCIPLE

Max Planck's Finding of Light Quanta: Much of Stephen Hawking's *A Brief History of Time*, explaining the origin and nature of the universe, deals with quantum theory. In his chapter on the uncertainty principle, Hawking cites the finding of Max Plank in 1900 that light, and other electromagnetic transmissions, could not be emitted at any arbitrary rate but only in certain 'packets' which he called 'quanta.' And each quantum had an amount of energy that was greater the higher the frequency of the waves. But, says Hawking, the "implication for determinism" from quantum theory was not realized until 1926 when physicist Werner Heisenberg announced his *uncertainty principle*.

The Uncertainty Principle: That principle, notes Hawking, deals with certain coupled uncertainties inherent in quantum mechanics, such as the fact that the more accurately you try to measure one of particle pair characteristics, like position and velocity of a particle, the less accurate will be the measurement of the other. So the more accurate you measure a particle's speed, the less certain will be its position. Heisenberg demonstrated that uncertainty in the position of the particle times the uncertainty in its velocity times the mass of the particle can never be smaller than a certain quantity, called the 'Planck's constant.'

Effects on Determinism: The uncertainty principle, he says, had "profound implications for the way in which we view the world," something still not fully appreciated by many philosophers-- and still controversial. The principle, he says, ended the dream of French physicist Laplace of a completely deterministic model of the universe. You just can't predict the future if you can't even know the present state of a single particle, he says.

Quantum State of Combination of Characteristics: These discoveries, notes Hawking, led Werner Heisenberg, Erwin Schrödinger and Paul Dirac, in the 1920's, to formulate quantum mechanics in which particles no longer had separate, well-defined observable positions and velocities but, instead, a quantum state of a combination of the two characteristics. Thus, quantum theory predicts a number of different possible outcomes and the likelihood of each, but not results of a particular measurement.

Quantum Theory As Underlying Most of Modern Science: Hawking notes that Einstein never fully accepted quantum uncertainty--that God would "play dice with the universe." But, he notes, quantum theory underlies "nearly all of modern science," governing the behavior of such technology as transistors and integrated circuits and, he notes, is also the foundation of modern chemistry and biology. It is only gravity, he says, that has not yet been quantized.

Particle-Wave Duality: Hawking points out that although light is made up of waves, Planck's quantum hypothesis tells us that in some ways it behaves as if it were composed of particles. Light can be emitted or absorbed only in packets, or *quanta*. So also, Heisenberg's uncertainty principle which implies that particles behave in some respects like *waves*. That is, they do not have a definite position but are "smeared out with a certain probability distribution."

Quantum mechanics, he says, is based on an "entirely new type of mathematics" that no longer describes the real world in terms of particles and waves. It is only the observations of the world, he says, that may be so described. So there is a "duality" between waves and particles in quantum mechanics where for some purposes they are best viewed as waves, and for other purposes, as particles.

Each Electron Passing Through Two Slits at the Same Time: He cites the 'two slit' experiment in which individual electrons are directed, one at a time, at a screen through two slits. We would expect them to go through one slit or the other giving a uniform distribution on the screen. What happens, though, is that fringes appear on the screen implying that each electron is passing through *both* slits at the same time. The new theory of quantum mechanics , he says, suggests that an electron orbiting around the nucleus could be thought of as a *wave* with a wave-length that depended on its velocity.

Feynman's Sum-over-Histories Technique: Hawking cites physicist Richard Feynman's "sum over histories" technique as a "nice way of visualizing the wave/particle duality." In this method, the particle does not have a single history or path in spacetime but would go from A to B by *every possible path.* With each path are associated a couple of numbers. One represents the size of a wave and the other, the position in the cycle, that is, whether it is at a crest or a trough.

The probability of going from A to B, he says, is found by adding up the waves for all the paths. In general, if one compares a set of neighboring paths, the phases or positions in the cycle will differ greatly. This means, he says, that the waves associated with these paths will "almost exactly cancel each other out." However, for some sets of neighboring paths the phase will not vary much between paths and the waves for these paths will not cancel out. Such paths, he notes, correspond to quantum physicist Bohr's allowed orbits.

Classical General Relativity Fails at Infinite Density and Zero Volume: Hawking notes that while Einstein's general relativity theory governs the large-scale structure of the universe, it does not factor in the quantum uncertainty principle. The singularity theorems, he notes, indicate that the gravitational fields should get very strong in at least two situations, black holes and the Big Bang.

In such strong fields quantum effects should be important. So, by predicting points of infinite density, classical general relativity, "predicts its own downfall," just as classical mechanics predicted its downfall by suggesting that atoms should collapse to infinite density.

Explore Recent Attempts to Quantize Gravity: Concluding the chapter, Hawking observes that we don't yet have a "complete consistent theory that unifies general relativity and quantum mechanics" but, he says, we do know some features it should have. So, he says, we will address recent attempts to correlate the forces of nature into a single unified quantum theory.

Quantum Mechanics and Freedom of Will: Edward Harrison, in his *Masks of the Universe*, after discussing the various "universes" constructed by society at different stages in history, goes on to discuss determinism and free will and comes to the "many worlds" interpretation of quantum theory in which all possible alternatives do in fact occur, in different universes.

Stapledon's 'Exfoliating Universe': Harrison cites Olaf Stapledon's *The Star Maker* in which *all alternative courses of action are taken*, creating many "distinct temporal dimensions and distinct histories of the cosmos." The multiple alternative courses of the many creations and their innumerable combinations, he says, leads to "an infinity of distinct universes exfoliated (separated) from every moment of every temporal sequence in this cosmos."

Freedom Choice As An Illusion: *Freedom of choice is an illusion in a rational universe*, Harrison says, for all is determined by the inviolable laws of the universe. So liberty to do or go is intolerable, for it contradicts the rational determinism of the universe. Hence, you take both paths. Says Harrison, this is the many-worlds interpretation of the freedom of will. *Free choice is an illusion, and all that can happen, must happen.*

The Universe As a Wave of Becoming: Harrison goes on to describe the interference effect of observation on particle behavior and speaks of the atomic world as a vast and intricate probability wave of becoming, *full of potentiality, empty of actuality.* Parts of the wave collapse (in the Copenhagen interpretation) into actual states of being, quantum jumps from one state to another, he says, whenever we make observations, and we can only predict probabilities.

All Alternative Courses Are Taken in Different Universes: But in the 'many-worlds' version of quantum uncertainty, *you take all alternative paths in different universes,* thus eliminating chance and free choice. You *must* take all offered courses. Harrison cites Hugh Everett's suggestion that the multiform wavelike world of quantum mechanics is the real world, all else being illusion, and *the atom of potential states becomes the atom of all the states, each state in a different universe.* "From any one universe exfoliates many universes each time a single atom is excited...each universe branches into a vast number of universes at every instant."

Frank Tipler and Quantum Mechanics: In *The Physics of Immortality*, dealing with his Omega Point theory of intelligence spreading through the universe to become God at the Omega Point, Tulane University mathematician and cosmologist Frank Tipler discusses the Copenhagen and Many Worlds interpretations of quantum mechanics. Since the Omega Point Theory is fundamentally a cosmological theory, that is, a quantum cosmology, Tipler says he is forced to adopt the Many-Worlds Interpretation. But he goes on to summarize both interpretations.

Schrödinger's Cat--Both A Live and a Dead Cat? The classically-cited quantum mechanics experiment, he notes, involves the famous Schrödinger cat, sealed in a chamber for an hour with a diabolical device. If an atom decays the cat dies and if it does not decay, the cat lives. But in the mathematics of quantum mechanics, the wave function of the cat is neither that of a dead cat or a live cat but the wave function of *both* a dead cat and a live cat. The true wave function, he says, is the *sum* of the dead cat and live cat wave functions.

Copenhagen Interpretation: But according to the Copenhagen Interpretation, there is a process called *wave function reduction* which reduces the sum of the dead and live cat wave functions to *either* the dead cat or the live cat wave function, not both, and the reduction is random. So half the time the cat will be alive and half the time it will be dead.

'Many Worlds' Interpretation: But according to the Many-Worlds Interpretation, there is no reduction of the wave function at all, Tipler says. After one hour in the chamber, the cat really is in the quantum state "dead cat plus live cat."

The Many-Worlds Interpretation, notes Tipler, resolves the obvious inconsistency by saying that the radioactive decay of the atom has forced the cat and all the pieces of equipment to split into two different worlds: the cat is alive in one and dead in the other. If, he cautions, we now try to see whether the cat is alive or dead, then we also split into two worlds. In one, we see the cat dead, and in the other we see the cat alive.

Omega Point Theory Adopts the Many-Worlds Interpretation: Since the Omega Point Theory is fundamentally a cosmological theory, that is, a quantum cosmology, Tipler says he is forced to adopt the Many-Worlds Interpretation. Since the Copenhagen Interpretation assumes that "wave function reduction" eliminated quantum effects on cosmological scales an exceedingly short time after the Big Bang, "the universe today is not quantum except on a very small scale."

No Reduction of Wave Function in Many-Worlds Interpretation: But the problem with this assumption, he notes, is that the wave function reduction process is almost entirely mysterious--we have no rules for deciding what material entity can reduce wave functions--so it is impossible to give a sharp analysis of contingency when this process is operating.

"The Many Worlds Interpretation,"on the other hand, does not suffer from this drawback since there is *no reduction of the wave function*. Physical reality is completely described by the wave function of the universe, he says, and the universe is "just as quantum now as it was in the beginning." The Many Worlds Interpretation may be wrong, he observes, but many physicists either accept the truth of the theory or employ it anyway because it delivers the best results in solving quantum physics problems.

> In the 'many-worlds' version of quantum uncertainty, *you take all alternative paths in different universes,* thus eliminating chance and free choice. You *must* take all offered courses, citing Hugh Everett, "From any one universe exfoliates many universes each time a single atom is excited...each universe branches into a vast number of universes at every instant," and *the atom of potential states becomes the atom of all the states, each state in a different universe* --Edward Harrison

CHAPTER 14
PARTICLES--THE FIELDS AND FORCES

The Small "Contamination" of Protons and Neutrons: In *Perfect Symmetry,* physicist Heinz Pagels observes that "The continuing interaction in the primordial radiation soup following the Big Bang created a small contamination of protons and neutrons. Hold on to that, he says, because from that small contamination, came the present structured universe--all the galaxies, stars, and the earth, and the human race.

Introducing Quantum Theory--Subatomic Mechanics: As noted earlier, quantum theory deals with physical mechanics at the subatomic level, the so-called quantum field theory, including the mechanics of the fundamental forces. As discussed earlier, the term 'quantum' is derived from physicist Max Planck's finding that radiation did not flow continuously but in packets or *quanta.*

Within the atom are the subatomic components, so-called fields, sometimes appear-ing as 'particles,' like the protons and neutrons of the nucleus or the surrounding electrons, other times as 'probability' or 'proxy' waves. Also functioning within the atom are the fundamental force 'exchange' or 'messenger' particles which apply or 'mediate' the fundamental forces (electromagnetism, the strong and weak nuclear forces, and gravity).

The Atom and Its Particle Components: Matter is made of molecules which are assemblages or groupings of atoms representing the various elements classed according to their atomic number. Within the atom is a nucleus with one or more 'hadron' particles (positively-charged 'protons' and neutral or no-charge 'neutrons.') For every (positive-charge) proton in the nucleus, the surrounding electron field includes one (negative-charge) electron.

The number of protons in the atom's nucleus determines the atomic number and the number of electrons define its chemical and electrical properties. The different numbers of neutrons in the nucleus identify the element's particular variation or *isotope*.

The Atom Mostly Empty Space: As Nick Herbert notes in *Quantum Reality,* the hydrogen atom with a one proton nucleus has one electron, and the atom has a diameter of 5 billionths of a centimeter, which also represents the electron's 'uncertainty' range-- its realm of spatial and position possibilities. *The atom is largely* *empty space*. If the atom were the size of a baseball field, the nucleus, though more than 99.9 percent of the atom's *mass*, would be only the approximate size of a baseball, so the nucleus is some 200,000 times smaller than the atom.

The seven-eighths-of-a-mile-long, circular Continuous Electron Beam Accelerator Facility, near Norfolk, Va., is used to accelerate electrons to almost the speed of light to smash them against the atomic nucleus and produce short-lived particles which have not existed since the very beginning of the universe. The nucleus, which represents *99.9% of all matter*, is held together by so-called *gluons*, representing the most powerful force in the universe, the strong nuclear force, which remains one of the mysteries of subatomic physics.

Particles with Components: While electrons are considered elementary (non-divisible) particles, protons and neutrons have components, groupings of 'quarks' which for most purposes are also considered elementary or nondivisible particles, although recent (1996) Chicago Fermi National Accelerator Laboratory Tevatron Accelerator collision experiments, between protons and antiprotons, suggest possible *constituents* within quarks, a concept which might undermine the entire so-called Standard Model of particle physics.

There are two classes of 'structure' or matter particles, classed as *fermions*, which include (1) 'leptons' (the family of electrons & neutrinos) and (2) 'hadrons' (the family of protons and neutrons). The hadrons contain combinations of quarks.

The Field Concept: In *The First Three Minutes,* physicist Steven Weinberg defines the universe as consisting of a set of fields--an electron field, a proton field, an electromagnetic field subject to the rules of special relativity and quantum mechanics.

Physicist Heinz Pagels, in *Perfect Symmetry,* observes that "All material reality...consists of interacting quantum fields, the transformation and organization of field quanta...the ultimate triumph of the field concept in the human attempt to comprehend reality."

Particle-Wave Duality: The so-called 'fields' representing the atomic constituents (protons, neutrons and electrons), when observed or measured, are described as existing in the form of 'particles.' Before observation or measurement, they are described as statistical 'probability' or 'proxy' waves,' represented by the symbol ψ. These atomic components are, in effect, positioned to provide the structure of matter by the 'mediation,' or 'exchange,' of the fundamental force particles.

Two Basic Types of Particle -- Structure and Force: So there are two basic categories of particles, sometimes referred to as 'interacting quantum fields': First are the *fermions,* the particles that provide 'structure,' consisting primarily of protons and neutrons in the nucleus of the atom, and electrons in the electron field surrounding the nucleus. Second are the *bosons,* the so-called 'messenger' or 'exchange' particles that mediate or apply the fundamental forces through an exchange process.

The Force Particles: So *fermions* function to provide *dimensional structure*--matter; the other particles, so-called *messenger* particles, known as *bosons*, apply or mediate the fundamental forces of electromagnetism, gravity, and the strong and weak nuclear forces. Thus, the nucleus of the atom consist of the fermions or structure particles called protons and neutrons. These, in turn, are made up of *quarks*, which are mediated or acted on by the so-called strong nuclear fundamental force.

The strong nuclear force is applied or *mediated* by exchange of the boson or messenger particles called *gluons* which, as the name implies, hold quarks together in protons and neutrons, and hold protons and neutrons in the nucleus. Light (and the other components of the electromagnetic force) is applied or mediated by the electromagnetic force *photon* particle. Gravity is applied or effected by the *graviton* particle.

Fermions and Bosons Once Interchangeable: Unification theory tells us that earlier in the history of the universe, the fermion or matter particles and the boson or force messenger or exchange particles, were *interchangeable*. The distinction between the two classes did not exist prior to 10^{-43} second in the universe's history, the point in time when gravity is thought to have separated from the other forces.

Overcoming the Obstructive Dualisms in Formation of Matter: In *The Cosmic Code--Quantum Physics As The Language of Nature*, physicist Heinz Pagels notes the two dualisms that had to be overcome before the modern concept of matter could unfold. The first dualism, *mass and energy,* was overcome by Einstein's special relativity theory, represented by the famous equation $E=MC^2$ which showed that mass and energy were equivalent and convertible forms of energy.

The second dualism was the *field and particle or wave-particle duality.* How, asks Pagels, could light be an electromagnetic wave field, as in Maxwell's theory, and a particle in Einstein's photoelectric effect? Well, Pagels cites Broglie's suggestion that particles had associated wave fields, and could behave like wave fields--and wave fields like particles. This dualism of fields and particles was finally overcome, says Pagels, with the advent and explanations of quantum theory, citing the work of physicist Steven Weinberg.

The Wave Nature of Particles: We noted earlier that particle-wave duality in which particles behave sometimes as waves and sometimes as particles, citing the two-slit experiment in which particles go through both slits at the same time, demonstrating their wave nature, that they do not have a definite position but are 'smeared out' with a certain probability distribution." So in quantum mechanics an electron orbiting around the nucleus could be thought of as a wave with a wave-length that depended on its velocity.

The Bose-Einstein Condensate: As we have seen, the wave nature of particles was demonstrated by the multiple slit experiment described earlier. In the January 1966 *Discover* magazine David Freedman reports on work at the Boulder lab of the University of Colorado and the National Institute of Standards and Technology, in which Carl Weimar and Eric Cornell succeeded in drastically slowing atoms close to absolute zero, the atoms in the test vial condensing into a single blob or "super-particle."

In this state, courtesy of the uncertainty principle, they *smear* so that the atom's location becomes wavelike, the wave being a probability distribution denoting the likelihood of the atom's being in any particular location. This state of matter had been predicted by Einstein using the work of Satyendra Nath Bose. Although the mutiple-slit experiment demonstrated the wave nature of particles, the Boulder lab experiment *permitted particle wave nature to be seen for the first time with the naked eye*.

Particles As the Quanta of Various Fields: Weinberg describes how physicists Jordan, Wigner, Heisenberg and Pauli, in the late nineteen twenties, showed that particles could be considered "the quanta of various fields," in the same way as the Einstein's photon is the quantum of the electromagnetic field, providing a field for each type of elementary particle.

Thus, says Weinberg, "the inhabitants of the universe were conceived to be a set of fields--an electron field, a proton field, an electromagnetic field. This represents the central dogma of quantum field theory: *the essential reality is a set of fields subject to the rules of special relativity and quantum mechanics.*" And everything else, he declares, represents "the quantum dynamics of those fields."

The Beginning of Relativistic Quantum Field Theory: Well, these ideas, says Pagels, signify the beginning of "relativistic quantum field theory--the merger of relativity and quantum theory," the universe viewed as a *vast arena of interacting fields*. Gone was the earlier view of matter in particles and energy in fields and "the dualisms of energy and matter, particle and field, were dissolved."

Pagels, with his unique facility for constructing classic descriptors, says that material reality thus consists of "interacting quantum fields," the "transformation and organization of field quanta....the ultimate triumph of the field concept in the human attempt to comprehend reality."

He goes on to describe the “central dogmas" of relativistic quantum field theory:

> Essential reality consists of nothing but a set of interacting fields--obeying the principles of special relativity and quantum theory--the intensity of which, at any point, gives the probability for finding their particle quanta. The interactions are themselves mediated by their associated quanta.

Stanford (SLAC) and Geneva (CERN) Announce Three Families of Matter: The nature and number of subatomic particles has, over the years, been the subject of much conjecture and speculation, as well as the object of extensive laboratory testing. On October 12, 1989 the Stanford Linear Accelerator Center (SLAC) in Palo Alto, California announced its conclusion that the various subatomic particle ‘families’ were limited to three.

Similarly, one day later, the European Laboratory for Particle Physics in Geneva (CERN) announced that the universe is made of twelve fundamental particles in not more than three families of particles. The two laboratories have been in high competition to be the first to define the basic constituents in the composition of the matter and energy of the universe.

Pauli's ‘Exclusion Principle’--Fermi Electron Pressure: Wolfgang Pauli in 1925 had discovered the ‘exclusion principle,’ that the electrically charged electron par-ticles which, until observed, exists as a cloud of probabilities around the atomic nucleus, repel each other.

This repulsion by electrons is not like the repulsion of similarly charged particles, like two positive-charge protons, but a far more powerful repulsion force, in quantum theory called an ‘exchange force,’ which keeps the electrons orbiting the atomic nuclei from collapsing. And the electrical repulsion between electrons is balanced by the resisting 'Fermi (electron) pressure.' It is only a matter of time, believes Pagels, before quantum theory will be applied to explain the most complicated molecules. Absent the repulsive exchange force of the electrons, he notes, a result of the "sleight of hand of quantum probabilities--atoms would collapse and we would not be here."

Exclusion Principle Explains Chemical Properties of the Elements: Recall the discussion of the stellar fusion process by which hydrogen and helium are fused into the heavier elements, and Russian chemist Mendeleev's construction of the periodic table for classifying the chemical elements. As electrons were added in orbit, as required by Pauli's exclusion principle, the electrons were put into different states, creating different elements and accounting for the periodic table of elements. So it took quantum theory to explain the theoretical foundation for the chemical properties of the elements in Mendeleev's periodic table.

Molecules Share Electrons: Not only did quantum theory explain the theory of the periodic table of elements, Pagels points out, but quantum theory explained how atoms combined to form molecules. By sharing electrons, atoms of a molecule bring into play "electronic wave shape...along with its associated exchange forces" that bond the atoms into "molecular configurations," and formulate the laws of chemistry. Linus Pauling's work in applying these rules of exchange forces, says Pagels, helped to achieve molecular bond theory, "the beginning of quantum chemistry."

Molecular Tapestries: The electrons surrounding the nucleus, are almost poetically described by Edward Harrison, in *Masks of the Universe*, as "waves spreading out and interweaving wherever possible [to] account for the structure of the atom. They lace together arrays of atoms into molecular tapestries and create the rich and varied patterns of our world of plants and flowerpots." As explained by quantum field theory, protons and neutrons of the nucleus also have wave-like characteristics.

Particles or Waves: Although we generally deal with particles as they are incorporated in atoms, as in the Earth, in ourselves, and in the objects around us, they do exist apart from atoms, in certain states. In the early universe energy existed only as radiation soup or plasma, until the universe cooled to allow atoms to form. Thus matter is said to have 'broken out' of the primordial 'thermal equilibrium' radiation state in the thermodynamic cooling process.

Originally thought to function sort of like small bricks, the material building blocks of the universe, particles are viewed now as 'interacting quantum fields,' shifting *quantum states* governed by the uncertainty principle, behaving sometimes as particles and sometimes as waves of statistical probabilities. Quantum fields are quantum waves until observed or measured, when, in the Copenhagen interpretation of quantum theory, the quantum wave collapses to a particle representing a single state.

The Wave-particle Duality and 'Fields' of Quantum Theory: So the subatomic world is seen to involve a quantum wave-particle duality, a quantum uncertainty involving statistical probabilities explainable in terms of certain principles and theories. These include: Heisenberg's *uncertainty principle* and wave mechanics; Bohr's resulting *complementarity principle;* Schrödinger's *matrix mechanics*, and Paul Dirac's *transformation theory* which demonstrates the equivalency of wave mechanics and matrix mechanics as two descriptions of the same behavior, known as 'quantum mechanics' or 'quantum theory.'

Particles: Ripples one moment, Particles the Next: Particles, says Edward Harrison, are one moment "like the ripples on the surface of a pond," and the next moment "like a cherrystone in the palm of your hand...both wavelike and corpuscular" but rarely acting "as a discrete entity."

The electron appears 'smeared out' over a large region of space but when detected it takes on the characteristics of a particle with precise location. Particle characteristics are subject to a lack of definition and predictability--Heinz Pagels' 'quantum slipperiness' inherent in the 'uncertainty principle.' The smaller the scale of examination of particles, the greater the uncertainty, the particle's behavior being expressible only as an 'array of probabilities.'

Space and Time as Fields: Physicist Paul Davies, in *Superforce,* notes the belief of most physicists that the spacetime concept ceases in the superforce unified era, that is, it cannot be continued back within that era. He notes the view that perhaps *space and time are to be treated as fields* which also froze out of the "pregeometric elements," so that the forces, fields and space and time were indistinguishable parts of the mix. In the superforce era, spacetime, he observes, would not have yet "jelled into a coherent form."

The Forces As Additional Dimensions: In unification models, the dimensions representing our geometry of three spatial and one time dimension are complemented by additional 'invisible' *space dimensions which manifest themselves as forces.* So in this theory, as physicist Paul Davies observes, both forces and energy-matter are related "manifestations of space and time," geometry put in place when the universe was incredibly less than one second old.

The Interdependence of Cosmology and Quantum Forces: In the unification or superforce period, the universe, says Paul Davies, would have been merely a "collection of ultra-simple components--the raw materials from which God fashioned all of space, time and matter."

Davies goes on to observe the growing scientific perception that the infinitely large, described by cosmology, and the infinitely small, described by quantum effects, coalesce to provide a unified description of the cosmos. It is, he says, a description in which "the ultra-microscopic structure of matter is intimately connected to the global structure of the universe, each influencing the development of the other."

Particles or Fields As An Energy Expression of Information: Some physicists and computer scientists think of our universe in terms of it being some kind of computer--being run by somebody or something--*some kind of ancient intelligence*--and for a purpose. They are not necessarily equating this controlling intelligence to God, preferring religious neutrality. But neither do they exclude God as the controlling mechanism. In this perspective, quantum energy forms are viewed as *information components,* rather than particles.

Information As the Ultimate Component of Everything: In his *Perfect Symmetry*, Heinz Pagels engages in metaphysical *Reflections.* His first is 'The Cosmic Computer,' in which he describes the metaphor of the universe as a cosmic computer for which the quantum particles, fields and forces are the 'hardware,' the laws of physics are the 'software,' and the evolution of the universe, including the transition of energy to matter, life and mind, is the execution of the 'program.'

But the ultimate fundamental components, say some, like computer scientist Edward Fredkin and physicist John Wheeler, are bits or units of information, part of the underlying informational mental subquantum reality, and part of physicist John Wheeler's 'meaning' universe consisting of 'meaning' software. But back now to particles.

First Level Atomic Components--Hadrons and Leptons: Reviewing particle nomenclature discussed earlier, at the *first level*, atoms are shown to consist of particles of *two types.* The *first type,* are the particles in the nucleus of the atom, called 'nucleons,' consisting of positively-charged, relatively heavy-mass, protons and no-charge neutrons. These are also classed both as heavy-mass 'baryons,' interacting with the strong force and, along with the gluon, intermediate-mass meson particles (including pions), as 'hadrons,' part of the family of heavy particles, interacting with the strong force.

The *second type* at this first level are the 'dancing' particles surrounding the nucleus, the electrons, one of the six-member 'lepton' electron-neutrino family of lightweight particles. Charged leptons, like electrons, interact with the electromagnetic (EM) and weak forces. Uncharged neutrinos are unaffected by EM. Electric current used by appliances for daily household chores is the streaming of electrons.

Interacting with the protons and neutrons to hold the nucleus together, are the intermediate-mass gluons, called 'mesons,' consisting, says physicist Steven Weinberg, of a quark and an antiquark.

The chemistry of atoms, their chemical identification as elements, is determined by the number of negatively-charged electrons and the equal number of positively charged protons together with the number of neutrons. The zero-charge neutron is unusual in that it has a positively-charged proton, a negatively charged electron and an antineutrino. Inside an atom, this 'combination' neutron particle is stable, but outside the atom, it simply goes to pieces, venting an electron and an antineutrino to become a proton.

Second Level: Quarks and Mediating Force 'Gluons': Protons and neutrons are now known not to be basic particles but to consist of a second level of particles, *quarks.* The quarks are held together so powerfully by the strong force gluons that the attempt to separate them results simply in *creating new quarks* in new hadrons. Physicist Paul Davies suggests that trying to separate quarks is somehow akin to *trying to separate matter from spacetime*. The idea draws on supersymmetry in which particles may represent dimensions which compactified in the symmetry-breaking process.

Quarks are thought to have existed as independent particles in the thermal equilibrium of the primordial Big Bang radiant energy, consisting of the "homogeneous gas of quarks, leptons and gluons," to quote Heinz Pagels in *Perfect Symmetry*.

Upon the decoupling of matter in reduced-energy-forms from the original higher temperature radiant energy, quarks lost their character as independent particles and exist today, for the most part at least, only in groups of twos (mesons-intermediate mass particles consisting of quark and antiquark) or threes (the baryons--protons and neutrons). The socially-inclined quarks co-inhabit protons in threes, consisting of up and down quarks, and in mesons in twos, as quark and antiquark. The bonding mechanism keeping quarks together in protons and neutrons is the 'strong' mediating force 'gluons,' intermediate-mass mesons (consisting of quark and antiquark).

Pions and Quarks at the Core of a Neutron Star? Some scientists, says Pagels, think that the center of a neutron star may be a pion condensate, pions being one class of gluons that hold the atomic nucleus together. Others think the very core of a neutron star consists of quark particles. Only in the primordial universe of the infinitely hot initial radiation, it is believed, did quarks exist as separate or unjoined particles.

Three Families of Matter: The October 1989 announcements by the competing Stanford Linear Accelerator Center (SLAC) and the Geneva European Laboratory for Particle Physics, CERN, that subatomic particle 'families' were limited to three, culminated years of speculation and experimentation.

The First Family of Everyday Matter: These findings suggest a universe consisting of twelve basic particles in three families and two classes: 'quarks,' of which combinations of three 'up' and 'down' quarks make up the protons and neutrons in the nucleus of the atom, and 'leptons'--the most common being electrons, the carriers of electricity--which occupy an electron field surrounding the atom's nucleus. These are the components of the everyday matter we encounter, the first of the three 'families.'

The Second and Third Families of the Earlier Universe: Other quarks and leptons, available in the early universe of high temperatures, are constituents of the *second family*, consisting of the 'charmed' and 'strange' quarks, and the muon and muon neutrino (leptons) and the *third family,* consisting of the 'bottom' (beauty) quarks, the 'top' (truth) quarks (thought to have been first human life-produced in April 1994 by the Batavia Ill. Fermilab accelerator-collider), and the tau and tau neutrino (leptons).

Up and down quarks exist normally, in sets of three, in every-day matter protons and neutrons. The other quarks--charm, strange, bottom and top existed during the earlier high energy, symmetry-breaking stages of the forces and fields.

The sixth or 'top' heavy-mass quark (200 times heavier than a proton, 35 times heavier than the fifth or bottom quark, and almost as heavy as a full atom of gold) was only theorized, until April of 1994, when the Chicago (Batavia) Fermi National Accelerator Laboratory announced finding evidence of the 'top' quark in its accelerator-collider detector chamber.

The name 'quark,' incidentally, was whimsically so dubbed by physicist Murray Gell-mann who first theorized the necessary existence of quarks, a term borrowed from James Joyce's novel, *Finnegan's Wake* where 'quarks' was a colloquialism for 'quarts' (of beer).

Fermilab's four-mile circular Tevatron supercollider collided highly-accelerated protons and anti-protons to produce the energy to create the top quark identified in its giant three-story detector. How big is the diameter of a quark? A thousand million times smaller than an atom or 0.0000000000000000001 of an inch. So what's the diameter of an atom? Much bigger--0.00000001 of an inch. In *Quantum Reality,* Nick Herbert describes the atom as "1 Bohr or about 5 billionths of a centimeter" which also represents the hydrogen atom electron's 'realm of spatial or position possibility,' also called its 'uncertainty.'

Primordial Quark Forms in a Newly Discovered Neutron Star Pulsar? As a momentary aside, within the debris of the 1987 supernova astronomers recently detected a neutron star, a 'pulsar,' spinning fast enough, that according to laws of physics, centrifugal force should have flung it apart--unless it had the extraordinary bonding power of quarks other than the 'up' or 'down' quarks. Well, physicists Joshua Friedman and Angela Olinto at Chicago's Fermilab, in the journal Nature, suggested that this pulsar is a form of quark 'strange' matter with this higher bonding power.

Use of Accelerator Colliders to Achieve Primordial Temperatures: Now back to the Second and Third Families of particles. Both laboratories have been seeking to create these primordial particles of the temperature conditions of the early universe by accelerating and colliding particles to produce quarks in the elusive and short-lived but massive Z particles, mediators of the electromagnetic and weak forces, as described in electro-weak unified-field theory.

The Paradox of Disappointment: Interestingly, the announcements of the findings at the two laboratories met with expressions both of elation--at identifying the components of the universe, and disappointment--at the prospect of reduced opportunity for new discovery. But the conclusions, about the limited number of particles, were based on the so-called 'standard model' of the organization of matter. However, there are other proposed models.

The Standard Model--Relativistic Quantum-Field Theory: The standard model is a 'relativistic quantum-field theory' in which the quanta are fundamental particles called quarks, leptons (both called 'fermions') and the messenger particle bosons, mediators of the four fundamental forces: the strong nuclear interaction, the weak interactions that result in the decay of atomic nuclei and quantum particles, the electromagnetic force, and gravity.

Three of these four forces, the strong nuclear and weak nuclear (both operating within the nucleus of the atom) and electromagnetic forces are included in the standard model. Excluded is gravity (affecting all particles and radiation), which happens to be both the weakest force, at a distance, and the strongest force, when concentrated--as in a black hole. This exclusion of gravity from the standard model represents the unsolved problem of a quantum theory of gravity.

The Three Forces Mediated by the Gluon Particles: Each of the three forces is mediated by a set of quantum particles called messenger, exchange, or force particles, which are quanta of a Yang-Mills gauge field. The strong force messenger particles, referred to as 'gluons,' are the adhesive force that keeps the quarks in protons and neutrons and both in the nucleus, with its surrounding electron field, without which the universe would have no structure--it would be just a gas of quarks and leptons.

A set of eight 'colored gluons' mediate or act as messengers for the strong force; a set of weak gluons called the W and Z particles, mediate the weak force, and the photon (light carrier) mediates or applies the electromagnetic force. These boson messenger particles interact with quarks and leptons. The quarks interact with all three sets of gluons. The leptons interact only with the weak gluons and the photon but not with the colored gluons of the strong interactions.

The Standard Model Illustrates Field Unification: This so-called standard model, says Pagels, integrates two relativistic quantum-field theories: The first is quantum chromodynamics, a field theory of quarks binding tightly together (by the strong force mediated by the colored gluons), the quarks being trapped in tiny bags--the strongly interacting hadron particles (including the stable protons and neutrons and other unstable hadrons). The second field theory is the Weinberg-Salam model of the electro-weak model, that is, the unified weak and electromagnetic interactions.

The electro-weak model unites quantum electrodynamics (the earlier theory of electrons and protons) with the Yang-Mills theory of the weak interactions, describing the decay of quarks and leptons.

The Standard Model As Step One in the Recipe for the Universe: The stable lepton (the electron) can combine with the nuclei (made of protons and neutrons), to make atoms. "Atoms can make stars, planets, molecules and life." And so, says Pagels, the standard model is the "first step in the recipe for cooking up a universe." Thus combined, these field theories seemingly account for everything in the material world, except the strong force and gravity.

The Standard Model Does Not Embrace the Strong Force or Gravity: Heinz Pagels, in *Perfect Symmetry*, points out that the standard model, while unifying the electromagnetic and weak fields, does not embrace the strong colored field nor gravity and general relativity.

But further field unification has been attempted to bring in the strong force, in the theory called *the Grand Unified Theory (GUT),* according to which the weak, electromagnetic and strong colored gluons are all quanta associated with a single multicomponent Yang-Mills gauge field.

Pagels describes the 'SU(5) Model' in which the single unified Yang-Mills field has 24 components, of which 12 correspond to the 3 weak gluons, 1 photon and 8 colored gluons of the standard model and 12 correspond to altogether new gluons. This multicomponent Yang-Mills field interacts with the quarks, leptons and also a new set of *Higgs fields* which break the 'single symmetry' distinguishing the strong, weak and electromagnetic interactions. Scientists may be on the verge of finding the Higgs boson particle thought to have functioned to shatter the original symmetry of unified matter and force particles.

Under the Standard Model the Proton Becomes Unstable: The 12 new gluons get a mass so enormous when the symmetry is broken, says Pagels, that these particles will never be detected by any particle accelerator because no such required energy is achievable here. These 12 new supermassive gluons have the unusual property of making quarks and leptons interchangeable and permit interactions that violate quark and lepton-number conservation laws.

In this model, the proton, previously stable under the standard model, becomes unstable. *Thus proton decay seems to be a natural consequence of the grand unification idea.* As described elsewhere here, efforts to detect proton decay have, so far, been without conclusive proof.

Eliminating Input Parameters and Dimension less Constants: The standard model, says Pagels, uses nineteen parameters: masses of quarks and leptons, coupling strengths which are input parameters from which one can mathematically describe all the strong, weak and eletromagnetic interactions. But physicists feel that the ultimate theory should have no input parameters and no Dimension less constants, and that all the masses of quarks, and all coupling strengths, should be predicted by such a master theory.

Did God Have Any Other Choice? Pagels quotes Einstein's musing whether God had any choice in creating our universe. Discussing Dimension less constants, Einstein said that his faith in simplicity (intelligibility) in nature precluded arbitrary constants; that only "rationally completely determined constants" occur--not constants whose numerical value could be changed without destroying the theory.

If Einstein was right, says Pagels, and no free parameters are permitted, then God had no choice, he could not have 'adjusted' the parameters, like the masses of quarks, to make different universes. Other scientists have suggested that the development of alternative models for workable universes is essential to our understanding of the meaning and implications of this universe.

Supersymmetry and Supergravity: But even beyond GUT, physicists have suggested theories of 'supersymmetry' and 'supergravity,' that incorporate gravity into the field unification efforts, involving energies only available before the first billionth-billionth of a second in the big bang. And so the study of the early universe becomes the laboratory for clues. But we are led to ask where did these atoms with these particles and forces come from?

Where Did This 'Material' of the Universe Come from? Earlier, in one theory, we speak of the creation of matter, the universe "spontaneously erupting" from a primordial quantum void, into an unstable false vacuum state which then decays, leading to the exponential inflation which provided the energy to "continue its expansion and to create particles." The still-continuing (big-bang-produced) expansion led to various phase transition 'freezing' states, and, in the process, releasing energy to create matter and the universe we observe today.

The History of the Universe As the History of Energy and Matter: The history of the material universe is the history of the evolutionary formation of matter and, essentially, a function of the symmetry breakdown and thermodynamics of the original primordial radiant energy. Temperatures are the method of measuring the energy involved in the motion of particles, and temperature, in effect, dictates our structured universe, the shift from 'motion energy' to 'rest energy' or matter.

The Period of Unified Forces and the Decoupling of Matter: Cooling temperatures, with lowered energies, following the Big Bang, brought loss of the original 'symmetry' in which energy, space, time, all the present forces of electromagnetism, the strong and weak forces, and, it is supposed, gravity, were unified. *With the cooling* *came the decoupling of matter from the thermal equilibrium,* then, in sequence, nuclear evolution, chemical evolution and finally biological evolution leading to life forms.

Present Matter a Function of Departure from the Original Symmetry: In *Perfect Symmetry,* Heinz Pagels describes the universe as a function of breach of the original 'symmetry.' The universe begins in a very hot state of utmost simplicity and symmetry and as it expands and cools, its perfect symmetry is broken, giving rise to the present complexities, the 'frozen asymmetric remnant' of its earlier hot state.

Physicist Hubert Reeves, in *Atoms Of Silence*, speaks of "Energy, Forces, and Elsewhere," in an effort to bring together certain ideas about energy, fundamental forces, mass changes in particle reactions, binding energy and energy release in bonding, chemical reactions, heat and gravity, and the role of the 'great elsewhere' conveniently provided by the expanding universe.

Light Dominated the Early Universe: The primordial light, cosmic background radiation now pervading the universe, originated in the Big Bang and represents the oldest photons in the universe. Its presence everywhere was discovered by scientists Penzias and Wilson who were studying electromagnetic microwave transmission.

This 'primeval afterglow,' was originally far more intense, carrying temperatures of "trillions of trillions" of degrees. In this early hot, dense universe, radiant or luminous energy--light, dominated the universe. As the universe cooled, radiant energies weakened until subatomic and atomic matter dominated, as it does now.

The Opaqueness of the Early Universe: Hubert Reeves says, in his *Atoms Of Silence*, that the primordial or fossil glow was "emitted at the moment of transition from opacity to transparency." This is a reference to the original infinitely dense, hot, universe as being 'opaque'--the emitted light immediately re-absorbed so that those informational photons could not reach us as light.

Consequently, unlike the light transmissions from early galactic formations reaching us now after millions of light years, we cannot use telescopes or electromagnetic wave receivers to 'see' (intercept the photons) reflecting the events of starting universe. They were hidden from us by their re-absorption in the original firestorm of creation.

Quantum Theory and Astrophysics--The Very Small and the Very Large: Quantum theory deals, as we have noted, with *the very small*, the composition of the micro-universe of subatomic particles, fields, and forces and how the quantum energy and matter states provide for the structure of the universe at large. But quantum theory also deals with applications at the cosmic scale in astrophysical phenomena such as stars, galaxies, black holes, and, as theoretical physicist Stephen Hawking suggests, perhaps even the creation of new universes. On a practical level, quantum mechanics is the foundation for such down-to-earth applications as nuclear power, lasers, electronic gadgetry, and the newly developing superconductors.

> All material reality...consists of interacting quantum fields, the transformation and organization of field quanta...the ultimate triumph of the field concept in the human attempt to comprehend reality.--Heinz Pagels (*Perfect Symmetry*)

CHAPTER 15
SUPERSTRING THEORY

Supersymmetry: We have seen how the grand unified theory (GUT) unified all the forces except the more elusive force of gravity, requiring an even larger framework of symmetry potential than Yang-Mills and GUT theory. Required to get to this higher unification, was a higher state of symmetry allowing the blending of gravity-relativity with quantum effects to dispense with the mathematical infinities introduced into the equations by the incorporation of gravity.

The Higher Symmetry Framework: The higher symmetry framework must provide both for: interchangeability of force and matter particles; a mechanism for *eliminating the infinities* that plagued unification approaches; and a theory of particles which *doesn't treat particles as mathematical points*, in order to allow for inclusion of gravity. The leading candidate at this time for this unification framework is so-called *superstring* theory derived from relationships between two separate theories involving 'superstrings.'

Supersymmetry and Superstrings: 'Superstrings' is a term derived from *two theories*. The first is the particle unification theory called *supersymmetry*, describing a condition thought to exist before the force of gravity separated from the state of total unification of force-carrying particles and matter particles.

In that unified state, not only were matter particles all unified, but *both matter particles and force particles were unified*, interchangeable. Each type of matter particle (*fermion*) would have a partner force-carrying particle (*boson*). So an electron matter-particle would have a 'selectron' force-carrying partner particle. An electromagnetic force-carrying photon would have a 'photino' matter-particle partner. In this unified state, these partner particles were described as 'interchangeable.'

The second theory involved in superstrings is *string theory.* String theory proposes that so-called 'point' particles, like protons, quarks or electrons, are not the basic components of matter but, instead, represent the *vibrational effects* of almost infinitely small *single dimensional strings* or *loops* which 'oscillate' or 'vibrate,' as they take on or give off energy, in compacted dimensions that co-exist with the four enlarged dimensions of our universe.

Supersymmetry--The Interchangeability of Force and Matter Particles: In the first of the two theories making up superstring theory, namely supersymmetry theory, each boson or *force*-carrying particle, like the graviton for the force of gravity, or the photon for electromagnetic force, has a fermion or *matter* 'supersymmetric' partner-particle, differing from its partner counterpart in its *spin* characteristic. The two 'matter' supersymmetric partners for the force-carrying graviton and photon, for example, would be the 'gravitino' and the 'photino.'

In the same way, each fermion or *matter* particle has a supersymmetric particle-partner representing a *force*. Thus, the electron (matter) particle has the 'selectron' (force) partner and the quark (matter) partner-particle, has the 'squark' (force) partner-particle. Although supersymmetry theory thus introduces additional unification through interchangeability of matter and force particles, that still leaves gravity out from unification.

The Universe Split into Two Parts: In supersymmetry theory, at the so-called Planck time, 10^{-43} seconds after Time = 0, when gravity is believed to have split off from the other still-unified forces, physicists theorize that the universe split into two parts, one, our world of particles, and supersymmetric partner-'sparticles,' and a *mirror or shadow world* of particles and supersymmetric partner-'sparticles.'

Our universe's particles are the 'sparticles' of the shadow universe, and the shadow universe's particles are our 'sparticles.' Each universe would have evolved according to its own laws, the forces in each universe interacting only within that universe, except for the force of gravity, common to both. So the two universes, in this theory, *interact with each other through the single force of gravity.*

We Are Candidates for Dark Matter in the Shadow Universe: Presumably, says Trefil, in *The Dark Side Of The Universe*, some shadow world astronomers look upon us as candidates for their dark matter. He muses whimsically that if both universes deterministically evolve in parallel, then a "shadow you" may be reading his book.

The Mirror Supersymmetric Sparticles Become Massive: The loss of this originally-existing boson-fermion equivalency or supersymmetry occurred in a process in which the mirror supersymmetric *sparticles became very massive*. The photino partner of the massless photon, for example, became perhaps forty times more massive even than a proton. While the shadow universe evolved according to its own laws, it continued to interact with our own universe but only through the force of gravity.

Early Use of Geometry to Unify Field Theory: Einstein had used geometry, the warping or bending of spacetime by matter, to explain the effect of gravity--what it does and how it does it. Einstein had envisioned using geometry as the mechanism to find unification mechanics. In *Superforce,* says Paul Davies, the best hope we have of achieving Einstein's dream stems from the work of Polish physicist Theodor Kaluza.

Kaluza "enlarged geometry just enough," says Davies, "to accommodate Maxwell's [electromagnetism] theory" by treating electromagnetism (EM) as a form of gravity, but that of an "unseen [fourth] dimension of space." This provides a five-dimensional universe in which "the gravitational field ...behaves just like ordinary gravity *plus* Maxwell's electromagnetic field, when viewed from the restricted perspective of four dimensions."

So, says Davies, in the Kaluza five-dimensional universe, in effect, there is only one force field, and that is gravity. EM "becomes part of the gravitational field, ... a ripple in the fifth dimension." Looked at in this way, *there are no forces at all*, only "warped five-dimensional geometry with particles meandering freely through a landscape of structured nothingness."

Where Is the Fourth Space Dimension? But where is this fourth space dimension? asks Davies, who goes on to discuss dimensionality. Swedish physicist Oscar Klein who worked on Kaluza's theory, suggested that we don't see this dimension because it is rolled up in a tiny circle *around the dimension*, beyond our sense perception, the loops being not *inside* space but "they extend space." Size of the dimension? Well, "10^{-32} cm or about 10^{-20} of the size of an atomic nucleus." We should think of this dimension, says Davies, as *somehow inside the atom*.

After some fifty years of dormancy since the Einstein-influenced interest in the theory, Kaluza-Klein theory was recently revived and 'remodeled' to adapt to the additional nuclear forces and applied to unification theory. Here the three forces can be described as guage fields with abstract symmetries which in Kaluza-Klein theory become concrete, as the "geometric symmetries associated with the extra space dimensions."

The Forces As More Space Dimensions: So the forces are accommodated by grafting more dimensions of space onto the known spacetime. Davies goes on to describe the process of adding additional dimensions to arrive at seven extra dimensions, making ten space dimensions plus time, or eleven spacetime dimensions. The resurrected Kaluza-Klein theory has stimulated physicists to recast physics in eleven dimensions, of which seven "curl up to invisibility."

Why Eleven Dimensions? Why eleven dimensions? asks Davies. Is this some numerology quirk, or does it represent some "deep mathematical significance?" Davies refers to the N=8 supergravity formulation which connects particles of different spins into a "superfamily of 163 particles," thought to be the largest possible. As we have seen, N=8 refers to the number of steps that "connect particles with different spins across the whole range of available spins under supersymmetry operation" and involving the eight species of gravitinos.

If the N=8 supergravity theory is recast in *eleven* dimensions, says Davies, "the eight separate symmetry operations of N=8 supergravity collapse to just one-we get 'N=1' supergravity." So, asks Davies, is this concurrence of the number eleven just coincidence or is there "a deep connection between supergravity and Kaluza-Klein theory?" He says that physicists hope that the supergravity and grand unified theories "will come together in a common description."

Nature as Geometry: Einstein had built relativity out of geometry and saw geometry as the path to unification. Relativity defined gravity as a distortion of four dimensional spacetime. The other forces, says Davies, are now "reduced to higher-dimensional spacewarps," so all the forces are "revealed as nothing more than geometry at work." So, says Davies, one is attracted to the idea of the universe as ultimately "convoluted nothingness...a sculpture of pure emptiness."

Interestingly, back in 1870, the British physicist W.K. Clifford had prophetically declared that the motion of matter is essentially the wave by which space curvature is passed from one portion of space to another, and matter, particles, represent "bumps or kinks" in empty space. In *Creation Revisited,* Atkins, as do some other physicists, refers to particles as spacetime "knots."

In *Superforce* Paul Davies declares that "With the superforce unleashed, we could change the structure of space and time, tie our own knots in nothingness, and build matter to order." The universe was explained by physicist Steven Weinberg as "inhabited by" nothing but "forces and fields." Now these forces and fields are, as Einstein had envisioned, viewed as being pure geometry, somewhat analogous to physicist John Wheeler's "geometrodynamics," particles and forces as geometric structure. A particle was a small space tunnel or wormhole through another dimension, the remote end being another particle of opposite charge.

String Theory: As stated earlier, besides supersymmetry theory we have been discussing, "superstring" theory involves also the second theory, so-called "string theory." According to string theory, so-called point particles, such as quarks, ordinarily dealt with in quantum mechanics, are not basic energy components as thought, but consist of constituents. These components are exponentially-minute, high-density, *single-dimensional lines, loops or strings* of basic Planck-length stuff, 10^{-33} of a centimeter long (the smallest measure that has any meaning).

These strings or loops vibrate in multidimensional space that has the geometry of either *ten or twenty-six dimensions*. The ten dimensions include our three space and one time dimension plus six other dimensions which got rolled up or 'compactified.' It is the very compactification process that permits the presence of 'shadow matter' particles which interact with our world through gravity.

A Curled Up Six-Dimensional World: So, say Gribbin and Rees in *Cosmic Coincidences*, "Each point of our ordinary space, at each moment of time, is really a tiny but complex curled-up six-dimensional world." The "structures and resonances in these 'extra' dimensions control how the superstrings behave, and thereby determine what particles exist at each point in ordinary space and how they interact." In *The Omega Point*, Gribbin refers to the "extra" dimensions beyond our three of space and one of time as wrapped up in "these tiny tubes," referring to the strings and loops.

Comparing a String and a Particle: A particle is pretty small and it is useful to get some grasp of these comparative 'smallnesses' between particles and strings. In *The Dark Side Of The Universe* physicist James Trefil depicts the size comparison between a 'string' and a 'particle,' such as a proton, as corresponding to that between "a person and a small galaxy!"

Energies Involved in Strings: And what are the energies involved in these 'strings?' Physicists John Gribbon and Martin Rees, in *Cosmic Coincidences*, speak of strings as having a typical "tension" that "stores energy equivalent to 10^{38} proton masses."

What Happened to the Other Dimensions? If the starting universe had this more complex geometry of these 10 or 26 dimensions, what happened to that multi-dimensional geometry? Where did the extra dimensions go?

Confronted with the concept of multi-dimensions, we draw a blank because we think of dimensions in terms of those in which we function, namely, relativity's three spatial dimensions and the fourth time dimension, being coordinates which identify a position in space. Dimensions other than those in which we function are difficult to conceive. For this reason some physicists suggest that rather than trying to conceptualize these dimensions geometrically -- or structurally, it may be simpler to consider these dimensions as a framework of mathematical concepts in which disparate aspects find a hidden symmetry.

Compactification: Well, when gravity separated at 10^{-43} seconds from the initial supersymmetry state, the extra dimensions rolled or curled up in a process called *compactification,* while our conventional three dimensions for space coordinates enlarged, giving us our familiar three dimensional space or four-dimensional spacetime geometry and structure.

Strings Vibrate in Compactified Dimensions: Thus, Princeton University physicist Edward Witten suggests that all matter and forces consist not of particle bits or 'points' but of *tiny strings which change identities to reflect different 'particle' characteristics according to their energy levels.*

The strings are said to 'oscillate' or 'vibrate' in the compacted dimensions that co-exist outside the four enlarged dimensions. Each of the vibrations of the strings correspond to different particles, leading to a theoretical infinity of possible particles. The strings vibrate in different particle forms as they take on or give off energy.

"The interconnected structures and resonances," say Gribbon and Rees, in *Cosmic Coincidences*, control superstring behavior, consequently determining "what particles exist at each point in ordinary space, and how they interact."

Single Dimensional Strings or Loops Rather Than Points: As described earlier, the strings rather than being mathematical 'points' are single-dimensional open lines or strings, or closed, looped. The looped strings, as we have noted, embrace both matter and force particles, including the gravitational force particles called 'gravitons.'

Gravity Is Already Built into String Theory: So, in string theory, as Martin and Rees point out in *Cosmic Coincidences*, not only do infinities disappear, but "rather than having to add gravity into the theory, they found it there already, specifically a mass zero, spin 2 graviton, is already a necessary part of any workable string theory of the particle world." Every consistent variation on the string theme has one, and only one, graviton--a massless, spin 2 particle that "leads inevitably to the spacetime curvature description of general relativity...."

The Quantum Approach to Relativity: As physicists point out, Einstein started with relativity and concluded with quantum consequences. He used general relativity to describe spacetime curved in the fourth dimension by "gravity waves, ripples in the fabric of spacetime, and calling for the associated quantum form, the graviton." String theory starts from the other direction, with a quantum theory of the graviton that winds up with "the usual form of general relativity, with curved spacetime."

For years, physicists have accepted general relativity theory with vague notions of its quantum underpinnings. String theory offers new insights in building the general relativity consequences, curved spacetime, from its quantum sources. So physicists are excited about the unification-potential for superstring theory providing the basis for mathematically harmonizing the forces, fields, space and time, in a theory of everything (TOE), even though the concepts which underlie the unifying mathematics are not yet fully understood.

Supersymmetry and Strings Give Us Shadow World Gravity Effects: So, getting back to the symmetry break at 10^{-43} second, which gave us 'sparticles,' the shadow world particles, these sparticles, more elusive and undetectable than neutrinos, interact in our universe only weakly with our matter and, says Trefil, in *The Dark Side Of The Universe*, are *perfect candidates for dark matter*.

While they interact with our matter only weakly, thus not interfering with our structures like stars or accelerators, being massive and exerting gravity, they would interact with matter in our world through gravity and thus *contribute toward the total mass of our universe*.

Other Comments on Superstrings: In *Superstrings, and the Search for the Theory of Everything*, physicist F. David Peat reviews the string theory work of physicist Yoichiro Nambu, and that of the prime movers in superstring theory, physicists Michael Green and John Schwarz. He also reviews the work of other prominents in this field like David Gross, Edward Witton, Chris Isham, and related 'twistor theory' work of Roger Penrose and his Oxford 'twistor group.'

Superstring Theory Approach to Unification: As we noted, superstring theory evolved from the search for the mathematics of unification that would go beyond quantum theory, relativity, and supersymmetry theories and combine them with string theory. This was the effort to bring gravity and spacetime into the unification scheme by quantizing them. Superstring theory, formulated in quantum terms, delivers a spin-two massless particle, the graviton; so *quantized gravity emerges naturally from superstring theory*, a major conceptual breakthrough.

As Hugh Ross points out in *Beyond the Cosmos*, string theory is the only theory that "self-consistently explains" all the known properties of the known fundamental particles, of the principles of quantum mechanics, both special and general relativity, all four forces of physics, and "the known details of the creation event."

Particles As Excitations of One-Dimensional Strings: Conventionally, particles are viewed as sizeless *points* having no extension. In string theory, particles are excitations or vibrations of one-dimensional, exponentially small extensions or infinitely thin strings, about 10^{-13} centimeters long, the fundamental building blocks of the universe.

Matter Dissolves into Field Energy Vibrations--Ten Dimensional Universe: So in this theory, say Paul Davies and John Gribbon in *The Matter Myth*, "solid matter dissolves away [into] excitations and vibrations of invisible field energy [and] little distinction remains between material substance and apparently empty space which itself seethes with ephemeral quantum activity." So superstring theory "seeks to unite space, time, and matter , and to build all of them from the vibrations of submicroscopic loops of invisible string inhabiting a ten-dimensional imaginary universe."

The superstring model works in ten dimensions (or twenty-six), six of which are 'compacted,' rolled up in tiny spheres or circles. Where movement of conventional particles represent world line trajectories, moving strings, notes John Gribbon in *The Omega Point,* form "sheets and surfaces in spacetime." The compactification process is seen as providing for 'shadow matter' particle families, existing in a related 'shadow universe' created when gravity separated from the other forces.

Two Symmetries, One Ours and One a Shadow World? In this theory, at the first symmetry break following the Big Bang--the gravity breakoff, the initial super-symmetry split into two symmetries, one ours, with our particle structure, and the second, the symmetry of a shadow world which may be "interpenetrating ours, expanding with it, but interacting with ours only through gravity."

Roger Penrose's Twistors Also Extended Objects: The shadow universe matter may conceivably make up part of the 'missing mass' from the envisioned 'critical mass' called for by Big Bang models. Roger Penrose's 'twistors' are also extended objects that "have the potential for describing both the internal structure of the elementary particles and the quantum nature of spacetime."

Reconciliation of the Twin Cornerstones of Modern Physics? So Peat observes that both the elementary particles and their interacting forces are a manifestation of superstrings, or twistors, and further, that "space-time itself has its origins in the shimmering world of superstrings or the interweaving of twistors." This holds out the hope, he says, that relativity and quantum theory, the "twin cornerstones of modern physics," will finally be reconciled.

Unresolved Problems in Superstring Theory: But Peat describes some of the unresolved problems with superstring and twistor theory. He had earlier discussed the potential revolution in physics in which the Cartesian ideas of continuity and dimensionless points would be replaced by new ways of thinking more in harmony with the "insights of the quantum theory."

He notes that while twistor theorists are committed to a truly revolutionary approach, the whole theory has proved more difficult than anticipated. But he reluctantly concludes that the initial promise of superstring theory has yet to be fulfilled. He goes on to cite more "philosophical" approaches like David Bohm's notion of the "implicate order," which is "nonlocal" and of an "enfolded, rather than an explicate nature," but suggests that concept still lacks "mathematical form," and, he says, attempts to "transcend quantum theory" call for "a new order."

The Deepest Questions Remain--Profound New Ideas Needed: And so, he says, while the deepest questions remain, physicists are realizing that a crisis does exist in physics calling for a profound new idea. Nevertheless, he says, superstring theory has generated very provocative questions about the nature of space and time at exponentially small distances. He hopes he will be around to write about the answers!

Since Peat's book, new mathematical breakthroughs have been suggested supporting the superstring theory approach to unification. The June, 1966 Time Magazine notes that the brilliant Princeton Institute for Advanced Studies physics professor Edward Witten has given new impetus to efforts in this theory by his forecast that the theory will dominate the next 50 years of physics. It was noted that it would take a particle collider much more powerful than any in existence or in prospect to test the theory.

Dimensionality: In *Beyond the Cosmos*, astronomer and minister Hugh Ross discusses unified field theories and various aspects of God's use of dimensionality in the creation, one aspect being string theory. For an exponentially small period of time after the moment of creation, he says, the universe had six extra dimensions needed for the creation process. He discusses the breakthrough by physicists Ed Witten and Nathan Seiberg who reduced an entire field of mathematics to a single concept, transforming very complex equations to a simple version.

Dimensions and Massless Black Holes: He cites the work of Andrew Strominger, Brian Green, and David Morrison to transform a hundred million plus different string models operating in various dimension numbers to a single theory, involving ten dimensions. This required the development of a theory of *massless black holes* which on its face is a manifest contradiction, black holes involving infinite gravity concentration.

Why Only Four Dimensions for Us? The basic element of string theory, Ross notes, is that at 10^{-43} seconds after the creation big bang event, a ten-dimensional universe of curled up dimensions split in two: a six-dimensional portion that ceased expanding (compactified dimensions) and a four-dimensional piece representing our present three space and one time dimension which continued expansion to produce matter, galaxies, and stars. Ross uses this description to observe that God must be operating in a minimum of eleven dimensions of space and time and we function in three space and one time dimension because those are the dimensions required for a universe to sustain life.

We stand, in a manner of speaking, midway between the unpredictability of matter and the unpredictability of God.--Freeman Dyson

CHAPTER 16
PARALLEL UNIVERSES

Many-Worlds Quantum Theory: In *Cosmogenesis,* physicist David Layzer discusses the "many-worlds' interpretation of quantum theory in which every measurement-like process causes the universe to split into an infinite number of parallel universes.

While its simplicity of use attracts physicists, like Steven Hawking in *A Brief History of Time,* Layzer feels that in addition to its "metaphysical extravagance" or inconsistency with the 'Occam's Razor' simplicity principle, it has the flaw of taking as a given "the distinction between reversible quantal processes and irreversible measurement-like processes." If we accept the many-worlds interpretation, he says, we must "renounce any attempt to understand the role of irreversibility." A fatal flaw, he suggests.

Universes with Other Histories: Physicists and scientific writers seem intrigued, as well they might, by the existence of other possible universes among the quantum "allowed histories," universes somewhere next to us in the superspace of the mega-universe. But these are universes representing quantum states that are complementary or mutually exclusive and unable to communicate.

The question has been asked whether these neighbor universes, sometimes described as other 'branches' of the universe, coming into being through quantum effects, are actually *real* universes or whether they exist only as mathematical entities. While we deal with this question in the chapter on creation theories, it may be worthwhile setting the stage for that discussion by a preview here since the question of parallel universes keeps arising apart from creation theories.

Are These Other Universes Real? In *Stephen Hawking's Universe*, author John Boslough asked Hawking whether these other universes derived from quantum theory were "actual physical regions that could be observed." In a masterpiece of non commitment, Hawking replied, "It just means that there is a nonzero probability for the universe to have a lot of different forms."

The idea of universes existing side by side is treading on metaphysical ground, he noted, quoting Cambridge University philosopher Ludwig Wittgenstein that the existence of other universes is "not a predicate," since their existence would have no meaning without some observable consequence or confirmability.

While other physicists have expressed similar views about other universes in quantum theory, there are some who remind us that most of the science we're now discussing was once considered metaphysics.

Of course, we would not want scientists in other such universes to dismiss our universe--perhaps occupying the same space as theirs--as merely metaphysics or a quantum mathematical quirk, even though they may, as we noted earlier, consider our's, if they could evaluate it, as one of philosopher David Humes "botched and bungled" universes.

Actual Parallel Universes: Interestingly, this question of whether these quantum theory 'other worlds' or universes are 'real' has recently generated a host of new approaches. In *Parallel Universes, The Search For Other Worlds*, physicist Fred Allen Wolf characterizes these possible quantum states as real, actually-existing, parallel universes. Stephen Hawking's comments are noncommittal, as we noted, in the absence of verifiability, speaking of them rather as mathematical states.

A Revolution in Thought about the 'Physical' Universe: In his Introduction to *Parallel Universes*, physicist Fred Alan Wolf observes that today, more so than in the past, we face a revolution in thinking about the physical universe, engendered by discoveries in physics which put into question previous concepts of "concrete solid reality," and point toward a "more abstract reality" which implies a need to "unify our picture of the world."

Other Universes Existing Alongside or Even Interpenetrating Our Own: The idea that the universe is not alone, that other universes exist alongside ours, says Wolf, is a concept which has emerged relatively recently as part of the effort to bridge conceptual gaps in the effort to unify our knowledge, no small part of which is the driving direction of physics toward Einstein's earlier goal, finding the mathematics of unification of space, time, matter, dimensionality, and the fields and forces.

Relatively New Ideas about the Universe: This has led to new ideas about the universe such as: that space is not infinite; that space and time are joined--a continuum, spacetime; that 'now' is not absolute but relative to observer movement; that gravity 'attraction' is geometry, the effect of mass to bend space and curve time; that matter represents knots in the fabric of spacetime.

Quantum Gravity Must Be Introduced: In tracing the universe back to an origin in the Big Bang, Wolf notes that it is now recognized that general relativity theory, postulating the starting point as one of infinite density, zero volume, and infinite temperature, must be augmented by a 'quantum gravity' theory. Required, he suggests, is "bringing quantum theory into the beginning of space, time, and matter."

Quantum Uncertainty--A New Vision of Reality--Parallel Universes: Wolf points out that the development of quantum theory brought a new perspective about matter in which 'observation' plays a previously-unsuspected role of influencing energy states.

In addressing the problem of bringing relativity and quantum theory together, Wolf explores a "bizarre" theory in which the solution to the conundrums of quantum uncertainty are approached by resort to "a new vision of reality." This new reality involves the existence of other, "parallel" universes which are "within our technologically extended senses that must connect or relate with our own."

What Is a Parallel Universe? So what is a 'parallel universe'? It is a "region of space and time," he says, "existing side-by-side with our own," perhaps even "occupying our same space," and "containing matter, galaxies, stars, planets, and living beings similar and *possibly even a duplicate of our own universe*." In a parallel universe not only must there be "other human beings," he notes, but they may even be "exact duplicates....connected to us by 'mechanisms' explainable only by quantum physics."

Included in his materials, he remarks, are ideas on quantum physics, unification, relativity theory, cosmology, time, psychology, and "the effects of the human mind on all of this."

Bringing in the Observer: Quantum physics, he comments, requires that the properties of matter depended on how it was observed, so that observation plays a quite unexpected role, affecting even "macroscopic matter in subtle ways" so as to "change cosmology and...our concept of just what a universe is." So quantum physics introduces the effect of an observer upon a physical system. This effect, Wolf feels, cannot be realistically understood without the existence of parallel universes.

Relativity, Black Holes--Topological Holes to Parallel Universes: Relativity, he notes, deals with relationships between matter, energy, space, and time. It includes "weird" ideas, such as light traveling across the universe without "spending any time on their own or going anywhere," from their vantage point.

Our universe contains regions of matter that strongly distort surrounding spacetime, called black holes. We now believe, he says, that the laws of physics hold everywhere and consequently, that these singular regions turn out to be "mappable" and "topological *holes* leading to *parallel universes*."

Cosmology: Search for the Beginning: Wolf notes that cosmology deals with the theory of the early universe and how it began. It is now recognized, he notes, that earlier cosmology theories were insufficient because they did not include quantum theory and, he says, when we include quantum theory, strong evidence emerges for the existence of parallel universes.

Parallel Universes and Communication with the Future: That the future may play a role in the present, he suggests, is a new prediction of quantum physics. Interpreted literally, he notes, the mathematical formulas suggest not only how the future blends with the present but also how our minds may sense the effect of parallel universes.

If Parallel Universes of Relativity and of Quantum Theory Are the Same: If the parallel universes of relativity are the same as those of quantum theory, the possibility exists, Wolf feels, that parallel universes may be "extremely close to us, perhaps only atomic dimensions away, perhaps in a higher dimension of space, an extension into what physicists call 'superspace.'

What Is Reality? What Is Existence? What is reality? What is existence? asks Wolf. From his quantum perspective, he says, all the parallel universes are *real* but not all *exist.* He explains that he wishes to distinguish "reality" from "existence" and describe how parallel universes fit into the scheme of things.

Reality--Mathematical Superspace of All Possibilities--The Mind of God: First, what do we mean by *reality*? He cites the American Heritage Dictionary definitions which, in effect, includes everything that is "actual or true," "exists objectively and in fact," and is "real, absolute and unchangeable." But for Wolf, reality consists of a "gigantic superspace--the mathematical space of all possibilities." We might think of this, says Wolf, as *the mind of God.*

Mind Floats in Superspace--Putting Together Two or More Dimensions: Each dimension in the superspace is like a dimension in our ordinary three-dimensional space. Each dimension of the space also points to or indicates a 'direction.' This direction is a possibility. In this superspace something called mind floats and freely associates. It puts together--for a fleeting moment, which could be an eternity, depending on the time scale one is looking at--any two or more dimensions, and it asks what about this combination?

The Most Significant Insight Since Advent of Quantum Theory: He believes that this insight into the workings of quantum physics, derived from the work of physicists like John Cramer, John Wheeler, Sir Fred Hoyle, David Albert, Yakir Aharonov, Susan D'Amato, Jack DSarfatti, and others, is the most significant insight into the "strange landscape" introduced by the advent of quantum physics.

Time Not A Barrier--Future and Past Exist Now--We Construct Existence: If it is a stable hypothesis, he says, it will revolutionize our view of the world. It will say in effect that time is not a barrier. The *future* exists now, and so does the *past*. Finally, from this perspective much is evident. We both choose "subsets of reality" and "construct existence." Because we are the outcome of the past and future, he muses, we are "stuck" in time.

God's Way to Have Anything Real Come into Existence: This "stickiness," he observes, is God's way of having anything real become something *in existence*. It is the ultimate middleman, the beneficiary "from both ends simultaneously." The more resonant the transaction, the higher the probability, the greater the number of "parallel universes contributing to the transactional existence." Since these universes are constantly overlapping and splitting through time, he says, we are ultimately the beneficiaries or "maleficiaries" in the parallel universe scene.

The Fog of Time Is Parallel Universes: The action of "piercing through the time fog," and the existence of parallel universes, appear to go hand in hand, he suggests. The "fog of time" is parallel universes. One of his views is that it will be possible to have experiences of the future, referring to psychics and seers. Many scoff at the idea while others accept them.

> A parallel universe is a region of space and time...existing side-by-side with our own...perhaps even occupying our same space...containing matter, galaxies, stars, planets, and living beings similar and *possibly even a duplicate of our own universe* [in which] ...not only must there be other human beings...but they may even be exact duplicates....connected to us by mechanisms explainable only by quantum physics.
>
> Fred Alan Wolf

CHAPTER 17
QUANTUM REALITY

The Macroscopic 'Physical' World: In every day life we see 'material' objects, things whose existence we verify through our sense organs, and so we are accus-tomed to thinking in terms of the macroscopic 'physical' universe. Ever since Einstein's famous special relativity theory equation, $E = MC^2$, defining the equivalency of energy and matter, we have known that so-called 'physical' things are actually just forms of energy that register in our senses as forms of solid matter, 'rest energy.'

However, these energy forms, touchable or not, were once viewed as physical 'building blocks,' atoms, which were eventually thought to consist of even smaller physical entities--particles. So, it seemed that 'reality' was physical, 'all the way down.' Some thought there might even be an infinity of ever-smaller entities, with physical structure.

Disappears into a Quantum State: But concepts in the physics of the very small--*quantum theory*, evolved. Quantum theory deals with the state of things at the subatomic level, the term derived from Max Planck's finding that radiation did not flow continuously but in packets or *quanta*.

Quantum theory, it turned out, suggested another and deeper reality, one in which, strangely, the 'physical' nature of the universe-at-large, at a deeper, underlying level, is transformed into something *nonphysical*.

Existence--Statistical Probability and Subjectivity: Even *more unsettling*, whatever 'exists' is governed by *chance*, the quantum uncertainty principle, the product of statistical *probability*. Thus the entire universe owes its existence to a chance quantum eruption from the primordial quantum void in which there is no matter, nor time or space, just the laws of physics which reduce to mathematics.

But the mystery of 'existence' deepens even more. What 'exists' is influenced by observation and measurement and, in at least one theory, influenced by human perception. In this view, *an object not perceived does not exist.* As one physicist noted, the moon does not exist until looked at.

Retroactive Creation of Reality: In this context, physicist John Wheeler, in *Genesis and Observership, Foundational Problems in the Special Sciences*, asserts that the precise nature of reality must await the participation of the conscious observer.

In his discussion of whether God created the universe, physicist Paul Davies discusses the question of God's existence and plan to create the universe, *including time*, before there *was* time. In order to facilitate the case for God's existence 'before' time, he asks whether it is possible for causes to act *backward* in time, to produce *prior effects*.

Picking up on Wheeler's proposition that reality must wait the conscious observer, Davies notes that in this way, the mind can be made responsible for the "retroactive creation of reality," suggesting the Big Bang as *the retroactive product of our own mental processes!*

What Does Quantum Theory Tell about Reality? There are, of course, different views of quantum reality--what actually exists in the absence of observation and measurement and the role of observation and human perception in firming up an objective reality. While Stephen Hawking speaks of observer effects and subjectivity in quantum reality, he notes in *his Black Holes and Baby Universes and Other Essays*, that he is not a *solipsist*, a believer that all but self is subjective. Yet there are some, scientists and philosophers, who assert essentially just that.

Long before quantum theory, Anglo-Irish bishop George Berkeley's subjective idealism asserted that matter does not exist independent of its subjective perception. In the Copenhagen interpretation of quantum theory, the observer or measuring system and the quantum system being measured or perceived are part of a single situation of interacting quantum systems.

On the occasion of a meeting with some visiting computer people, Stephen Hawking was asked, out of the clear, how he would define 'reality,' ostensibly in the quantum theory sense. He responded, somewhat testily, that he didn't know how to respond to such a question. The point is, of course, that a quickie answer is either inaccurate or meaningless.

SPACE AND TIME AS MENTAL CONSTRUCTS

Re-examining the Universe: Earlier philosophic and scientific perspectives about the universe and, to an extent, the nature of our existence, seemed to call for some re-examination following the discovered concepts of relativity and quantum theory.

What are Space And Time? As baffling as space and time seemed when physicists first explored their implications in the context of relativity theory in the early part of this century, their meaning seemed even more uncertain in the light of the emerging physics of quantum theory dealing with subatomic physics.

This strange theory suggested: that *reality itself is the product of observer participation;* that while space and time may have (unknowable) physical properties apart from our perceptions, in our quantum-observer reality, space and time consist of what the thinking mechanisms of our minds construct them to represent.

Space and Time As a Virtual Reality? Earlier in this century, physicists gave expression to their speculations that quantum theory supports the strange concept that our own subjective observations may dictate our reality. Scientist George Zebrowski, in his fascinating *The Enigma of Distance,* Omni Magazine for November 1992, deals with the nature of the reality of time and space.

He notes that ours may be a "*virtual reality...a mental construct*, built up within minds to the point where we experience it as a literal reality of three dimensions and time....*Its true nature may be quite different from what we experience....*" Perhaps, he muses, these interpretations of space, distance, dimensionality, and time are, as Einstein expressed it, "free creations of the human imagination, means devised for easier comprehension of our sense experience."

Quantum Genesis and Pre-universe Unity: Quantum genesis and unification theories suggest *what may have existed before the Big Bang*, and describe the symmetry-breaking process through which we derived space, time, matter, the forces and fields, and the present dimensionality.

The theory postulates that out of the primordial void (the *superspace* of the primordial mega-universe), empty of space, time, and matter, erupted a cosmic particle, exponentially smaller than a proton. *In this particle was unified everything common to our present universe* of "broken symmetry," including space, time, matter, dimensionality, fields and forces. This cosmic particle represented a single unified state governed, it is thought, by the master law of all physics.

Shatter a Unity and Inflate It: In his discussion of 'distance,' Zebrowski speaks of this "grand monism...where ordinary intuitions about space and time just don't apply...which suggests that *everything is in the same place and does happen all at once."* Perhaps, he notes, space [three-dimensions of space and the single dimension of time with its one-way arrow permitting cause and effect]...is what you get when you shatter a unity and then inflate it so that space is what exists between the pieces that still seem to 'know' each other."

Space and Time As Mental Constructs: Physicists like John Wheeler, Paul Davies, and others, have discussed the notion that our ideas of space and time are creations of the human mind, *constructed realities representing the interaction between two quantum systems,* the universe on the one hand, the human mind on the other, something described by physicist George Greenstein as the *symbiotic universe*.

The suggestion here is that while space and time represent some theoretical, but for us undiscernible, abstract state in physics, their demonstrable or experiential reality arises from the participation of an observing or measuring quantum system like human mind in its corporeal setting.

Philosophical Asides: And so Debrowski speculates on space and time as *mental constructs* and the possibilities that *everything is in the same place and does happen all at once,* and that there is "no *literal reality* without the participation of observers." This reminds us, of course, that all reality has *subjective underpinnings* which, we are reminded, have been the subject of philosophic discussion, especially by the eighteenth-century philosophers.

Early Philosophy Proposed This Underlying Mental Reality: As startling as is this contemporary idea in quantum physics that reality is nonphysical beyond the observing mind, the idea is actually a very early metaphysical concept. Plato suggested (idealism) that only ideas are real, with objective standards that are independent of personal involvement.

Irish Anglican Bishop George Berkeley asserted that everything was a monistic or single-substance mental representation, a mental substance, ideas and objects did not exist unless perceived. In Immanuel Kant's views of transcendental or critical idealism, objects are simply the constructs of our minds and what may exist "out there" can not be known. Georg Hegel's universe was of pervasive mind and unfolding thought.

Metaphysics First to Grasp a Nonphysical Underlying Reality: And so, strangely, philosophic thought, metaphysics, seems to have been able to grasp intuitively, long before quantum physics, the concept of a nonphysical reality underlying the physical trappings of the universe. And theology, of course, has always dealt with a universe whose importance lay in a nonmaterial, ethereal, spirituality.

These notions bear strange resemblance to recent *concepts of subjective reality in quantum theory*. In fact, some physicists have suggested that the quantum observer effect implies, as we discuss elsewhere, *that the Big Bang itself is a product of our retroactive quantum observations.* The arrow of time from present to the future, sometimes referred to as a half dimension, applies to the macro world at large not to the subatomic quantum world where time is fully dimensional, the arrow pointing back and forward.

Such a notion, of course, raises questions about the ability of the quantum system representing the human mind to function not only in the present but in the past and future, reminiscent of Einstein's mysterious caveat that distinctions between the past, present and future are an illusion.

Kant and Objective Reality: The eighteenth century German philosopher, astronomer and scientist Immanuel Kant, regarded as having initiated the school of philosophy known as German Idealism, authored his famous critiques of Pure Reason, Practical Reason, and Judgment.

These expressed the fundamentals of his remarkable philosophy dealing with epistemology, what the human mind can know, especially scientific knowledge, relationships between cause and effect, and the nature of reality. Kant, perhaps the most striking figure in 18th century philosophy, whose powerful influence on succeeding philosophers exists to this day, expressed some thoughts remarkably similar to certain ideas in contemporary physics, while other of his views no longer hold.

Interestingly, Debrowski's musings about time and space as mental constructs,not dissimilar to views expressed by physicists like John Wheeler, Paul Davies, and others, bears striking resemblance to certain aspects of Kant's philosophical conclusions.

Thus Kant asserted that the nature of objective reality--of objects, "things in themselves"-- can be known to us only to the limited extent permitted by the filter of our mental processing apparatus. So earlier philosophy and contemporary quantum physics somehow coalesce.

Understanding the Human Mind to Understand the Universe: It is interesting to observe that science often focuses on the need to understand the human mind in order to understand the nature of the universe. Physicists suggest that it is the aggregate of human minds, within society groupings that shapes the many different “universes” of human perception from time to time and from society to society.

This focus is well illustrated in an erudite and exhilarating view of ancient and modern cosmology presented by University of Mass. Physics and Astronomy Professor Edward Harrison, in *Masks Of The* Universe, in which he describes the sequence of "universes" representing human life's changing understanding of the Cosmos. His views of the societal universes provides an interesting comparison with those described in Astronomer Gerald S. Hawkings' *Mindsteps To The Cosmos*.

Other Intelligence--Different Mental Constructs of Time and Space: This raises the interesting question of whether there may be other intelligences, observing quantum systems, in the universe which may construct different mental constructs of time and space, and are able to function accordingly--perhaps without our distance limitations.

We generally assume that other sentient beings in the universe must reside too far away (speed-of-light limits) to reach planet Earth in the "time" represented by likely organic-life durations. The case is suggested of intergalactic messages received long after the transmitting civilization has ceased to exist.

In "The Search For Signals From Space" (*Parade* of September 19, 1993), astronomer Carl Sagan discusses the NASA SETI (no longer funded)and other programs for the search for extraterrestrial intelligence, "other civilizations out there in the dark, sprinkled through the vast Milky Way Galaxy." He speaks of the role of the discovery of other intelligent civilizations (perhaps far more advanced than ours), as "a rite of passion for our species, a transforming event in the ancient quest to discover our place in the universe."

Eight Quantum Realities and the Missing Universe: One of the easier-reading books on quantum reality is Nick Herbert's *Quantum Reality* in which he discusses eight quantum 'realities' based on eight theories of what exists and happens at the quantum level and what these approaches mean. Another fascinating approach are physicist Fred Alan Wolf's *Taking a Quantum Leap* and *Parallel Universes* which explore the interrelationships of quantum mechanics, God and human thought, perception, and decision making.

At the quantum level our so-called objective 'physical' universe disappears into some sort of indescribable nonphysical 'quantum reality.' About quantum theory, physicist John Wheeler observed that "Nothing is more important than this: it has destroyed the concept of the world as 'sitting out there.' The universe will never afterwards be the same." Wolf introduces Wheeler's statement as "The Case of the Missing Universe."

Our physical world is considered by some physicists and science writers as some sort of *holographic projection of underlying [quantum] information.* As discussed later, space and time may be viewed as mental constructs. Einstein comforted the sister of his dear departed friend by asserting that the distinction between past and present was an illusion, implying the coexistence of all time.

The Shock of Quantum Theory: *Anyone who is not shocked by quantum theory has not understood it,* observed Danish quantum theory pioneer Niels Bohr. As contemporaries, Einstein and Niels Bohr engaged in extensive dialogues on the meaning of quantum theory. Einstein, like St. Augustine, was a firm believer in a deterministic universe in which everything from the creation unfolds in a cause-and-effect manner.

Bizarre Philosophical Implications: Einstein was so unnerved by quantum uncertainty theory, as proposed by quantum physicists like Bohr and Heisenberg, that for the rest of his life he was unable to fully accept the philosophic indeterministic implications of quantum uncertainty, and of *observation as influencing reality,* according to the Copenhagen interpretation of quantum theory.

To the very end Einstein never accepted the notion that God would resort to mechanisms of chance, *play dice,* to build and operate the universe. Although there are some theories that lend some support to Einstein's views, the weight of scientific authority favors Bohr's view of the absence of objectivity at the quantum level of the universe.

Physicists perceive quantum theory as having implications so profound as to affect the very concepts we employ in thinking of God and our very existence. Physicist Paul Davies, in *God and the New Physics*, describes the quantum revolution as having *bizarre philosophical implications*, implications which warrant our close attention in the search for an understanding of God and existence.

The Weirdness of Energy and the Quantum Revolution: Davies speaks of energy as a *slippery concept* and physicist Freeman Dyson says that *energy-matter is weird enough that God can do anything He wants with it*. In the beginning, at the incredibly high primordial temperatures in the early universe, energy existed only as pure radiation, in a state of 'thermal equilibrium.' As the universe expanded and cooled in its thermodynamic sequence, matter-energy decoupled from radiation, first to become particles, then atoms, then molecules, eventually becoming the 'frozen' or 'rest' energy of solids, liquids, galaxies, planets, and people.

Everything in the Universe Is Energy: Einstein's famous special relativity equation declaring the equivalency of energy and mass or matter, $E = MC^2$ (Energy equals mass times the speed of light, squared), is described as providing the defining equation of the universe. Everything which our senses make us aware of in our universe, and more, consists ultimately of different manifestations of a single medium, *energy*, which, as physicist Steven Weinberg notes, in *The First Three Minutes*, consists of a set of *interacting quantum fields*.

Moreover, according to the latest in contemporary unification theory, all space, time, and energy-matter were once a single unified state, described as the *superforce*. Although we generally deal with energy components as subatomic 'particles' these particles exist also, of course, outside the atom, as cosmic radiation, for example.

A Momentary Aside--Physical Forms of Energy: The physical universe we observe, and live in, thus seemingly consists of various forms of motion energy and rest energy or 'structured' energy--matter (galaxies, stars, planets, people, etc.)

Energy is said to represents the capacity for work or change by producing motion (kinetic energy), heat (thermal energy), sunburn (radiant energy), chemical energy (growth), nuclear energy (fission and fusion). Chemical energy, of course, involves the rearrangement of atoms in molecules, while nuclear energy involves rearrangement of atomic component particles, protons and neutrons, within the atomic nuclei.

Energy Becomes Life--To Ask about Energy and Life in the Cosmos: Earlier we noted the fundamental question which physicists Hawking, Heinz, and others pose: Why there is something rather than nothing--why does the universe exist? And how does it come to pass that the universe consists of a single medium, energy, which metaphorphizes into matter, then life, consciousness, subjectivity, and exquisite abstractions--like love, hope, justice and contrition? And how does an infinitesimally minute piece of the universe, the human mind--a living energy form--contemplate the entire cosmos--and ask these kinds of questions?

Physical Things Slip into Quantum Immateriality: Quantum theory deals essentially with natural phenomena at the *subatomic level*, energy phenomena within or smaller than the atom, manifested through 'fields and forces.' At that level, the definite, sense-identified physical things of everyday life in our universe--stars, planets, trees, automobiles, and living creatures--fade and disappear into mysterious uncertain and shifting, *ethereal energy states*, particles or probability waves, seemingly more mathematical concepts than the physical reality to which we are accustomed.

Shifting Quantum States: Before the incredibly hot starting universe cooled enough for matter to *break out of thermal equilibrium* and combine to form atoms and molecules, these particles existed independently as radiation, as a soup or plasma.

Originally thought to be the material 'building blocks' of the universe, the 'brick and mortar' of material things, particles are viewed now as 'interacting quantum fields,' *shifting quantum states* governed by uncertainty, behaving sometimes as particles and sometimes as waves, *waves of statistical, observer-affected probabilities.*

Quantum Theory Required New Thought Categories: Quantum mechanics involves effects which are so unportrayable by ordinary scientific descriptions that new *thought categories* had to be devised as analogy communication techniques. Employed are such quantum theory terms as *strangeness* and *charm, color* and *chromatics, and flavor.* Actually the particles or quantum force or field states have no actual color, charm, or strangeness as those terms reflect every-day meaning.

Unity and Diversity: But these seemingly whimsical terms, communication mechanisms for categorization, have practical utility in expressing the ideas involved, through *analogy.* Which reminds us, of course, that we describe or define any one thing by reference to a second thing which, in turn is described in terms of a third thing, and so on. In unification theory, all these 'different' things (and the diversity that we ourselves represent) are part of a single master unity. As physicist Freeman Dyson observes, *the unity of the universe is matched only by its incredible and growing diversity.*

People and Stars As Quantum States: We have noted the strange reality that emerges at the quantum level where the physical universe of galaxies, stars, planets and people, disappears into uncertainty states of nonphysical statistical probability.

This state is referred to as something that is nonphysical, nonmaterial, sometimes described as representing *information, mind, consciousness*, all part of an interconnected cosmic unity. A person, or a star, or any other object that we 'see' as part of our big, *macroscopic* so-called 'physical' world is, in quantum theory, inextricably interrelated with the examining (in quantum terms, observing or measuring) mind.

One Quantum System of the Universe Contemplating Another: In the *microscopic* or quantum world, a person, a table, a car, is essentially a quantum system consisting mostly of empty space, and the physical structure we visualize represents the interrelationship of the aggregate of forces and fields being viewed by the (living) observing quantum system that we represent. In quantum theory, the observer quantum system and the observed quantum system are part of a single situation and applying the so-called Bell theorem, part of the interconnected cosmic quantum.

In quantum terms, a human being is a quantum system piece of the universe with the power to contemplate other quantum system pieces of the universe or, for that matter, to contemplate the quantum system representing the entire universe of which it is so infinitesimally small a subsystem.

Consciousness As a Quantum Effect: In *Quantum Reality*, Nick Herbert observes that "science's biggest mystery is the nature of consciousness" and he asks whether all human experience is a product of classical physics or whether is it "some sort of quantum effect?" If ordinary awareness is a direct connection to quantum reality, and if external knowledge of quantum entities is "quantum ignorance," then, he suggests, we might call our awareness or experience of the world's real nature as "quantum knowledge," a relationship between the "waveform alphabets of quantum theory and certain human states of consciousness."

The Invisible Quantum Connection and the Illusion of Separateness: Following Bell's theorem, while the world's phenomena seem 'strictly local,' underlying this localness, "the world's deep reality is maintained by an invisible quantum connection whose ubiquitous influence is unmediated, unmitigated, and immediate." This is illustrated, possibly, by such phenomena as telepathy and other unexplained powers of the mind, as evidenced in persons thought to have 'psychic' powers, such as finding objects through mental association, affecting materials at a distance, prescience, and other unexplained mental feats.

Biologists say that we are all 'brothers and sisters,' genetically intertwined. Physicists tell us, Herbert says, that "the very atoms of our bodies are woven out of a common superluminal fabric" (communicating faster than the speed of light). He cites Einstein's references to our 'illusion of separateness.'

The Quantum *Nonmaterial* Reality: We discuss elsewhere the mysterious, nonmaterial, informational and mental nature of our underlying Quantum Reality, in which *our physical world seems to be some sort of holographic projection* of an underlying interconnected nonmaterial network of information and mind. This is *one in which physical reality is somehow related to mental processes.* As physicist Paul Davies reminds us, reality cannot be separated from the observer and his choice of measurement strategy.

Physical Reality and the Observer: So while we 'exist' in our observed physical universe, quantum physics postulates another more mysterious universe of the ephemeral, the ethereal. At the quantum level, our objective, material world somehow dissolves into ghostly quantum states in which *mind-consciousness, plays a key and catalytic role in bringing quantum indeterminate states to forms of perceived physical reality.*

Quantum theory, according to physicist John Wheeler, says that *the physical universe requires observer participance for its very existence;* the 'observer particles' that constitute us, and those particles representing objects or phenomena observed, being all part of the same interconnected whole.

Quantum Reality States of the Universe: The 17th-18th century philosopher Immanuel Kant in his *Critique of Pure Reason*, argued that we cannot know the real universe, what's really 'out there,' we can only conjure what our mental construct permits us to grasp. Anglican Irish Bishop-philosopher George Berkeley saw a subjective world, one in which material objects had no existence except as they are perceived.

Strangely, related concepts are wrapped up in *subjective probabilistic quantum reality,* where, at least according to the Copenhagen quantum interpretation, multiple possible quantum states are collapsed to a single state by observation or measurement and the observer quantum state and the observed quantum state are a 'single situation.'

Physicists speak of time and space, too, as possibly 'mental constructs,' part of a *virtual reality.* Quantum theory posits the entire universe, and all its quantum system components, as the product of quantum statistical probabilities--influenced, in most theories, by observation.

As we will discuss, in *A Brief History of Time*, British theoretical physicist Stephen Hawking proposes a *quantum theory of gravity*, and in the process, explores all the likely quantum probability states to determine the initial conditions of the universe. This requires calculating the *quantum wave function of the entire universe,* using the mathematics of 'imaginary time,' in which the distinction between time and space dimensions disappears and time becomes a fourth *space* dimension.

The Boundary between the Physical and Quantum: In *The Philosopher's Atom* in November 1995 Discover Magazine, Hans Christian Von Baeyer notes that one of the most stubborn enigmas of modern physics is the missing link, the 'metaphysical' gap between the atom and molecular matter, between the unimaginable quantum world of uncertainty and potentiality and the physical actuality of the macro world. Stepping across this as yet unreconciled gap, says Von Baeyer, physicists rely on instinct rather than reason.

But 'Rydberg atoms' may shed some light in this area. These are delicate atoms in which the outermost electron follows an enlarged irregular orbit that periodically takes it *outside the atom's internal quantum world,* moving in a real orbit through space and time of the classical world, then back into the atomic world of quantum mechanics, an "atomic amphibian."

Its trajectory through classical physics world produces a highly irregular complex wave which, on return to the atom's quantum venue, produces wave interference ripples caused by the returning wave electron no longer being a perfect quantum three dimensional sphere. The variables related to the Rydberg orbits are energy and time rather than the quantum variables of position and speed.

While there are particle trajectories through space and time of classical physics, there are no such trajectories in quantum mechanics. The functioning of the Rydberg atom's outer electron behavior as it weaves in and out of quantum land gives some insight into this metaphysical gap between the quantum and classical worlds.

> Distance...[the] grand monism...where ordinary intuitions about space and time just don't apply...which suggests that *everything is in the same place and does happen all at once.* Perhaps, space [three-dimensions of space and the single dimension of time with its one-way arrow permitting cause and effect]...is what you get when you shatter a unity and then inflate it so that space is what exists between the pieces that still seem to 'know' each other.--George Zebrowski

PART VI--COSMOLOGY--CREATION THEORIES

CHAPTER 18
CREATION THEORIES--QUANTUM COSMOLOGY

Four Approaches to the Start of the Universe: Since we have been talking about theories of the origin of the universe and some of their implications, it will be helpful briefly to describe some of these theories.

***First*: The Eternal Universe:** In this theory, the universe is eternal, so there was no 'creation' point where the universe of time, space and matter actually began. This is the so-called 'steady state' universe, of which there are several variations, including the more recent 'continuous creation' steady-state theory in which new matter is continuously being created, via a gravitational energy field.

***Second*: The 'Big Bang' Creation Concept:** In this theory, based on general relativity theory, the universe starts, from an unknown environment, with the so-called Big-Bang event. Einstein's general relativity equations and the laws of thermodynamics implied a universe that was expanding (or contracting). Thus the universe is thermodynamically traceable back in correlated size, time, and temperature, to a cataclysmic, so-called Big-Bang expansion from an infinitely small, hot, and massive singular state, termed a *singularity*.

The 'singularity' state represents in classical physics a so-called 'boundary condition' signifying a beginning in time, space, and matter. In the words of physicist Stephen Hawking, it is "a point of infinite density and infinite curvature of spacetime," a point incorporating all spacetime, forces, and matter potential.

Left undefined in the original theory, was the question of what may have existed *before* this state of singularity since time before the Big Bang had no meaning. In terms of the laws of classical physics there was no 'before.'

What Existed Before? Left open in Big Bang theory is what existed 'before' the Big Bang, although the question is sometimes said to be meaningless since the dimension of time originated in the Big Bang--so there was no technical 'before.' Did the universe come out of sheer nothing, *ex nihilo,* as is said to represent the Genesis scenario? And if not, then how did the universe arise, out of what--and is the question scientific or theological?

The conventional Big Bang theory, of course, left open the cause and so allowed metaphysical speculations, including the possibility, at least, of divine causation. God might just have *thought* the universe into existence, as Genesis suggests. The Big Bang model does not invoke quantum theory or presuppose a pre-existing quantum state.

***Third*: The Quantum Vacuum Fluctuation Model:** The time was when theories of the start of the universe were based on the classical deterministic physics of the Marquis de Laplace, Newton, and even Einstein. In classical mechanics, all events were explainable in terms of *causal relationships* that could be traced back to the creation. Even Einstein's relativity theories conform to this causal characteristic of classical physics.

But the advent of the theory of quantum mechanics forever altered our perceptions of reality, and, according to mainstream physics, at the subatomic or quantum level, essentially negated the cause-and-effect, deterministic universe of St. Augustine Laplace, and Einstein.

The vacuum genesis theory introduces quantum theory, a theory which emerged in the nineteen twenties to deal with then (and even now) unexplainable subatomic phenomena of particle physics, the subatomic world. Quantum mechanics would, it was hoped, define the *reality* underlying the macroscopic universe of galaxies, planets and people.

A Universe-triggering Event: The Big-Bang theory was based on Einstein's classical-physics, general-relativity equations, which postulated an infinitely small starting point, exponentially smaller even than the particles inhabiting the atom. So physicists constructed a hypothesis which would allow a universe-triggering event from a pre-Big-Bang quantum void or vacuum state, a 'vacuum genesis' creation theory in which the Big Bang was generated by a quantum-uncertainty-induced random fluctuation in the hypothesized quantum vacuum of a primordial *mega-universe.*

In this theory, from that strange fluctuation there erupted an exponentially small *mini-universe point of super-energized spacetime,* infused, it is speculated, with the superforce of unified physics, a point, exponentially smaller than a proton, constituting an unstable high-density energy state.

This supercharged 'false vacuum' energy state, inherently unstable, 'decayed' in a 'phase transition,' unleashing an *exponential inflation* of the mini-universe point of spacetime, into which the negative gravitational energy, locked up in the expanding quantum vacuum, was released as positive radiation to become the energy-matter of the universe.

The exponential or *faster-than-light* inflation of space then subsided into the conventional Big-Bang sequence of linear, that is, less-than-speed-of-light, expansion process, an expansion, which continues to this day. There are several alternative versions of the phase transition and inflation sequence.

***Fourth*: The Quantum Genesis or Quantum-wave-function Approach:** This is often described as the 'quantum genesis' theory, requiring, according to British theoretical physicist Stephen Hawking, the use of a 'quantum theory of gravity' to try to determine the 'quantum wave function of the universe.'

In doing this, thought Hawking, one must explore the 'many worlds' quantum uncertainty interpretation, involving *all the possible quantum- probability states* from which the universe we observe may have started, and in the process, *define the quantum wave function* of the universe we now inhabit.

Looking at the starting universe as a quantum state, calls for calculating the *total quantum wave function* of the entire universe to determine the *initial conditions*, the so-called 'boundary conditions' of the starting universe in terms of the multitude of quantum probability states inherent in the "primordial quantum flux." Hawking's version of the initial universe also calls for an exponential inflation stage.

Hawking's Startling Conclusions: Hawking's quantum-genesis approach leads him to *stunning conclusions*: In this approach, the starting universe conforms to Einstein's *finite but unbounded* characteristics, and, since, technically, quantum fuzziness explains away the singularity hypothesized in the Big-Bang model, so in Hawking's model there were no time or space boundaries--*no technical beginning of time*.

Although the universe might fluctuate between cycles of expansion and contraction, Hawking notes, the universe would be unaffected by any external mechanism, the universe simply exists, it would just *be,* says Hawking.

So Hawking's quantum-genesis model posits a universe that has no boundaries, no beginning, eternal but cyclically expanding and contracting so that it signals a periodic end to, and perhaps, a new beginning of, life in the universe, perhaps in an eternal repeating process. This approach, of course, initially raised many interesting questions, such as whether the arrow of time reverses when the universe starts to contract. Hawking and others finally concluded that it does not. We will further review Hawking's quantum cosmology below.

Hawking Asks Did the Universe Have a Beginning? It will be helpful, in the context of self-generated universes, to explore further the work of physicist Stephen Hawking as reflected in his *A Brief History of Time* and some of his sequel and related essays published in his *Black Holes and Baby Universes and Other Essays.*

Hawking asks fundamental questions. "What do we know about the universe and how do we know it? Where did the universe come from, and where is it going? Did the universe have a beginning, and if so, what happened *before* then?"

What is the nature of time, he asks, and will it ever come to an end? Recent breakthroughs in physics suggest possible answers to these questions, and some day, he observes, the answers may be as obvious to us as the fact of the earth's orbiting the sun.

Nothing for a Creator to Do? In the Introduction to *A Brief History of Time*, astronomer and space scientist Carl Sagan observes that Hawking embarks on a quest to answer Einstein's famous question of whether God had any choice in creating the universe. He suggests that *Hawking's effort is no less than an attempt "to understand the mind of God,"* reminiscent, of course, of Einstein's oft-quoted expression of his desire to "know how God created this world.....to know His thoughts...."

But Sagan goes on: "And this makes all the more unexpected the conclusion of the effort, at least so far, *a universe with no edge in space, no beginning or end of time, and nothing for a Creator to do.*"

Hawking's Theory: Profound Implications for God's Creation Role: In some of his work, Hawking, ordinarily preoccupied with the *natural* state of things, occasionally talks about physics & God in relation to theories of quantum genesis and quantum gravity, the role of supersymmetry and unification and superforce theories.

Before quantum theory was introduced to address cosmological questions beyond the scope of Einstein's classical physics, general-relativity theory, the mathematics of general relativity theory predicted an expanding universe. It thus required a starting point, a beginning of time, a creation event--the *Big Bang*, from an incredible state of exponential singleness called a *singularity*, out of which emerged all energy-matter, space, time, and, of course, dimensionality.

New Possibilities from Use of Quantum Histories and Imaginary Time: By introducing quantum theory, says Hawking, a "new possibility" arises, one not envisionable from the classic physics of general relativity theory. It is that in quantum theory, "space and time...might form a [self-contained] finite, *four-dimensional space* without singularities or boundaries, like the surface of the earth but with more dimensions," and even accounting for the observed arrow of time.

The mechanics of this possibility derives from being able to apply physicist Richard Feynman's technique of quantum uncertainty *sum-over-histories* of the universe, and resorting to use of so-called spatial or *imaginary time* where the time dimension is, in effect, a fourth space dimension. Hawking depicts 'imaginary time' as time which lies 'at right angles' to conventional or what he calls 'real' time.

Treating Time as a Fourth Space Dimension: Hawking notes that *this treats the time dimension like a space dimension*, and permits three space directions and a single time dimension to meet to form a spacetime that was a "closed surface like the surface of the earth...without boundaries or edges...without a point that could be called a beginning or end."

Time Turns into Space in the Early Universe: Hawking's quantum cosmology 'creation' suggests that there is no real beginning of time. This is because, as physicist Paul Davies puts it, *time*, in a sense, *turns into space* in the uncertainty fuzziness of the early quantum universe.

Instead of a space of three dimensions at a point in time, there is a four-dimensional, infinitely-curved space which represents an 'unbroken continuum.' To use Hawking's illustration, the earth's north pole is not an edge or beginning of the earth's surface.

All Possible Quantum Wave Function Probability States: As noted, because the infinitely small mini-universe starting point in the vacuum fluctuation was a quantum particle state, Hawking sees the need for exploring *all the possible quantum wave function probability states* from which the universe we observe may have started. This hypothesis, he notes, requires the use of an appropriate 'quantum theory of gravity,' calculating the *quantum wave function of the entire universe* to determine the 'initial conditions.'

These would be the boundary conditions of the starting universe in terms of the multitude of quantum probability waves inherent in the primordial quantum flux. Hawking's quantum theory approach, as noted, also incorporates the so-called 'chaotic inflation' theory.

Calculating the Quantum State of the Universe: To derive the quantum state of the universe, by calculating the 'quantum wave function of the universe,' Hawking employs several advanced quantum mechanical approaches. These include use of physicist Richard Feynman's *sum-over-histories* strategy of calculating all possible quantum probabilities or wave functions to reach the *most probable* path by which the universe reached its presently observed state. It also involves use of the mathematics of what Hawking terms "imaginary numbers" to arrive at so-called "imaginary time."

Imaginary Time Eliminates Distinction between Time and Space: In the use of this imaginary time, the distinction between time and space disappears, says Hawking, noting that space-time, using imaginary values for time, is known as Euclidian space-time, a mathematical device to compute answers in so-called *real* space-time.

Using the quantum theory of gravity, if we know how the Euclidian curved space-times in the sum-over-histories behaved at early times, he notes, *we would identify the quantum state of the universe* out of the many possible alternative quantum states.

In Quantum Gravity Theory, Space-Time May Be Finite but Unbounded: Hawking observes that in classical gravity theory, there are *two possibilities*: either the universe has existed for an infinite time, or else it had a beginning at a singularity at some finite time in the past. But, in the quantum theory of gravity, a *third* possibility arises.

Because one is using Euclidean space-times, in which the time direction is on the same footing as directions in space, it is possible for space-time to be finite in extent and yet to have no singularities that formed a boundary or edge--like the two-dimensional surface of the earth, only with two more dimensions.

No Boundaries, No Beginning or End--The Universe Would Just *Be*: Since, in this quantum theory of gravity, there would be no boundary to space-time, there would be no need to specify the behavior at the boundary, *no singularities at which the laws of science break down*, no edge of spacetime. Here, "the boundary condition of the universe is that *it has no boundary*," declares Hawking.

So Hawking's quantum-gravity theory arrives at *startling results*: The Universe is finite but unbounded. There is no singularity, no boundaries--no beginning or end of time. Although cyclically expanding and contracting, the universe would be completely self-contained and not affected by anything outside itself. It would neither be created nor destroyed. It would just *be*.

The Many-Worlds and Sum-over-Histories Approach: In applying quantum mechanics, as we noted, Hawking used a number of mathematical devices, including so-called 'imaginary time' derived from the use of 'imaginary numbers,' and physicist Richard Feynman's 'sum-over-histories' method of calculating most likely quantum probabilities.

Hawking calculated the quantum wave function of the universe as a whole and drew certain tentative conclusions about the initial state of the universe, about the initial dimensional nature of time and space, whether the initial universe had singularities or boundaries, all *bearing on whether the universe was created or eternal.*

The addition of quantum theory to general relativity to explore the starting universe had profound effects, of course, on earlier perspectives of how the universe may have started. For one thing, this required the development of a model of the initial quantum state of the universe to determine its initial conditions.

Attacks on Imaginary Time: Hawking defends himself against denigrating attacks for his conjuring up of something seemingly unreal, namely, 'imaginary time.' These were attacks characterizing him as a 'nominalist,' an 'instrumentalist,' a 'positivist,' or a 'realist.'

Those denigrators, he feels, just don't understand the quantum environment in which these terms are used, that it is the use of various mathematical devices that permit all sorts of progress as in infinity 'normalization' devices permitting certain advances in unification theory. He asserts that the search for self-consistency has always been more important in making advances than experimental results; that theory has always come first and made predictions tested by later observation.

This occurred with Einstein's now amply-demonstrated relativity theory predictions which were not based on experiment but derived from concept efforts to reconcile the laws relating to electromagnetism and the laws governing the motion of bodies.

Calculating the Quantum Mathematical Description of Our Universe: At the 10th annual Conference on Relativity, in Padua, Italy on July 7, 1983, Stephen Hawking noted that the "quantum state of the universe will determine the initial conditions for the classical universe." So today," he announced, *I want to make a proposal for the quantum state of the universe*."

In the *Copenhagen Interpretation* of quantum uncertainty, before we observe a system, it exists in all possible quantum states, and, upon the act of observation, the quantum wave function 'collapses' into one of its possible states, following physicist John Wheeler's hypothesis. That hypothesis is that the existence of the universe requires subjective 'observer participation.' Since we are a part of the universe--inside it, notes John Gribbin, *in In Search of The Big Bang*, we have no 'outside observer' to trigger its quantum collapse to a single state.

Calculating Probabilities: 'Many Worlds' Plus 'Sum-Over-Histories': So how do you go about calculating the probabilities that define the universe? Well, Stephen Hawking suggests that the way to arrive at a quantum mathematical description of the universe is to employ a combination of the *Many Worlds Interpretation* and physicist Feynman's path integral or *sum-over-histories* method, involving all possible quantum alternative ways the universe could emerge.

Combining Quantum Theory and Relativity Eliminates Boundaries: Since physicists can't calculate *one* such history, never mind *all* possible histories, Hawking arbitrarily adopts a 'more likely' set of initial conditions with fields for gravity and matter, from which the evolution of the universe might then be extrapolated. These 'boundary' conditions require the combination of general relativity and quantum theory, and, he feels, *the model seems to work.*

The Big Bang Singularity Is Questioned --Time Is 'Smeared Out': We have observed how the introduction of quantum theory can be construed as questioning the Big Bang singularity, the inception of time being stretched or 'smeared out' over a period of time, albeit very small, namely, 10^{-43} second. Hawking's model combines general relativity and quantum mechanics to produce a finite, closed-system universe that has no boundaries or edges, no singularities in space or time--no beginning or end of time or space, and destined to undergoing cycles of inflation and collapse.

He uses the 'sum-over-histories' method for all histories which have finite size and no time or space boundaries. Use of this method, it is noted, narrows the alternative histories to a *few that are more probable*.

But There Was Still a Big Bang, Says Hawking: However, despite his intriguing 'no boundary' proposition, Hawking believes that there was in fact some singularity event implying a beginning of time; and that the universe did have a unique quantum history.

As we noted earlier, Hawking, is concerned that he not be misconstrued by his 'no-boundary' quantum history theory, and he believes that the overwhelming evidence favors a Big Bang event, especially since the 1992 NASA COBE satellite findings of early background radiation temperature fluctuations, ripples in the fabric of spacetime, the gravitational seeds for present galactic structures.

It's just that quantum theory, he suggests, requires the resort to all possible histories of the universe; that a quantum system doesn't have "just a single history in spacetime, as one would normally assume it did in a classical nonquantum theory. Rather, it had *every possible history,*" as reflected in the 'many worlds' quantum theory interpretation.

Use of 'imaginary time' is needed to make computations for those histories to determine the universe's starting conditions. Using imaginary time and the sum- over-histories technique, *the laws of physics then apply* and the starting conditions of the universe are theoretically calculatable.

Hawking Cites the Black Hole Analogy: He goes on to explain the probability of a particle traveling from A to B by adding up the numbers associated with all the paths; that *all the paths include some faster than light* (back in time); thus permitting some particles to escape from a black hole. So the particles in a black hole *need not have the end of their histories at a singularity.* Instead, he says, they could "continue their histories outside."

Applying Quantum Principles to the Structure of Space and Time: Maybe, he adds, the quantum principle would mean that *one could also avoid the histories having a beginning in time,* a point of creation, at the big bang. This, he observes, involves *applying the quantum principle to the structure of time and space themselves* and not just to particle paths in a given space-time background. What one needs, he says, is a way of "doing the sum-over-histories *for the whole fabric of space and time.*"

Though we don't know how to do the sum-over-histories properly, he says that we do know certain required features. We know, he says, that it is easier to do the summing by using histories in 'imaginary' time rather than in ordinary, 'real' time. So while quantum uncertainty in the primordial quantum void triggered the emergence of our 'branch' of the universe, technically, there may be *no real creation*, no true beginning (or end) of the universe, and no true beginning of time. This set of quantum uncertainty possibilities, he observes, *has profound implications for the role of God as Creator.*

Universes with Other Histories: Physicists and scientific writers seem intrigued, as well they might, by the existence of other possible universes among the 'allowed histories,' *universes somewhere next to us in the superspace of the primordial mega-universe*. But these are universes representing quantum states that are complementary in the sense of being mutually exclusive and unable to interact.

The question has been raised as to whether these neighbor universes, sometimes described as other 'branches' of the universe, coming into being through quantum effects, are actual or 'real' universes or whether they exist only as mathematical entities.

Are These Other Universes Real? In *Stephen Hawking's Universe*, author John Boslough asked Hawking whether these other universes resulting from quantum theory were "actual physical regions that could be observed." It just means, Hawking replies, that "there is a 'nonzero probability' for the universe to have many different forms."

The idea of universes existing side by side is treading on metaphysical ground, he said, quoting Cambridge University philosopher Ludwig Wittgenstein that the existence of other universes is *not a predicate*, since their existence would have no meaning without some observable consequence or confirmability. While other physicists have expressed similar views about other universes in quantum theory, there are some who remind us that most of the science we're now discussing was once considered metaphysics.

Actual Parallel Universes? Interestingly, this question of whether these quantum theory 'other worlds' or universes are 'real' has recently generated a host of new approaches. In *Parallel Universes, The Search For Other Worlds*, physicist Fred Allen Wolf characterizes these possible quantum states as real, actually-existing, parallel universes. Stephen Hawking's comments are a bit more conservative speaking of them rather as mathematical states.

Phase Transition Sets Off Exponential Inflation: In the quantum vacuum fluctuation model, described earlier, a quantum uncertainty event in the void of the primordial mega-universe created an infinitely small, dense, and hot mini-universe (one-thousand-millionth the size of a proton) point or 'bubble' of spacetime, sometimes referred to as characterized by the unified *superforce,* with the potential for all the fields, forces, and fortuitous dimensionality of our universe.

In one theory, this point undergoes a *phase transition,* triggering an *exponential inflation* at a rate faster than the permitted velocity of force transmission (like the speed of light). Into this exponentially fast-expanding space there was released from the excited negative gravitational energy, positive energy-matter, geometricizing into our spacetime dimensions, and breaking down into component forces and fields, leading to the present energy-matter distribution.

Breaking of the Perfect Symmetry of Unification: This inflation subsumes into the Big Bang expansion which still continues today. *The Big Bang phase starts a 'symmetry-breaking' sequence* in which the unification's potential components separate.

First gravity separates, then the strong force, followed by the then-unified weak and electromagnetic forces, which in turn, separate to give us our present spacetime structure and dimensionality--three dimensions of space and one of time. [To be noted is the fact that for certain purposes, like superstring theory, there are actually *six or more additional dimensions* that exist in 'compactified' state.]

Manipulation of Physics by Ancient Intelligence? Physicist Paul Davies, in *God And The New Physics*, wonders whether perhaps some ancient *natural* god intelligence, having been handed the superforce, perhaps by a *supernatural entity*, presumably God, proceeds to play out different evolutionary routines, as computer scientist Fredkin discusses, to see what will happen.

Along this line, physicists like Paul Davies and Freeman Dyson have suggested the idea that even advanced extraterrestrials, accustomed to manipulating quantum phenomena like black holes and quantum-connection, time-bypass mechanisms, might trigger new Big-Bang sequences for new universes from quantum umbilicals in our own universe.

Creating New Universes in the Laboratory: Physicists Alan Guth and Edward Fahri wrote a paper entitled *An obstacle to Creating a Universe in the Laboratory*, in which they describe how even our relatively primitive scientific skill might harness quantum effects to provide the mass-energy needed to trigger a new universe. This involves supplying the very *high density and high temperature condition* of 10^{24} K to trigger the required inflationary sequence which subsides into a Big Bang symmetry-breaking sequence. We already have hydrogen bombs to supply the needed energy. All we need do is to learn how to concentrate that energy in a small enough space, a space much smaller than that of a proton.

In *Cosmic Coincidences*, John Gribbin and Martin Rees say that applying Einstein's general relativity equations what you would get is a black hole or, given the 'right initial conditions,' an exponential inflation that subsides into the standard Big Bang linear expansion scenario of our own universe. One pictures a mad scientist triggering universes from a basement laboratory.

How much matter is needed to appear in a quantum fluctuation to start the universe creation expansion process? The answer, says James Trefil, in *Reading the Mind of God,* is twenty- two pounds of matter. The problem, says Trefil, is how to achieve the density required, a volume much less than a proton, a state of matter described as a 'false vacuum.'

Says Trefil, "If you could create a bit of false vacuum, it would distort the space-time network of our universe, form a kind of neck, and then seperate. In effect it would *leave our space and become another universe in its own right.*"

What would such a procedure do to our own universe? Hopefully nothing, because the expansion of the new mini-universe would, it is thought, pinch off "at right angles" to our own universe's dimensions of spacetime, "into a universe of its own." *This is exactly what would happen, they note, to 'any inflationary seeds' from quantum fluctuations in our own universe's quantum vacuum.*

Anthropic Cosmology: Symmetry Breaking Differently in Other Universes: So, quantum cosmology allows for "an infinite number of universes out of nothing at all." In fact, *our own universe may simply be an umbilical of spacetime* "pinched off from another bubble." Symmetry may break differently in different bubble universes, leading to altogether different physical constants in those bubble universes.

But given an infinity of such universes, *anything that can will happen* and some must undergo symmetry breaks that dictate life, something further discussed later on the "anthropic principle."

Is the Universe Subject to Another Quantum Leap? Well, there are physicists who suggest the possibility that *the present universe may represent an intermediate, unstable false vacuum* (energy-reducible) state subject to further phase transition and quantum decay. We wouldn't be here to cope with this, of course. Would we continue to exist in parallel universes? Might *they* also exist in an unstable false vacuum state?

Only One Self-consistent Set of Unified Theories? If Hawking's self-contained 'no boundary' universe proposal is valid, then, he notes, there may be only one 'self-consistent' set of unified theories allowing structures like human beings "who can investigate the laws of the universe and ask about the nature of God."

This is reminiscent, of course, of Einstein's so-often asked question--whether God had any other choices (that would sustain life). Unfortunately, Einstein's deliberations preceded our progress in quantum theory. And so it was left to later physicists like Stephen Hawking to develop the techniques to explore such questions. The consensus seems to be that the slightest alterations to the values in the constants of physics would rule out life--at least as we know it.

But, even if there is only one possible set of rules and equations, Hawking asks his oft-quoted question, *What is it that breathes fire into the equations and makes a universe for them to describe?*

The Mediocre Universe: In the February, 1966 Discover magazine, David Freedman discusses Tufts University Russian-American professor Alexander Vilenkin's *mediocre genesis* theory in which the universe springs from a primordial quantum state, possibly a space-time energy field, the average value of which is zero. Our own void or vacuum of space is actually an energy-loaded state, constantly generating *virtual* particles which flicker into and out of existence, courtesy of quantum mechanics which permits the average of zero energy level to fluctuate.

Any fluctuation can reflect a negative or positive value and be significant enough to erupt as a virtual particle, only to disappear back into the 'quantum flux.' Vilenkin's thought is that if the quantum void can generate particles why could it not generate whole universe particles? The mega-universe quantum void may be thought of as a spacetime-energy field with zero-average value, able to generate universe particles of unified space-time-energy able to escape annihilation, instantly expanding in a Big-Bang configuration.

Freedman notes that with a set of quantum equations, Vilenkin turns a metaphysical creation into a physical event. But to transform a philosophical hypothesis into science with a 'testable prediction,' Vilenkin uses physicist Alan Guth's theory of a starting universe in which gravity is *repulsive* and leads to an exponential 'inflation' which subsides into the Big Bang sequence. The universe seems poised or balanced exactly between a runaway expansion and a Big Crunch contraction in which energy-matter-space-time curvature is said to be flat, undetectable.

The Mystery of the Cosmological Constant: The 'cosmological constant' is a term derived from general relativity equations which, Stephen Hawking reminds us, provides a cosmic attraction or repulsion force. It gives spacetime an inherent tendency to expand or contract depending on whether the constant is positive or negative. We are reminded of Einstein's self-described blunder of assigning a positive or anti-gravity constant to prevent gravitational contraction of what was then considered an eternal steady-state universe.

The constant is thought to represent the amount of *energy inherent in so-called empty space*. Empty space is spoken of as having a zero average value energy field, although there is no reason other than the aesthetics of physics for the value being exactly zero. In theory, the cosmological constant should be exponentially large, it is thought, but the behavior of a universe with a large positive or negative constant would prevent life formation.

Some unusual phenomenon must work to drive the constant down. For want of a better explanation, resort is made to the 'anthropic principle,' that the constants of physics, like the speed of light and the electron charge, require values consistent with the existence of life. The principle invokes the range of different values for constants in multiple varied universes and the relative likelihood of occurrence of our universe's particular constants. Most universes would not permit life but chance occurrence of some would permit life.

Seeking a More Rational Basis for the Constants: Physicist Steven Weinberg seeks to find a rational explanation for our own constants other than the anthropic principle but is willing to *accept anthropic reasoning as a temporary expedient*. Vilenkin reasons that a small portion of universes would account for most civilizations and we are likely to exist in one of these high civilization-producing 'fecund' universes.

The Principle of Mediocrity: So our civilization would be nothing very special, and so he calls this criterion the 'principle of mediocrity' in which the characteristics of the universe lead to an abundance of life. One such characteristic would be size, because the larger the universe, the more likely would be life. So our universe is likely one of the largest universes--one that produces the largest inflation consistent with life support.

The Cosmological Constant: And to support such life, the cosmological constant should lie in a narrow range, one which allows life and high inflation, which he calculates would have to be near zero, somewhere between .9 and zero. Physicists are getting closer to predicting the value of the cosmological constant and verifying Vilenkin's mediocrity theory prediction and, Freedman notes, perhaps the mediocracy principle will play a role in verifying other difficult-to-verify predictions.

There may be only one 'self-consistent' set of unified theories allowing structures like human beings who can investigate the laws of the universe and ask about the nature of God.--Stephen Hawking

CHAPTER 19
THE PRIMORDIAL VOID AND QUANTUM BEGINNINGS

The Universe As Triggered By a Quantum Fluctuation: Many seek a better understanding of the pre-existing environment from which the Big Bang occurred. Most reviews of the scientific preconditions for creation generally start with a primordial, pre-existing quantum uncertainty state. That seething *uncertainty* state generates a *quantum fluctuation* in the primordial quantum vacuum of an eternal chaotic mega-universe without space, time or matter, but latent with the laws of physics.

The quantum fluctuation generates an incredibly small mini-universe point of spacetime impressed with a superforce of unified physics which, because it is not *in* space, exponentially inflates at a speed greater than the velocity of light, and then subsides into the Big Bang linear expansion sequence still in progress today. It is this incredibly small mini-universe point, *exponentially smaller than a proton*, from which, in quantum-genesis theories, our present universe evolved through the 'unfolding process' triggered by the Big Bang.

A Point Containing All Space, Time, and Matter: In *Steven Hawking's Universe*, John Boslough describes the process this way: Into a void, empty beyond human conception, he notes, came a "point bearing all matter, all dimension, all energy and all time...spewing forth its contents...[in which] all matter..and forces were indistinguishable [but]...split apart...as the universe expanded and cooled..." to form our present universe.

Spontaneously Erupting from Nothing: In his *Search for a Super-theory--From Atoms to Superstrings*, Idaho State University physicist astronomer Barry Parker speaks of the ultimate question, "Where did the universe come from and how do we get something from nothing?"

So, he says, we have the universe spontaneously erupting from "nothing," into an unstable false vacuum state, which decays, leading to the inflation which provided the energy to "continue its expansion and to create particles." The continued big-bang-expansion, produces, in sequential cooling states, today's universe of still-expanding space, and multiple energy forms.

Picturing The Pre-Big-Bang Void: Before we leave this question of the role of God versus quantum uncertainty as the source of creation some further discussion of the pre-big-bang primordial quantum void 'nothingness' is warranted. We have discussed several descriptions of what existed 'before' the Big Bang. The various versions boil down to a 'superspace' or mega-universe state of quantum nothingness, without space, time, matter or dimensionality but fraught with physical potentiality.

But what processes went on in the void? How did the quantum void, a chaotic state of quantum randomness, deliver such a rich and unusual, neatly wound-up combination of unified components capable of unfolding into our incredible life-sustaining universe? In *Creation Revisited,* Oxford chemistry professor P. W. Atkins engagingly undertakes to describe the process.

Thinking Rationally about the Processes of Creation: Atkins expresses the view that it is possible to think rationally about what some see as "beyond explanation," such as the processes involved in the creation of the universe. He argues that there is no need to invoke the notion of a Supreme Being in any of its "numerous manifestations." His concept is that abstract *mathematics alone* can create the universe, *without outside intervention.*

He wishes to make clear at the outset that he does not intend offense, either to devout religious believers on the one hand, or on the other, to scientists who might consider his views an "incautious extension of scientific thought burst through the membrane of propriety."

In the Beginning, Nothing, Not Even Space or Time: In his last chapter, "Things Created," be states that in the beginning, there was the primordial "absolute void," connoting a state without space or time, but characterized by the pregeomtery of "unstructured points."

Chance fluctuations in this void produced "dimensional patterns" of space, or time, or even space and time--momentary minute universes, too weak in "stabilizing interrelationships" to persist, and so they subsided back into the void.

The chance formation of a pattern of points resulted in the emergence from nothing, and "without intervention," of time, from what he calls "coalesced opposites." "The emergence of the dust of points and their chance organization into time was, he says, the haphazard, unmotivated action that brought them into being.

Universes Not Rich Enough for Persistence: But, says Atkins, these lines of time, "incipient universes," evaporated, not rich enough for "persistence." Patterns of time or space, or even time and space, emerged repeatedly, points "inducing their own existence," only to disintegrate--victim of their instability. These defective universes, inadequate for survival, remind us of Scottish philosopher David Hume's description of the "botched and bungled" universes struck by God before our own imperfect but life-sustaining universe.

Our Fluctuation: But on one occasion, notes Atkins, came *our fluctuation*, a pattern of points configuring three dimensions of space and one of time, an amalgamation of geometric complexity adequate to give it sufficient survival features to generate matter, energy, forces, and, later, "elements" and, still later, "elephants" (Atkin's metaphor for life).

Energy as Coiled Spacetime--the Universe an Unwinding Spring: The initial spacetime was left "coiled and twisted," and the "local twists into persistent knots" are the particles that make up the physical objects in the universe. Atkins says that different knots (twists) represent different collections of the binary entities coming into existence at creation.

Different particles, he suggests, represent the different "topological structures of spacetime." Embedding these local structures in spacetime induces the phenomenon of gravitation, representing the *global twist of spacetime.* Energy [& matter] he describes as *coiled spacetime*, and all activity, he suggests, represents "aspects of its unwinding."

The Benevolent Job Lot of Strengths of Forces: Presently, the universe is alive, Atkins notes, courtesy of the balance of strengths and forces. The emergence of just the *right blend* of forces may give the impression of a miracle, and suggest "intervention," he notes. But, he concludes, while we cannot see far enough to decide on the correct explanation, we can be reassured that intervention was not required. It is chance that may have created a "benevolent job lot of strengths of forces." If they had not, he says, we would have been none the wiser-- or sadder.

Chance Stumbles on Fortune: Chance might have "stumbled on fortune" in due course, because, he says, a universe "looping through existence stages might have rebounded with a different strengths of forces, millions of times since the "true creation," until the "recycled structure of spacetime" supported a "web of forces" that just happened to be suitable for consciousness to emerge.

The Universe Will Sleep Again: But even if the universe is a one-shot affair, he muses, a single *coiling of the spring* and "one irreversible drift into uncoiled uniformity and global flatness," such a universe still has no purpose behind the happenstance "benevolence" of the forces.

We are, he says, chance beneficiaries. The universe has "awakened in this benev-olent epoch," and consciousness emerged. Why? Simply because it happened, he declares, and when the epoch passes, the universe will sleep again. We and the universe are awake, he observes, necessarily "amid benevolence."

Benevolent Constants Do Not Imply Purpose: It is possible, Atkins says, that the generation of spacetime from the void must lead to our present force strengths because they are aspects of the structure of spacetime. But, he insists, that this does not imply a purpose. We can still be the "children of aimless chance" and "wonder at their values" no more than we wonder at the speed of light, or of π.

Complete Knowledge Just within Our Grasp: When we have seen that the fundamental constant values are "unavoidably so and...irrelevant," he notes, we will have achieved a complete understanding. Fundamental science, he declares, "can then rest." Complete knowledge, he ventures, "is just within our grasp [and] comprehension is moving across the face of the Earth."

Dimensionality and Gravitation: Atkins cites the need for three space dimensions. Without three spatial dimensions, he notes, there could not exist our life-supporting gravitational physics.

There could be no matter, he says, and space would have random curvature. Gravitation, he says, reflects the three dimensions of space and one of time, in which space and matter can exist. Gravitation propagates its effect locally and at a distance and functions compatibly with electromagnetism and the other forces in a "coordinated entity."

Time, the Causal Universe: So also with time, says Atkins. More than one dimension of time would permit shifting in time as in space, eliminating our causal universe in which the consequences of present action lie in the future. And spacetime is stabilized against collapse by the conservation of energy, he says.

But, he notes, with another time dimension, this limitation crumbles and with it crumbles matter. Add another time dimension, and the universe “would be over in a flash....[It] could not survive its own creation....Without time in our sense, there is no beginning."

> The emergence of just the *right blend* of forces [to generate our life-supporting universe] may give the impression of a miracle, and suggest outside intervention. But while we cannot see far enough to decide on the correct explanation, we can be reassured that intervention was not required. It is chance that may have created a benevolent job lot of strengths of forces. If they had not, he says, we would have been none the wiser or sadder. --P.W. Atkins

PART VII--THE UNIVERSE AS MATHEMATICS

CHAPTER 20
MATHEMATICS AND MEANING

God and Mathematics: Some have pointed out that the universe is essentially a mathematical construct; that there is a remarkable relationship between God, mathematics and human life; that everything is ultimately mathematics, and that God is ultimately mathematics, or a mathematician. Physicist Eugene Wigner explored the relationships of mathematical abstractions to physical phenomena writing on "The unreasonable effectiveness of mathematics in the natural sciences."

Grasping the Nature of the Universe through Mathematics: Does pure human thought alone, without physical experiment and observation permit us to construct an accurate description of reality? In his 1933 Herbert Spencer Lecture Einstein expressed his view that pure thought alone, "free invention" of the mind, is capable of grasping reality as reflected in the laws of physics.

Greatest Feat of Human Thinking: The distinguished physicist Edward Teller had declared that Einstein's general relativity represented "The single greatest human creative thought achievement in the history of science," and physicist Max Born had described Relativity as "the greatest feat of human thinking." Nigal Calder said that "Between 1904 and 1917, or between his twenty-fifth birthday and his thirty-eighth, [Einstein] had rebuilt the universe in his head."

But what is it that permits the discovery of such physical concepts without test verification? It is, said Einstein, *mathematical construction,* the "correct path" within "our power to find." Einstein is credited with having figuratively "rebuilt the universe in his head" through mathematics and related "thought experiments" of his relativity theories, with their profound unification implications.

Mathematics and Physical Reality: Earlier, Einstein noted that scientists have long observed how mathematics, the pure construct of human thought and essentially unrelated to actual human experience, "fits so excellently the objects of physical reality." As we have seen, this led him to go on to conclude that pure human thought can construct the true nature of our reality, "the properties of real things."

The example frequently given is how the physicist Subrahayam Chandrasekhar pursued mathematical abstractions. He searched for aesthetical physical concepts--the mathematically beautiful, discovering the ellipsoid, an apparently useless purely mathematical geometric creation. It was years later that astrophysics and astronomy resorted to these concepts to explain galactic gravity dynamics.

Mathematical Beauty, Guage Fields And Riemann Geometry: Similarly, guage fields, now so important in describing fundamental forces in particle physics, attracted physicists Chen Ning Yang and Robert Mills because of their peculiar mathematical aesthetics.

Einstein was, by his own acknowledgment, heavily indebted for his relativity spacetime geometry to Bernhard Riemann for the latter's work in exploring theories in mathematical physics--non-euclidian geometry, and to Carl Gauss for his differential geometry.

Einstein Envisions Unification: Since the mathematics of relativity theory showed correlations between space, time, electromagnetism and gravity, Einstein had conjured up visions of mathematics as underlying the total unification of all forces and fields. After relativity theory, Einstein spent much of the rest of his life in the vain pursuit of the mathematical formulation of unification of electromagnetism and gravity. His progress in unification theory was blocked, of course, by the then unknowns of quantum mechanics, particularly the lack of information on the strong and weak nuclear forces.

But Pursuit of Unification Theory Was Premature: As it turned out, you must mathematically solve unification of electromagnetism and the weak force before you could wrap in the strong force and then tackle gravity-- for which you needed a quantum theory of gravity. And you also had to solve the mathematics of the interchangeability of force and matter particles. You had to construct the mathematics of unifying space, time, matter and dimensionality, and how matter particles may represent 'compacted' dimensions.

The Universe, Mathematics, Mind and Human Life: Physicists as well as biologists have noted the strange connectedness of mind, mathematics, and the physical universe. The intuitively experienced beauty of abstract mathematics, physical reality, and the very processes of human thought seem strangely interwoven. There is growing consensus that human thought turns out to be an expression of the laws of physics and their mathematical underpinnings. The way we think and what we think are the very application of the laws of physics. God or the cosmic intelligence has somehow constructed both.

Mathematics As the Heart of Science: Physicist Paul Davies notes, in *God And The New Physics*, that a "neurotic fear of mathematics" is what turns most people off from science. Yet the criticality of mathematics in our universe has been noted down through the ages. Davies quotes Leonardo da Vinci's view that "no human investigation can be called real science if it cannot be demonstrated mathematically."

Impressed with the mathematical elegance of the laws of physics, physicists assert that *mathematics reveals a fundamental feature of existence*. "Mathematics is the poetry of logic," says Davies. "Logic is the only branch of mathematics that can think about itself," said physicist John Wheeler, and "God is a mathematician," asserted Sir James Jeans.

Physics As Mathematics: Physicist John Wheeler employs the Aristotelian syllogism to conclude that at the heart of physics is the mathematics of logic. He observes the premises that if all mathematics reduces to the mathematics of logic and all physics reduces to mathematics, then one must draw the conclusion that all physics reduces to the mathematics of logic.

Deducing the Nature of the Universe from Logical Inference: Along this line, Davies muses that much of nature, *the physical laws, might be deduced from logical inference rather than empirical evidence*, a view shared by other physicists like Arthur Eddington and, to an extent, Einstein. Citing the French scientist Jean d'Alembert, Davies asks whether the totality of existence, the nature of the universe, could be *the inevitable consequence of logical necessity.*

A Single Logically Consistent Universe--God's Only Choice? Western religion asserts this particular universe as God's *choice* from an infinity of alternatives. Einstein wondered whether God had any other alternatives of physical laws. Davies observes that the fact, if such it is, of the nature of our universe as the "inevitable consequence of logical necessity," touches curiously on the issue of God's *omnipotence*. He notes that *even God cannot break the rules of logic* and if there is only a single 'logically consistent' universe, then that's it, God has no alternatives.

Cited in *Einstein: A Centenary Volume*, interestingly, Einstein asked this very question, " ...whether God could have made the world in a different way; that is, whether the necessity of logical simplicity leaves any free will at all."

A Single Mathematical Principle: Under the relativity models of the Big Bang, fields or particles, on the one hand, and the fundamental forces (electromagnetism, the strong and weak nuclear forces, and gravity), on the other hand, were considered distinct phenomena. But newer quantum physics unification theories suggested, as we've noted, a time when the universe was in a unified state of 'supersymmetry,' for short, SUSY, in which the four fundamental forces and the energy-matter fields were unified and operating under what Davies calls a "single simple, breathtakingly elegant mathematical principle."

Mathematics As the Language of Nature: As we have seen, Davies speaks of *the resonance of mathematics with reality*. Why does mathematics work? asks Oxford Professor P. W. Atkins in *Creation Revisited*. Why is it that mathematics describe features of physical reality? That Newton's equations predict certain particle behavior? That Einstein's field equations have observable physical counterparts? That quantum theory describes subatomic phenomena, despite its mysteries?

Why Mathematics Works--An Insight into Cognition: If we can understand why that "supreme construct of the human intellect, that arch-disembodiment of intellect, mathematics, works as a description of the world," says Atkins, "then maybe we shall have an insight into cognition."

Perhaps it is not too much to suggest, Atkins observes, that if we answer the question of why mathematics works, we will have answered *all questions of being*. But, he is reminded that experimental science is needed to guide the way through the world's "labyrinth of complexity."

Unpacking the Complexity of the Universe: Atkins points out that our current level of study of the substructure of the world has brought us to the "bedrock of being;" that we are *unpacking the complexity of the universe,* revealing what may be the underlying structure of the universe and revealing the relationship of that structure to mathematics.

Does everything mathematical have its counterpart in the real world, Atkins asks. If the physical world is mathematics, then should not every mathematical structure exist in it? He wonders whether the correlation of *all mathematics* with the physical world is waiting to be discovered.

The 'Hypothesis of Deep Structuralism': In exploring why mathematics works, Atkins resorts to employing a hypothesis as a "framework" within which he can deposit elements to develop his model of mathematics as the underlying foundation for the physical universe. The hypothesis will, he declares, generate speculation, carry the indicia of authority, and bear a name, *the hypothesis of deep strucuralism.* This term is derived, he remarks, from *Chomskyan linguistics,* suggesting that the brain has a built in "hardware-like capacity for language," and speech is a sort of "surface structure" generated from an "innate structure."

Physical Reality *Is* Mathematics--and Mathematics *Is* Physical Reality: The purpose of the hypothesis, he states, is better to convey the idea that the physical world has the same logical structure as mathematics. By implication, he says, the reason why mathematics works as a description of physical reality is that both either *share* or actually *are* the same logical structure. And here he distinguishes between a *strong* and a *weak* version of the hypothesis.

By *weak deep structuralism,* he means that "mathematics and physical reality mere-ly *share* the same logical structure." In *strong deep structuralism,* he says, mathematics and physical reality do not merely *share* the same logical structure but *are* actually the same. Physical reality *is* mathematics and mathematics *is* physical reality.

Resonance between Mathematics and the Physical Universe: The speculations in both versions of the hypothesis, he notes, account crudely for the ability of mathematics to describe the physical world--because there is in some sense a *resonance* between the structure of mathematics and what will turn out to be the fundamental structure of the physical world.

The reason why we may be conscious of the world, including the inner, introspective world of emotion and intellect, he observes, may be that our brains are physical representations of the same fundamental structure-- and the reason why brains can conjure the mathematics needed to understand the world.

So, in a philosophic aside, Atkins observes, there may be an aspect of truth in the concept of the *logos* (reference to Jesus at the start of the gospel of St. John), but "in place of shared divinity there may be shared structure."

The Mathematical Description of the World: The mathematical description of the world, Atkins notes, depends on assembling numbers in a certain way. Thus the formulas of theoretical physics are "generalized statements about quantities," illustrated by Newton's equation for motion, Force = mass x acceleration--which he describes as "a class of behavior."

A formula in theoretical physics, he notes, is a universalized summary of relations between measurements, and numbers are "generalizations of lists." So the number '2' is a 'generalization' of any two objects. Numbers, he notes, arise from sets, and mathematics is an application of the logic of sets.

So, in addition to summarizing the essence of sets, numbers are also "the universal currency of calculation and exposition." Physics began, he says, when people started expressing observations numerically and searching for relationships between numbers from different classes of observation, such as between force, mass, and acceleration.

The Manipulation of Numbers As Statements about Logic: Manipulations of numbers in physics, he notes, are actually "statements about logic," and the writing of 6 = 3 x 2, which Atkins describes as a particular case of Newton's equation in a "certain choice of units," constitutes a *condensed version of a statement in general logic*. So a relationship between observations and logic exists, and he proceeds to describe the connecting chain.

Aspects of the universe, he says, are described by mathematical formulas; formulas are in turn universalized statements about relationships between quantities numerically expressed; hence, formulas are statements about the kinship between numbers; and statements about numbers are really statements in general logic.

Thus, Atkins observes, we discern the hint that the formulas of physics are *expressions about some underlying logical structure of the universe.* It is this set of interrelationships which, he notes, represents the substance of deep structuralism.

Hints Confirming Strong Deep Structuralism: He refers to existing hints which suggest the efficacy of the strong version of deep structuralism, that the physical world *is* mathematics-- not merely the two as sharing the same logical structure.

The biggest hint, he says, is that "we seem to exist" and therefore the mathematical description of the origin of the world must "accomodate the emergence of *the apparently something from the certainly nothing*. This happens to be one expression of the 'anthropic principle,' even though Atkins disavows the principle.

'Something' As Elegantly Reorganized 'Nothing'--Pure Numbers: We have seen, Atkins notes, that the seemingly something is actually "elegantly reorganized nothing;" that the net content of the universe is now the same as it has always been, ...namely, nothing. So, he asks, what mathematics can "emulate the emergence of something from nothing?" Why, he answers, nothing but *pure numbers*!

Atkins Illustrates: To see numbers emerge from the empty set Ø, the set with absolutely nothing, we name it '0' and denote it as **0 = Ø**. This set that contains the empty set is named '1' written as 1 = {Ø}. By number 2 we mean the set that contains both the empty set and the set that contains the empty set or 2 = {Ø, {Ø}}.

By "3" Atkins notes that the word description gets difficult but 3 is the name of {Ø, {Ø}, {Ø, {Ø}}, and so forth through 4, 5, and continuing indefinitely. Thus numbers, he observes, are the *currency of physics*.

The Universe Arguably Nothing at All: So, says Atkins, we see that in a universe that is arguably nothing at all, "numbers are still essentially collections of the empty set...assemblages of absolutely nothing...the illusion of something." In an ego-diminishing conclusion, he remarks, that "the origin of mathematics is a mirror... of the origin of the cosmos." *We, like mathematics...are elegant, self-consistent, reorganizations of nothing!*

If the physical world is indeed mathematics, Atkins asks, should not every mathematical structure exist in it? And he asks again: Is all mathematics waiting to be discovered in the physical world?

Brains As Portrayals of Underlying Logical Structure: Atkins goes on to explore the question, leading to the matter of brains and their functions. We remarked earlier that in the speculative framework of his "deep structuralism," brains may be viewed as physical, that is, chemical or biological, depictions of an underlying logical structure--physical realizations of logic, consciousness being a "manifestation of the resonance of the deep structure of a brain with the deep structure of the universe."

Atkins notes that by virtue of their nonlinearity, brains have gone on to transcend evolutionary requirements for simple survival and pursued the processes of maturation of neural networks by selective stabilization of certain synapses and the elimination of others, as occurs following infant acquiring of certain language skills.

The brain has evolved as a highly redundant network, Atkins notes, as a "generalized, semi-soft computer, that can have aspects of its environment impressed on an exuberantly-sprouted and only partially genetically-predetermined network."

Brains As 'Material Realizations of Logic': Atkins looks at some of the implications of deep structuralism. He notes that the fact that we can be conscious of the world is, at its deepest level, a manifestation of *the resonance of the deep structure of a brain with the deep structure of the universe.*

Brains can generate mathematics because, in a sense, they are "chemical portrayals of an underlying logical structure: brains are *material* realizations of logic." The deep structure of the brain may be in resonance with the deep structure not only of mathematics but also of the physical universe.

Until recently, the major thrust of mathematical physics, notes Atkins, has been the "formulation of differential equations and the discovery of their analytical solutions." The wonder of physics , he declares, is the power with which an analytical expression summarizes a "great swatch of observation and prediction."

Profound Complexity as Starting from Simple Equations: The new paradigm of emerging explanation is based on the view that even simple equations might have consequences of profound complexity. Although we may know the seed of a pattern and can specify the very simple rules of transformation of the seed, he observes, we *cannot predict the outcome of the repeated application of those rules of the initial seed.*

Chaos Theory and Mathematical Descriptions of Consciousness: So, he says, extraordinary complexity can evolve from this underlying simplicity," as shown by examining the "mind-stirring images" conjured from "fractals, the Mandelbrot set, and cellular automata." Thus, he says, mathematical physics and aesthetics have blended, and "science is ready to become whole." Not only has the chaos theory provided a new "paradigm of understanding," it may be just what is needed to describe the *mathematics of consciousness.*

Binary Computers May Mirror the Deep Structure of the Universe: The fact that we need a computer to explore the ramifications of complexity, and that a binary computer is the simplist type of computer, may point to a deep analogy between Boolean logic and the workings of our brains. Atkins says that *binary computers may precisely mirror* the deep structure of the universe.

Space and Time As the Central Event of the Creation: So Atkins has developed his theme of how a natural process of events, explainable within a framework of underlying mathematics starting from nothing, can account for the present universe. The necessity for a creator will be "seen to fade." And as the creator disappears from any involvement, "the universe comes into being without external intervention, out of absolutely nothing." But the emergence of space and time, he says, is "the central event of the creation."

Mathematics--The Ultimate Reality : In *The Mind of God*, Davies speaks of mathematics as the "language of nature," as contrasted with the ordinary person's view of mathematics as an esoteric mystery, and, he observes, physicists "invest the ultimate reality of the physical world in mathematics."

But the fact that mathematics is the key to unlocking cosmic secrets, he notes, goes back to ancient times, citing Greek preoccupation with geometry and numerology, Pythagoras' theorem, and Pythagorean philosophy. He cites Pythagoras' belief that "Number is the measure of all things" and notes the role of numbers in music. The octave corresponds to the ratio 2:1 and "rational" numbers involve the ratio of whole numbers, like 3/4.

Ancient Preoccupations with Numbers: Davies cites Greek consternation with the inability to express the square root of 2 as the ratio of whole numbers, no matter how big the numerator and denominator, such numbers being called "irrational." Plato in his *Timaeus*, goes on to develop a "musical and numerical" model of the cosmos, treating the physical world as "a manifestation of concordant mathematical relationships."

The idea carried over to the scientific era when Kepler described God as a geometer, and even today when physicists, like Sir James Jeans, refer to God as a mathematician, or as mathematics. The old testament, he points out, places special significance on the number 40, the devil is associated with number 666, and William Blake's *The Ancient of Days* shows God as measuring the universe.

Mechanizing Mathematics: Today, notes Davies, there are suggestions that we regard nature as a computational process and, Newton's mechanical or clockwork universe as having been displaced by the concept of the universe as a cosmic information-processing system--a comuputer.

He cites the pioneering work of Charles Babbage, the eccentric British inventor who pioneered the modern concept of computers, constructing a "difference engine," a type of adding machine, then a general-purpose "analytical engine," a forerunner of the modern computer. Babbage's insight into the computation process, notes Davies, led to his speculation on the universe as a computer with energy as the hardware and the laws of physics governing behavior as the software.

Manipulating Symbols: Davies also cites the work of British Alan Turing and American mathematician John von Neumann, both credited with "laying the logical foundations for the modern computer." Back in 1900, mathematician David Hilbert set out the 23 most important outstanding math problems to be tackled.

One of these problems was whether a general procedure could be found for proving math theorems, that is, with a finite number of steps, proving a math statement true or false. Mathematics becomes a system for "manipulating symbols" according to "certain specified rules and establishing tautological relationships," without any required relevance to physical reality.

Gödel's Theorem: But this concept was questioned in 1931 when Austrian mathematician and logician Kurt Gödel proved a sweeping theorem to the effect that there existed mathematical statements the truth of which could not be determined by any systematic procedure. It provided indisputable proof that something in mathematics is actually impossible--that there are "undecidable propositions."

Gödel's theorem, says Davies, springs from an admixture of "self-reference paradoxes" that have confused for centuries. He cites the following: "This statement is a lie." If true, then the statement is false; and if false, then it is true. By identifying the subject with the object, says Davies, mapping the description of the mathematics onto the mathematics, he uncovered a "Russelian paradoxical loop that led directly to the inevitability of undecidable propositions."

Why Does Arithmetic Work? Machines do what they do because the laws of physics permit, says Davies. He quotes Oxford mathematician David Deutsch that computability is an empirical property, depending on "the way the world happens to be" rather than on necessary logical truth. So the laws of physics just happen to permit the existence of physical models for the operation of arithmetic, such as addition and subtraction, and if they did not, these operations would not be computable.

A Deeper Resonance between Mathematics and Reality? Davies cites mathematician R.W. Hamming as finding it "strange and inexplicable that the abstraction of integers for counting is both possible and useful." Is it not a miracle, he asks, that the universe is so constructed that such a simple abstraction as number is possible? This fact--that the universe reflects the computational properties of arithmetic-- has a profound implication, observes Davies. *It means that the physical world is a computer; that computers can not only simulate each other, but simulate the physical world.*

Apparently, he suggests, there is a *crucial concordance* between the laws of physics and the computability of the mathematical functions that describe those laws. Is this "loopy self-consistency just a coincidence?" he asks, or is such consistency somehow required by some "deeper resonance between mathematics and reality?"

The Universe As Numbers: "Can a String of Numbers Tie the Universe Together?" asks George Johnson in New York Times *Ideas & Trends* (Feb 5, 1995). Physicist Eugene Wigner is most quoted for his fascination for "the unreasonable effectiveness of mathematics" to explain the nature of the universe.

And, as we have seen, Einstein believed that through the mental gymnastics of mathematics a person can unravel the mysteries of the universe. As Johnson points out, the existence of many of the particles populating the universe have been discovered, not in the laboratory but "imagined into existence, conjured from thin air" through mathematics.

Paul Dirac and Enrico Fermi Conjured up Particles Mathematically: Johnson cites Paul Dirac's discovery of the positively-charged electron, the positron, as a solution of one of his equations, later confirmed in the laboratory. So also, Enrico Fermi, borrowing from Wolfgang Pauli, conjured up the existence of the neutrino.

According to the equations, the neutrino (little neutron) *had to exist* to represent the unaccounted for energy left upon decay of the neutron into a proton and an electron, requiring the neutrino to balance the energy equations. More than two decades later the neutrino was found, by Frederick Reines and Clyde Cowan.

Neutrinos Discovered to Figure in the Mathematics of Universal Mass: Now the neutrino is part of the recent standard model of the three families of particles, each including two types of quark, one type of electron and one type of neutrino. Neutrinos, considered virtually massless, countless numbers of them constantly passing through the earth--and our brains--were recently discoverd by the Los Alamos National Laboratories as having minute mass, and therefore accounting for some of the universe's 'missing mass,' the mass required for eventual gravitational contraction.

Mathematics, Logic, and the Creator: Physicists associate physics and logic, via mathematics. Physicist John Wheeler employs the Aristotelian syllogism to conclude that *all physics reduces to the mathematics of logic;* that "Logic is the only branch of mathematics that can think about itself." Physicist Paul Davies observes that "Mathematics is the poetry of logic," something inviolate-- even by God. Some are led to conclude that God *is* Logic--applied through mathematics.

But What Breathes Fire into the Equations? In varying degrees, at least, physicists like Stephen Hawking suggest that the laws of physics and their underlying mathematics may be "so compelling" as to self-generate the creation--the ultimate first cause?

In *A Brief History of Time*, Stephen Hawking wonders, though, what it is that actually "breathes fire" into the equations to generate a physical universe from mathematical abstractions--which, of course, leaves room for God's intervention as the triggering mechanism. He goes on, though, to ask whether physics is *so compelling* that it is sufficient by itself to generate the universe.

Atkins Goes All the Way--The Universe Is Mathematics! But in *Creation Revisited*, British scientist P.W. Atkins is willing to dispense with intermediaries and assert that creation is self-initiated *without outside intervention,* because underlying everything, including the primordial 'nothing,' is *mathematics, a universe self-generating mechanism.* In Atkins view, nothing else is needed to "breathe fire into the equations," in the context of Hawking's terms. So Atkins' view is that *mathematics does it all!*

[Discussing "deep structuralism"--that the universe *is* mathematics] Brains may be viewed as material, that is, chemical or biological, portrayals of deep underlying logical structure--material realizations of logic, consciousness being a manifestation of the resonance of the deep structure of a brain with the deep structure of the universe. (Paraphrased)--P. W. Atkins

CHAPTER 21
MIND, MATHEMATICS, AND THE UNIVERSE

Resonance Between Human Mind and the Structure of the Universe: Through-out the *The Mind of God*, Davies reverts to a "tantalyzing mystery" which has attracted increasing scientific and philosophic interest: Why human beings should have the ability, through science and mathematics, to understand how the universe works.

Is it just a "lucky fluke," he asks, or is it something deeper? Does the spectacular progress of our science point to a "deep and meaningful resonance" between human mind and the universe's underlying organization?

The Mind of God and Human Mind--Matter and Mathematics: Davies, as we have seen, expresses the idea that the cognitive capabilities of the human mind somehow reflect a 'resonance' with the structure of the universe which functions through the medium of energy. We find various references to human mind as an extension of cosmic mind or the mind of God, references to the absence of distinction between mind and matter, the reconcilability of scientific materialism and religious transcendentalism, and references to the overlap of God and mathematics as source of creation.

In his delightful *Infinite in All Directions,* physicist Freeman Dyson, observes that "Between matter as we observe it in the laboratory, and mind as we observe it in our own consciousness, there seems to be only a difference in degree but not in kind."

Dyson declares, "If God exists and is accessible to us, then his mind and ours may likewise differ from each other only in degree and not in kind.... Our minds may receive inputs equally from matter and from God....Therefore, I say, speaking as a physicist, scientific materialism and religious transcendentalism are neither incompatible nor mutually exclusive." *We have learned that matter [energy] is weird stuff. It is weird enough, so that it does not limit God's freedom to make it do what he pleases.*

Mathematics--The Language of Nature: Similarly, we note many references to the relationship between nature, mathematics, and the human mind. Davies speaks of mathematics as the "language of nature," as contrasted with the ordinary person's view of mathematics as an esoteric mystery, and, in a provocative reflection, *physicists invest the ultimate reality of the physical world in mathematics.*

But the fact that mathematics is the key to unlocking cosmic secrets, he notes, goes back to earlier times. Astronomer Kepler described God as a geometer, and today physicists, like Sir James Jeans, refer to God as a mathematician or as mathematics.

The Universe As a Computational Processing System: Today, notes Davies, there are suggestions that we regard nature as a *computational process* and, Newton's mechanical or clockwork universe as having been displaced by the concept of the universe as a cosmic information-processing system--a computer. He suggests, there is a "crucial concordance" between the laws of physics and the computability of the mathematical functions that describe those laws. Is this somehow required by some "deeper resonance between mathematics and reality?"

Mathematics Describing Reality? Why does mathematics work? asks Oxford Professor P. W. Atkins in *Creation Revisited.* As we noted earlier, Atkins observes that If we can understand why that "supreme construct of the human intellect, that arch-disembodiment of intellect, mathematics, works as a description of the world," then perhaps we will have an insight into cognition. Answering that question will have answered *all questions of being.*

Does everything mathematical have its counterpart in the real world? he asks. If the physical world is mathematics, then should not every mathematical structure exist in it? And, he asks, "Is all mathematics waiting to be discovered in the physical world?"

Physical Reality and Mathematics As the Same Logical Structure: The physical world, says Atkins, has the same logical structure as mathematics. By implication, he says, the reason why mathematics works as a description of physical reality is that both either *share* or actually *are* the same logical structure.

The reason why we may be conscious of the world, including the inner, introspective world of emotion and intellect, he observes, may be that our brains are "physical depictions of the same deep structure and the reason why the human brain can grasp the mathematics needed to understand the universe. Thus, Atkins observes, we discern the hint that the formulas of physics are *expressions about some underlying logical structure of the universe.* So, in a philosophic aside, Atkins observes, there may be some truth in the concept of the *logos* (reference to Jesus at the start of the gospel of St. John), but in place of shared *divinity* there may be shared *structure*.

Brains As Portrayals of Underlying Logical Structure": Atkins goes on to explore the matter of brains and their functions. We remarked earlier that brains may be viewed as chemical or biological renderings of an underlying logical structure--material realizations of logic, consciousness being a "manifestation of the resonance of the deep structure of a brain with the deep structure of the universe."

Mind and Matter: In *Infinite in all Directions*, Freeman Dyson then crosses over into theology and metaphysics, giving expression to the vision of future mind over matter. *Mind*, he says, "has infiltrated and taken control of matter on planet earth, fulfilling a law of nature that requires this and makes the process unstoppable." And, he adds, "if we falter, others will--or may have already done so." He observes that mind has taken three eons to control this planet and three more billion years may pass before it covers the galaxy.

Will Life and Intelligence Survive Forever? If, as Dyson hypothesizes, the universe is open and expanding forever, is it possible, he asks, for life and intelligence to survive forever and transmit information across the constantly expanding galactic distances? Life and intelligence, he feels are potentially immortal, with available information growing as the universe's temperature declines and free energy reserves diminish. He thinks that intelligence forms throughout the universe can forever preserve a communication network, exchanging ideas.

Life and Intelligence Molding the Universe to Its Own Purposes: The energy requirments for survival and communication, he calculates, are "surprisingly modest." New things will be happening, he says, and no matter how far we look into the future, there will be new information, new worlds to explore, a constantly expanding domain of life, consciousness, and memory.

He sees a universe of life, "growing without limit in richness and complexity, surviving forever and making itself known to its neighbors across inimaginable gulfs of space and time." There is a substantial scientific basis, he notes, for hypothesizing that "life and intelligence can succeed in molding this universe of ours to their own purposes."

Mind Controlling Matter As a Law of Nature: Through the long course of biological evolution, he observes, mind has established itself as a moving force in our little corner of the universe. Here on this small planet mind has infiltrated and taken control of matter. It appears to him that the tendency of mind to infiltrate and control matter is a law of nature.

If our species is extinguished, he remarks, others will be wiser or luckier. Mind is patient, he notes, having waited 3 billion years before starting its spread over the galaxy.

The Point at Which Science Ends and Theology Begins: What, he asks, will mind choose to do when it controls the universe? We cannot hope to answer he says. When mind has expanded its physical reach and its biological organizations many "powers of ten beyond the human scale," he muses wistfully, "we can no more expect to understand its thoughts and dreams than a Monarch butterfly can understand ours."

Mind can answer our questions only as God enigmatically answered Job out of the whirlwind. In contemplating the future of mind in the universe, he says, we have exhausted our puny human science. This, he declares, is *the point at which science ends and theology begins*.

Dyson's God As the Evolving God of the Socinian Heresy: He cites a meeting with theologian Charles Hartshorne. Dyson had explained his views about the evolving God and Hartshorne told Dyson that his theological outlook was that of the 'Socinian heresy.' Socinus was an Italian heretic in the sixteenth century whose principle tenet was that God is neither omniscient nor omnipotent but "learns and grows as the universe unfolds."

Socinianism, a religious movement originally organized in Poland, was founded by Faustus Socinus, the Italian Catholic reformer, based on the teachings of his uncle Laelius Socinus. Not only did Socinus have trouble with 'perfection' aspects of God, given the world's manifest imperfections from human experience, but, like others, he had trouble with the doctrine of the Trinity in the face of the unity of God emphasized in the Bible.

The Evolving God Concept: Church intervention on the grounds of Socinianism as heresy effectively brought the Socinianism movement to a close. Interestingly, Socinianism, the evolving God, has, in a sense, taken on new vitality not only in the writings of scientists like Freeman Dyson but in the mathematics of Omega point Theory as spelled out in physicist Frank Tippler's *The Physics of Immortality,* supported by the mathematics display in his remarkable Appendix for Scientists.

Not Making Any Clear Distinction Between Mind and God: Dyson's view of the developing God is one that he feels is consistent with "scientific common sense, in which God is not omniscient and omnipotent but "learns and grows" as the universe unfolds. He merely finds it "congenial and consistent with scientific common sense." In his thinking, he reflects that he does not make any clear distinction between mind and God.

We Are the Chief Inlets of God on This Planet: "God, he says, is what mind becomes when it passes beyond the range of our understanding. God, he suggests, may be thought of as a world-soul or a collection of world souls. We are the "chief inlets of God on this planet at the present stage of his development" and, he says, we may "grow with him as he grows, or we may be left behind."

Dyson's Theology Leaves Room for Diversity: The virtue of his version of the Socianian theology, Dyson says, is that it leaves room at the top for diversity. Just as the greatness of the creation lies in its diversity, he observes, so also does the greatness of the creator.

When mind grows to fill the universe, he says, it comes as a "diversifier as well as a unifier." He also speaks of the possibility that the very laws of physics are still evolving with us, reminiscent of the anthropic principle. He refers to St. Paul's concern with 'unity' in First Corinthians, but says he cares more deeply for the diversity of God's workings than for God's unity.

Whispers of Immortality: Dyson speaks of the play, Spring 1600, by Emily Williams, which ends with the actors meditating in the dark of the London playhouse, after the Queen and the crowds have gone. One asks the other, do you hear anything? The other answers. "Yes, whispers of immortality."

Despite our science and our visions of the future, we know very little yet about the potentialities and the destiny of life in the universe. In speculating about these matters, he notes, we follow a great tradition of men like Bernal and Newton. And "Letting our imagination wander among the stars, we too may hear whispers of immortality."

A Revolution in Thought about the 'Physical' Universe: In his Introduction to *Parallel Universes*, physicist Fred Alan Wolf observes that today, more so than in the past, we face a revolution in thinking about the physical universe, engendered by discoveries in physics which question previous concepts of *concrete* reality, and suggest a more *abstract* reality which implies a need to unify our view of the world.

Other Universes Existing Alongside or Even Interpenetrating Our Own: The idea that the universe is not alone, that other universes exist alongside ours, notes Wolf, is a concept which has emerged relatively recently as part of the effort to bridge conceptual gaps in the effort to unify our knowledge, no small part of which is the driving direction of physics toward Einstein's earlier goal, finding the mathematics of unification of space, time, matter, dimensionality, and the fields and forces.

Relatively New Ideas about the Universe: This has led to new ideas about the universe such as: that space is not infinite; that space and time are joined-- a continuum, spacetime; that 'now' is not absolute but relative to observer movement; that gravity 'attraction' is geometry, the effect of mass to bend space and curve time; that matter is produced by the universe as *knots in the fabric of spacetime*.

Quantum Gravity Must Be Introduced: In tracing the universe back to an origin in the big bang, Wolf notes that it is now recognized that general relativity theory, postulating the starting point as one of infinite density, zero volume, and infinite temperature, must be augmented by a 'quantum gravity' theory. Required, is introducing quantum theory into the "beginning of space, time, and matter."

Quantum Uncertainty --A New Vision of Reality--Parallel Universes: Wolf points out that the development of quantum theory brought a new perspective about matter in which 'observation' plays a previously-unsuspected role of influencing energy states.

In addressing the problem of bringing relativity and quantum theory together, Wolf explores a "bizarre" theory in which the solution to the conundrums of quantum uncertainty are approached by resort to "a new vision of reality." This new reality involves the existence of other, parallel universes which are "within our technologically extended senses that must connect or relate with our own."

What Is a Parallel Universe? So what is a 'parallel universe'? It is a region of space and time which exists alongside ours, perhaps even "occupying our same space," he says, with galaxies, stars, planets, and living beings similar and *possibly even a duplicate* of our own universe. In a parallel universe not only must there be "other human beings," he notes, but they may even be *exact* duplicates, connected to us by "mechanisms" explainable only by quantum physics. Included in his materials, he says, are ideas on quantum physics, unification, relativity theory, cosmology, time, psychology, and the effects of the human mind on all of this.

Bringing in the Observer: Quantum physics, he notes, required that the properties of matter depended "on how it was observed," so that observation plays a quite unexpected role, affecting even "macroscopic matter in subtle ways" so as to "change cosmology and...our concept of just what a univere is."

So quantum physics introduces the effect of an observer upon a physical system. This effect, Wolf feels, cannot be properly understood" without the existence of parallel universes.

Psychology: Consciousness & Machine Intelligence--Quantum Computer: Psychology deals with human consciousness, human behavior and the nature of observation and he feels that the parallel universe hypothesis enriches psychology, suggesting that it may help the understanding of major disorders such as multiple personality and schizophrenia.

He notes also that psychology deals with machine intelligence and explains that his book explores the possibility that parallel universe theory will generate a new kind of quantum computer that "could not exist if parallel universes were not real."

Taking Data from the Future As Well As the Past: This computer will exhibit intelligence that present computers only anticipate, they will be able to make the kinds of decisions that we ourselves can make. These decisions, he says, would be *based on data not only from the past but also from the future*. It is his view that parallel universe theory takes into account the coexistence of all time, and shows that the future can influence the present just as much as does the past.

Can Mathematics Describe Reality? Can mathematics describe reality? He cites Einstein's affirmative response. He agrees, provided that we factor into account the new view provided by parallel universe theory. The laboratory of parallel universe experimentation, he suggests, may not lie in a mechanical time machine, but could exist "between our ears," the human mind.

If Parallel Universes of Relativity and of Quantum Theory are the Same: If the parallel universes of relativity are the same as those of quantum theory, the possibility exists, Wolf feels, that parallel universes may be "extremely close to us, perhaps only atomic dimensions away, perhaps in a higher dimension of space," an extension into what physicists call 'superspace.'

What Is Reality? What Is Existence? asks Wolf. From his "quantum viewpoint," says Wolf, the universe, any universe, *all the parallel universes are 'real' but not all 'exist.'*

He explains that he wishes to separate 'reality' from 'existence' and describe how parallel universes fit into the picture.

Reality--The Mathematical Superpace of All Possibilities--The Mind of God: First, what do we mean by reality? He cites the American Heritage Dictionary definitions which, in effect, includes everything that is "actual or true," "exists objectively and in fact," and is "real, absolute and unchangeable." But for Wolf, reality consists of a "gigantic superspace--the mathemtical space of all possibilities." We might think of this, says Wolf, as "the mind of God."

Mind Floats in Superspace--Putting Together Two or More Dimensions: Each dimension in the superspace is like a dimension in our ordinary three-dimensional space. Each dimension of the space also points to or indicates a "direction." This direction is a posssibility. In this superspace something called mind floats and freely associates. It puts together--for a fleeting moment, which could be an eternity, depending on the time scale one is looking at--any two or more dimensions, and it asks what about this combination?

Minds--Sensing Possibility Waves from Past & Future to Create Existence: Our minds are thus tuned or are tunable to multiple dimensions, multiple realities. This freely associating mind, he says, is able to cross time barriers, "sensing the future and reappraising the past." As "time machines," our minds sense the flow of possibility waves from both the past and the future. And, Wolf says, in his view you can't have "anything like existence without this higher form of quantum reality."

The Fog of Time Is Parallel Universes: The action of "piercing through the time fog," and the existence of parallel universes, appear to go hand in hand, he suggests. The fog of time is parallel universes. One of his views is that it will be possible to have experiences of the future, referring to 'psychics' and seers. Many scoff at the idea while others accept it.

Is It Possible to View a Future Parallel Universe? So, he asks, if time is "present and unredeemable," might it be possible to observe a future parallel universe? The only way to have time unredeemable, he says, is to have it exist as a maze of parallel universes. Perhaps a better approach, he thinks, is the hologram of parallel universes.

Viewing the Hologram Alters the Probabilities: What one sees when viewing the hologram depends on the viewpoint one takes. And that viewpoint alters the hologram because it alters the probabilities by moving in time.

After all, if the hologram is constructed from transacting quantum waves moving in all time directions, then who assigns the probabilities? Somehow, he says, a "field of consciousness" must illuminate the hologram and IT assigns the probabilities.

The Human Mind as the Laboratory of the New Physics: *The human mind is the laboratory of the new physics*, he notes. It already is tuned to the past and the future, making "existential certainty out of probable realities." It does this, Wolf says, by simply observing . Observing oneself in a dream or when awake--observing the action of observing.

But What is Consciousness--Mind? Defined by What it Does: But what is self--what am I? Wolf asks. "Is my mind an illusion...am I John Lilly's 'a human liquid bio-computer'?" The answer to the question of what distinguishes a person from a washing machine, he observes, is his consciousness, his mind. And how do we define consciousness? In the same way we define an electron-- by what it *does*. Wolf says mind is "both the awareness and the creation of experience....the *being* and *knowing* of experience."

Knowing As Mind and Being As Matter: With the "stroke of the twentieth century quantum eraser," he says, the dividing line between ontology (theory of being) and epistemology (theory of knowing) is rubbed out. In short, he declares, *knowing is mind and being is matter*. And how the two separate is represented by "the magical process we call consciousness."

The Mind-Body Problem--How Does the Mind-Body Interact? How the mind-body interact, he notes, has perplexed philosophy for centuries under the name of the "mind-body" problem. How does a "decision" to do something, which is mental, trigger a responding physical act? Wolf quotes University of Queensland, Australia, Professor L. Bass in his *A Quantum Mechanical Mind-Body Interaction* in the *Foundations of Physics.*

That book, says Wolf, offers a "quantum physics solution." It is "knowing that the job is being done" that is the grabber, Wolf comments. When I chose to do something, how does it get done and how do I know I'm doing it? The problem has to do with *indeterminacy* and *will*, he continues.

Reality Noticed Is Reality Created: It is quantum indeterminacy that leads to deterministic choices on the normal level of perceived experience. And how and where does this choosing take place? It takes place because "special channels open in the walls of my neurons,...highly adapted electrochemical , excitable, elongated cells that make up my central nervous system (CNS)."

The Quantum Mechanics of Consciousness: This device within our CNS, Bass says, exists as an independently operating agent in each and every neuron, *perhaps even in each molecule or even in each atom*, he offers. "And at these levels of reality, reality noticed is reality created." In a single act of noticing by one of my CNS agents, he states, a pure qwiff [quantum wave function] pops [act of observation where the quiff vanishes] and a dream becomes a reality.

This reality, he continues, takes place inside a single neuron where the neuron is left in an indeterminate state until the "act of consciousness" takes place. This is the "conscious noting of the positional configuration of the active atomic group on the enzyme molecule."

The neuron is then no longer in the indeterminate state, it has fired, he states, and "I am aware of it." So it all comes down to the "popping of a quiff in our central nervous system" by a "magical device within that nervous system," evolved through complexity theory's evolutionary process.

Where Does the Uncertainty-Resolving Device Exist? So "quantum mechanics is necessary for life," he says, and "Out of uncertainty comes freedom. From the uncertain atoms comes free will." But Wolf asks, What device pops the quiff in the first place? Where does the device exist? The answer seems to be *nowhere*, he declares.

"The device is our awareness of our own existence." So, he says, consciousness is choice--"which comes to us from quiff land (the physical phenomenon of the quantum wave function), the land of our imagination." Then comes the big one: "Before...we are aware of anything, the universe is paradoxical and not "out there." Something called "mind" is present. Mind, he says, is the world of quantum reality.

Collective Consciousness--Process by Which Potentiality Becomes Reality: Wolf explains that Bass's "device" changes the quiff, pops the wave function. But the device has no location, it is the living neuron, the enzyme operating in the neuron, the molecule operating at the end of the tail of the enzyme; it is the conscious atom noticing itself and thus creating itself. Wolf concludes that *consciousness is the* process from which *potential reality* becomes *actual reality*, in quantum terms, the quiff popping, the wave function collapsing.

The "act of consciousness," says Wolf, is the creation of reality at the atomic and molecular level. Wolf goes on to explain how countless numbers of individual neurons become the "collective consciousness." And this integrated consciousness is our "consciousness of consciousness," an "overseer," and all the other acts of consciousness form a "mixed state of popped quiffs."

The Mind Has No Location in Space--Appears to Be Everywhere: Wolf quotes neurosurgeon Roger Penfield in *The Mystery of the Mind*, to the effect that the mind has no unique location within the body. So where is the mind? The mind, says Wolf "appears to be everywhere," observing on the scale of atoms, molecules, neurons, cells, tissues, muscles, bones, organs, observing on all scales of physical existence."

The atomic minds pop quiffs and operate at the level of quantum mechanics. But the one mind "adds up all of the pictures gathered by all the atomic minds" and "the one mind makes up its mind." So the mind, declares Wolf, "has no location in space."

Atomically Conscious: Wolf goes on to point out that consciousness is able to sense things on an atomic scale, the mind-boggling possibility that new or novel events can be accepted at the atomic level, "where potential reality becomes real reality." Wolf declares that humans found out who they were when they became "atomically conscious" and "learned how to live in quantum reality."

Back to the Source and Meaning of Human Reasoning: So we have reviewed some energy and quantum mechanisms by which the human mind functions and resonates with the universe, at the level of the quantum and the universe at large.

> Consciousness is able to sense things on an atomic scale, the mind-boggling possibility that new or novel events can be accepted at the atomic level, "where potential reality becomes real reality.... humans found out who they were when they became atomically conscious and learned how to live in quantum reality. --Fred Alan Wolf

PART VIII--COSMIC PURPOSE

CHAPTER 22
PURPOSE AND MEANING IN COSMIC TERMS

Cosmology--Theological and Scientific Implications: Theologians, philosophers and scientists all explore cosmology, the study of the origin and nature of the universe. Theoretical physicists, astrophysicists, space scientists and astronomers are all interested in cosmology as it relates to their respective fields. Philosophers and theologians also traditionally study cosmology for its relationship to the existence and nature of God and for insights into purpose and meaning of the universe.

Scientific Discoveries as Proof of God? In his *The Creator and the Cosmos*, astronomer, cosmologist, and Congregational Minister Hugh Ross notes the deep philosophical and theological implications of this study of cosmology. He refers to certain recent scientific discoveries as proof of the existence of God and of the truth of Biblical references.

He cites the physics of the Big Bang, recent confirmations by NASA satellites of primordial ripples in the fabric of spacetime occurring immediately after the Big Bang. He also cites the anthropic principle, the extraordinary coincidences in the exponential precision-tuning of the constants of physics so as to assure life in the universe, as evidence of "design" which he takes as supporting the argument for the existence of God.

'Before' the Big Bang: The equations of Einstein's theory of general relativity implied a universe that was expanding from a starting point, a beginning of space and time, called *a singularity*, giving rise to the physics and cosmology of the Big Bang. The Vatican and most of Western theology were comfortable with the Big Bang theory as conceptually compatible with the Genesis creation scenario--the existence of an actual beginning, a 'Creation' event.

What existed *before* the Big Bang was viewed largely as metaphysics or perhaps theology, since pre-big-bang events were, in science philosopher Karl Popper's terms, unverifiable and therefore unable to meet the 'falsifiability' criterion to qualify as science.

But the classical physics of space and time in Einstein's general relativity theory breaks down at the singularity, requiring resort to quantum theory, the mechanics of matter at the subatomic level which characterized the beginning, and related theories of 'quantum gravity' and unification.

These conditions invoke new concepts, so-called 'quantum cosmology,' in which the universe might be described as having its origin in a unified-force-particle generated by the primordial quantum void, described as a mega-universe pregeometry state, without space, time, matter or dimensionality, but a state which physicist Heinz Pagels, in *Perfect Symmetry* describes as fraught with the potentiality of all the laws of physics.

The Genesis Creation Provides a Universe of Meaning: Western theology reflects early anthropocentric perceptions of the Earth as the center and purpose of the universe. As chronicled in Genesis, and described by 4th-5th century St. Augustine in his *Confessions*, a timeless God, existing outside of time and space, created the universe to provide for human and other life in the universe. Virtually all agree that such a God-designed universe could justifiably be said to have meaning and purpose, associated with the existence of mind, consciousness, and life.

But a universe self-generated through quantum effects can have "no objective meaning," insists astronomer-minister Hugh Ross in *The Creator and the Cosmos.* Where quantum effects spontaneously self-generate the universe, says Ross, "morality and religion" in such a chance universe, while "temporarily useful" are "ultimately irrelevant," he declares.

To Study Cosmology Is to Explore the Basis for Life's Meaning: If the universe is (God-) created, observes Ross, there must be "reality beyond the universe" and "the Creator is that ultimate reality," the source of life, which "establishes its meaning and purpose." God's personality, says Ross, "defines personality" and His character "defines morality." Thus, he concludes, to study cosmology, the origin and development of the universe, is to explore "the basis for any meaning and purpose to life."

Cosmological Chauvinisms: Ross is concerned about three forms of what he terms "cosmological chauvinism," each, he says, vying for "supreme authority." First is *theological*cosmological chauvinism, giving theologians the authority since theol-ogy, in his view at least, is the "mother of sciences." Second, is *philosophical* cos-mological chauvinism, involving scientists' "bungling intrusion" into cosmology. And third, is *scientific* cosmological chauvinism, in which scientists do everything "better than anyone else" and confound the problem by employing mathematics in the process. What is needed, he says, is the interdisciplinary rather than the intim-idating-exclusionary approach. As to what is meant by 'meaning,' we explore below, some different approaches to meaning or lack of it in the creation.

P.W. Atkins' Proposition: Creation Required "No External Intervention": In his *Creation Revisited,* Oxford physical chemistry professor P.W. Atkins seeks to demonstrate the provocative proposition that underlying the apparent complexity of the universe lies a core of "organized simplicity"; that the universe emerged from "nothing"; and further, that no "external intervention" was required, either for the creation or for the later appearance of planets, "elephants" or sentient beings.

A Universe from Nothing, Inexorably Collapsing into Chaos: Atkins advances certain basic themes: that the universe's seeming complexity is traceable to ultimate simplicities; that all forces, fields, and physical phenomena are simply forms of "natural behavior" for which we provide code words expressing this behavior as "laws of physics."

"Everything" came from "nothing," he declares, and, mandated by the second law of thermodynamics, the universe, which is initially wound-up with the high-quality energy of creation, unwinds. The unwinding process is represented by energy being "degraded" by the *increased entropy or reduced order* induced by change, as the universe is irreversibly driven toward ultimate total *chaos.*

A Purposeless Process? Thus all events are driven, Atkins declares, by a *purpose-less* process--the thermodynamic collapse into chaos. We are reminded here of physicist David Layzer's observation in *Cosmogenesis* that, in a sense, "Our world may be just a fantastically improbable fragment of perfect chaos."

The 'Purposeless' Universe: Earlier, we noted Paul Davies' conclusion, in *The Mind of God*, that the exquisitely fine-tuned creation could not have been purposeless and, in *Creation Revisited*, Atkins' contrary conclusion--that the creation was a chance and purposeless event. Atkins' use of the term 'purposeless' runs contrary to the idea implicit in the 'anthropic principle' since he contemplates a universe dominated by inexorable physical processes that overwhelm notions of design or purpose, implying some sort of volitional act.

Fundamental Questions about the Existence of Meaning: Thus Atkins' concept of the universe's chance inception as 'purposeless,' raises fundamental questions about the existence of meaning of, and in, the universe, although Atkins, like Steven

Weinberg in *The First Three Minutes* and later, in *Dreams of a Final Theory*, does not deny that we can construct meaning--even in a universe which comes into being through chance natural processes.

There are those, of course, like physicist Fred Alan Wolf, in *Parallel Universes*, who suggest that quantum mechanics, including the uncertainty feature, is God's method of creating and running the universe. And, of course, there are the suggestions that the uncertainty principle is at the root of free will.

Can There Be Meaning in a Chance-Created and Temporary Universe? Supposedly, however, Atkins uses 'purposeless' in a quasi-scientific context; namely, that 'purpose' is difficult to construct for a universe born of uncertainty and inexorably and irreversibly directed toward total energy degradation--ultimate chaos, in which all events and achievements will be irretrievable erased as though they had never been.

Thus, in Atkins perspective, any appearance or implication of purpose, implied from the extraordinary life-supportive environment, is subsumed in the chance beginning and likely end of the universe and of every vestige of consciousness. The last evidence of human suffering and failure, of human progress and achievement, all will come to an end so absolute as to deny even the fact of its once existence.

Two 'Terminations of the Future': Speaking of the end of the universe and its relevance to purpose, Peter W. Atkins, in *Creation Revisited*, addresses two *terminations of the future:* One, the knots of matter loosen, "their local ripples dissipating until in due course the spring is unwound and the universe is uniformly and irretrievably dead," and all evidence of our achievement eradicated.

In the second, there may be sufficient knots (mass) that their combined twisting (gravitational critical mass) *winds back the universe* to a point, perhaps to re-expand, an event he describes as a "renewal" rather than a creation. We may be living now, he notes, in just such a renewal and "reconstituted universes may constitute a perpetuity for the future," but they must once have sprung from a true creation, unless, he adds, time lies in a circle.

Finally, the Present: Presently, the universe is alive, Atkins notes, courtesy of the balance of strengths and forces. The emergence of just the "right blend" of forces may give the impression of a miracle, and suggest "intervention," he notes. But, he concludes, while we cannot see far enough to select the correct explanation, we can be confident that intervention was not required. It is chance that may have created a "benevolent job lot of strengths of forces." If they had not, he says, we would have been "none the wiser" and "none the sadder."

Chance Stumbles on Fortune: Chance might have encountered fortune in due course, because, he says, a universe "looping through cycles of existence" might have rebounded with a "different bag of strengths of forces" millions and millions of times since the "true creation," until the "recycled structure of spacetime." was sufficient to support a structure of forces "exactly right for the emergence of consciousness."

The Universe Will Sleep Again: But even if the universe is a one-shot affair, he muses, "one coiling of the spring" and "one irreversible drift into uncoiled uniformity and global flatness," such a universe still has no purpose behind the happenstance favorable amalgam of forces. We are chance beneficiaries, he declares. The universe has awakened in this " benevolent epoch," and consciousness emerged. Why? "Simply because it happened." And when the epoch passes, he declares, the universe will "return to its sleep."

Benevolent Constants Do Not Imply Purpose: It is possible, Atkins says, that the generation of spacetime from the void must lead to our present force strengths because they are essential facets of the structure of spacetime. But, he insists, that does not translate into purpose. We can still remain the progeny of chance. And, he adds, we should not "wonder" at the values of the constants (the anthropic principle) any more than we do at the speed of light, or the value of π.

Complete Knowledge Just within Our Grasp: Continuing this line of thought, he notes that when we have seen that the fundamental constant values are inevitably so and extraneous, we will have reached complete understanding and science can " rest." And, he declares, with a Genesis metaphor, "Complete knowledge is just within our grasp [and] comprehension is moving across the face of the Earth."

The Universe Emerges from an Incipient State of Pregeometry: The simple-entity 'somethings' of the starting universe, according to Atkins, originate from the unstructured points of the universe's incipient state of pregeometry. But ultimately, all is traceable back to the fundamental power inherent in the mathematics of nothingness to bootstrap rudimentary opposites into being and then given "persistence" by the fortuitous event of the concurrent emergence of time.

Driven by natural "behavior," essentially processes inherent in or emerging from the primordial quantum state, Atkins proposes that all these events are self-generating and none require or involve "outside intervention" (Atkins' code words for a Creator or God) and that the process can therefore be said to be *purposeless.*

Physicists and Purpose: Scientists and philosophers have dwelt on purpose of and in the universe down through the ages. More recently, some scientists, such as Paul Davies in *God and the New Physics* and *The Mind of God* have considered the earlier-discussed anthropic principle, at least the stronger versions, as implying purpose, purpose derived from identifying the universe as somehow the product of incredibly intelligent precision fine-tuning.

In *Superforce*, physicist Paul Davies, like Atkins, Hawking, and others, points out that the new physics and cosmology explain how the physical structures came to exist *automatically*, as a result of natural processes, and would then "no longer have need for a Creator in the traditional sense."

Design and Purpose: In *Superforce*, Davies goes the other way. "Nevertheless," says Davies, "the laws which enable the universe to come into being spontaneously seem themselves to be the product of exceedingly ingenious design." If physics is the product of design, he notes, *the universe must have a purpose*, and the evidence of modern physics suggests strongly to him that the *purpose includes us.* So Davies suggests that purpose is somehow intrinsic to physics which can self-generate the universe.

Refuting Reductionist Philosophy--Mind As Apart from the Physical World: In *God and the New Physics,* Davies concludes that while the new physics requires a re-examination of conventional ideas of the role of God in the creation, that the human mind is something apart from our physical world--that we are something more meaningful than just a "co-location of atoms," as philosopher-mathematician Bertrand Russell had so drearily declared.

Davies observes that "It would be foolish to deny that many of the traditional religious ideas about God, human life, and the nature of the universe have been swept away by the new physics." But, he says, our search has turned up many positive signs. He notes that the existence of mind, for example, as an *abstract, holistic organizational pattern*, capable even of disembodiment, "refutes the reductionist philosophy that we are all nothing but moving mounds of atoms."

Design As Implying Purpose, Cosmic Intelligence, and ? Davies is saying that the remarkable coincidences in the precision-configured constantsof physics and underlying mathematics, which are such as to promote the emergence of consciousness in the universe, necessarily imply cosmic intelligence and purpose. For some scientists, like astronomer-minister Hugh Ross, such cosmic intelligence, design and purpose, sometimes equated to Cosmic Mind, demonstrates or at least implies the existence of God.

But Paul Davies, like George Greenstein in *The Symbiotic Universe*, is content to settle for the existence of cosmic intelligence, design and purpose having to do with our presence--that somehow the precision design of the universe suggests that our presence is something other than pure chance.

But these different points of view seem to resolve themselves into two categories, on the one hand, Ross's theistic God of Genesis, thinking the universe into existence through divine volition, resulting in the Big Bang creation event, and on the other hand, some more ambiguous form of cosmic intelligence, represented by physics and mathematics which 'self-generate' a universe, precision-configured, designed, to assure life. Is there a fundamemtal difference between these two views? Or is this a 'distinction without a difference'?

Natural or Supernatural? One possible significant difference may relate to whether God is a *natural* or a *supernatural* phenomenon, and whether the attributes of 'personhood' attributed by Western theology to God represents something 'supernatural,' or whether it can be explained as a 'natural' phenomenon. Sometimes this subject of the role of God versus the role of physics is addressed within the framework of the persistent tension between scientific materialism and religious transcendentalism--and whether these are'mutually exclusive' areas.

In *Infinite in All Directions*, Princeton physicist Freeman Dyson argues that scientific materialism and religious transcendentalism are not mutually exclusive. It is perhaps in this context that physicist-astronomer Edward Harrison, in *Masks of the Universe*, invokes the *Unigod*, the God who is both God and physics.

Physicist Steven Weinberg--A Pointless Universe? In a burst of frustrating impatience with the apparent scheme of things, in his *The First Thee Minutes*, physicist Steven Weinberg, after describing the end of the universe, strikes a note of philosophic plaintiveness: "The more the universe seems comprehensible, the more it also seems pointless." The empathetical reader might get the feeling that this expression is, *sub silentio*, a theological outreach--that God not countenance the unthinkable--the end of the universe; that God not impose a cold, random universe one that will have been without purpose or meaning.

Lifting Human Life above Farce: And, in a final moment of pensive contemplation, Weinberg observes that "The effort to understand the universe is one of the very few things that lifts human life a little above the level of farce, and gives it some of the grace of tragedy." Again, one might read into this nostalgic final note, the feeling that this must be a dream we experience, for the nature of God would not allow such an incomprehensibly massive tragedy to be played out on the stage of reality. Instinctively, we abhor and rejects such a purposeless end.

Steven Weinberg's 'Pointless Universe' Revisited: We note with concern Steven Weinberg's melancholy observation that "The more the universe seems comprehensible, the more it also seems pointless." Somehow this observation, from a scientist of such eminence, has captured the attention of other scientists, philosophers and theologians and is often cited, sometimes with acceptance but other times with disappointment if not outright disapproval.

His 'Pointless Universe' Sentence Got Him in Trouble: In his essay, *Confronting O'Brien*, in Clifton Fadiman's "Living Philosophies," Weinberg expresses certain philosophic thoughts. In the process, he observes that this single 'pointless universe' sentence got him into "more trouble with readers than anything else [he'd] ever written."

All He Meant: He observes that "All I meant was that if we search in the discoveries of science for some point to our lives, we will not find it. This does not mean that we can't find things that give point to our lives. If science can't provide us with values, neither can it invalidate them." He goes on to describe his own values as a "disorganized hodgepodge." But high on his list of values are loving one another, enjoying beauty, and learning about the universe scientifically. However systematic our values may be, he notes, we must live with the fact that in our value assumptions, we are "not responding to some cosmic imperative," but "inventing values...as we go along."

Melancholy Reaction: My own reaction to his 'explanation' leaves me even more melancholy than the original reading of *The First Three Minutes.* He reiterates that we must invent our own meaning, something few of us may wish to trust. In the essay referred to earlier, Steven Weinberg continues his melancholy reflections by noting that "It would be wonderful if, in uncovering the principles underlying the universe, we had discovered ourselves, we had found that a grand cosmic drama was going on, in which human beings played a starring role. *We haven't*, he declares.

Any Special Place for Human Beings? "Nothing that scientists have discovered suggests to me that human beings have any special place in the laws of physics or in the initial conditions of the universe." He makes one concession, though. He does acknowledge that there is "one possible exception, the anthropic cosmological principle," discussed elsewhere here.

But the views of this distinguished physicist are not shared by all. In *Infinite In All Directions*, Physicist Freeman Dyson speaks of the extraordinary role of the human mind, a part of the mental apparatus of God, destined to spread through the universe. We may note also, in *God And The New Physics*, physicist Paul Davies' discussions of quantum theory and relativity as pointing the way to a new appreciation of the human role in the universe.

Scientists Dwell on Weinberg's 'Pointless Universe': Weinberg was certainly right about his observation on the pointless universe getting him into trouble. In his *The Dark Side Of The Universe*, physicist James Trefil notes the gloominess brought on by Steven Weinberg's plaintive reference to the end of the universe.

He asks, "If billions of years in the future, there is to be no life, no memory of the struggles of humanity, what point is there to existence?" He long wrestled with this, he says, acknowledging that "in a quadrillion years...There may be no one to know how I behave...no one to remember what any of us did." "But this is irrelevant," he says. "The point is...that all that matters...is that *I* will know tomorrow what I have done, I will know whether I was the best person I could be."

Steven Hawking's quantum cosmology model of a universe which has no beginning or end, so that time does not end, lends new meaning. Even though his endless universe is one in which all life will periodically be wiped out, it nevertheless means that the laws of physics, the presence of time, guarantees that our existence, and all its trials and victories will have been a fact for all eternity.

We noted earlier the comments of physicist Heinz Pagels that he is not convinced of the grounds for this ultimate pessimism, noting that our view of the universe may be wrong; that no current cosmological model has factored into account the effect on the future of the universe of intelligent life which may influence the cosmos in an as-yet-unforeseen way. Similar views have been expressed by physicist Freeman Dyson in *Infinite in All Directions* and mathematician Frank Tipler in *The Physics of Immortality.*

But some scientists shy away from Weinberg's 'pointless' universe. Paul Davies' view of a universe of design and purpose having to do with our presence here is clearly at variance from Weinberg's outlook on this fundamental issue.

'Deciding' What to Believe? In the same "Living Philosophies" essay, Weinberg addresses the question of *how should we decide what we ought to believe.* He speaks of the "commitment to reason" under "scientific method" often crystallized, he notes, "as mathematics, and a deference to observation and experiment."

Do We Determine Our Own Beliefs? He asks whether *we* determine our own beliefs? Do we somehow will them or are they the coincidental product of unidentifiable external influences? Many accept their beliefs as naturally-generated, perhaps a product of their inheritance and environment, having little or nothing to do with volition. The question is a serious one, especially in Christian theology where faith in the Christian path may be viewed as critical in terms of salvation.

Like many others, Weinberg acknowledges his limitations and says that he does not really know the extent to which his own beliefs are really under his own personal control. So, he asks, what *does* he believe? Well, he believes, among other things, in "order in the universe." [In] "adaptive orderliness of living things...and an order in the logical structure of the universe." But ask theologians, where did that order come from?

Value Postulates As Not Responding to Some "Cosmic Imperative": There are, Weinberg is reminded, "some value judgments that we all want to see imposed on everyone, such as prohibitions against murder or slavery....But however systematic our values may be, we have to live with the fact that at bottom, *in our value postulates, we are not responding to some cosmic imperative,* but rather inventing values for ourselves as we go along."

Of course, this further explanation, too, is unlikely to satisfy 'purpose-oriented' philosophy or theology. The Bibles of the great religions--pronouncements like the Ten Commandments-- purport to be just exactly what he questions--universal value postulates, cosmic imperatives. Even so-called non-believers resort to moral [cosmic] imperatives as pragmatic necessity--required for orderliness in society.

What About God? Will a 'Final Theory' Answer Our Deepest Questions? In his later *Dreams of A Final Theory*, in his chapter called *What About God?*, Steven Weinberg notes that if there were anything we could discover in nature that would give us some special insight into the handiwork of God, it "would have to be the final laws of nature." And knowing these laws, we would have in our possession the "book of rules that governs stars and stones and everything else." So, he acknowledges that it is understandable that Stephen Hawking should refer to the laws of nature as "the mind of God."

But Weinberg does not share Hawking's optimism about final explanations. He cannot envision finding "ultimate principles" that do not require explanation in "deeper principles," as he puts it. And is it possible to find any "answer to our deepest questions," any "sign of the workings of an interested God in a final theory"? "I think that we will not," he reluctantly concludes.

Tolstoy , Frankl, Camus on Meaning: Most believe that the universe, and especially life, has transcendent rather than self-constructed purpose and meaning. And, as psychiatrist Viktor Frankl points out in the best seller *Man's Search for Meaning*, consciously or otherwise, they spend their lives seeking that meaning.

But down through the ages, as we have seen, there are some, challenging thinkers, who question the existence of meaning in life, and, as we have seen from the work of Weinberg and Atkins, question the existence of meaning of the universe, at least as derived from *some cosmic imperative*, as Steven Weinberg puts it.

The great Russian philosopher-novelist Leo Tolstoy asserted that "The meaningless absurdity of life is the only incontestable knowledge accessible to man." Albert Camus, the French existentialist novelist and playwright, observed that "We turn toward God only to obtain the impossible."

Bertrand Russell: The Soul's Foundation in Despair: British philosopher and logician Bertrand Russell, in *A Free Man's Worship*, observed that human life is the product of random events; that "the whole temple of Man's achievement must inevitably be buried beneath the debris of a universe in ruins;" that the "soul's habitation" can be safely built "only on the firm foundation of unyielding despair."

Heinz Pagels: The Destiny of Complete Oblivion Seen by Science: In *Perfect Symmetry* physicist Heinz Pagels observed that "... The incomprehensible suffering of individuals, the hope of religion, the insights and explorations of art and science, the madness, the glory and joy of our species are destined to an oblivion so complete that even the act of destruction will not be remembered....this is...the only rational inference that emerges from our scientific view of the cosmos." But despite this devastating pessimism, Pagels goes on to find meaning--which includes, just as Weinberg suggested, science's search for truth.

Voyager's Journey to the Stars: Launched by NASA in 1977, the spacecraft Voyager 2 rounded out its grand tour of the giant outer planets by a visit to the mysterious Neptune and its fascinating moon, Triton, and is now headed into interstellar space toward the bright star Serius, carrying the story of planet Earth, our voices, languages, music, way of life-- just in case it is examined by some interested extraterrestrial space travelers along the way.

Carl Sagan Muses Wistfully about Voyager's Infinite Journey: Carl Sagan mused wistfully that "perhaps they would recognize the tentativeness of our society, the mismatch between our technology and our wisdom." Perhaps, he says, there is no one in 5 billion years who will ever come upon Voyager from planet Earth.

"In 5 billion years...all humans will have become extinct or evolved into other beings, no artifacts of the human species will remain on Earth...the Earth itself will have been reduced by the evolution of the Sun to a charred cinder." So, in this view, Voyager 2 may some day be the only artifact and memorialization of our presence in the universe.

What would have been the meaning of human life, you ask, if such is our fate--the only memorial to existence of the human race and planet Earth recorded in a lonely robot spacecraft endlessly wandering the galactic emptiness.

Human Relevance-- and the Absurd Universe--Harold Fritzsch: In *The Creation of Matter, The Universe from Beginning to End*, physicist Harald Fritzsch of the University of Munich, the Max Planck Institute for Physics, and the European Nuclear Research Center, CERN, describes modern cosmology, the history of the universe which he divides into 'epochs.'

In his 17th and last chapter, he makes the startling observation that in all of human life's search of the universe he has not found the reason for our presence and a way of using science to find ethical values. He quotes Einstein's reference in *Out Of My Later Years*, to the reminder of the "limits of our rational understanding of our existence," and our inability to substantiate the "justification and value of the pursuit of truth."

Says Fritzsch, we don't know why the earth exists, the reason for the Big Bang--or even whether there was a reason. "The universe makes no sense: it is absurd," and he then quotes the essays of Albert Camus awarded the Nobel Prize for literature.

Camus and the Absurdity of the Human Condition: The French philosopher and writer, Albert Camus, thought of by some as an existentialist, others, as a humanist, wrote of the absurdity of the human condition. His widely read novels: the Stranger, The Plague, The Fall, The Rebel, and his essays, the Myth of Sisyphus, earned him the Nobel Prize in Literature.

Camus speaks, in *The Myth of Sisyphus*, of man feeling as an alien, a stranger in a universe divested of illusions. "His exile is without remedy since he is deprived of the memory of a lost home or the hope of a promised land...[and this] divorce...is properly the feeling of absurdity." If man perceived the universe as being able to love and suffer, "he would be reconciled," and if phenomena could be summed up in a "single principle," then there would be "intellectual joy."

From Lost Illusions and Need for Meaning Come the Idea of God: And so, in the face of lost illusions and desperate need for meaning, says Fritzsch, the idea of God arose to dispense with meaninglessness, to deal with the question of the meaning of life. He is the cause of all being, the mediator, the assurance of life after death. He quotes French Nobel Prize-winning biologist author, Jacques Monod, known for his views on human life's presence as the result of genetic mutation. *Monod cites man's search for meaning as an 'imperious need,' 'inborn' and 'inscribed somewhere in the genetic code.'*

Meaning and a Divine Creation: As astronomer-minister Hugh Ross points out in his *The Creator And The Cosmos*, the meaning *of* the universe and meaning *in* life will naturally be influenced by one's view of the origin and nature of the universe. Western thought, of course, is accustomed to pursuing 'meaning' of the universe in the context of the existence of a theistic, personally-involved God who purposely created, and continues to watch over, the universe, suitably designed for the human race, as chronicled in the Biblical Genesis sequence.

As theologians point out, since the universe, in the Biblical account at least, was *intentionally* created, it was reasonable to assume that it was *specially designed* by the Creator to achieve certain *purposes* or goals. Accordingly, life here must have special meaning which, of course, we are left to construct through resort to reasoning, intuition, scriptural guidelines, and writings of the world's great thinkers.

But Many Insist on the Purposeful Universe: But, as we shall see, there are many who insist that the universe, however created, has purpose and meaning apart from differing religious materials; that there are universal objective values; that individuals and societies do construct other justifiable values, goals, and meanings.

Physicist Heinz Pagels, as we have seen, noting the views of serious thinkers, like Bertrand Russell and Steven Weinberg, nevertheless saw meaning in the occurrence of our universe, even if it is destined, as ours apparently is, to end in obliteration so absolute that memory of it, and even time and our other dimensions may no longer exist.

'Many Worlds' Quantum Interpretation As Evidencing Purpose? As we noted earlier, in *Parallel Universes,* physicist Fred Alan Wolf defies conventional argument by using quantum uncertainty not to assert purposelessness but to assert the contrary, that the multiplicity of universes inherent in the 'many worlds' interpretation of quantum theory constitutes *majestic evidence of cosmic purpose*. In effect, *it is the very uncertainty-derived infinity of universes implicit in quantum theory that assures the existence of consciousness somewhere in that multiplicity of universes.*

So, while others suggest our functional universe as intrinsically meaningless, the product of quantum uncertainty, Wolf turns this use of uncertainty and randomness completely around, employing the many-worlds interpretation of quantum theory to demonstrate purpose. In effect, God would go to any lengths--an infinity of universes--to assure life somewhere in that infinity.

Purpose is so compelling a force in nature, he declares, that *nature creates an infinity of all possible universes to guarantee that one (or more) will necessarily produce consciousness.* But once here, Wolf feels, nature's job is done; purpose then becomes our own job to construct. This leaves open the question of where universally-experienced 'values' come from. Organized religions assert, of course, that they have their source in biblical revelation.

Theologians Dissent from Explanations That Do Not Invoke God: In *The Existence of God*, theologian Richard Swinburne, like Hugh Ross in *The Creator And The Cosmos*, takes issue with these ideas of a naturally-caused universe, one not necessarily requiring the invocation of God. It would be error, he says "to suppose that if the universe is infinitely old, or has a complete explanation through prior states, that the existence of the universe through eternity has a complete or even full explanation. It has not, it has neither." *It is totally inexplicable*, he declares.

Creation and Biblical Confirmations: The earlier reference to the views of astronomer-minister Hugh Ross on the Big Bang require some elaboration. An astronomer, astrophysicist, Congregational pastor, Ross lectures on Christian apologetics, and writes fascinatingly on the scientific creation as fully conforming to biblical description. Ross is founder of "Reasons to Believe," an organization which seeks to reconcile the broad structure of the scientific view of creation on the one hand, and on the other, religious doctrinal statements of both the National Association of Evangelicals and the International Council on Biblical Inerrancy.

In his *The Creation and the Cosmos*, Ross has no problem reconciling the laws of physics and biblical inerrancy, as he describes recent scientific Big Bang confirmations, like the 1992 NASA COBE satellite findings of cosmic background radiation thermal remnants of the Big Bang and related experiments.

Keeping God "Out of the Picture": Ross laments the interpretations of scientists in a manner so as to "keep God out of the picture" and "to protect certain [non-theistic] belief systems." He questions certain aspects of 'quantum cosmology' involving theories of physics supposedly applicable at the universe's beginning, when "gravity may possibly be modified by quantum mechanical effects." Echoing science philosopher Karl Popper's falsifiability criterion, he views these as beyond "observational verification" and therefore really "metaphysics."

Transcendent Creator is Required: Ross proposes that a "transcendent creator," not subject to dimensional limitations including that of time, must exist "for all matter, energy, length, height, and even time, suddenly and simultaneously came into being from some source beyond itself." For the most part he avoids resort to the 'supernatural' but points out that God has access to other dimensions of time and, in any event, is not subject to dimensions including that of time and "is capable of cause and effect operations before the time dimension of the universe," which, as he views it, is confirmed by the Bible.

Biblical Confirmation of "Extra-Dimensional Capacities": As biblical confirmation of God's "Extra-Dimensional Capacities" Ross cites Genesis 1:1 and Hebrews 11:3. For God's 'transcendence,' he cites 2 Timothy 1:9, and Titus 1:2. For a beginning of time, he cites John 17:24, Ephesians 1:4 and 1 Peter 1:20. As to the existence of God's causal capability before time, he cites John 1:3, Colossians 1:16-17, John 20:19 and Luke 24:37-43. While his reading of these verses is on the optimistic side, he brings to the subject an appealing rationality.

But Who Created God? He points out that religious 'alternatives' to Christianity deny, at least in part, God's transcendence and extra-dimensional attributes. But Who created God? He asks. He suggests that time of our universe is only a "half-dimension," therefore requiring a starting point and consequently a creation.

But, he says, God is not confined to such a half-time dimension and "both the Bible and general relativity speak of at least one additional time dimension." And, he says, in two or more time dimensions, "an entity is free from the necessity of being created." If God were so to chose, suggests Ross, "He could move and operate along an infinite time line that never touches or crosses the time line of our universe."

Spontaneous Self-Generation As Speculative: Ross proceeds to review and, in his words, "tackle" various arguments advanced by scientists who assert that God theories 'collapse' when certain serious questions are asked. Ross asserts that "any one-, two-, or three-dimensional shape for the origin of the universe that has zero volume yields a theistic conclusion."

In such cases, he notes, "the universe exhibits a big bang and an ultimate origin for the dimensions...and time." He argues that appeals to "spontaneous self-generation at the moment the universe began, followed by billions of years of self-organization...through the present, is purely speculative."

Ross Questions Hawking's Quantum 'No Beginning of Time' Theories: Ross goes on to analyze and critique Stephen Hawking's *A Brief History of Time* in which Hawking, employing 'imaginary time,' suggests that time might have had no true beginning. Although Ross vigorously attacks mathematician Frank Tipler's Omega Point God theory, Ross approvingly quotes Tipler as saying that Hawking might be simply substituting one singularity for another, one involving general relativity, the other, quantum theory.

Ross questions the Hawking 'imaginary time' explanations for 'no boundary in time' as essentially speculation, non-falsifiable in science-philosopher Karl Popper's science criteria, and therefore metaphysics.

Gödel's Theorem--Can't Know the Universe from within the Universe: Ross goes on also to reject Hawking's proposition of coming to know the physics of unification and therefore knowing the Mind of God, citing Gödel's theorem to the effect that "no non-trivial set of arithmetical propositions can have its proof of consistency within itself."

This means, Ross says, that from within the universe, it is impossible to know "that the universe can only be what it is." Ross continues, taking up other aspects of quantum theory, and denying the validity of the idea that human observership might have created the universe out of quantum uncertainty, in the same way as observation determines a particle from an earlier uncertainty state.

Differing Perceptions of the Universe: Edward Harrison writes an intellectually stimulating work in *Masks of the Universe.* He describes, with great erudition, the subjective universes of different societies in succeeding time periods, and the distinctive "world picture" constructed by each human being. As have others, of course, he suggests that *each person constructs his own universe.*

And if, as thought by many scientists, there are extraterrestrials in more advanced civilizations which have, perhaps, learned to travel in the dimension of time, or utilize the incredible power of black holes, then, he notes, there are countless other subjective personal universes beyond our more primitive power of perception.

Wright's Profiles: In *Three Scientists And Their Gods*, Robert Wright provides analytical profiles of computer scientist Edward Fredkin, entomologist and socio-biologist Edward O. Wilson, and economist Kenneth Boulding, whose background was chemistry. The central theme of the book deals with the information nature of the universe.

He discusses information and communication, randomness, entropy, organization, order, and complexity, genetic and cultural evolution, and evolution of consciousness and awareness, noting that entropy and growing disorder in the 2nd law of thermo-dynamics is accompanied by the growth of islands of order, complexities--like human beings.

Meaning in a Universe Driving toward Disorder and Chaos: There seems to be meaning in this very fact, that in a universe driving toward disorder and chaos, and in which cosmic complexity is unraveling, there are islands of increasing order and evolving complexity. It goes even further--the growth of this order is possible only at the cost of increased disorder elsewhere, what physicist Hubert Reeves, in *Atoms Of Silence*, terms 'the great elsewhere,' conveniently provided by the expanding universe.

In his book, physicist Hubert Reeves interrelates 'energy, forces, and elsewhere,' to bring together certain ideas about energy, the fundamental forces, mass changes in particle reactions, binding energy and energy release in bonding, chemical reactions, heat and gravity, and the 'great elsewhere' to absorb disorder from the creation of new forms of order.

The Cosmic Information System: As we have seen, views of purpose and meaning of the universe vary widely, and of course, are approached from different contexts. Science writer Robert Wright, in his *Three Scientists and Their Gods*, speaks of his interviews with MIT computer scientist Edward Fredkin. Fredkin is a quantum physics iconoclast who thinks physicists should focus more attention on particles, fields and forces as parts of the cosmic information system. The ultimate components of everything, he feels, are bits of information.

Purpose and the Computer Universe: Wright and Fredkin get to discussing meaning of life. Fredkin views the universe as a computer run by *somebody or something* [he's neutral on whether it is God]. He notes that society as a whole is an information-processing system and life itself is a kind of computer; and DNA functions like a cellular automaton, everything running on information.

At the end of the interview, Wright asks Fredkin about the meaning of life in the context of Fredkin's thoughts on the universe as a computer. Fredkin responds that meaning has to do "with intelligence and information," perhaps, he declares, the "mission is to create artificial intelligence."

Interestingly, this is a thought somehow paralleled by physicist Freeman Dyson in his *Infinite in All Directions*, where he speaks of the role of human life to extend Mind throughout the universe, in different biological forms and perhaps even in non-biological formats.

The same idea is expressed by Frank Tipler in *The Physics of Immortality* which deals with Omega Point Theory in which intelligence spreads throughout the Milky Way and then the entire universe as part of the universe's drive toward the Omega Point which corresponds roughly to the evolution of the Personal God. Freeman Dyson also speaks of this evolving imperfect Socinian God.

Meaning, God, and Beneficial Behavior: Aside from the computer universe subject and speaking of meaning and belief in God, Wright makes reference to the philosophic school of thought, pragmatism, promoted by Charles Peirce and William James, in which beliefs and meaning are related to the practical consequences and beneficial behavior which they induce.

So, he notes, if belief in God enhances the quality of life--which it generally does, then, in the pragmatic context, it represents truth. So meaning involves the processing of information which induces behavior suitable to the particular environment. In *Dreams Of A Final Theory*, Steven Weinberg questions this view of "pragmatic truth" in which opposite views are equally valid if they are practically successful.

Dyson's Understanding of Purpose: In his Aberdeen, Scotland, Gifford Lectures, updated in *Infinite In All Directions*, physicist Freeman Dyson gives expression to the thought that there is 'some purpose' for our presence here, something beyond our comprehension, something "having to do with the future." If you like, he says, you can "call the transcendent purpose God," but, he muses, it is the "God of the Socinian Heresy" who, though not omniscient or omnipotent, is "growing in power and knowledge as the universe unfolds," a process in wish we share.

There is a nostalgic analogy between Dyson's evolving and not-yet-perfect God, and philosopher Liebniz's God who gave us his best in our imperfect but "best possible world."

Not Just Mounds of Atoms--Coming to Understand the Universe's Meaning: In *God and the New Physics*, Paul Davies elaborately explains the chance, self-starting universe, seemingly eliminating God's creation role. A self-generating universe, of course, tends to weaken the idea of the universe as having purpose and meaning. Davies notes that "It would be foolish to deny that many of the traditional religious ideas have been swept away by the new physics."

But, he says, our search has turned up many 'positive signs' too. "The existence of mind, for example, as an abstract, holistic, organizational pattern, capable even of disembodiment, refutes the reductionist philosophy that we are all nothing but moving mounds of atoms." Thus questioning Bertrand Russell's despairing assertion that we are meaningless conglomerations of co-located atoms, Davies goes on to endorse a universe of meaning. with the caveat that this can only be grasped through an elaborate analysis.

"It is my deep conviction," he observes, "that only by understanding the world in its many aspects -- reductionist and holist, mathematical and poetic, through forces, fields and particles, as well as through good and evil--that we will come to understand ourselves and the meaning behind this universe, our home."

Ingenious Design Implies Purpose--Which Includes Us: In *Superforce*, Davies notes that "though science may be able to explain the world, we still have to explain science." The laws which enable the universe to come into being spontaneously seem themselves to be the product of exceedingly ingenious design. If physics is the product of design, the universe must have a purpose, and the evidence of modern physics suggests strongly to me that the purpose includes us."

Purpose to Modify the Universe? Physicists, like Freeman Dyson in *Infinite in All Directions*, and John Gribbin and Martin Rees in *Cosmic Coincidences,* suggest that we have an important role and goal to modify the galactic environment. "If there is no life elsewhere," say Gribbin and Rees, "the Earth acquires universal significance as the spark with the unique potential to spread life and consciousness through the cosmos...the greening of our Galaxy and beyond by forms of life and intelligence...seeded from Earth."

It is my deep conviction that only by understanding the world in its many aspects --reductionist and holist, mathematical and poetic, through forces, fields and particles, as well as through good and evil--that we will come to understand ourselves and the meaning behind this universe, our home.--Paul Davies

CHAPTER 23
PURPOSE AND THE ANTHROPIC PRINCIPLE

Why Discuss the Anthropic Principle? As a part of our concern with 'meaning' we need to start, perhaps, with meaning in cosmic terms. If the universe is a product of 'design,' this implies a Designer, that the universe was 'intentionally' created to fulfill some purpose, from which we may go on to build meaning in life. But if the universe's existence and characteristics are the product of pure chance, then, as scientists like Oxford's Peter Atkins and Nobel Laureate Steven Weinberg suggest, it has no objective purpose--though we are still free to rationalize our own purpose and associated meanings.

Interrelationship between the Universe and Life? Because the enormous complexities inherent in the physical characteristics of the universe just happen to be such as to promote the development of life and mind, there has evolved a principle suggesting that the universe and human life have some sort of necessary or symbiotic interrelationship which, for want of a better term, has been designated the *anthropic principle*--that the physics of the universe are 'anthropic,' somehow *human life-related.*

Differing Interpretations of the Anthropic Principle: As we shall see, attitudes toward the principle vary from Heinz Pagels' view of the principle as *needless clutter* to physicist Paul Davies' view of the principle as evidence of a universe of design and purpose, to minister-astronomer Hugh Ross's view of the principle as confirming God's existence and the correctness of certain biblical statements.

The Fitness of the Universe for Life: Harvard Biochemist Lawrence Henderson, back in 1913, had written on the peculiar "fitness of the environment" for biological organisms, using conceptual material suggestive of what became known as the 'anthropic principle.'

Similarly, a deep connection between astrophysics and quantum mechanics--between cosmology and atomic theory, was highlighted back in the thirties by physicist Paul Dirac who noted the strange similarity in numbers between certain phenomena related to the fundamental forces, ratios between electromagnetic and gravitational force and between certain basic time measurements.

The Anthropic Principle: In the nineteen-seventies British physicist Brandon Carter invoked the anthropic principle, dealing with our universe's proclivity for life and the suggestion that the laws of physics *required* life. His was the first real effort to catalog those laws of physics which represented the critical coincidences needed for life.

But there remains a curious ambivalence among those who use the term. Some mean only to say that since life is here, whatever the reasons, we must conclude that the laws of physics in this universe happen to be structured to allow for our existence.

But others go further, suggesting that the incredible aggregate of coincidences represented by the myriad of critical life-supporting values which exist in the laws of physics cannot be explained by pure chance, and therefore suggest some external controlling design influence. Although this implies the function of the universe as life-promoting it does not necessarily suggest a fix on our particular form of life, human life--or human life in its present form--as the ultimate design.

The Requirement of Observer Participance: Physicist John Wheeler suggests that life may be of crucial importance to the universe--that the universe requires "observer participance" for its further development, a suggestion with anthropocentric implications not altogether popular in all scientific circles.

Greenstein's Symbiotic Universe: An elaborate, and fascinating treatment of the anthropic principle is contained in physicist George Greenstein's *The Symbiotic Universe*. It deals with the remarkable coincidences in the law of physics describing the symbiosis or necessary relationship between the universe and the existence of life which we will be discussing.

And So If Any Condition Is Needed for Life It Exists: If, for instance, supernovae stars with certain characteristics are needed to generate elements required for life, then they exist. If three microscopic particles in the stars must have 'resonated' together eons ago for us to exist, then they did just that. And if stars have to be far apart for life to exist, then they are. If a certain type of planetary orbit around the sun is necessary to avoid extremes of heat and cold--and it is, then that orbit must exist--and it does!

The Copenhagen interpretation of quantum theory, requiring 'observation' to bring reality to quantum possibilities to reality, is itself suggestive of a symbiotic relationship between the universe and life, both being quantum systems. Although this suggests the function of the universe as life-promoting it does not necessarily suggest a fix on our particular form of life as the ultimate design.

Biologists point out that we have been and still are part of an evolving, and in some ways, experimental process that may lead to more sophisticated life versions--which may already exist elsewhere in this life-promoting universe.

God Had No Alternative but to Create Human Life? Interestingly, Greenstein asks whether the exploration of alternative possible models of the universe will demonstrate that the very sets of laws that are necessary to create universes are the same laws that will result in creating human life, or at least life. Greenstein generates a sense of urgency for the study of the potentiality of life in alternative universe models, so we can judge the full import of the fact of the habitabilty of our own universe--whether it is peculiar and limited to our particular universe model.

If There Are Lots of Inhabitable Universe Models? If study of these alternative universe models demonstrate that there are lots of other potential habitable universe environments, then, he declares, in a fit of melancholy, throw out his book because it will have been meaningless; there would then be no need to invoke "some weird and speculative symbiosis between mind and the cosmos." The idea of the fitness of the environment for life as a monumental achievement in the process of thought would be mere legerdemain. Life, he says, would have been shown to be a necessity rather than some remarkable human manifestation of purpose.

If There Are No Other Habitable Universe Models: But if studying these alternative possible sets of laws of physics leads us to the conclusion that none of them will provide life--Henderson's "fitness for the environment," then we really have latched on to something. How did the universe get to know precisely the right laws to bring life into being?

An examination of alternative models of the universe might show that perhaps there was no alternative but to set in motion the process by which creatures capable of contemplating the universe and pondering the mysteries of creation, would come into actuality. Or it might show that habitability and life are a matter of chance.

Human life As Unavoidable? So, life may, on one hand, be simply chance, or, on the other hand, inescapable. We might find that in any logical universe, God might have *had* to accept the presence of human life--or give up inventing universes! But while Greenstein says we must pursue the answer to whether we are either purposefully unique or simply unavoidable, he personally opts for uniqueness--despite its implication of some kind of outside intervention.

At this point, one can't help remembering computer scientist Fredkin's references to this universe as exhibiting the characteristics of a "laboratory," a computer program being run "to see what would happen," implying that there is something "out there," running the program to see what results.

Rules of the Game: Greenstein speaks of the rules of the game and the greatest game of all the "Game of the World," the laws of nature. He asks, why the rules should be so *life-nurturing.* Nothing in physics, he suggests, explains their conformance to the requirements for life.

Greenstein describes in persistent detail the strange coincidences which suggest purpose. And when he is on the threshold of Q.E.D.-demonstrating a 'designed' universe--and when his readers are trapped in his carefully constructed web of suggestion of some kind of divine intervention, he promptly disavows authorship of any such conclusion!

This is, of course, the unhappy lot of every scientist who would retain the respect of his fellow scientists. After all, scientists can't afford to find themselves as spokesmen for conclusions with clear theological implications. It must be science all the way.

The Bridgewater Treatises: The early 1800's *Bridgewater Treatises* of Paley and Whewell had examined the question of a Grand Design from the disadvantage of the much-earlier base of scientific knowledge and, in conclusion, pointed to a Designer. A watch found on the heath, and a flower, both have intricacy of design. And so Paley attributes the watch to a <u>w</u>atchmaker and the flower to a <u>F</u>lowermaker.

Greenstein asks whether, more than a century later, we have found Paley's "watchmaker" within the very laws of nature? And then he finally gets to the Big one, abandons caution and uses the Big word--Was it "God Himself" who crafted those laws so precisely for our benefit? Backing off a little (but still pretty heady) he asks, Have we scientifically proved the existence of the "Supreme Being?"

He Does It Again: We, the readers, are all excited by now. Like in every whodunit, we know the answer, or hope we do--or at least we know the answer we would like. So what follows? He does it again! Science discoveries have nothing to do with religion, he says. But (fortunately) he hastens to add: "No matter what your choice I have no way of proving you right or wrong."

Our Own Choice--About God's Role in Our Presence: Thank God for small favors, we have our 'own choice'--and no one can disprove us. (In science philosopher Karl Popper's terms, the conclusion of a designed universe is not 'falsifiable,' and accordingly, not science.) But Greenstein continues in this now disappointing vein. *Any attempt to prove God scientifically is going to fail*, he says. Science and religion are "different realms," he insists. But hold on a minute, we have thoughts of other scientists that don't look at these two realms as so mutually exclusive; they don't come out quite where he does. They leave a little room.

Natural Explanations Tend to Surface: But Greenstein reminds us that time and again a purely natural explanation has surfaced for events thought to have "religious" answers. A miracle is an event the natural explanation of which is not yet available. Or so the explanation goes. We just can't keep giving religious answers to everything, like the Sun rises because God wants it that way.

And so we get back down to earth, a little deflated but none the worse--and still looking for scientific evidence in support of ...... The problems of religion are "spiritual, not empirical," he insists. But he then acknowledges that "Science is the art of doubt." And there are so many who insist that religion is the art of belief.

Triumphs of Religion--But Life May Not Be the Source of Rejoicing: He's not content to leave things here and he waxes eloquent. Religions, he says, are the source of great triumphs of humanity, the noblest acts of self-sacrifice. Religion relates to the ultimately human--but he does not consider the cosmos as having conspired to being us about, any more than it did the barnacle clinging to the rock.

In any event, he observes, the appearance of life in the cosmos may not be the great cause for rejoicing. "No one among us has lead a life free from grief, loneliness, and despair....The more cosmic evolution has progressed, the more suffering has come forth in the world."

Pagels Takes a Dim View of the Anthropic Principle: Often cited is the view of physicist Heinz Pagels on the significance of the anthropic principle. Pagels cites physicist John Wheeler's view of the "participatory universe" requiring the participatory act of observers for the existence of the universe. But other scientists, he points out, go on to calculate mathematical formulations for the wave function that triggers creation, quite apart from the relevance of human life's existence, a reference to the many-worlds quantum interpretation.

Life Does Not Act on the Laws of Nature, says Pagels: Unlike some scientists that are intrigued with the principle, Pagels is unimpressed with the idea that the fact of life in the universe demonstrates a determining or controlling set of criteria for the physical laws of the universe--the 'anthropic principle.' He backs away from the anthropic principle as having independent causative, scientific effect.

It has explained nothing, he says, and has had a negative influence; it is "needless clutter in the conceptual repertoire of science." The existence of life in the universe is not a "principle acting upon the laws of nature," but simply the coincidental *result* of the laws of nature. Despite Pagel's misgivings, though, even Stephan Hawking and Steven Weinberg give it serious thought despite their own reservations.

It seems to the reader that we have the 'which comes first, chicken-or-the-egg' question here. If one takes the Pagels' 'life results from the laws of physics' proposition and then asks why that should be the case--that physics leads to life--then we've somehow worked our way back to the anthropic principle. Certainly this would be true if the universe is 'determined' at creation. Then physics *must* produce life and, arguably at least, life is the or at least a purpose or function of the universe.

Interestingly, despite his stated reservations about the value of the anthropic principle, Pagels, in his *The Cosmic Code,* seems, nonetheless, to reaffirm the principle, *sub silentio,* when he suggests that *life is written into the cosmic code.*

Harrison and Anthropicism-Why Is the Universe the Way It Is? In *Masks of the Universe*, Edward Harrison poses several introductory questions: Why is the universe the way it is? Why is the universe organized with planets, stars and galaxies; why is it furnished with certain laws of nature and not others; and why have the fundamental constants, such as the mass and charge of the proton, their particular (life-supporting) values?

The Anthropic Principle Inverts the Deistic Design Argument: One answer given, he says, is that the universe is as it is because God designed it that way; another answer is the anthropic principle. Life exists, not because the universe by chance happens to be a fit place, but that the universe necessarily is a proper place because life exists, a fact that places severe constraints on what can be the nature of the universe, *turning the deistic argument upside down*. We must, he says, *postulate the existence of human beings, not God*, to explain the universe. The anthropic principle, he says, inverts the deistic design argument.

The Universe Is the Way It Is Because We're Here to Ask the Question: He asks again. Why is the universe the way it is? Because, he responds, we are here to ask the question. And he says, the answer to many of the why-type questions is the same. Things are as they are because we are here; if things were different, we would not be here to ask about them. Thus our presence explains the physics of this universe. Harrison goes on to discuss universes with other characteristics which make them totally dark, or having insufficient luminous life for stellar and biological evolution, or unable to get from hydrogen to carbon--and to galaxies.

But Don't Get Too Far Out on the Limb of the Anthropic Principle: But Harrison, like Greenstein in *The Symbiotic Universe,* takes pleasure in speculating on the anthropic principle though careful not to get too far out on a limb with it. He points out that the presence of human beings that can ask these questions is not really essential to the principle. A Wild flower, a hazelnut, or a butterfly, will do, he notes. Each involves the same mystical cosmic numbers and qualifies for the same universal physical constants. Consequently, says Harrison, the principle is *biopic, and biometric* rather than anthropic or anthropometric.

Two Views of the Universe--By Random Chance and by Design: Harrison suggests two views of the universe. One is the 'Aristotelian' view of accidental design--requiring a "stupendous ensemble...impotent to spawn life, ...cosmogenesis on a most profligate scale,...in effect, a law of cosmic selection,...the anthropic principle that discards hordes of universes as unfit to bear life."

The other, the deistic universe precisely configured by deity, which does away with the waste of countless barren universes. Why this life-supporting universe? In the universe by design, because God wished a universe of life. Which is our universe, a random chance universe, the product of countless useless universes, or a single universe from design?

Harrison goes on to comment on our tendency to create other universes to compensate for the deficiencies and injustices of our own. "With the exfoliating universe we repair the rationality of the physical world; with the anthropic principle we cover up its inexplicable accidentals." Each of the societal universes is "a simulacrum of the Universe," he says.

The Mystical Cosmic Numbers: Unable to resist the impulse of reflecting on the mystical cosmic numbers, Harrison quotes Arthur Eddington and Paul Dirac as showing the remarkable relationship between the size of the observable universe and the constants of physics, such as the speed of light, particle electric charge and mass, and Planck's constant.

The ratio of the proton and electron masses, the ratio of the electrical and gravitational forces between a proton and an electron, and the size of the observable universe in reference to the size of a subatomic particle, are all represented by a *single huge number, 10 followed by 40 zeros*, a fact made even more mystical by the fact that the size-of-the-universe figure is time-oriented, that is, the observable universe was smaller in the past and will be larger in the future.

Caveat: Life Requires the Present Size of the Expanding Universe: With the more expanded future universe, will come a change in the size of the universe numbers--and, of course, the extinction of the life on which those numbers depend. Thus life requires the present size of the universe. We are here only because the universe has reached its present size. *Somehow, the human race is a reflection of the mystery of coincidental numbers*. The universe had been viewed by some as mathematics. Now that view is perhaps supported by the incredible coincidence of these cosmic numbers reflecting the universe's physical constants.

Dyson's View of the Anthropic Principle As 'Meta-Science': Physicist Freeman Dyson describes the anthropic principle as *teleological*, dealing with purpose, something forbidden to science but acceptable as "meta-science." He points out that questions of *choice of the initial conditions* for the universe are outside the realm of pure science, and purpose gets into anthropocentric views of the universe--with which science is uncomfortable. His own view is to accept the principle as one of two complementary styles of explanation.

Hawkins' Mindsteps to the Cosmos and the Anthropic Principle: In several chapters we refer to or discuss the significance of the anthropic principle, in connection with the strange coincidences in the laws of physics, which, almost in defiance of logic and what one would expect of the universe, one can find answers to seemingly insoluble problems in physics by 'reverse engineering' the universe's formation to find those arrangements necessary for the introduction of life.

As we have seen, physicist George Greenstein labels it the "Symbiotic Universe," to highlight the apparent interdependence of two "organisms," the physical structure of the universe and life in the universe--a universe which is extremely hostile in general but, strangely, characterized by the fitness of the environment for habitability.

Physicists Differ on Use of the Anthropic Principle: As we have seen, physicists differ in their definitions and use of the anthropic principle. In his *Mindsteps To The Cosmos,* astronomer Gerald S. Hawkins breaks the concept down into three principles.

I: The Universe Must Be 'Known' To Exist (Quantum theory)

The universe's existence is dependent on the existence of an intelligent awareness of it--the philosophic proposition, elsewhere discussed, that existence requires a 'knower' or observer (presumably, apart from divine knowledge) without which, existence is meaningless.

II: The Laws Of Physics Are Designed To Produce Life

The laws of physics are not accidental, they are specifically designed arranged and sequenced to produce the biochemistry of life.

Thus gravity allows safe planetary orbit, and the electric fields are strong enough to bind particles and atoms to produce necessary structure. The red giant stars react to make the jump from hydrogen and helium to carbon which is then, through the supernova explosions, poured out into space to provide for coalescence of the materials needed for life-supporting planets. Here, he says, the line between proof and belief, science and religion, seems to have *softly and suddenly vanished.*

III: The Presence Of Life Has Changed The Universe

The universe is altered because life exists. Intelligent observation *alters matter*. In quantum mechanics, human's measurements of particles resolves quantum uncertainty--changes the characteristics of the particles. Human life is itself capable of changing the working of the laws of nature--as through genetic engineering and the making of new materials. Human beings have freed themselves from the binds of gravity to leave planet earth and walk the moon or send satellites to Mars and Venus, or surveillance spacecraft past the great outer planets--and out of the solar system. Space colonies in near earth orbit, on the moon and on Mars are in contemplation if not actual planning.

Hawkins Summary: Thus Hawkins combines (1) the requirement of a "knower" for a universe to exist, (2) the laws of physics as designed to produce life, and (3) human life's role in altering the universe via quantum mechanics, as aspects of the anthropic principle.

The Theological Implications: In a sense, coupled this way, these elements sug-gest: the traditional Design argument for the proof of God, the scientific justification of Creation, and Human life--or at least Life--as the central purpose or meaning of the universe. This God- human relationship, of course, is at the core of all theology. In that sense, the anthropic principle may be thought of perhaps as meta-scientific restatement of theology.

A New Vitality to Reverence for Life? This would seem to be the case because then, the importance of life--Albert Schweitzer's intuitively-derived *Reverence For Life*--is the key to the meaning of the universe, a concept which by its very nature is essentially theological. Science, in this view, has seemingly stumbled across a quasi-scientific explanation for the meaning of the universe, from which we can surmise the meaning of everything else.

In *The Creator and the Cosmos* and in *The Fingerprint of God,* astronomer-minister Hugh Ross assembles a massive collection of the requirements for life in the universe and even more massive set of requirements for human life on a planet.

After ruling out explanations like many-worlds quantum theory, Ross, unlike his more reserved scientific compatriots, unabashedly concludes by noting that "the only explanation left for how living organisms received their complex and ordered configurations is that an intelligent, transcendent Creator personally infused this information....A personal transcendent Creator must have designed planet Earth. A personal, transcendent Creator must have designed life."

> With the exfoliating universe we repair the rationality of the physical world; with the anthropic principle we cover up its inexplicable accidentals.--Edward Harrison

PART IX--GOD AND SCIENCE

RANDOM THOUGHTS ON GOD AND THE UNIVERSE

To believe in God is impossible--not to believe in Him is absurd. --Voltaire

The Fool has said in his heart, "There is no God." Psalm 14:1

If the work of God could be comprehended by reason, it would be no longer wonderful. --Pope Gregory I

God's eternal power and divine nature have been clearly seen, being understood from what has been made. -- St. Paul in Romans 1:20.

We stand, in a manner of speaking, midway between the unpredictability of matter and the unpredictability of God.--Freeman Dyson

It would be foolish to deny that many of the traditional ideas about God, human life and the nature of the universe have been swept away by the new physics.--Paul Davies

Science without religion is lame, religion without science, blind.--Albert Einstein

It is easy to understand God as long as you don't try to explain him.-- Joseph Joubert

The world is God's epistle to mankind--his thoughts are flashing upon us from every direction. --Plato

Two books in which the actions of God are recorded, the Bible--theology, and the 'Book of Nature' --Natural Theology. These are two windows from which we can view the world, two human enterprises, sharing many common features.—Freeman Dyson

For the scientist who has lived by his faith in the power of reason, the story ends like a bad dream. He has scaled the mountains of ignorance; he is about to conquer the highest peak; as he pulls himself over the final rock, he is greeted by a band of theologians who have been sitting there for centuries. --Robert Jastrow

When God lets loose a great thinker on this planet, then all things are at risk....There is not a piece of science, but its flank may be turned tomorrow. --Emerson

Faith has need of the whole truth. --Pierre Teilhard de Chardin

Truth in science can be defined as the working hypothesis best suited to open the way to the next better one. --Konrad Lorenz

I had to set limits to knowledge in order to make place for faith. --Immanuel Kant

I believe that in actual fact, philosophy ranks before and above the natural sciences. --Thomas Mann

Any statement which attempted to express a consensus of scientists about religious and philosophical questions would miss the main point. There is no consensus among us.--Freeman Dyson

One thing I have learned in a long life: that our science, measured against reality, is primitive and childlike--and yet it is the most precious thing we have. --Albert Einstein

What is it: is man only a blunder of God, or God only a blunder of man? Nietzsche--The Twilight of the Idols

The universe is a message written in code, a cosmic code.--Heinz Pagels

Full account must be taken of the quantum revolution in the search for an understanding of God and existence....So many common sense notions of space, time, and matter have been overturned that no serious religious thinker can ignore it." -- Paul Davies

Natural Theology is the reading of God's mind as expressed in the works of Nature.--Adam Gifford

Between matter as we observe it in the laboratory and mind as we observe it in our own consciousness, there seems to be only a difference in degree but not in kind. If God exists and is accessible to us, then his mind and ours may likewise differ from each other only in degree and not in kind. --Freeman Dyson

The heavens and the earth alike speak of God, and the great natural world is but another Bible, which clasps and binds the written one; for nature and grace are one--grace the heart of the flower, and nature its surrounding petals. --H.W. Beecher

Theology is but the science of mind applied to God....Truth is everlasting but our ideas of truth are not. Theology is but our ideas of truth classified and arranged....Theologies are well in their place but repentance and love must come before all other experiences. --H. W. Beecher

Our world may be just a fantastically improbable fragment of perfect chaos. --David Layzer

It may be that our role on this planet is not to worship God--but to create him. --Arthur C. Clarke

There is a reason in nature why something should exist rather than not.-- Gottfried Wilhelm von Leibniz

Even God cannot change the past. --Agathon

The universe is but one vast symbol of God.--Thomas Carlyle

The universe is one of God's thoughts. -- Friedrich Schiller

So God created man in his own image. --Genesis 1:27

The [universe] is alone, having nothing outside, resting on itself immobile as a whole; and it alone is all things. --Johannes Kepler

The inhabitants of the universe were conceived to be a set of fields--an electron field, a proton field, an electromagnetic field.-- Steven Weinberg

The existence of mind, for example, as an abstract, holistic, organizational pattern, capable even of disembodiment, refutes the reductionist philosophy that we are all nothing but moving mounds of atoms.--Paul Davies

When God lets loose a great thinker on this planet, then all things are at risk. --Ralph Waldo Emerson

An awareness of our smallness may help to redeem us from the arrogance which is the besetting sin of scientists. --Freeman Dyson

Scientific materialism and religious transcendentalism are neither incompatible nor mutually exclusive. We have learned that matter is ....weird enough, so that it does not limit God's freedom to make it do what he pleases. --Freeman Dyson

[Upon the advent of non-Euclidian geometry]: I acknowledge...that I have a Euclidian earthly mind and how can I solve problems not of this world? ...I advise you never to think about it...especially about God, whether he exists or not. All such questions are utterly inappropriate for a mind created with an idea of only three dimensions. And so I accept God...His purpose...utterly beyond our ken. --Feodor Dostoevsky

What is it: is man only a blunder of God, or God only a blunder of man? Nietzsche--The Twilight of the Idols

Hawking embarks on a quest to answer Einstein's famous question about whether God had any choice in creating the universe. Hawking is attempting, as he explicitly states, to understand the mind of God. And this makes all the more unexpected the conclusion...a universe with no edge in space, no beginning or end in time, and nothing for the creator to do. --Carl Sagan

CHAPTER 24
SCIENCE-THEOLOGY BORDER--CONTRASTING VIEWS

Exploration of Meaning Must Consider Physics and Theology: From our earlier discussion we have already seen that any rational discussion of the meaning of the existence of the universe, and our presence in it, must be addressed in the context of science and theology.

Cosmology--The Border between Physics and Theology: Accordingly, this material explores areas of accord and discord between physicists, theologians, and philosophers on cosmology, the nature and source of the creation of the universe and relationships between God, the universe, and physics & mathematics.

In their writings on physics and cosmology, physicists and mathematicians like Stephen Hawking, Paul Davies, Freeman Dyson, and Frank Tipler, have seen fit to address problems and concepts on the borderline between physics and theology-religion.

Divergent Views on Quantum Cosmology: Uniquely, Davies' entire book, *The Mind of God,* is addressed boldly to this 'interface' between science and theology. Hawking's *A Brief History of Time*, while not devoted primarily to this interface as is Davies' work, is oriented to quantum cosmology with frequent questions about the role of God and physics in the creation, and unification theory as related to the reading of the Mind of God.

Although Davies sees quantum cosmology and the 'new physics' as having a key role in the creation, astronomer-minister Hugh Ross, in his *Creator And The Cosmos*, sees science, especially the relativistic interpretation of the Big Bang, as confirming the Creation described in Genesis and as evidencing the supernatural God of the Bible, and he also finds a basis in physics to account for some of the biblical events considered miraculous.

Scientific Materialism & Religious Transcendentalism: In any event, differences in viewpoint between scientists and theologians are often subsumed in the question of whether scientific materialism (everything explainable as physical, part of nature) and religious transcendentalism (favoring spiritual or religious explanations) are compatible or are mutually exclusive.

While theologians are increasingly accepting new scientific discoveries affecting cosmological outlook, there are those who continue to feel that regardless of creation mechanisms, everything is traceable ultimately to the supernatural. Physicists and theologians increasingly feel there is room for both science and theology.

As physicist Freeman Dyson observes in *Infinite in All Directions*, "scientific materialism and religious transcendentalism are not mutually exclusive." Other physicists, like Paul Davies as expressed in *God and the New Physics* and *The Mind of God*, suggest similar views, albeit more cautiously expressed.

P.W. Atkins Sees Mathematics As Creating the Universe: On the other hand, Oxford chemistry professor Peter W. Atkins, in *Creation Revisited*, sees mathematics as creating the universe without supernatural intervention. The universe, he declares, can come into existence "without intervention"-- without need to "invoke the idea of a Supreme Being in one of its numerous manifestations." Atkins recognizes at the outset that any religious person is unlikely to be swayed by such arguments.

Atkins proceeds to embark on the track of zero intervention and to examine the universe for the indicators of its underlying simplicity. What we view as complexity, he notes, may be the outcome of "chains of simplicity."

Atkins speaks of the "recipe" for the universe as a few types of particles and forces the descriptions of which are just "codes" or words for behavior and structure. These descriptions, he observes, having no meaning at the "deepest level of comprehension...can be allowed to evaporate." And when they evaporate, he says, they will leave behind *virtually nothing.*

Astronomer-Minister Hugh Ross Sees Science As Revealing the Biblical God: Diametrically opposed to Atkins' concept of a universe creation without external intervention are the views of some scientists like Dr. Hugh Ross, an astronomer, cosmologist, and Congregational Minister. Dr Ross initiated and directs the 'Reasons To Believe Institute' and writes on science as supporting rather than contradicting the Biblical God.

In The Creator and the Cosmos, Ross presents his fascinating explanation of how scientific accounts of the Big Bang, as supported, for example, by the NASA COBE satellite findings of primordial temperature variations, ripples in primordial spacetime, are interpretable as supporting the Big Bang and Biblical accounts of the creation.

Ross portrays Hawking's quantum cosmology, not requiring God's direct involvement, as metaphysics, and physicist Frank Tipler's Omega Point Theory God, who represents the spread of intelligence throughout the universe, as essentially fantasy. Ross makes a very different and deeply religious case for science as demonstrating the existence of the Biblical Creator who is personal and transcendent. He also addresses the physics involved in the New Testament accounts of the role of Christ, and that of the pre-existent Christ referred to in New Testament passages like those in John and Colossians.

Purpose and Meaning Defeated by Quantum Cosmology? Ross points out that a universe randomly created by the laws of physics has no meaning and leaves us lost in a sea of artificially-created and subjective meanings. He despairs at scientific blindness to what he regards as bible-confirming scientific discoveries but is encouraged by the expressed thoughts of some "non-theist" physicists, like Paul Davies, regarding the remarkable coincidences reflected in the exponential-precision design of a universe which has every immensely-complicated feature required to assure life sometime in its duration.

Ross questions much of recent quantum cosmology, including Stephen Hawking's quantum states of the universe suggesting the universe as having no beginning (boundary in time) as being unverifiable and therefore metaphysical speculation. He views the physics of Frank Tipler's evolving Omega Point God as fanciful.

'Logical Necessity'--Some Might Call This God--As Creating the Universe: Paul Davies, in *The Mind of God*, sees the universe's creation as from a *regression of causation*, resting ultimately, perhaps, on 'logical necessity' which, he observes, some might wish to define as God. Other physicists, like Freeman Dyson and Heinz Pagels, express somewhat similar ideas.

In *A Brief History of Time*, Stephen Hawking tracks the creation to quantum physics and mathematics. But, he asks, what is it that "breathes fire into the [abstract] equations" to create a physical universe? And he too regresses to the concept of "logical necessity," which, to the theologian, is just one whit better than the pure chance creation of quantum uncertainty.

Meaning and Frank Tipler's Omega Point God: Mathematician Frank Tipler, in *The Physics of Immortality*, develops his Omega Point Theory of God as evolving -- like Freeman Dyson's changing God of the Socinian heresy. Tipler's Omega Point God represents intelligence as information-processing, now in our human form, spreading throughout the universe in different forms, to the end or Omega point at which this intelligence represents Tipler's Omega Point God.

This spreading intelligence becomes total cosmic mind--the omniscient, omnipotent, personal Omega Point God, who, through computer simulation, resurrects all life that ever lived, or life which may be said to have existed under the 'many-worlds' interpretation of quantum theory.

His Omega Point Theory, of course, requires a number of hypotheses, such as a closed universe (sufficient mass to force ultimate contraction) which contracts but permits life in the form of processed information to survive by preserving the required energy differential through the 'Taub' universe 'shear' effects, sufficient to store the computer information as radiation during the final contraction state.

Tipler's Omega Point God, which, in some respects at least, resembles Freeman Dyson's Socinian evolving God--though considered by Ross as fanciful metaphysics, nevertheless would have meaning, because God exists and influences the nature and progress of the universe. In fact, even Tipler acknowledges that *he is a theist if his Omega Point Theory proves out.*

Meaning and Kant's Fundamental Problems: Eighteenth century philosopher-mathematician Immanuel Kant, in his deliberations on meaning and the nature of the universe, posed three 'fundamental' metaphysical problems: Does God exist? Do we have free will? Is there life after death?

Because these questions were viewed as 'metaphysical,' unverifiable, Kant con-cluded that we could only speculate about such things, although he reasoned his way to affirmative conclusions on the questions. Moreover, reality as measured by whatever is actually "out there," declared Kant, is forever constrained by the limitations of our mental equipment, our thinking processes.

Frank Tipler Addresses Kant's Questions: In *The Physics of Immortality* Tulane University mathematician Frank J. Tipler uses his *Omega Point Theory* to provide specially-defined "yes" answers to all of Kant's three questions, offering "mathematical proof" of the physics involved. So Tipler regards the questions as subject to physical rather than just metaphysical conclusions. He distinguishes, for example, between relativistic and quantum theory versions of free will. Tipler points out that the history of science has typically turned insoluble metaphysical problems into problems of physics "and then solved them."

Princeton physicist Freeman Dyson, in his *Infinite in All Directions,* contemplates a similar 'Socinian' evolving God, as did Jesuit evolutionist & paleontologist Pierre Teilhard de Chardin, in his *Phenomenon of Man,* and, Tipler says, his own use of the term 'Omega Point' is from Teilhard's work although Teilhard's God more closely resembles Freeman Dyson's evolving God.

Dyson's *Infinite In All Directions* adopts the imperfect God of the Socinian heresy, growing with the universe. Dyson hypothesizes the spread of life through an expanding open universe in which life influences physics to continue through eternity.

Critical to Meaning--Kant's Number One Question--Does God Exist? In The *Mind of God,* Physicist Paul Davies, ignoring Monod's admonition that scientists stick to science, resolutely explores the interface between science and religion-theology. He points out that many earlier 'metaphysical' questions now have physics answers. He asks a number of God-related questions. Included are questions involving meanings of the term 'God.' In addition to Kant's first question of whether God [of Western religion] exists, Davies, like Kant, asks whether God *must* exist.

Must the 'Infinite Regression' of Explanations Rest on Theology? Like most physicists, Davies notes in his *The mind of God* that he does not regard himself as a believer in the theistic Judaic-Christian supernatural, personal God. But he does acknowledge that perhaps the "infinite regression" of universe causation must come to rest on some foundation of what he assumes are natural phenomena involving "logical necessity."

Davies notes that these ideas may *cross into theology* and some may wish to identify this point of ultimate regression as God, although he prefers to couch these things in terms of 'design' and 'purpose.' Those ideas on felicitous design for life are associated with the 'anthropic principle,' discussed earlier, which addresses the extraordinary coincidences in the constants of physics which permit life in the universe.

Breathing Fire into the Equations: Even if the universe can be said to erupt from the mathematics of quantum uncertainty, Stephen Hawking asks, in A *Brief History of Time*, what is it that *breathes fire into the equations* to convert such abstractions as mathematics and unified laws of physics into a physical universe? Perhaps, suggests Hawking, the causation rests on "logical necessity." Davies goes on to explore forms of self-causation.

Profound Theological Implications from Quantum Cosmology: Physicists, like Heinz Pagels, David Bohm, Freeman Dyson, Frank Tipler, Paul Davies and computer scientists like Edward Fredkin, see the universe as an *information system*, an interconnected cosmic fabric of information and mind that underlies the physical universe. And in this context, they see life as a reactive 'information-processing system,' as we will be discussing.

In emerging quantum theory concepts, physicists, like Paul Davies, in *God And The New Physics* and *The Mind of God*, suggest that we face startling new realities in quantum theory, quantum cosmology, and self-organizing complexity physics, requiring a searching reappraisal of earlier notions about the nature of God and the universe--and ourselves as well.

Science and Theology Finding New Areas of Common Interest: As we will be discussing, though relationships between science and religion have never been altogether harmonious, they may be viewed as improving as science and theology find new areas of reciprocal interest. Theologians now regularly study physics, astrophysics and the evolving cosmologies. Religious organizations are interested in astronomy and astrophysics, and the Vatican is significantly contributing to discoveries in astronomy and cosmology.

While there are still spokesmen for religion expressing the view that the nature and existence of God are explained through inspiration and revelation, there are others who are receptive to broadened understanding through new scientific developments in cosmology.

As science has plumbed the subatomic quantum reality in which our material world disappears into a mystical etheriality of forces and fields, there are those who feel that this underlying non-material quantum world may lead us some day to *a spiritual field or energy form.* In astronomer Gerald S. Hawkings' *Mindsteps To The Cosmos,* human life proceeds, via stages or "mindsteps," to the discovery of a spiritual field, the "nonphysical cosmos," and an "age of the spirit...."

Science As a Way of Probing the Ultimate, a Path to God: The New Testament speaks of perspectives of God revealed through his Creation. St. Paul in Romans noted that we may also understand God by reflecting on His physical handiwork in the creation. Some physicists, like Davies and Tipler, suggest science as an alternative or even preferred avenue for bringing us closer to ultimate truth, to God.

Science a Surer Path to God? "It may seem bizarre, but in my opinion," says physicist Paul Davies in *God And The New Physics*, "science offers a surer path to God than religion. Right or wrong, the fact that science has actually advanced to the point where what were formerly religious questions can be seriously tackled, itself indicates the far-reaching consequences of the new physics...."

Mathematician and astrophysicist Frank Tipler, an acknowledged non-theist, declares in his *The Physics of Immortality,* that if the physics and mathematics in his Omega Point Theory (in which evolving intelligence becomes cosmic mind, the omnipotent, omniscient, personal God at the end or Omega Point) is found correct, then, as discussed earlier, he must declare himself a theist.

More to the Universe Than Meets the Eye? Davies acknowledges, of course, that his is "simply one man's perception of the universe; there are many others....[but] There is more to the world than meets the eye." Of course, there will be many, he recognizes, who will not share this view, who perceive science and theology as two independent and perhaps even inconsistent approaches to God.

John Polkinghorne--A Creation Leading to Saints and Scientists: British theoretical physicist, clergyman and theologist John Polkinghorne is often quoted on his thoughts of science-theology relationships. In *Alone is Never Enough,* the October 1994 Omni Magazine, Polkinghorne observes that he does not have to chose sides in the commonly perceived science-theology conflict--that "insights of both science and religion" are needed.

Both are concerned with the truth, he notes, and, he suggests, neither necessarily deals with immutable truth. Why are our minds such that we can not only understand the "everyday world" but the world of quantum mechanics which, he declares, is "shot through with signs of mind," interpretable as "an encounter with the mind of God?"

The Universe as the Home of Saints and Scientists: Science is possible, Polkinghorne declares, *because the universe is a creation and we are made in the image of the Creator.* He suggests that the history of the universe, which "has turned an expanding ball of energy into the home of saints and scientists" implies "purpose at work" as it "actualizes...endowed potentiality."

Scientific vs. Theological Surfaces in the Context of Biology: Boyce Rensberger's *Discovering New Worlds In The Realm Of The Cell,* prominently displayed in several issues of the Dec. 1988 Washington Post, suggested that early philosophers and biologists once invoked a "mystical vital force" that breathed life into ordinary matter

(Genesis 2:7--God breathes life into Adam), but that present biological physics and chemistry serve to explain phenomena of life in "purely materialistic terms." He was, of course, simply reaffirming the traditional scientific approach to the origin of life in terms of conventional scientific materialism, a view, however, not always popular with persons of strong religious orientation.

This Brought Interesting Letters to the Editor: Rensberger's reference to there being no need for a 'mystical vital force' breathing life into matter, brought editorial responses. A Letters to the Editor reaction by Christine Hanson Mercer said: "I hope he didn't mean to suggest that the God-concept is unscientific. My memory banks yield numerous examples of deeply religious scientists," citing her devout Catholic father, a scientist at the Johns Hopkins APL; Nobel prize-winning physicist Werner Heisenberg, author of the uncertainty principle in physics; and Albert Einstein: "... in this materialistic cage of ours the serious scientific workers are the only profoundly religious people."

She goes on to suggest that "Arguments that try to explain away God by using science, do a disservice to both." She suggests that people who engage in such sophistry instead spend a little time "examining the private agonies of their repressed souls."

The How and Why of the Universe As Philosophical and Theological: In another Letters to the Editor response, Robert A. Boisture asserts "...how and why our universe came to be....has always been, and remains, a philosophical and theological inquiry, not a scientific one.

As we grapple with this most important of all questions, the marvelous order and complexity of life which science describes becomes a strong circumstantial clue pointing to a conscious creator rather than random chance as the final answer." Implied here, of course, is a strong endorsement of the anthropic principle that there is a strong connection between the laws of physics and the universe's proclivity toward life.

> For the scientist who has lived by his faith in the power of reason, the story ends like a bad dream. He has scaled the mountains of ignorance; he is about to conquer the highest peak; as he pulls himself over the final rock, he is greeted by a band of theologians who have been sitting there for centuries. --Robert Jastrow

CHAPTER 25
SCIENCE AND THEOLOGY--
CHANGING RELATIONSHIPS

Science and Theology As Two Sides of the Coin: Implied in the work of some physicists, Freeman Dyson and Paul Davies for example, is the concept of science and theology as two sides of the same coin and the proposition that definitions of God need to be reexamined in the light of new knowledge of the role of mathe-matics and physics in the creation, the emergence of quantum cosmology, theories of quantum genesis, quantum-gravity, unification theory, and the physics of chaos and emerging complexity.

The Metaphysical Becoming Science? Do developments in science now suggest scientific as distinguished from theological or philosophical aspects of God? Are conventional metaphysical concepts increasingly being swept up into science? Must theology become a branch of science or lose its credibility and relevancy, as math-ematician Frank Tipler suggests in his *The Physics of Immortality*?

As we will discuss, there are many who recognize the growing role of science in cosmology. Unlike the days of Galileo, religious institutions like the Vatican recog-nize the growing need for buttressing theological concepts with developing scientific underpinnings.

Two Great Systems of Human Thought: In *God and the New Physics*, physicist Paul Davies observes that science and religion represent "two great systems of human thought." Religion, he notes, predominates in human lives while people encounter science not at an intellectual level but primarily through everyday technology.

Biblical and Natural Theology--Two Books of God: In *Infinite in all Directions,* Princeton physicist Freeman Dyson notes that according to Christian doctrine, God gave us two books in which his actions are recorded, one, the Bible, the other, the "Book of Nature." And Scottish philanthropist Adam Gifford, sponsor of the Edinburgh, Scotland scientific lectures of that name, observed that "natural theology is the reading of God's mind as expressed in the works of Nature."

The Earlier Religious Suppression of Scientific Discovery: Science today is still mindful of earlier suppressions of science by the Church: of the Polish astronomer and cleric, Copernicus, for his theory that the earth orbits the sun; of the Italian astronomer-physicist, Galileo, silenced by the Inquisition for similar thoughts; and of the 16th century Italian Dominican Monk, Gordiano Bruno, burned at the stake in Rome for the Copernican "heresy" of the planets revolving the sun, and of the glory of God as reflected in the infinity of other suns and planets.

Galileo's Dispute with the Church: In Galileo's dispute with the Church over banning the advocacy of the Copernican theory that the earth revolved around the sun, Galileo appealed to Church officials to permit publishing an impartial account of both the Church's and scientific theories. When the published 'impartial' book was recognized as making the case for Copernicanism, Galileo was sentenced to house arrest and enjoined to disavow the theory. As a faithful Catholic, he obeyed. But Copernican astronomy, abandoning Earth as center of the universe, was off and running. Officially forgiven in 1981, Galileo's heliocentrism was banned until 1835.

The Vatican's Interest in Science: In the centuries since the 1633 trial of Galileo, the conflict between science and religion has diminished to a considerable extent although the tension is still expressed, more as polemics or academic debate, though not without some passion. Although many currently view science and religion as increasingly *interdependent* rather than adversarial, some still draw lines in the sand.

But both physicists and theologians have noted the diminishing tension. In *A Brief History Of Time*, Stephen Hawking notes the recent church-sponsored dialogue between the Catholic Church and scientists concerning Big Bang theories of creation. Church denominational, educational, and journalistic materials show an open and active interest in new developments in science though they may differ on the implications.

The Vatican's Arizona Advanced Technology Telescope: Vatican interest in science is not limited to sponsoring world-class science conferences such as on cosmology. For centuries the Vatican astronomers have studied the heavens. The work in Rome is now supplemented by hands-on work of Vatican astronomers at its Arizona Mount Graham Advanced Technology Telescope constructed by the University of Arizona. Reverend Chris Corbally, a Vatican astronomer, speaks of the spiritual dimension in the work of great minds in science, something encouraged by the Church in its dialogue with the scientific community.

Such Catholic involvement in cosmology and astronomy, in the Copernican and Lemaître tradition, is now an established part of the Vatican's exploration of God's handiwork. In the Vatican perspective, physics and cosmology is viewed as essential to the theological search for truth.

John Polkinghorne--Physicist, Clergyman and Theologian: Because of his unusual science-theology interdisciplinary credentials, the British theoretical physicist, clergyman and theologian John Polkinghorne, President of Queens College, at Cambridge, England, is often quoted on his thoughts of science-theology rela-tionships. In *Alone is Never Enough,* October 1994 Omni Magazine, John Polk-inghorne observes that he does not have to "chose sides" in the commonly perceived science-theology conflict--that the insights of both are needed.

While each is concerned with the truth, he suggests, *neither necessarily deals with immutable truth.* Science, he says, interweaves fact (experiment) with opinion (theory). And religion is the search for understanding human experience "rooted in worship, hope, and the history of holiness" reflected in "great religious figures of world history."

Science Asks How? And Religion Why? Science asks *How*, and religion *Why*, Polkinghorne says. While science treats the world as an object, religion invokes the personal encounter where "testing" gives way to "trusting." He cites the example of music which, in physics, represents vibrational energy and in religion provides emotion and meaning, including spiritual uplift. Of course, emotional effects can, in turn, be explained in terms of the chemistry of brain activity, an extension of physics.

The Universe a Creation Leading to Saints and Scientists: Some questions in physics, Polkinghorne says, seem to point in a religious direction. Why are our minds so constructed that we can understand not only the "everyday world" but the strange quantum world, "shot through with signs of mind," and interpretable as "encounter with the mind of God?" Science is possible, Polkinghorne declares, *because the universe is a creation and we are made in the image of the Creator.*

The Science of the Sacred: In *The Search for the Sacred,* Newsweek, Nov. 28, 1994, Sharon Begley addresses *Science of the Sacred--From Quantum Mechanics to the Evolution of Life, Rationalists Discover Awe.* She triggers an immediate alert by referring to science's "usurpation of the place of faith in the last years of the millennium."

How Science Can Nourish the Spirit: Begley cites the NASA 1992 COBE Satellite findings of early gravity distortions in the Big Bang cosmic radiation, and other new discoveries in physics, astronomy and biology. These discoveries, she notes, are leading physicists and theologians to express the metaphor of these discoveries as disclosing the *handwriting of God*, inspiring "a sense of cosmic piety, of serene holism nd even a moral code." She observes that these scientific contributions to knowledge of the universe exemplify how science can "nourish the spirit."

Omega-Point Theory--Proof of Personal God & Resurrection: Begley cites Tulane mathematician Frank Tipler's mathematical 'proof' in his Omega Point Theory, updated in Tipler's *The Physics of Immortality,* that all creatures who ever lived will, in the "end"--the Omega Point representing the evolving God--be resurrected--in a better environment--through computer simulation.

In the *Science and Religion* issue of Omni (Oct. 1994) Frank Tipler further explains his *The Physics of Immortality*, citing the New Testament Book of Revelation reference to God as the Alpha and Omega, and further commenting on how physics proves the ultimate (future) existence of the personal God, and the physics of how, perhaps, at the contraction of the universe in the Big Crunch, the information necessary for the resurrection is preserved.

Begley's point, of course, is that we have contemporary examples of science as more compatible, and in some ways at least, even *corroborative* of religious beliefs--something earlier studiously avoided by science. This point, of science as corroborative rather than incompatible with religion, is vigorously pursued by astronomer-minister Hugh Ross in his writings and lectures and especially in his *The Creator and the Cosmos*; *Beyond the Cosmos*; and *The Fingerprint of God.*

All that Existed Were the [Generative] Laws of Physics: Orienting her discussion to the creation event, Begley refers to [quantum genesis] notions that in the beginning, time, space and matter did not exist, only the [generative] laws of physics. She cites the "sense of oneness with the cosmos," as *beyond awe*.

Humanity As Critical Component of the Universe: Begley cites some thoughts of physicist, ordained-minister and theologian Robert Russell, professor of theology and science at the Graduate Theological Union (GTU). Russell, who is also director of the Center for Theology and the Natural Sciences (CTNS) at Berkeley, observes that humanity, once thought of as "an isolated fact in nature," is now perceived in cosmology as a "critical component" in cosmic unity.

"Existential" Physics--Observer-Created Reality? In this same vein, she quotes physicist-theologian Ian Barbour of Carleton College to the effect that we are related to all creatures past and present as well as to the non-living. Barbour notes that the view of human life and the universe as an "interconnected whole" is supported by quantum theory.

Barbour cites Princeton physicist John Wheeler's references to certain concepts she calls *existential physics*, in which the universe as a quantum system requires *observer participation.* Reality, in this view, requires interaction of another quantum system, the mind, to bring the universe to physical actuality, something, she observes, which some lay people take as a matter of faith.

Natural Law and Chance As Equal Instruments of God's Intentions: In conventional physics, including relativity theory, the universe appears deterministic--everything governed by cause-and-effect, starting from the creation. However, at the quantum or subatomic level, events are governed by a particular form of chance--quantum uncertainty, the very principle which, in quantum-genesis theory, generates the universe by a chance eruption from the quantum void.

Through quantum uncertainty, there erupted an incredibly small particle or point of supersymmetry and unified forces, a *superforce*, which, in inflation theory, exponentially inflated to trigger our still expanding universe. And so, says Begley, Barber perceives natural law and chance as equal instruments of God's intentions. Thus there can be "purpose without an exact predetermined plan." This, says Barbour, may offer comfort to a humanity “vulnerable to unpredictability and chance."

With this background of scientific discoveries bearing so profoundly on the nature of God and the creation, it would be foolish, muses Begley, to ignore science because, she asserts, *few secular quests approach the sacred as closely.* In *The Physics of Immortality*, Frank Tipler goes further. He urges that theology “join science lest it [theology] become meaningless.”

The Common Ground Shared By Science and Religion: In *Science & Religion, Blurring the Boundaries*, Omni Magazine, Oct. 1994, Margaret Wertheim reviews the work of physicist, theologian and ordained-minister Robert Russell, who, she observes, is at the forefront of an increasing segment of theologians and scientists whose religious faith and scientific interests are not incompatible.

Russell is founder of the Center for Theology and the Natural Sciences (CTNS) in Berkeley, California. This is on the campus of the Graduate Theological Union (GTU), near UCAL, Berkeley, where Russell is professor of theology and science. He is also an ordained United Church of Christ minister, and a member of the Society of Ordained Scientists.

Complementary Aspects of Science and Religion: Werthheim notes that Russell, who emphasizes "the common ground shared by science and religion," was early advised to disassociate his science from religion and de-emphasize his preoccupation with Christianity. He challenged this advice, proceeding to seek masters degrees in both physics and divinity, in the pursuit of the *conjunction* of cosmology & science and theology.

But unlike earlier views, such as those of Thomas Aquinas, that the emphasis on science should be to serve Christianity, and as Newton suggested, used to demonstrate the existence of deity, Wertheim notes that contemporary thought moves in the direction of showing the "mutual" and "complementary" aspects of science and religion--two sides of the same coin.

Quaker Physicist George Ellis and Priest Georges Lemaître: Quaker, and University of Capetown professor of applied mathematics, George Ellis is a former president of the international Society of General Relativity and Gravitation and coauthor, with British theoretical physicist Stephen Hawking, of *The Large Scale Structures of Spacetime.*

Wertheim cites Ellis' views that *the foundations of the latest physics manifest signs of a providential deity* and point to the fact that Einstein's most ardent supporter of general relativity's principle that space and time are a continuum and gravity is geometry--spacetime curvature, was physicist Arthur Eddington, an ardent Quaker. So also, notes Wertheim, Abbé Georges Lemaître, a Catholic priest and brilliant physicist, to Einstein's delight, demonstrated Einstein's equations as supporting an expanding universe originating from a Big Bang.

Wertheim goes on to cite the "fruitful relationships" between the Center for Theology and the Natural Sciences (CTNS) and the Vatican Observatory in Rome, involving biannual conferences of scientists, theologians and philosophers. CTNS is also exploring the theological implications of the Human Genome Project to decode and catalog the genes in human chromosomes.

Jesuit William Stoeger: Unless Current, Religion Becomes Anachronistic: Russell says that theology must keep current with the times, a point also made by Jesuit priest, and astrophysicist at the Vatican Observatory, William Stoeger, who points out that if religious people continue to rely on "categories of thought from the Renaissance or the Middle Ages," then religion will be perceived as outdated.

If "God as Creator"concepts are to continue to make sense, he says, then they need to be expressed within the broader cultural context, part of which is "modern science and cosmology." We need to be able, notes Stoeger, to see *just how God could be working within the natural processes revealed by contemporary science*. Here we note the growing tendency of theology to shy away from supernatural explanations.

God Acting through Natural Processes: Stoeger's reference to God functioning through "natural processes" echoes the views of many that God's action are consistent with the laws of physics--without need for resort to supernatural mechanisms (miracles). We discuss later, physicist Richard Feynman's analysis of how exquisite abstractions like love, compassion, and holiness evolve through a natural developing process of increasing order and complexity.

In the past, this has led some physicists toward the conceptual notion of the *UniGod*, where God represents nature rather than some remote supernatural deity who somehow resorts to magic or miracles rather than classic cause-and-effect rules and quantum uncertainty mechanics.

Mathematics as Underlying the Universe: Discussed earlier are views of God as representing the laws of physics and their underlying mathematics, and suggestions, like those of P.W. Atkins in *Creation Revisited*, that underlying the laws of physics is mathematics which alone brings the universe into existence and administers its evolution through benign "directionality," all *without "external intervention."*

Contemporary Science Can Provide Scope and Insight for Faith: But getting back to both Russell and Ellis, they argue, says Wertheim, that physics has criticized and restructured traditional theological positions and, in the process, says Russell, "contemporary science can provide scope and insight for faith." Particularly important are the biological sciences, genetics, involving a collection of technologies with "profound theological consequences."

As a professor of systematic theology at the Pacific Lutheran Theological Seminary, Russell is head of CTNS's Human Genome Project research team and he points to the problems derived from the fact that *DNA has already acquired a sort of tacit sacrality in our culture.*

The Problem of Genetic Determinism: Russell goes on to discuss the problem of genetic determinism, human freedom and free will and whether we can be held accountable in the eyes of God, or the State, if our traits and propensities are genetically determined.

He concludes that *there must be three parts to this: your genes, your environment, and finally, yourself.* He notes that it is the "self" which makes the decisions. But, he asks, where does this self come from? That's where the goal must be to understand real "freedom" rather than just "indeterminacy."

We Need Both Forces Co-existing in Our Lives: And so, says Wertheim, Russell emphasizes the need for mutual respect. The aim of a religion-science dialogue is to learn how both forces may "coexist in our lives." Russell hopes that the CTNS is able to increase public outreach to both the religious and the scientific communities.

Cooperation in Coping with the Environmental Crisis: Space scientist Carl Sagan (Washington Post "Parade" magazine, Mar. 1, 1992), pointed to the extraordinary and growing cooperation between science and religion in coping with the environmental crisis.

He points out that this was evidenced by the remarkable display of unity by distinguished representatives from almost a hundred nations at the Global Forum of Spiritual and Parliamentary Leaders. The forum which met at Oxford in April 1988 and in Moscow in January, 1990, emphasized the theme of the "interconnectedness of all human beings."

Joint Appeal by Science and Religion: At the Moscow conference, an appeal, signed by distinguished scientists, was presented to world religious leaders. The response was overwhelming, says Sagan. One response was the establishment of the "Joint Appeal by Science and Religion for the Environment." Such joining of spiritual and scientific forces, on a global level, is remarkable evidence of how much the earlier schism between science and religion has diminished in favor of concepts of "joint stewardship."

Sagan quotes Pope John Paul II in a remarkable statement which, not so long ago, would have been viewed as sheer heresy: "Science can purify religion from error and superstition; religion can purify science from idolatry and false absolutes. Each can draw the other into a wider world, a world in which both can flourish....Such bridging ministries must be nurtured and encouraged."

"Mind of God"--Words Made Famous by Stephen Hawking: But back to the title of Davies' book *The Mind of God.* These words have been made famous in their use by a number of physicists, in particular, British theoretical physicist Stephen Hawking, through his *A Brief History of Time*. In his concluding chapter Hawking had observed that there may be "only one or a small number [of unified theories]...that ...allow...structures as complicated as human beings who can investigate the laws...and ask about the nature of God."

But, even if there is only one theory, he noted, it is just rules and equations. "What is it that breathes fire into the equations for them to describe?....Why does the universe go to the bother of existing? Is the unified theory so compelling that it brings about its own existence? Or does it need a creator, and, if so, does he have any other effect on the universe? And who created him?"

Hawking concludes with his now-famous observation that "if we do discover a complete theory.... we shall all ...be able to take part in the discussion of the question of why it is that we and the universe exist. If we find the answer to that, it would be the ultimate triumph of human reason--*for then we would know the mind of God.*"

The Clash of Ideologies and the Science-Religion Interface: Hawking's oft-quoted proposition is as challenging as it is contentious--that a complete unified theory might explain why we and the universe exist, and , in turn, reveal "the Mind of God." It has probably attracted more attention, among scientists and theologians alike, than any other such statement in our time. So use of this title, *The Mind of God*, by Davies in his fascinating plunge into the "clash of ideologies" at the "science-religion interface," is understandable at the same time that it is provocative.

Will a "Final Theory" Answer Our Deepest Questions? In *The First Three Minutes,* Nobel-prize-winning physicist Steven Weinberg, a major contributor to unification theory, commented that the more we learn of the universe the more it seems *pointless*, for which observation he received considerable negative publicity. In his later *Dreams of A Final Theory*, in a chapter called "What About God?" he observes that science has found no "cosmic imperatives" in its discoveries about the universe.

The Laws of Nature and the Mind of God: But, he says, if there were anything we could discover in nature that would give us some special insight into the handiwork of God, it "would have to be the final laws of nature which would represent the book of rules that governs stars and stones and everything else." So, it is "natural," he says, that Stephen Hawking should refer to the laws of nature in the context of "The Mind of God."

Weinberg Does Not Share Hawking's Optimism--Always Deeper Principles: But it becomes clear that Weinberg does not share Hawking's optimism about any final theory as really giving us insights into the mind of God. Whatever one's religion or lack of it, he acknowledges that it may be "an irresistible metaphor to speak of the final laws of nature in terms of the mind of God." But as other physicists like Paul Davies have observed, Weinberg envisages no "ultimate principles" that do not require further regression, or some "deeper principles," as he puts it.

Conflict between Science and Religion? Weinberg believes that science will find no "hint of God" or "standards of value or morality," although, he notes, we may find these elsewhere. Weinberg is unaware of any scientist who finds scientific evidence for a divine being, although there are scientists, like astronomer-minister Hugh Ross, as we have discussed, who see scientific discoveries, like the NASA COBE satellite findings of primordial temperature variations in the cosmic background radiation as doing just that.

The Universe, God, and Mathematics: Works like Davies' *The Mind of God* and *God* and *The New Physics* suggest that the universe, its laws of physics, mathematics, logic---and God--- are somehow inextricably entwined, and ultimately represent mathematics and "logical necessity" which, he notes, one may wish to call God.

We are reminded that, like Sir James Jeans and Einstein, physicists tend to think of God in terms of mathematics, that is, as a mathematician or as the logic of mathematics. So, in this context, "The Mind of God" title may be viewed as a metaphor for the proposition that mathematical logic underlies the universe.

The "Mind of God" Emphasizes Mathematics: Davies speaks of the chain of causality of the universe perhaps coming to rest in God, logic, laws, "or some other foundation," all somehow related to mathematics. His book, explains Davies, represents his "more considered attempt" at exploring this "science-religion interface." A central theme is the tracing of the universe and its laws to mathematics and logic.

Mathematics, Logic, and the Creator: Physicists associate physics and logic, via mathematics. Physicist John Wheeler employs the Aristotelian syllogism to conclude that *all physics reduces to the mathematics of logic;* that "Logic is the only branch of mathematics that can think about itself." Paul Davies observes that "Mathematics is the poetry of logic," something inviolate-- even by God. Some are led to conclude that God is Logic--applied through mathematics.

Mathematics As the Structure of the Universe: In *Creation Revisited*, P.W. Atkins goes further than most, suggesting that mathematics, "the language of nature," not only "accounts for" the structure of the universe but *is* the universe's structure--that *the answer to why mathematics works, answers "all questions of being."*

Scientists and Theologians--Still the Schism: History shows cosmology as among man's earliest writing among the Egyptians, Mesopotamians, Greeks and Romans. It preoccupied the early Church as we have seen from the proof of God writings of St.

Augustine and St. Thomas Aquinas. Early discoveries in astronomy signaled the start of questioning of early church cosmology.

Differing Perspectives: We have seen how science and theology in recent times are finding accommodating concepts in their search for answers to the enigma of the creation. But in *The Fingerprint of God*, astronomer-minister Hugh Ross reminds us that "bitter animosity [still] divides scientists and theologians.... who, despite their education, ...are woefully uninformed of each other's disciplines."

He attributes this in part to theological reliance on revelation, faith, exegetical methods, and belief in a transcendental creator, on the one hand, as contrasted with scientific reliance on observations, human reasoning, scientific method, scientific objectivity and receptivity to self-creation ideas. Ross goes on to detail how relativity theory confirmed the biblical creation concept but requires long astronomical and geological days rather than the "creationist" literal reading of Genesis days as 24-hour earth days.

> Hawking embarks on a quest to answer Einstein's famous question about whether God had any choice in creating the universe. Hawking is attempting, as he explicitly states, to understand the mind of God. And this makes all the more unexpected the conclusion...a universe with no edge in space, no beginning or end in time, and nothing for the creator to do. --Carl Sagan

CHAPTER 26
SCIENCE, RELIGIOSITY& HUMAN LIFE

Fritzsch: Modern Cosmology, Cosmic Evolution, Science and Meaning: In *The Creation of Matter, The Universe from Beginning to End,* physicist Harald Fritzsch of the University of Munich, the Max Planck Institute for Physics, and the European Nuclear Research Center, CERN, describes modern cosmology, the history of the universe, which he divides into "epochs," and the origin of matter. Of particular interest here are some of his thoughts in his search through physics and metaphysics for meaning.

On a Journey from Nowhere to Nowhere? Are we on a journey "from nowhere to nowhere," he asks, *or is there meaning?* Does religion still make sense where science posits creation without divine intervention? *We may solve the riddles of science but who sets the values to provide meaning?* Not intent on human life's belittlement, he hastens to add that we are not only part of the universe's history but "makers" of that history, seemingly, *an integral causative part of the evolutionary process itself.*

The Role of Intuition and Faith in Science: Fritzsch speaks of the role of intuition in science. Einstein had finished his work on general relativity; the predicted bending of starlight by the sun's gravity had been demonstrated. When asked how he would have responded if his prediction had not been valid, the humble Einstein, who relied heavily on intuition, impulsively replied, in the ultimate expression of faith, that he would "feel sorry for God had He missed this opportunity."

The Universe Won't Last Forever: Speaking of eternity, Fritzsch suggests the universe has long to go but that either quarks and atomic nuclei--protons and neutrons--will disappear, decay, in an endless expansion or, in a closed universe, which will end in the contraction implosion--the Big Crunch.

Of course, recent mathematical models, like those of physicist Stephen Hawking, also provide for "rebirth" or oscillation--expansion and contraction. But in any of these alternatives our familiar configured universe--and any life forms it harbors, will vanish. This is so, of course, unless, as suggested by Freeman Dyson in *Infinite in All Directions*, and Frank Tipler in *The Physics of Immortality*, intelligence spreads throughout the universe, comes to control physics, and finds a way to preserve life forever.

In Heinz Pagel's *Perfect Symmetry*, he discusses whether the production of intelligent life--capable of contemplating meaning, the nature and purpose of the universe, will or not have been meaningful if all is to end in the primordial chaos of a dying universe.

The Galaxies: To set the stage for discussing "unity with the universe," Fritzsch speaks of the structure of the universe; the endless great galaxies and our own "wagon wheel" galaxy, the Milky Way which, from a distance, would look like the Andromeda Galaxy, two-million-light-years distant.

We note that recent findings reported to the American Astronomical Society show the Milky Way as a 'barred spiral' rather than a typical spiral galaxy like Andromeda. For comparison, the Sun is but eight light-minutes from Earth. By looking at Andromeda from Earth, he says, we know what Andromeda looked like two-million light years ago, before the earth was inhabited. *We don't know what Andromeda looks like now--or even whether it still exists.*

The Search for Meaning of the Universe Inbred in Us--The Eighth Epoch: Fritzsch has a purpose in creating this sense of oneness with something so vast. He makes the point that neither we, nor the entire universe for that matter, are eternal.

Whatever may be the meaning of it all--and the search for the meaning of it all is inbred in us, we are all in it together--we, our ancestors and our descendants, all part of his "eighth epoch" of the universe. The eighth epoch, he says, is our time, the period, 1 million to 20 billion years, in which matter condenses, the galaxies, stars, planets form, and life develops.

The End of Our Sun: He describes the predictable end of our sun. Things will rock along for the next five billion years or so and then the sun will start getting larger and brighter until, in eight billion years, it will be a hundred times bigger and two thousand times brighter, a red giant star which has consumed its planets. Ultimately (an estimated 10^{14} years) it will cool into a white earth-size dwarf star and finally into a black lifeless dwarf star.

Humankind Has Not Found the Reason for Its Presence: In his 17th and last chapter, Fritzsch makes the arresting observation that *in all of man's search of the universe he has not found the reason for our presence and a way of using science to find ethical values.* He quotes Einstein's reference in *Out Of My Later Years*, to the reminder of the "limits of our rational understanding of our existence," and our inability to substantiate the "justification and value of the pursuit of truth."

Says Fritzsch, we don't know why the earth exists, the reason for the Big Bang--or even whether there was a reason. "The universe makes no sense: it is absurd," and he then quotes the essays on this subject of Albert Camus, awarded the Nobel Prize for literature.

Interdependency Rather Than Opposition: Speaking of the "complementarity of science and religion," by which he means their mutual exclusivity as in the particle or wave states in quantum uncertainty, Fritzsch insists that religion and science should not be viewed as adversarial but as interdependent. Fritzsch echoes Einstein's concern in behalf of interdependence: "Science without religion is lame, religion without science, blind," he insisted.

Religiosity Provides a Framework for Meaning: Fritzsch notes *the special role of religiosity in filling the gap in meaning.* Religiosity, says Fritzsch, provides a framework for ethical values. We are "imbedded in the inexhaustible continuum of possibilities" of our, the eighth, epoch, including the not inconsiderable power to "shape the course of history." He emphasizes uniqueness--our uniqueness, the uniqueness of our actions and the uniqueness of every second of time that unfolds.

Meaning and Experiencing Cosmic Unity: The meaning, he says, lies in living, the experience of the cosmic unity that transcends the individual, individual consciousness--and acceptance--of one's potential and limits--one's acceptance of life here. And so God is the unity of the universe in its remarkable diversity, muses Fritzsch.

Idea of God Brings Meaning: And so, in the face of "lost illusions" and the desperate need for meaning, says Fritzsch, *the idea of God arose to dispense with meaninglessness*, to deal with the question of the meaning of life. He is the cause of all being, the mediator, the assurance of life after death. He quotes French Nobel Prize-winning biologist author, Jacques Monod, known for his views on human life's presence as the result of genetic mutation. *Monod cites man's search for meaning as an 'imperious need,' 'inborn' and 'inscribed somewhere in the genetic code.'*

Irreconcilable Contradiction Between Science and Religion? Meaning, as we have discussed earlier, is generally associated with God and purpose. But, asks Fritzsch, is there room for God in a universe that, seemingly, can dispense with His intervention in creating the universe? And what about God, he asks, if the universe of the mind disappears? Are religion and science an "irreconcilable contradiction?"

Fritzsch does not accept irreconcilability. He speaks of scientific method establishing connections, the "road of ideas," chains of causality, between concepts of scientific phenomena. But he cautions against confusing this scientific concept network with the "real world" which is a "continuum" that can't be defined by a network of single-dimension "roads of causality."

He speaks of religiosity as *the intuitive sense of cosmic unity* experienced by Spinoza and Einstein and ingrained in Eastern religion. In our discussion of concepts of God we discuss theism--the personal, involved God, deism--the God who set up the system to be self operative, and pantheism--the idea of everything, including human life, as part of the totality of Nature, of God. Like Henry David Thoreau at Walden Pond, many see God in the beauty and serenity of the wilderness--and in the miracle of each other.

We see here the inclination of scientific thought trending toward concepts of deism and pantheism, as do the Eastern religions of Hinduism and Buddhism and Taoism. Buddhism's eightfold path to a divine state of consciousness, Nirvana, says Fritzsch, is like the eightfold path of the evolution of the universe. This sense of separation from the individual and oneness with the unity of the cosmos is the essence of Buddhism, he notes.

"Such a religiosity has no room for a personal god who rewards and punishes. It is part of our universe of the mind, a product of the human spirit like music and art, a product of our culture." But in our discussion of God and religion, we have seen that linguistics plays a fundamental role in expressing such concepts; that aspects of theism, deism and especially pantheism, involving the feeling of one with the cosmos, are not mutually exclusive.

The Complementarity of Science and Religion: Speaking of the "complementar-ity of science and religion," Fritzsch insists that religion and science are not adversarial but interdependent. Fritzsch echoes Einstein: "Science without religion is lame, religion without science, blind."

Religiosity, Uniqueness, and Acceptance of Life Here: Fritzsch sees meaning in religiosity, our uniqueness and our acceptance of our life potential. Religiosity, he says, provides a framework for ethical values. We are "imbedded in the inexhaustible continuum of possibilities" of our, the eighth, epoch, including the not inconsiderable power to "shape the course of history."

He emphasizes uniqueness, our uniqueness, the uniqueness of our actions and the uniqueness of every second. Meaning, he says, lies in living, the experience of the cosmic unity that transcends the individual, in individual consciousness, and acceptance of life here with both its potential and limits. And so God is the unity of the universe in its remarkable diversity, says Fritzsch.

The Reciprocal Exclusion between Science and Religion is Past: He cites the four centuries, since the Vatican's trial of Galileo, that science and religion have been at odds and asserts that *the time of reciprocal exclusion between science and religion is past*; we find ourselves in "interdependent worlds," and he regrets the sharp dividing line still drawn by some. Steven Hawking, in his *A Brief History Of Time*, notes the recent dialogue between the Catholic Church and Science concerning Big Bang theories of creation and we noted earlier, the Vatican's active role in astronomy and astrophysics.

Scientific and Biblical Compatibility: The Bible, he says, is primarily a historical document "subject to reinterpretation," in the light of the discoveries of the whole field of science, physics, chemistry, geology, molecular biology, genetics, astro-physics, particle physics and so forth. Knowledge of the cosmos, he says, has "made us humble but has freed us from the captivity of myth" and given us rightful pride in our unique albeit minor roles in the universe.

Early Theology Interpreted Scripture Consistent with Reality: We noted that early theology was able to interpret scripture to avoid incompatibility with reality. As far back as the fifth century, for example, St. Augustine did not accept the "captivity of myth" and interpreted the first six days of Genesis as long periods, and the planet's life species as having originated from the evolution of primordial matter.

An Irrelevance to All of This? And isn't there an irrelevance to all this since some day in the distant future we will all be gone, without trace. Fritzsch rebels against the thought that there is no point to all of this. Our planet, glowing in the dark of the cosmos, is "our home." We are, he says, all enmeshed in the complex network of relationships and dependencies of the eight epoch, without which there could be "no life, no awareness, no human spirit, no God." The universe imposes the duty on us to preserve this world, he says, a view expressed by other scientists like physicist Freeman Dyson and Heintz Pagels.

Some Intuitional Insights: Interestingly, these concluding observations reveal something of Fritzsch. Despite his erudite and persuasive description of a pantheistic universe of unity in the cosmos, his last cry is to struggle to give continued meaning to our individuality, to preserve this haven in the cosmos where human spirit exhibits awareness of the universe and dependency on God. Here we see the intuitional longing in the human spirit for the personal caring God to whom each of us in the human race can directly relate.

Theologian Paul Tillich points out that one's deeply held intellectual views, even concerning the merits of the atheistic explanation of the universe, can only be expressed as an affirmation of and in the name of the personal God of theism. Even the Nobel Prize-winning mathematician, philosopher and logician, Bertrand Russell, evidences a sadness and disappointment in his denial that human life and our planet is the product of Divine Plan and his assertion that we are the result of a universe of chance.

Two Realities, One of Physics and One of Our Experience: In explaining his philosophic approach to describing the history of the universe, Fritzsch describes his conversation with a student at the University of California at Santa Barbara. The student felt truth in being one with nature, and being content with one's experience. In essence, the idea of scientists picking everything apart, somehow detracted from the perception of reality gained from the constant stream of impressions of daily experience.

Cited is the atomic scientist Niels Bohr's reference to wholeness and detail, the two extremes. The idea here is that there are two different realities: the holistic, what we perceive through the totality and aggregate of our experience in the world and the reductionist view of just the physical characteristics of the physical structure of the universe.

Max Planck on Realities Greater Than Experience: In his essay on "Physics and World Philosophy," in *The Universe in the Light of Modern Physics*, Max Planck says that scientists have learned that science must go beyond the "perceptions of the senses," that science cannot exist without "some small portion of metaphysics," that there are realities apart from such sense perceptions, realities "of greater value for us than the richest treasures of the world of experience."

Theology and Science See Two Distinct Realities: To an extent, it is this basic distinction that separates theology and, to some extent, all of metaphysics, from the requirements of a purely "scientific reality." We eat and enjoy ice cream, bowl, play or listen to great music, love, pity, and believe in God, or abstractions of some sort, all important to us. We have no intrinsic interest in quantum physics which shapes these things into physical structure. In a sense, we deal with the tangible end products on the one hand, and the intangibles which enter into everything that has meaning.

The 'Next World' Not the World of Physics: The "next world" of theology, or at least of theistic religion, is not dependant on physics or materiality and even the meaning of this world, in a sense, is unrelated to laws of physics--even if those laws are what makes possible our very existence. In fact it is often this very materiality that is blamed for the evils of the world.

Why didn't God chose a nonmaterial way to bring us into existence? Then there would be no evil and supposedly we wouldn't deteriorate--and die. No doubt the New Testament idea of the "spiritual body" of the next life was influenced by this distrust for the physical structure of which we are made and from which we live out our lives.

Planck Speaks of the Younger Generation Groping for Meaning: Max Planck, of quantum theory fame, pointed out that the vast and never ending stream of new discoveries and kaleidoscopic changes lead the younger generation to grope in every direction for some lasting element of permanence, for peace for a weary spirit. While it is the Church's function, he states, to meet these needs, "its demands for an unquestioning belief serve rather to repel the doubters" who then have "recourse to rather dubious substitutes...."

The Role of Hypothesis and Faith in Scientific Investigation: Planck goes on to suggest, however, that every effort in scientific endeavor is "guided by an unpretentious but essential principle." Scientific material is always incomplete and the gaps must be filled in by the association of ideas from the investigator's activity which may be described as hypothesis or faith. "The disjointed data of experience can never furnish a veritable science without the intelligent interference of a spirit actuated by faith..."

The universe is but one vast symbol of God.--Thomas Carlyle
The universe is one of God's thoughts. -- Friedrich Schiller

CHAPTER 27
GOD AS NECESSARY FOR MEANING--
THE CHANGING GOD

God As a Necessity for Meaning: Many people, especially religious persons, see a universe without God as meaningless and foolish, reminding us of Psalms 14:1: The fool has said in his heart, "There is no God." In *The Creator And the Cosmos*, astronomer and ordained-minister Hugh Ross indulges in a clear and cogent expression of these sentiments.

He declares that if the universe is "accidental" (emerging by chance from quantum-uncertainty), then the universe has "no objective meaning," and so it follows that life, "has no meaning." Everything is determined by a "mechanical chain of events." And, he declares, morality and religion, though "temporarily useful," are "ultimately irrelevant."

The Creator As Establishing the Meaning and Purpose of Life: But if the universe is "created," says Ross, then there must be "reality beyond the confines of the universe" and it is the Creator that is that ultimate reality. Thus the Creator is not only the source of life, but "establishes its meaning and purpose." And he adds, it is the "Creator's personality that defines personality" and "the creator's character that defines morality." Ross would endorse Hebrews 11:6 that "He who comes to God must believe that He is a rewarder of those who diligently seek Him."

So to study the origin and development of the universe, he observes, is to investigate "the basis for any meaning and purpose to life." Thus, he concludes, "cosmology has deep theological and philosophical ramifications."

Mainstream Physics Goes Only As Far As the Anthropic Principle: But not all physicists share a view of urgent necessity of Ross's theistic God to assure meaning and purpose. Some physicists, like Paul Davies in *The Mind of God* and George Greenstein in *The Symbiotic Universe*, are willing to give some credence to versions of the anthropic principle and, from the principle, to imply some aspect of providential design and purpose.

The "strong" version of that principle holds that there is evidence of "design" or purpose in the remarkable coincidences reflected in the exponential precision in the constants of physics which assure life in the universe. Physicists Davies, Greenberg, Rothman, Kistiakowski, and others, seem inclined in this direction.

An Infinity of Universes to Assure Life? While some physicists suggest that the "many-worlds" interpretation of quantum theory rules out a universe of "purpose," physicist Fred Alan Wolf, a devotee of the "many worlds" interpretation of quantum uncertainty, in his *Parallel Universes*, goes so far as to suggest that the creation of infinite universes is God's way of assuring life somewhere in that infinity.

A Universe Without Purpose? Some scientists, like Stephen Hawking, Heinz Pagels, Steven Weinberg, and P.W. Atkins may on occasion resort to the "weak" anthropic principle to the effect that the fact that we are here necessarily means that the constants of physics must be such as to allow for our presence.

Steven Weinberg's "Pointless" Universe: The distinguished, unification theory Nobel-prize-winning physicist Steven Weinberg got into real trouble with his best seller *The First Three Minutes,* in which he had declared that *The more the universe seems comprehensible, the more it also seems pointless.*

His statement had immediate and dramatic responses, some with approval but more with disapproval, dismay, and even despair. Physicist Heinz Pagels, while agreeing that self-generating universes do not suggest purpose, nevertheless felt that there is still room for purpose and that science must continue its pursuit of truth.

Of course, these views on the limited efficacy of the anthropic principle are at variance from Paul Davies' finding of "design" and "purpose," or Hugh Ross's necessary-for-meaning supernatural God which, he suggests, is evidenced by the 1992 NASA COBE satellite findings of primordial temperature ripples in the fabric of spacetime following the Big Bang.

"Deciding" What to Believe? Do *we* determine our own beliefs? Do we somehow will them or are they the coincidental product of unidentifiable external influences? Many accept their beliefs as naturally-generated, perhaps a product of their inheritance and environment, having little or nothing to do with volition. The question is a serious one, especially in Christian theology where faith in the Christian path may be viewed as critical in terms of salvation.

The God of Grandeur: For most religious, church-going people, looking up at the night sky, contemplating the grand stellar display and the vastness of the universe, somehow triggers thoughts of a distant God representing the grandeur and the immensity of the universe. Looking about us at the beauties of nature here on earth, some see God as reflected in that beauty and diversity, just as Henry David Thoreau saw God in the beauty of the forest at Walden Pond. At times, the idea of God may seem remote and aloof, cold as the winter wind; at other times, close and embracing as the warmth of the summer sun.

The Immanent and Transcendent God: For many, God is not "out there" somewhere, but within themselves, reachable in the very depths of their being. In *Man's Search for Meaning*, author Viktor Frankl, father of logotherapy, felt that when in his solitude he was engaged in a deep soliloquy, *with whom he was in meditation was God*. The idea of the God within us has theological significance, sometimes expressed as the God immanent in man, the God Who gave of himself to create the universe and provide for our presence here.

But some, like dialectical theologian Karl Barth, saw God as transcendent, above and apart from our materialistic world, reachable through aspiration and prayer. To many, including members of the Eastern religions, God is the unity of the cosmos of which we are a living, sentient part.

Tolstoy and Camus Question the Existence of a Universe of Meaning: Most believe that the universe, and especially life, has transcendent rather than self-constructed purpose and meaning, and they consciously, sometimes desperately, seek that meaning, as Viktor Frankl points out in *Man's Search for Meaning.* But, as we have seen, down through the ages, there are some--challenging thinkers--who question the existence of meaning in life, and for that matter, meaning of the universe.

The great Russian philosopher-novelist Leo Tolstoy asserted that "The meaningless absurdity of life is the only incontestable knowledge accessible to man." Albert Camus, the French existentialist novelist and playwright, observed that "We turn toward God only to obtain the impossible."

Bertrand Russell: The Soul's Foundation in Despair: Similar expressions of pessimism and a world without meaning are not uncommon. British philosopher and logician Bertrand Russell, in *A Free Man's Worship*, observed that human life is the product of random events; that "the whole temple of Man's achievement must inevitably be buried beneath the debris of a universe in ruins;" that the "soul's habitation" can be safely built "only on the firm foundation of unyielding despair."

Heinz Pagels: The Destiny of Complete Oblivion Seen by Science: In another expression of pessimism, physicist Heinz Pagels, in *Perfect Symmetry,* observed that

"... The incomprehensible suffering of individuals, the hope of religion, the insights and explorations of art and science, the madness, the glory and joy of our species are destined to an oblivion so complete that even the act of destruction will not be remembered....this is...the only rational inference that emerges from our scientific view of the cosmos." Fortunately, Pagels does not leave us totally in the doldrums. He goes on to find meaning--which, he says, includes science's search for truth.

Meaning and a Divine Creation: The meaning *of* the universe and meaning *in* life will naturally be influenced by one's view of the origin and nature of the universe. Western thought, of course, is accustomed to pursuing "meaning" of the universe in the context of the existence of a theistic, personally-involved God who purposely created, and continues to watch over, the universe, suitably designed for the human race, as chronicled in the Biblical Genesis sequence.

Since the universe, in the Biblical account at least, was *intentionally* created, it was reasonable to assume that it was *specially designed* by the Creator to achieve certain *purposes* or goals. Accordingly, life here must have special meaning which, of course, we are left to construct through reasoning, intuition, and resort to scripture.

But Many Insist on the Purposeful Universe: But, as we shall see, there are many who insist that the universe, *however created*, has purpose and meaning apart from differing religious chronicles; that there are universal objective values; that individuals and societies do construct other justifiable values, goals, and meanings. As we have seen, physicist Heinz Pagels, while noting the pessimistic views of some thinkers, like Bertrand Russell and Steven Weinberg, nevertheless saw meaning in the occurrence of our universe, even if it may be destined to end in obliteration so absolute that memory of it, and even time and our other dimensions, may no longer exist.

Defining God--Better Sensed Than Described? The problems inherent in the expression of ideas about God, and defining God are all too apparent. As theologians like Paul Tillich point out, attempts to use descriptors are fraught with pitfalls and at best may inappropriately imply certain aspects of finiteness.

It has been suggested that God cannot be described or defined; that the subject matter transcends the power of rhetorical or linguistic expression; that God can only be experienced or *sensed*--known through feelings. Rhetorical or linguistic descriptions are, by their very nature, imprecise and confining.

'God' Has Many Meanings: The term 'God' involves many conceptual descrip-tions and, as we have seen, has many meanings. Major distinctions are illustrated, for example, by the differences between the great Eastern and Western religions. Most societies from very early times have shared belief in forms of deity, God concepts, which we will explore. But the term is perhaps most frequently invoked in relationship to the human role in the world, his predicament, his existential concerns and dependence on something outside of himself, and in the context of meaning of life and the universe, which we explore.

Depicting God: In religious art and training materials God is often depicted as the patriarchal figure, like that in Michelangelo's famous painting of Bible scenes on the ceiling of the Vatican's Sistine Chapel. Despite the plethora of descriptions of God offered down through the years by the great artists, philosophers, theologians, and even scientists, these descriptions are, after all, the product of human frailties.

They are also influenced by wish, nostalgia, and imagination, and, of course, influenced by Genesis 1:27, man created in God's image which, in depicting man as in God's image, also depicts God.

"Cannot Agree on...Who God Is": In *Groping In The Light*, a delightful publi-cation of the Worldwide Church of God, author John Halford traces the "new way of looking at the world," brought about by the startling findings of science in this century. After summarizing some of the remarkable scientific advances, such as quantum and relativity theories, he gives plaintive expression to frustrations over lack of any corresponding progress in revealing information of God.

"It is a sad commentary, he observes, that, after 2,000 years of Christianity and some 3,500 years of biblical tradition, philosophers and theologians still *cannot agree on even the most basic explanations of who God is, and his purpose for creating man*." Halford goes on to note that what is known of God derives from Biblical revelation, a view of course, shared by Western theology.

The Ultimate Question: The ultimate question--about God, says theologian Abraham Heschel, in *Man Is Not Alone*, begins with the ineffable--the inexpressible. The question about God inquires what all things stand for, the transcendental; the mystery shared by all men and a dimension of all existence.

The Mind of God, and the Human Mind: The theological concept of God's immanence in human life is thought by some to find support in evolving concepts of quantum physics. Quantum theory says that our physical universe melts into a quantum world of uncertain energy states; that underlying this quantum world is a fabric of interconnected cosmic information, something more *mental* than physical. Some relate this to the Eastern concepts of cosmic mind; others, including physicists like Freeman Dyson speak of this underlying mentalness as *the Mind of God.*

In *Infinite In All Directions*, Princeton physicist Freeman Dyson observes that if God exists and is accessible to us, "his mind and ours may differ from each other only in *degree* and not in kind." He speaks of *the human mind as an actual part of "God's mental apparatus."*

Of course, there are others, as we shall see, who say that such assertions are adventures into the speculative; that God will always be the eternal mystery, beyond our ken, grasp, and definition, that it's safer to place our confidence in accepting established concepts of organized religion derived from revelation and centuries of acceptance.

Defining Religion: Religion is sometimes described as a system of thought, feeling, and action, toward an object of reverence and devotion, and commonly characterized by group or congregant practice. Religion involves shared beliefs, and spiritual and emotional attitudes toward the mysteries of the unknown, that which is not explainable by reference to natural observable events, and so described as the *supernatural*. It is summed up and, in Western monotheism, at least, generally personified, in the term *God.*

Practicing Religion: Religion in our society is conspicuous for the social environment which it generates and in which it thrives. Some congregants may feel confused by the mysteries or complexities of religious theory or dogma. Or, finding little interest in doctrinal nuances or niceties, they prefer to "live" their religion, that is, preoccupy themselves with social, practical, and practicing aspects of religion.

For many, "practicing" religion involves not only worship and observing biblical injunctions and admonitions, but constantly living, managing and idealizing the whole spectrum of God-human and human-human relationships, as reflected in every facet and activity of everyday life.

Religion Supplies Values, Promotes Acceptance of Death: In this context, the adoption and practice of religion as represented by any of the major religions, supplies values, purpose, and meaning of and for life. It also promotes the ultimate reconciliation with and acceptance of death within a recognized theological framework. But despite these approaches to an understanding of religion, to a great extent, religion, like God, in many ways remains the great mystery and enigma.

Religion Eludes Apprehension: For many, religion has always had an aura of mystery-- dealing with the unknown. In *Masks Of The Universe*, Edward Harrison observes that it is religion which, down through the ages, has been the primary recourse in the search for the unexplainable and the unattainable.

The distinguished mathematician-philosopher Alfred North Whitehead described religion as "the vision of something that stands beyond, behind, and within...yet eludes apprehension; something whose possession is the final goal, and yet is beyond all reach; something which is the ultimate ideal, and the hopeless quest."

Religion As the Mystery of the Universe: Many view religion as something within themselves which identifies with the ineffable, the unexplainable. We encounter an indefinable sense of mystery when we look out toward the stars and back in time. The experience triggers profound existential questions about the miracle of our awareness of "our place," the cosmos, so vast, inscrutinable, enigmatic, and fraught with uncertainty in quantum and human terms.

For many, the indefinableness of God is wrapped up in the profound mysteries of the universe. "The most beautiful and deepest experience a man can have," thought Einstein, "is the sense of the mysterious." He mused that perhaps religiousness could be expressed as the idea that behind what can be experienced there is "a something that our mind cannot grasp and whose beauty and sublimity reaches us only indirectly and as feeble reflection." For him, "it suffices to attempt humbly to grasp...a mere image of the lofty structure of all that there is."

God Comprehended by Reason? Pope Gregory I said that "If the work of God could be comprehended by reason, it would be no longer wonderful." To believe in God is impossible--not to believe in Him is absurd, said French philosopher-satirist François Voltaire. There are many expressions of the concept that God exists as a spiritual force or a physical reality akin to mother nature, or perhaps as some kind of an abstraction.

But however He may exist, there is a deep subjective need for the personal (theistic) God who watches over us, judges, punishes and rewards, the source of salvation. Down through the ages most have perceived this personal God as essential to human existence. And if this God did not exist human life would need to create him. As Arthur C. Clark observed, "It may be that our role on this planet is not to worship God but to create him."

The Changing God: In *A History of God,* Karen Armstrong, a Londoner and former Roman Catholic nun, who views herself as an "unaffiliated monotheist," describes the evolution of the present God of the world's three great monotheistic religions. *Yahweh.*, She suggests, started as a savage, partisan god of war, perhaps one of the Israelites' deities.

All Religions Change: All religions change, she notes, and *each generation has to create its own imaginative conception of God.* An ecumenical devotee who regularly prays with Jewish and Muslim congregations, Armstrong speaks of the Muslim movement devoted to *Falsafah* which recognizes the chaos around us and sees God as "reason itself."

The Search for God as the Search for Meaning: The book chronicles the changing views of the great Christian, Muslim and Jewish religious thinkers down through the ages. As do many others, she sees the effort to understand the meaning of God as part of the human search for meaning in human life and in the existence of the universe--why there is something rather than nothing.

God as Knowable Only by Intuition: Armstrong discusses great figures in the Western monotheistic religions, and writes of reciprocal Muslim-Jewish theological influences as reflected, for example, in the writings of Spanish physician-philosopher Moses Maimonides. She discusses the Muslim *Falsafah* movement, in which rather than viewing God "as a mystery," the "Faylasufs" saw God as "reason itself," reminiscent of Greek philosophy and some contemporary theology.

Knowing God by Intuition--The "Duty of Compassion": Another thought, also reflecting the views of some theologians, who or what God is and his role in the universe, is beyond human comprehension; that we can only know God by intuition--by experiencing him, not by reasoning. One theme she develops is "duty of compassion," that to be authentic, religious experiences must be integrated into everyday life or they are meaningless abstractions, a thought which echoes certain New Testament passages.

Secularism and Fundamentalism: She discusses the present competition between the growth of secularism, which, she feels, tends to diminish God's contemporary relevance, and the growth of fundamentalism, which she views as relying on the biblical God to rationalize "loves and hates," and in the process, perhaps unwittingly, problem-contributing rather than problem-solving.

"God: A Biography": We are used to the definition of God as omnipotent, omniscient, ubiquitous, perfect, noncontingent, immutable, etc. But how does the Bible depict God? In *God: A Biography*, Jack Miles, journalist and former Jesuit, reviews the Bible as great literature from which he constructs the life story of God, A Biography reflecting God's nature.

The God Miles finds in the Bible starts as a lone figure, omnipotent at the start in Genesis. As the Biblical account unwinds, and after Adam's breach in Eden, God becomes adversarial to humanity and his interactions with human frailties somehow introduce changes in Him--as though He created human life so He could experience change in Himself and find freedom from immutability--and perfection.

God Witnesses and Is Affected by Human Frailties: Miles' God exhibits anger, frustration, love, consolation, patience, impatience, destructiveness, restorativeness, in an almost endless sequence of changes in mood and outlook as he witnesses and experiences human growth, folly, intemperance and immorality. He guides, leads, punishes, rewards, loves, consoles, protects and nurtures, is lifted up then discouraged and disillusioned as he witnesses, in turn, human glories and achievements, and human weaknesses and transgressions.

There are frequent arguments over the weight to be attributed to the Old versus the New Testament. The Old Testament, of course, covers a time span of some two thousand years of the history not only of the peoples involved but of how God conducted divinity in those years of primitive society. So the God of the old Testament--said to exhibit anger and holiness, meting out tough justice and forgiveness, wrath and compassion, is compared with the God of the New Testament--the God of love and grace.

As Christian theologians point out, the comparison is unfair and faulty because the two are a continuum and interdependent, and the Old regarded as an essential foundation for the New. Much of the wisdom of the New finds its foundations and counterparts in the Old.

Does God Experience Emotions and Frustrations? Mile's God exhibits the range of emotional and behavioral features characteristic of the humanity He created in His own image. He becomes the savage partisan God of war. In Job, He engages in whimsy, going along with Satan in the test of Job's loyalty and devotion by inflicting on him unspeakable tragedy and degradation.

Appearing out of the whirlwind, He thunders of his power in creation of the heavens and the Earth, in defiance of Job's demand for an explanation of Job's desperate plight in the light of Job's record of goodness and exemplary behavior. The Book of Job, of course, has been the challenge to much great literature, such as Leibniz's *Theodicy*, as we discuss.

As Miles wends his way through Biblical adventure and experience, he observes God's changing outlook, His moodiness, and His increasing detachment as God withdraws into the passive mode reflected in Psalms, Proverbs, and the books that follow.

"God Is No Saint" : "God is no saint," says Miles. "There is much to object to in him, and many attempts have been made to improve him." Miles notes that "much that the Bible says about him is rarely preached from the pulpit because, examined too closely, it becomes a scandal." This brings us back to Karen Armstrong's *A History of God* which notes that "All religions change and develop...[and] each generation has to create its own imaginative conception of God."

The Evolving God: In Miles' *God: A Biography*, God evolves throughout history in the Bible. In physicist Freeman Dyson's *Infinite In All Directions*, we note that Dyson's God is the Socinian evolving God, and in Frank Tipler's *The Physics of Immortality*, cosmic intelligence spreads to pervade the universe, becoming the loving personal God at the end or Omega Point, where the cosmic intelligence, the Omega Point God resurrects all actual or potential life, in an improved higher state through computer simulation.

How Widespread Is Belief in God? Belief in supernatural powers, spirits, and gods, antedates recorded history. The most primitive societies had their beliefs in gods or spirits associated with the sun, the moon, war, love, fertility, and other specially-feared or revered phenomena. But old and widespread as it is, belief in God is not a societal universal. Statistics are made less meaningful by problems of definition. Leo Rosten's "A Guide to the Religions of America," contains a wealth of facts and figures on religions.

One older Gallop international poll showed that, in England, some 45% of the population believed in a personal God, 39% believed in the existence of some sort of spirit or vital force and 16% were uncertain. The belief in a God in rural Denmark was greater than in urban areas. Belief in a theistic or personally-involved God was somewhat higher than God more generally defined as a vital force in the universe.

Belief in God, in the U.S., was considered to vary by age--less belief by the younger, more by the older. A Gallup poll in 1990 showed, oddly, that belief in heaven and hell was higher in lower age groups and lowest in those over fifty.

Interestingly, church attendance does not necessarily equate to belief in a personally-involved God, or at least a God who watches over our every move and thought, reacting immediately and judgmentally to praise or admonish, reprove and punish. But some see church attendance as part of the search for faith or belief, perhaps as part of the more general search for meaning.

Belief in Heaven and Hell: According to a Gallup poll in 1990, reported by the Religious News Service, the percentage of Americans believing in heaven and hell was the highest in 39 years. *Seventy-eight percent* believe that good behavior here brings a heaven of eternal reward.

Sixty percent went along with a hell of eternal damnation for those whose conduct warrant. Most of those expressing belief in Heaven and Hell, incidentally, would judge themselves heaven-qualified. In fact, only four percent felt they were hell-bent. And women, it seems, are more likely to be believers than men.

A 1995 Gallup poll found that in Great Britain some one-fourth believed in a literal hell of fire and brimstone. The Doctrine Commission of the Church of England, though, recently disavowed the hell of fire and brimstone, viewing it as standing for God as a "sadistic monster," and favoring the view of hell as an annihilation for those who reject the love of God.

Faith and Religious Affiliation: A 1995 religion in U.S. poll reported in Time magazine, showed that, with the greatest religious diversity in history, 65% of those polled experienced the answering of prayer, 53% felt the direct presence of God, 37% witnessed what was believed to be a miracle, 8% were visited by an angel, the Virgin Mary, or a saint. Eighty-eight percent believed that the ethical and moral standards should be higher. Of charitable donations, 45% went for religion.

Reported were 1600 denominations, 44% of them non-Christian (half of them started since 1960). In relative size, they were (rounded): Evangelical Protestants 26%, Roman Catholics 24%, Mainline Protestants, 18%, Black Protestants 8%, Other Christians 3%, Jews 2%, Other Religions 1% (Muslims, Hindus, Buddhists, others, in that order), 19% reported as "nonreligious."

The most striking development was the shifting balance between decreasing mainline Protestant (Episcopal, Methodist, Presbyterian) and increasing Evangelical (Fundamentalists, Pentecostals, Charismatics), the largest single evangelical group being the Southern Baptist Convention.

The Evangelicals are reported as more inclined to modify worship practices such as use of soft-rock bands, song sheets instead of hymnals, and Saturday night worship. Services offered include counseling and fitness centers. Over 400 churches have weekly attendance of 2,000 or more and many have a wider T.V. "electronic church" family. While Catholicism has a wider official membership, The Evangelicals now have more "active churchgoers."

God As a Reality beyond Human Expression: While the search for meaning will continue, says Armstrong, the personal God of human creation is giving way to the mystic God experienced "as a reality beyond human concepts." This idea of human inability to communicate the nature and meaning of God has been expressed by theologians of different faiths. Cited in some Catholic theological literature is Jewish theologian A. J. Heschel's writings, for example, in *Man is Not Alone: A Philosophy of Religion.*

Heschel gives erudite expression to the concept of the ultimate inexpressibility of what human "senses" and "experiences" respecting God, which he terms the *ineffable*. The imperative, says Heschel, is to seek the sublime, "that which lies within our reach but beyond our grasp." Thus God is conceptual, can be *thought*, sensed, and experienced, but is beyond communicable description, given the limitations on our ability to verbalize our thoughts in the tools of language.

Human Beings Continue to Mold God According to His Needs: In *A History of God,*Karen Armstrong dwells on the theme that it is human life's destiny to continue to mold God to human life's needs. She cautions that "If we are to create a vibrant new faith for the twenty-first century, we should, perhaps, ponder the history of God for some lessons and warnings." And along this line, physicists and mathematicians like Paul Davies and Frank Tipler suggest that theology and religious concepts will wither away unless theological outlook is receptive to startling new scientific discoveries about the nature of the universe and related perceptions of the nature and role of God.

Science As a Way of Probing the Ultimate, a Path to God: The Bible itself speaks of perspectives of God as revealed through his Creation. In Romans, St. Paul noted that we might better understand God by reflecting on His great works in the creation.

In their excitement over the remarkable scientific progress in unwrapping the mysteries of quantum theory, some suggest science as an alternative or preferred instrument for bringing us closer to the ultimate, to God.

"It may seem bizarre, but in my opinion," says physicist Paul Davies in *God And The New Physics*, "science offers a surer path to God than religion. Right or wrong, the fact that science has actually advanced to the point where what were formerly religious questions can be seriously tackled, itself indicates the far-reaching consequences of the new physics...." Astrophysicist mathematician Frank Tipler expresses similar views in his *The Physics of Immortality.*

Davies acknowledge, of course, that "This is simply one man's perception of the universe; there are many others....[but] There is more to the world than meets the eye." Of course, there will be many who do not now and will not in the future share this view, who perceive science and theology as two, if not irreconcilable then at least independent, approaches to God.

Which Is Closer to God? In discussing the source of "abstractions" as being derived from increasing order and complexity arising from the quantum, physicist Richard Feynman asks, *which is closer to God, atoms or abstractions?* They are *equidistant from God,* he observes. Although Feynman wasn't discussing it from that aspect, one may imply his agreement with physicist Freeman Dyson that science and theology represent two books of God, one, the Bible, dealing with the spiritual, and the other, the book of Nature, which, of course, confronts scientific materialism.

Jesuit philosopher-paleontologist Pierre Teilhard de Chardin is often cited as a defender of materialism; that the physical world and human daily activity in the material world on the one hand, and the spirituality on the other hand, are equally-sacred manifestations of God. He is noted, of course, for his impassioned defense of the sacredness of the materialistic aspects of transubstantiation in the Catholic Eucharist.

Intensifying the God-Human Interrelationship-- Human life Creates God: God is regarded by some to exist only if there exists an "awareness" of his Being. Sometimes the God-creation concept is reversed. This idea is expressed as God being the "creation" or "conception" of human life; that God's existence comes only in the form of his coexistence with human life--the two being inextricably interdependent. Thus, the argument goes, God must have human life and human life have God--or neither has anything.

Religiosity , Uniqueness, and Acceptance of Life Here: Religiosity, says Fritzsch in *The Creation of Matter*, provides a framework for ethical values. We are "imbedded in the inexhaustible continuum of possibilities" of our, the eighth, epoch, including the not inconsiderable power to "shape the course of history." He emphasizes uniqueness--our uniqueness, the uniqueness of our actions and the uniqueness of every second.

The meaning, he says, lies in living the experience of the cosmic unity that transcends the individual, individual consciousness, and in acceptance of one's potential and limits according to the nature of life here. And so God is the unity of the universe in its remarkable diversity, says Fritzsch.

"God and the Absurd Universe": Physicist Harald Fritzsch makes the depressing observation that *in all of man's search of the universe he has not found the reason for our presence and a way of using science to find ethical values.* Says Fritzsch, we don't know why the earth exists, the reason for the Big Bang--or even whether there was a reason.

He quotes French Nobel prize writer Albert Camus, in his *Myth of Sisyphus*, of man in exile, a stranger in a universe, "deprived of the memory of a lost home or the hope of a promised land, imbued with the feeling of absurdity." But if man could perceive the universe as being able to love and suffer, he muses, "he would be reconciled." If the universe could be summed up in a "single principle," then, he observes, there would be "intellectual joy."

Similarly sad and depressing views on the apparent meaninglessness of life and the nature and fate of the universe are reflected also in Bertrand Russell's "foundation of despair" and physicist Steven Weinberg's wistful account, in his *The First Three Minutes*, both contemplating the likely end of the universe without a trace of human life's existence or its successes and tragedies.

Human Search for Meaning As Inscribed in the Genetic Code: Fritzsch observes that in human life's evolutionary progress the idea of God arose to dispense with just such ideas of meaninglessness of the universe and of life. God, he says, becomes the cause of all being, the mediator, and the assurance of life after death.

He quotes French Nobel Prize-winning biologist author, Jacques Monod, known for his views on human life's presence as the result of genetic mutation. Monod cites *human life's search for meaning as an 'imperious need,' 'inborn' and 'inscribed somewhere in the genetic code.'*

Physicist Heinz Pagels says that the universe is a message written in code, the *'cosmic code,'* it being science's duty and destiny to decipher that code. His implication, of course is that such a message or code must have *Authorship.*

Defining the Term 'Universe': The relationship of the term universe to the term 'God' is tricky and leads to much confusion. Frank Tipler, in *The Physics of Immortality*, is careful to delineate what he means by universe, which is everything. And everything means everything, including God. But many, including astronomer & Minister Hugh Ross treat the transcendent God as outside the universe and its space, time, and dimensionality and so Ross's reference to the 'universe' does not include 'God.'

Use of the term universe is further complicated since the advent of 'other' and 'parallel' universes, and of quantum cosmology which speaks of parallel universes, and the 'Many-Worlds Interpretation' of quantum theory.

God, Linguistics and Semantics: But in our later discussion seeking to define God and religion, we see that linguistics and semantics play a fundamental role in the seeming diversity of religious concepts; that aspects of theism, deism, pantheism, materialism, transcendentalism, immanence, and even humanism, are not mutually exclusive.

Scientific Materialism and Religious Transcendentalism: Physicist Freeman Dyson gives strong assertion to the proposition that so-called scientific materialism and religious transcendentalism are not mutually exclusive. It is worth noting that the new physics, postulating a quantum nonmaterial mental cosmic fabric of information and mind as underlying our physical universe, may be viewed as a fundamental conceptual change from earlier perspectives of 'irreligious' scientific materialism.

> If the universe is created, there must be reality beyond the confines of the universe and it is the Creator that is that ultimate reality. Thus the Creator is not only the source of life, but establishes its meaning and purpose. It is the Creator's personality that defines personality and the creator's character that defines morality.-- Hugh Ross

PART X--DEFINING GOD

CHAPTER 28
EFFORTS TO DEFINE GOD

Background--Efforts to Define God: How do we define 'God'? This is an often-asked question, not only in church study but in everyday personal encounters. Books like Paul Davies' *The Mind of God*, Stephen Hawking's' *A Brief History of Time*, and numerous others dealing with the interface between physics and God, can be better explored by reviewing some background on efforts to define God, a term which may invoke differing perceptions.

Defining God also calls for defining the term 'universe,' especially in the face of recent quantum cosmology contemplating a *primordial mega-universe* consisting of an uncertainty-dominated quantum void from which, according to quantum cosmology, this universe erupted as a random event.

How Do Scientists Think of God? As physicist Freeman Dyson notes in his *Infinite in all Directions*, there is no monolithic scientific perception of the existence or nature of God. Some will display intense piety and faith in a *theistic*, personal God. Others might use the term God simply to describe a primordial creative force which triggers into existence our universe equipped with behavioral characteristics which permit or promote the eventual emergence of life and mind.

Those who perceive this universe as having self-regulating mechanisms, including those governing human behavior, without need for continuing outside intervention, might be seen as *deists*. Others who see God as coterminous with all that exists, might be considered *pantheists*. Still others who see everything as chance occurrence in a random universe involving no supernatural intervention, might be called *non-theists*. The problem of the definition of God is made more difficult because of different uses of the term *universe*.

Complications from Use of Term 'Universe': Sometimes the word 'universe' refers to the expanding universe we inhabit, the product of the Big Bang. But newer quantum vacuum and quantum genesis models hypothesize a primordial quantum mega-universe void from which untold numbers of universes are constantly erupting, as well as the infinite number of universes as described by the Many-Worlds interpretation of quantum theory. Also, relativity-black-hole theory posits black holes as possible connections to other universes or other branches of the mega-universe, sometimes referred to as *superspace*, a sort of laws-of-physics maternity ward for universes.

The Universe and Transcendent God: We are accustomed to use "universe" to refer to our own universe, which we used to think of as *everything that exists*, the old dictionary definition. As physicist Frank Tipler points out, "everything" would include God. Yet we are also used to thinking of God as *transcendent* or outside of time and space and outside of the universe, at the same time we think of God as also *immanent* in our universe--and in us.

But with the advent of the various theories just described, involving universe 'states' outside or apart from our own universe, at least for now, we probably should use God's 'transcendence' to mean outside of all universes or states, quantum, relativistic, or otherwise. Yet physicists like Freeman Dyson and Frank Tipler see God as still evolving with our particular universe, growing as our knowledge grows, a concept that seems rather confining if our particular universe is just one of an infinite array.

But if, by 'universe,' we mean the common dictionary definition--all and everything that 'exists'--this may be interpreted as including all other universes and *even God* if He may be said to 'exist.' Of course, theologian Paul Tillich wouldn't demean God by attributing 'existence' to Him since in his view God represents the very state of 'being itself,' a concept, incidentally, in which some see pantheistic implications.

God--The Supreme Being As Defined by Religions: But getting away from defining 'universe' and back to defining God, the name 'God' is usually applied to designate the supreme being whose nature is characterized consistent with the meaning assigned by one or more of the various religions. And, of course, even within any single religion there likely exists a diversity of views as to meaning and doctrine.

Martin Luther almost lost his life refusing to recant his doctrinal views on justification by faith. Descriptions of God are influenced also by evolving perceptions of God, as reflected, for instance, in Karen Armstrong's *A History of God* and Jack Miles' *God: A Biography.*

Knowledge of God: 'Knowledge' of God is considered to be based on a combination of intuition, revelation, and reasoning, rather than direct sense perception. Revelation is, in turn, often defined to include those ideas deducible from creation itself and those the Bibles particular religions. And as knowledge of the universe, its origins, nature and behavior grows, perceptions of God and His role undergo change.

Frustration over Inability to Agree on Basic Explanations of God: We noted earlier the Worldwide Church of God publication, *Groping In The* Light, where author John Halford traces the "new way of looking at the world," brought about by 20th century scientific discovery, expressing frustration over lack of progress in discovery about God. He noted that after 2,000 years of Christianity and some 3,500 years of biblical tradition, *philosophers and theologians still cannot agree on even the most basic explanations of who God is*, and his purpose for creating human life. He went on to note that our knowledge of God derives primarily from Biblical revelation. So what does the Bible tell us?

God as Described in the Western Bible: Most readers of the Bible interpret the Bible as revealing God to be self-existing outside of time or space, omnipotent, omniscient, and transcendent, existing outside the universe, and yet immanent in the universe. Other attributes derived from the Bible include loving and compassionate, forgiving and human life-oriented.

The Worldwide Church of God publication, *God Is...* observes that the Bible shows God to be: Self-existing, unlimited in time, place, power, and knowledge, and transcendent (beyond the universe) and immanent (involved with the universe).

Each Society and Age Defines God: In her *A History of God,* former nun Karen Armstrong notes that it is human life's destiny to continue to mold God to human life's needs. And along this line, physicists and mathematicians like Paul Davies and Frank Tipler suggest that theology and religion must be open to new scientific discoveries about the nature of the universe which in turn change perceptions of God.

Science As a Path to God: In Romans, St. Paul noted that we might better understand God by reflecting on His works, and some physicists, like Paul Davies and Frank Tipler suggest science as an alternative or *preferred instrument for understanding the nature of God.* Davies acknowledges that his is but "one man's perception of the universe; there are many others....[but] There is more to the world than meets the eye." Of course, there will be many who do not now and will not in the future share this view, who perceive science and theology as two, if not irreconcilable then at least independent approaches to God.

More on the Transcendent God and the Universe: Some speak of God as "transcendent," meaning outside the universe but supporting its existence. As we have discussed, this term comes into conflict with the term 'universe' if universe is taken to mean everything that exists since "everything that exists"would include an "existing God." This is the definition which Frank Tipler chooses to adopt in his *The Physics of*

Immortality, in which intelligence in the universe grows until it becomes the God of his Omega Point theory. That God also exists now, Tipler explains, is the case because of the quantum and relativity coexistence of all time.

God as a Being or As Being Itself: As we have seen, some see God as a 'being,' an 'existing' entity. Others, like theologian Paul Tillich, see God as 'being itself.' So as the Worldwide Church of God literature suggests, there is really no single 'scientific,' lay, or even theological view of what or who God is or what the term means.

While some scientists make a point of not discussing God, others are increasingly willing to explore a range of possibilities--even to an extreme, *suggesting scientific proof of God*, as derived, for example, from the *anthropic principle* involving the extraordinary coincidences in the mathematics of the constants of physics which are *precision-tuned* to dictate life in the universe.

Scientists with Religious Backgrounds: It is not mere coincidence that many physicists have backgrounds in religion as priests and ministers, and there are many persons active in religion who have abiding interests in physics and chemistry, in astronomy and astrophysics, in paleontology, in the biological sciences, and in the sciences generally. Many cosmologists of the past have been religious figures, especially holding ecclesiastical positions in the Catholic Church, and the Vatican to this day actively supports, pursues and participates in the study of astronomy and astrophysics.

It is probably true, however, that scientists, being trained to deal in subjects with carefully defined terms, are inclined to be more guarded in their frame of reference for statements of belief. More and more phenomena thought to have only metaphysical explanations are found to have scientific explanations.

The New Physics--Quantum Cosmology Suggests a Different Role for God: Understandably, scientists express ideas about God which sound more deistic (a self-sufficient world created by God, requiring no further divine involvement) or pantheistic (the universe, representing everything in existence, equals God). Of course there are some, like astronomer-minister Hugh Ross, who, in *The Creator and The Cosmos*, find no incompatibility between science and the *theistic* personal God (who continues supernatural involvement in the workings of the universe, including day-to-day human events).

But there is growing acceptance of the recent models of our universe, suggesting the universe as being 'self-created' by the laws of physics and their underlying mathematics. These models imply that, assuming His existence, God's role has to do more with being the source of the mathematics concepts which underlie the laws of physics and describe the behavior of the universe.

This perception of the universe as self-generating may be bolstered by theories that the self-generating mechanism is augmented by laws governing the growth of complexity and order, taking such forms as life, consciousness, and going on to personal abstractions, like justice, love, compassion, reverence and holiness.

Metaphysical Considerations in Physics: The great 19th century Victorian battle over mingling values with science, then waged between biologists and churchmen, personified by evolutionary biologist Thomas Huxley and Bishop Wilberforce, may have been won by Huxley, but if this was a victory it has been swallowed up in the passage of time. For many at least, the conflict continues, albeit on a more subdued level.

But physicists and theologians now work more closely together. As physicist Freeman Dyson observes, "scientific materialism and religious transcendentalism" are “not mutually exclusive” and some, like the Jesuit paleontologist Pierre Teilhard de Chardin, see the God-created material universe and the spiritual world as equal recipients of God's grace.

Using Science to Explain Away God: But tempers still flare. As a spirited "letter to the editors," Washington Post, once noted, "Arguments that try to explain away God by using science, do a disservice to both." The writer suggested that people who engage in [such] "sophistry" instead spend a little time "examining the private agonies of their repressed souls." And so the subject still triggers emotional controversy. Many scientists quote French molecular biologist Jacques Monod, in *Chance and Necessity*, who reasserts that "Any mingling of knowledge with human values is unlawful, forbidden."

God's Existence Affirmed in the Very Denial of God: Speaking of explaining away God, theologian Paul Tillich asserted that God, our ultimate concern, is always present in any act of faith, even if the act of faith constitutes the very denial of God; that in any expression of ultimate concern, God can only be denied in the name of God.

The Origin of the Universe, Scientific or Theological Question: To many, reference to ‘the universe’ will trigger ideas of God. There are many who feel that the nature, origin, and fate of the universe, while subject to scientific exploration, is ultimately subsumed in theology. But in *The Physics of Immortality*, physicist Frank Tipler asserts that theology must ultimately recognize itself as a branch of physics or wither away and lose credibility.

Omega Point Theory: Tipler goes further, suggesting his *Omega Point Theory*, a name taken from Revelation 1:8 and the work of Jesuit paleontologist and Christian evolutionist Pierre Teilhard de Chardin, dealing with mystical concepts of immortality. Tipler's Omega Point theory is an extension of physicist Freeman Dyson's concept of the evolving God and human destiny to spread life in different forms throughout the universe.

Omega Point theory provides 'mathematical' proof of the ultimate existence of the personal God to be, and the necessary ultimate resurrection of all human life--past, present, and future--through computer simulation. Omega Point theory sees life spreading throughout the entire universe to become the Omega Point God and saving itself through manipulation of the contracting universe in the Big Crunch.

The resurrection is permitted by computer simulation of all data involved in the universe's evolution. The data is savable at the final singularity by intelligent control of the forces to produce an irregular contraction (shear) which alters the thermo-dynamics to provide sufficient temperature differential to store the data to reproduce life through computer simulation.

An Ancient Intelligence--A Natural God? Physicist Paul Davies, in *God And The New Physics*, in a fascinating flight of scientific fancy, discusses the ideas of the existence of an *ancient intelligence*, a "natural god," or gods, manipulating the laws of physics, leaving the supernatural God pretty much to the single function of creating the laws of physics, or "pushing the button," unless even the laws themselves are, "so-compelling" as to be self-generated. So in this view, quantum beginnings lead to self-generated universes, God's role reduced, perhaps, to representing the underlying mathematical logic, a concept consistent with the idea of God as a mathematician or as mathematics.

Freeman Dyson's View of Purpose and Human life's Role to Share: In his Aberdeen, Scotland *Gifford Lectures,* updated in his *Infinite In All Directions,* Princeton physicist Freeman Dyson gives expression to the thought that there is "some purpose" for our presence here, something beyond our comprehension, something "having to do with the future."

Dyson's Socinian God: If you like, he says, you can "call the transcendent purpose God," but, he muses, it is the God of the Socinian Heresy, the God who, though not yet omniscient or omnipotent, is "growing in power and knowledge as the universe unfolds"--an evolutionary process in which we share.

Socinianism, a religious movement originally organized in Poland, was founded by Faustus Socinus, the Italian Catholic reformer, based on the teachings of his uncle Laelius Socinus. Not only did Socinus have trouble with 'perfection' aspects of God, given the world's manifest imperfections experienced by human beings, but, like others, he had trouble with the doctrine of the Trinity in the face of the unity of God emphasized in the Old Testament.

Church intervention on the grounds of Socinianism as being heresy, effectively brought the Socinianism movement to a close. Interestingly, Socinianism has, in a sense, taken on new vitality not only in the writings of scientists like Freeman Dyson but in the mathematics of Omega point theory as spelled out in physicist Frank Tipler's *The Physics of Immortality,* supported by the mathematics display in his Appendices for Scientists.

Mathematician-Philosopher Liebniz's Less Than Omnipotent God: In contemplating the less than perfect but growing God, we are reminded of mathematician-philosopher Gottfried Wilhelm von Liebniz's famous *Theodicy,* exploring the existence of evil. Leibnitz asserted that God gave us the "best possible world," in which there are unavoidable imperfections.

Liebniz, of course, sought to sidestep the pitfall of the less than omnipotent God by discussing the *inescapability* from the world's imperfections rather than dealing with Divine imperfection. But critics of Liebniz nevertheless atttacked Liebniz's theory as heresy. Dyson's Socinian evolving God, coping with Leibniz's 'best possible (but defective) world,' may seem more credible than the omniscient, omnipotent God permitting the enormously tragic events which human life experiences.

Einstein's God As Nonpersonal--The Harmony of the Universe: In answer to one of the many questions of him whether he believed in God, Einstein replied that he was drawn to Spinoza's view of God who "reveals himself in the harmony of all that exists, but not in a God who concerns himself with the fate and action of men." Einstein searched for a "a simple and lucid image of the world," to explain the confusing differences between the fundamental forces. He speculated "whether God had any choice in the creation of the universe" and expressed his faith that "God is subtle, but not malicious," and beauty and simplicity characterized the universe.

Einstein had difficulty with a personal God dominated by concern for men because of the incompatibility of God's omnipotence and omniscience on the one hand and free will including human propensity for evil. If God is an omnipotent-omniscient personal God who predestines all time and events (as classical physics implies), including human acts of folly, then, Einstein felt, such a God rather than human beings is ultimately responsible. Nigel Calder, in *Einstein's Universe*, speaking of Einstein's views as being "soft" and "moderate" in religion," interestingly describes Spinoza's writings as anticipating modern cosmology.

The Mind of God and the Human Mind: Both Steven Weinberg in *Dreams of a Final Theory* and Steven Hawking in *A Brief History of Time*, discuss "The Mind of God" as a metaphor for the final (unification) theory--the source of all physics and of the universe. Well, Princeton physicist Freeman Dyson, in his book *Infinite in All Directions*, speaks of the role of human life in the universe and the relationship of human mind to the mind of God. If God exists and is accessible to us, then, he says, his mind and ours may differ from each other only in degree and not in kind (human beings made in God's image).

God As Mind: In *God and the New Physics*, physicist Paul Davies observes that whatever their differences about the nature of God, *the major religions teach of God as mind.* The Christian God is viewed as omniscient, infinitely knowledgable and omnipotent, free to act, and "there can be no mind greater than God's, for God is the supreme being."

Davies speaks of scientific research as shedding new light on the "age-old enigma of the mind" and its relation to the material world. *The consequences for religion are profound*, he notes. His vision embraces the role of mind in establishing reality, theories on consciousness, and the interconnected mental, cosmic fabric of information underlying the physical universe.

Physicist Freeman Dyson's preoccupation in *Infinite In All Directions*, with mind in the universe, and human mind as an extension of God's mental apparatus is based in part at least on this concept of the quantum phenomenon of *mind* as underlying the physical universe.

Various Concepts of God Explored: We have explored various concepts of God: theism--the personal God; deism--God sets up the universe as self-functioning; pantheism--God as the universe; God as mathematics and physics; God as the quantum forces; God as outside the universe(s); The supernatural vs. the natural God, God and Time, God as evolving, God as the creator, Humanity as creating God, The crucial interdependency of God and human life, and other ideas.

Can these varying notions support some set of minimum demonstrable characteristics that may be attributed to God? Astronomer-minister Hugh Ross believes that scientific discoveries reveals much about God's nature. Mathematician Frank Tipler, in *The Physics of Immortality*, actually provides definitions in physics of (the Omega Point) God's characteristics of omnipotence, omniscience, perfection, and personhood.

Is God Definable? Human life's primitive conceptions of multiple deities have undergone change, along with growing knowledge of the universe and the broadening intellectual perspectives. Although human life has taken the quantum leap in knowledge of nature of the universe and its origins, humanity still struggles with the ultimate enigma, the nature of God--What or Who He is--how to define God.

Search, scrutiny, and reflective contemplation leads us to believe, and some perhaps sense, God's existence, an awareness of something *there* but *inexplicable*. Some feel that in attempting to explain God we're dealing with the ineffable, the indefinable, real but perhaps not in our reality; something here but yet beyond reach. There are others who perceive God as unexplainable but as an *abstraction*.

God Big Enough to Embrace New Theories of Universes? We have explored new models of the universe which hypothesize an eternal mega-universe, a quantum state which quantum-randomly self-generates our universe. The model proposes the existence of perhaps an infinite number of universes. British theoretical physicist Stephen Hawking, in *Black Holes And Baby Universes*, envisions not only an infinity of universes, but universes being constantly generated, many through "umbilicals" from our own universe.

With all of these mind-boggling notions of a timeless and dimensionless quantum void mega-universe from which an infinite number of universes have been and are constantly erupting, we should make sure our view of God is conceptually grand enough to embrace these imponderables.

Human Effort to Define God: "It is easy to understand God as long as you don't try to explain him," noted Joseph Joubert. Probably no two persons on the planet are likely to give identical definitions of God. But the same might be true if you questioned extraterrestrials, unless, of course, being more advanced, they have finally broken through to a 'spiritual field' in which God is more 'accessible.'

But on Earth, the 'who or what is God' question is likely to get more religiously-oriented responses, descriptions of God in terms of Jesus, Mohammed, or the Buddha. But if one is searching for something universal, without sectarian personifications or religious orientations, as we have seen, one is more likely to get a model of God extrapolated from what we know about the physical universe.

What Do We Mean By God? In *The Quest for God, A Personal Pilgrimage*, British historian, and journalist Paul Johnson, notes that we believe in "some kind of God," and asks *what we mean by this*. He cites Jack Miles in *God: a Biography*, showing God changing as the Old Testament goes from the early anthropomorphised, personal, activist to the later mysterious, remote God--almost "written out of the script." Then comes the God-son, Christ person, tangible- visual, real, understandable in human terms for contemporaries--and the millions to follow. Father God becomes remote, "indefinable, invisible, unknowable, and...almost unimaginable."

Johnson suggests that a God with limitations is not God, though he'll accept the all-powerful God choosing not to exercise all powers, even against evil. It may be better, he suggests, to use the analogy of the infinite computer to convey God's grasp of information which includes every facet of all human lives. God is outside creation and therefore not to be confused with the pantheistic God--everything that exists.

Why should the infinitely-powerful God be viewed as the suppliant while we infinitely-weak mortals ignore God, then, in need, welcome him as the prodigal son? The keyword answer, he declares, is *love*, the only reason for the universe's existence, the creative, sustaining and energising principle, without need for reciprocation or "symmetry between his love and ours."

> After 2,000 years of Christianity and some 3,500 years of biblical tradition, *philosophers and theologians still cannot agree on even the most basic explanations of who God is*, and his purpose for creating human life.--John Halford

CHAPTER 29
CHANGING PERCEPTIONS OF GOD

Changing Perceptions of God: In *A History of God,* Karen Armstrong, former Roman Catholic nun and now an "unaffiliated monotheist," describes the changing human conception of God of the world's three great monotheistic religions, starting as a savage, partisan god of war.

Religions Evolve: Each generation has to create its own imaginative conception of God, she observes. Her book chronicles the changing views of the great Christian, Muslim and Jewish religious thinkers down through the ages. As do many others, she sees the effort to understand the meaning of God as part of the human search for meaning in life and in the existence of the universe.

"God: A Biography": In *God: A Biography,* Jack Miles, journalist and former Jesuit, reviews the Bible as great literature from which he constructs the life story of God, a biography reflecting God's nature.

As we noted earlier, the God Miles finds in the Bible starts as a lone figure, omnipotent at the start in Genesis. After Adam's breach in Eden, says Miles, God becomes adversarial to humanity and his interactions with human frailties somehow introduce changes in Him--as though He created human life so He could experience change in Himself and find freedom from immutability--and perfection.

God As Reacting to Human Frailties: Miles' God exhibits anger, frustration, love, consolation, patience, impatience. God becomes destructive, then restorative, in an almost endless sequence of changes in mood and outlook as he witnesses and experiences human growth, folly, intemperance and immorality. He guides, leads, punishes, rewards, loves, consoles, protects and nurtures, is lifted up then discouraged and disillusioned as he witnesses, in turn, human life's glories and achievements, and human life's weaknesses and perversity.

God's Changing Outlook: As Miles wends his way through Biblical adventure and experience, he observes God's changing outlook, His moodiness, and His increasing detachment as God withdraws into the passive mode as reflected in Psalms, Proverbs, and the books that follow. "God is no saint," says Miles. There is much to object to in him, he states, and "many attempts have been made to improve him."

The Evolving God: We have seen that Miles' God evolves throughout the Bible. In physicist Freeman Dyson's *Infinite In All Directions*, we note that Dyson's God is the Socinian evolving God, and in Frank Tipler's *The Physics of Immortality*, cosmic intelligence spreads to pervade the universe, becoming the loving, personal God at the end or Omega Point, where the cosmic intelligence resurrects all actual or potential life, in an improved higher state, through computer simulation.

Knowledge of God from His Handiwork: St. Paul, in Romans 1:20, suggests that God's nature can be understood not only through scripture but by examining God's physical handiwork as reflected in physical manifestations of the universe. God's eternal power and divine nature are "clearly seen, being understood from what has been made." We have now been entrusted, of course, with much more information about the origin and nature of the universe, its laws of physics and underlying mathematics, knowledge undreamed of in St. Paul's time.

The Immaterial Underlying the Physical Universe: We know, for example, that at least in relativity theory, time had a beginning in the Big Bang; that time is a physical dimension of the universe; that space and time are an interrelated continuum; that gravity is the geometry of spacetime curvature; and that at the subatomic level, the deterministic cause-and-effect universe is lost in quantum uncertainty; and matter--particles disappear into mysterious wave-particle manifestations.

We also know that primordial radiation became particles, particles become atoms which combine to form molecules. Through a process of increasing order and complexity, these combine to form inanimate material which evolve to become animate material (DNA) to become life, generate mind, consciousness, self-awareness, subjectivity, curiosity, with the capacity for dealing with exquisite abstractions and explaining how mind may pervade the universe.

A Different Perception of God? A model of God extrapolated from scientific observations now available may well give us a different perception of the God derived from scripture and religious insight. This is illustrated by Einstein's hangup against Heisenberg's and Bohr's quantum uncertainty--that God would employ mechanisms of chance (dice) to construct and sustain the universe.

Reconciling the Bible with Scientific Accounts of the Creation: Astronomer and cosmology lecturer Dr. Hugh Ross, is a Congregational minister who, besides his astronomy and cosmology lectures, regularly lectures on Christian apologetics, directs the Reasons To Believe Institute, and hosts science-religion broadcasts. *In The Creator and the Cosmos*, Ross, who is *not* a creationist, presents an unusual explanation of how scientific accounts of the Big Bang are interpretable as supporting Biblical accounts of the creation.

Making the Case for God As the Creator: Although he discounts physicist Frank Tipler's Omega Point theory in which cosmic intelligence evolves toward the Omega Point God, Ross makes a case for physics as demonstrating the existence of the Creator as personal and transcendent. He also addresses the physics of the pre-existent Christ such as cited in New Testament passages from John and Colossians.

Access to Additional Dimensions: Some of the acts of Jesus, in apparent defiance of the laws of physics, are explainable, Ross suggests, by Jesus' access to additional dimensions which contemporary physics now recognizes, namely, six or twelve more original dimensions which still exist today but are now 'compacted.'

While we humans presently have no practical access to those dimensions, God's immanence in Jesus would, he argues, have provided such access. Biblical events, thought of as 'miracles' in the past, have increasingly plausible scientific or natural explanations, though, in Ross's view, God represents the supernatural, rather than the natural, as some physicists imply, although this may turn out to be a play on words, a distinction without a difference. As our information grows on what is possible within the laws of physics, the field of miracles diminishes.

The Universe As Evidence of Providential Desaign: Speaking of dimensions, science fiction writers and some scientists, including astronomer and cosmologist Carl Sagan, suggest that advanced extraterrestrials may have acquired access to other dimensions and, to promote practical space and time travel, found the way to bypass the travel and communication limitations of the speed of light.

While decrying the proposals of some scientists that the universe is explainable by the laws of physics without intervention of God, Ross is encouraged by increased willingness of some physicists, even nontheists, to accept the extraordinary balance of the constants of physics as evidence that the universe somehow reflects 'design' and 'purpose.'

Biblical Correlations with Contemporary Physics: Especially helpful is Ross's cataloging of biblical materials which, in his view, correlate with contemporary physics, and his excellent summary of the anthropic principle coincidences in the constants of physics which imply a 'designed' universe.

Dostoyevsky and God's Existence: Through the ages, great thinkers have been frustrated in the search to understand God. Nineteenth century Russian novelist Fyodor Dostoyevsky, confused by evolving ideas in dimensionality, and the advent of non-Euclidian geometry, declared: "I acknowledge...that I have a Euclidian earthly mind and how can I solve problems not of this world?...I advise you never to think about it...especially about God, whether he exists or not. All such questions are utterly inappropriate for a mind created with an idea of only three dimensions. And so I accept God...His purpose...utterly beyond our ken."

Revelation a Continuing Process: Despite our scientific breakthroughs in understanding the origin and workings of the universe, our definitions of God are suggested as premature. Not only is the human species in its relative infancy but revelation, whether through science or inspiration, is a continuing process. A more advanced civilization may provide us altogether new insights--if we are not des-troyed by cosmic events or by our own environmental folly, or, as science fiction suggests, pursuant to galactic judicial decree for environmental profligacy.

In Greek Philosophy: In *A Handbook of Theological Terms*, Van. A. Harvey enumerates various def-initions of God that have evolved through history. In the Greek philosophic tradition God, he notes, was either the "supreme and ineffable principle of truth, beauty, and goodness, ...the timeless origin of all things" [or] "a sovereign soul or demiurge who...fashions the world in the light of ultimate principles." The early Christians, he observes, adopted the former idea and so God, he remarks, was conceived as the "utterly transcendent, timeless, changeless absolute...beyond being in dignity and power."

Thus, Harvey continues, God was "absolutely simple containing no complexity... [except that of the Trinity], immutable, eternal, ubiquitous or omnipresent..., omnipotent...and uncaused...." He goes on to point out some of the problems of this definition--such as insistence on His "absoluteness" resulting in His "unknowability."

"God Is.....": In *God Is...*, one of the many a delightful publications of the Worldwide Church of God, the author observes that "Who and what God is, what God wants-- these are aspects of God's nature. And *the nature of God, after all, underlies everything else*--why the universe is the way it is, who we are as humans, why our lives are the way they are, and what we should be doing with our time."

The author goes on to suggest that "we can even come to take part in that divine nature....through God's own revelation of himself." Atheism, Nihilism, Existentialism, he suggests, attempt to explain the world the way we find it, "without looking to an almighty." Similar observations have been addressed to humanism and scientific materialism. The author of *God is...* cites the John 21:25 chronicles of certain aspects of Jesus' life, observing that "the world itself could not contain the books that would be written" to describe the actions and nature of God as manifested through the life of Jesus.

The Bible as Revealing God's Nature: As noted earlier, the Worldwide Church of God publication *God is...* described biblical revelation as showing God as self-existing, timeless, omnipresent, of unlimited power and knowledge, transcendent and immanent. It further notes that many creeds and statements of faith also try to define what God is, such as the Nicene and Apostles' Creeds, the Anglican Thirty-Nine Articles, the Augsburg Confession, and the Shorter Catechism.

The difficulty in describing God , he says, is that our knowledge is acquired through our five senses, and being *conceptual*, God is not fully available through the senses. He points out that the Bible uses allegory and metaphors to communicate God's nature.

Parts of God's Body As Metaphors: Thus, God is a rock and fortress (Psalm 144:2), or a consuming fire (Hebrews 12:29). Although the biblical God is essentially spirit, there are metaphorical references to God's head and hair (Revelation 1:14), face (Genesis 32:20 and Exodus 33:11), eyes and ears (Deuteronomy 11:12, Psalm 34:15, and Revelation 1:14), nose and nostrils (Genesis 8:21, Exodus 15:8), mouth (Matthew 4:4, Revelation 1:16), lips (Job 11:5), voice (Psalm 68:33, Revelation 1:15), tongue and breath (Isaiah 30:27-28). Other references include arms, hands and fingers, shoulders, chest, back, waist, and feet.

The Singleness of God: Despite the fourth-century adoption of the doctrine of the Trinity, representing three God manifestations of Father, Son and Holy Ghost or Spirit, the essential singleness of God is recognized as central to Judaic-Christian monotheism (Isaiah 43:10-11 and 45:5). Judaism's most frequent *Shema* declaration: "Hear Oh Israel, the Lord our God, the Lord is One" (Deuteronomy 6:4), is cited by Jesus in Mark 12:29-30 and by his declaration in Mark 10:17-18 that "No one is good but One, that is, God."

Dealing with Sectarian Exclusions: As society adjusts to modern life, and begins to question doctrinal aspects which are exclusionary or tend to separate the religions, new ideas evolve in the effort to accommodate sensitive differences. We discuss eligibility for salvation through commitment to Christ in the chapter on Religion, Denominations and Meaning, but we might note here, in the context of evolving religion and changing perceptions of God, the effort to rationalize the salvation exclusion.

God and the Universality of the Pre-existent Christ: Members of the ministry interested in the study and practice of religion, and in New Testament cosmology, speak of the universality and pre-existence of Christ (John 1:1-18; Colossians 1:15-20). God's Christ spirit here is envisioned as a universal expression of God in the timeless expanse void before our universe and human life's advent on Earth.

This spirit is viewed by some as immanent in the entire human race, regardless of one's assumed religious preferences. We see here an abiding concern for all of humankind regardless of religious persuasion or absence of persuasion. In this broad sense, John 14:1-6 "No one comes to the Father, but by me," is interpretable as "everyone comes to God through the universal Christ because his spirit is present (immanent) in everyone everywhere, regardless of their acquired persuasions."

> The heavens and the earth alike speak of God, and the great natural world is but another Bible, which clasps and binds the written one; for nature and grace are one--grace the heart of the flower, and nature its surrounding petals. --H.W. Beecher

CHAPTER 30
OMNIPOTENCE & OMNISCIENCE
SOME QUESTIONS
FINITE ASPECTS OF GOD

God As Perfection: We have seen that there are widely differing perspectives on the nature of God--what or who He is, how God functions in the universe. Implicit in the traditionally-enumerated qualities is the characteristic of the absolute perfection of God, human life being made in the image of but an imperfect reflection of God, in fact, "a little lower than the angles," as Hebrews 2:6-11 puts it.

Omnipotence, Omniscience--Determinism, and Evil: But these "God characteristics" suggested by traditional theology, especially omnipotence and omniscience, have been questioned and agonized over by theologians and philosophers down through the ages.

Best known, perhaps, is philosopher-mathematician Gottfried Leibniz's rationalization in *Theodicy* of God's imperfect but "best possible world" for which Leibniz was, on the one hand, criticized as questioning God's omnipotence, and, on the other hand, satirized by Voltaire, for his "best possible world" proposition.

The predetermined, cause-and-effect universe of St. Augustine and John Calvert, earlier thought to have been supported by Einstein's general relativity theory, had put into question either or both God's omnipotence and his omniscience. God's predestined universe with imperfection, disorder and evil, all precast at the creation, brought consternation to many.

They worried that an omniscient God who knew all the terrible things that would happen and failed to prevent them, must be less than omnipotent. If he did not prevent them because he could not, or because he could not predict future events, then his omnipotence and omniscience are in question.

Quantum Theory and Omniscience: Then came the mysteries of quantum theory dealing with the universe at the subatomic level. Quantum theory made clear that absolute cause-and-effect did not describe the underlying universe. At the subatomic level the universe is governed by the quantum uncertainty, and to some extent, at least, this uncertainty carries over to the macroscopic universe at large which is also a quantum system.

Computer scientist Edward Fredkin asserts that quantum theory tells us that the universe is a computer program being run by someone or something outside the universe; that the program has to run to determine the outcome--which cannot be known, *even by God.* So, in this view, *quantum theory puts God's omniscience at risk.* As we have seen, scientists have suggested that the new physics, especially quantum theory, requires re-examination of the nature of God.

God Can't Change the Past--Defy Logic--Or Change Mathematical Values: We also know that omnipotence has its inherent limitations. "Even God cannot change the past," observed Agathon, and in *The Mind of God* physicist Paul Davies, notes that even *God is limited by logic* and its underlying mathematics. Even God cannot make a triangle with four sides--or for that matter, change the value of π.

Einstein--Omnipotence and Free Will: Fascinated by Einstein's extraordinary intelligence in his mathematical demonstration of the creation implied by general relativity, clergymen of various faiths pursued Einstein for his view that God the Creator could not be a personal God continuing to function and relate to every-day human life. How or why, they asked, could an intelligent and creative God not also be a personal God?

Einstein responded, in *Out of My Later Years,* to the effect that if God is omnipotent then human feelings, aspirations and deeds are also His work, and how could men be held responsible for their thoughts and deeds? In passing judgment for punishment or reward, God would, "to a certain extent be passing judgment on himself. How can this be combined with the goodness and righteousness ascribed to Him?"

So while many of his peers held different views, Einstein remained a staunch believer in God the Creator--but not a personal reigning, rewarding and punishing God. Einstein's God is described by some as closest to deistic.

God's Omnipotence and Human Free Will--The Timeless Conundrum: This timeless conundrum of the paradox of God's omnipotence, omniscience, and the predestination implied in the creation, on the one hand, and on the other hand, human free will and therefore related responsibility, has been addressed by philosophers and theologians down through the centuries.

Many scientists, and others of course, tend toward Einstein's perception that God created the universe to be self-functioning and can neither be praised for human triumphs or criticized for human or cosmic tragedies. And, in the view of many, neither does He violate the laws of physics to invoke miracles. It has been suggested, of course, that the laws of physics themselves have built-in self-modulating mechanisms which control, reward, and punish.

Physics As Providing New Evidence for the Existence of a Personal God: We noted Einstein's questioning of the nature of God as a personal God because of the omnipotence and free will problem. In *The Creator and the Cosmos*, astronomer-cosmologist ordained minister Hugh Ross observes: "I am grieved that no one ever offered Einstein a clear biblical resolution to the paradox he posed [and] that Einstein did not live long enough to see the accumulation of scientific evidence for a personal caring Creator." Ross proposes a biblical resolution of the paradox in his book, *The Finger of God.*

Miracles Reconciled: tists dislike dealing with miracles, preferring scientific explanations where the laws of physics are not violated. In *The Creator and the Cosmos,* Hugh Ross, devoutly religious, reconciles some of the apparent miracles performed by Jesus as consistent with the laws of physics by resort to the universe's multi-dimensionality of space and time beyond our conventional three space and one time dimensions. Interestingly, while invoking the supernatural God, Ross is convincing in his demonstrating God's use of natural phenomena to accomplish his purposes.

Possible Finite Aspects of God: "Where wast though when I laid the foundations of the earth? Declare if thou hast understanding!" So replied God to Job's challenge for an explanation of Job's undeserved punishments. Job's dramatic dialogue with God has given fruit to many speculations on the meaning of God's unsettling reply to Job's understandable demand for an explanation of his undeserved afflictions, considering his model human conduct.

Frustrated by God's enigmatic reply to Job, a number of theses have speculated on possible 'finite aspects' of God, that is, the existence of limitations on God's omnipotence or perfection. As Scottish philosopher David Hume observed, "If the evil in the world is from the intention of the Deity, then he is not benevolent. If the evil is contrary to his intention, he is not omnipotent. He cannot be both omnipotent and benevolent."

Behind some of these theses is the subliminal caring for God-- excusing Him for allowing evil by *sub silentio* acceptance of his finiteness. A compassionate solution to the problem of evil -- God allows evil because the world could not be constructed without it. Scientists have postulated other universe models and Einstein wondered whether God had other choices. David Hume spoke of God's "botched and “bungled” universes before he struck this one.

These theses on so-called ‘finite aspects’ of God are advanced to explain certain imperfections inherent in our universe. Of course, some would suggest the Book of Job as an aberration, God in a weak moment, perhaps a bit of divine mischief, or "a parade of power, devoid of moral content...positively immoral by any human standards" (Kenneth Rexroth). Many wonder why those who influenced the content of the Bible chose to include the Book of Job, except perhaps to show God as inexplicable, the ultimate mystery.

In his *Infinite In All Directions,* Princeton physicist Freeman Dyson speaks of the God of the Socinian heresy, a God who though *not omniscient or omnipotent*, is *evolving*, like human life, and consequently a God of finite characteristics.

[Where is God] When Bad Things Happen to Good People: In *When Bad Things Happen To, Good People*, Harold Kushner cites Proverbs 12:21 that "No ills befall the righteous, but the wicked are filled with trouble." He cites Isaiah 3:10-11 "Tell the righteous it shall be well with them, for they shall eat the fruit of their deeds. Woe to the wicked, it shall be ill with him, for what his hands have done shall be done to him."

He then goes on to say that *the idea that God gives us what we deserve and our misfortunes are the result of our misdeeds, is too simplistic* and leads to undesirable consequences in that it creates guilt and blame when there is no cause, and it results in *people being disappointed in God and themselves*.

Our View of God--A Misdescription of Reality? But most importantly, he observes, it is a misdescription of our reality. He cites the world's sufferings on a mass scale--indiscriminate or deliberate killings, derelictions in nursing homes, etc. as demonstrating the unfairness and suffering by the innocent, the death of the righteous unfulfilled, as we wait for the emergence of God's plan. Alas, he says, "the world...is not so neat a place as the prophets would have us believe." We’re back to the question of “best possible world.”

Does God Want Us to Suffer? We tend to assume, he says, that God causes us all our sufferings, deserved or not, for reasons as unexplainable now as they were to Job, whose question of God about his undeserved afflictions met with the most enigmatic response in human history. Well, Kushner suggests another approach--that perhaps God is not the source of our suffering, that perhaps there is some reason "other than the will of God."

Imperfections--Does God Have Finite Limitations? Well here we are back again to the work of Gottfried Leibniz, philosopher mathematician, and founder of symbolic logic. In his *Theodicy,* he asserts that despite its manifest imperfections, including evil, this is the "best possible world." Built into this concept, of course, is the idea that God does after all have finite limitations; that human life's very existence is a product of the fall from the perfect symmetry of the pre-universe void, or the original unified superforce; and that the very existence of the universe required the introduction of imperfection.

Physicists like Steven Weinberg and Heinz Pagels have noted that our universe of time, space, matter, geometry and dimensionality, permitting structure essential to life, derives from breakdown of the original "perfect symmetry" inherent in the unified superforce from which our universe started.

It was through the breach of the initial perfection, the symmetry-breaking thermodynamic cooling process, that our universe was made possible, our universe of time, space, and the separated fundamental forces permitting atoms, then molecular aggregates to evolve into living structures which could develop into mind capable of exploring these questions. So, in this perspective, *existence of life requires departure from perfection*.

Eternal Life As a Redeeming Feature of the Suffering World: If God's omnipotence is constrained by some limitations, such as the unavoidable imperfections required for a universe with life, is there some rational redeeming or compensating aspect? Philosopher Immanuel Kant, and others, pointed to the defect of the pre-sence of injustice in the world. Since justice is mandatory, an *a priori* characteristic of God and the universe, then, said Kant, a justice unfulfilled in our earthly life assures us of a rectifying afterlife. This is a comforting thought--although it still does not explain why a little less suffering and depravity would not suffice to satisfy the principle.

David Hume's "Botched and Bungled" Universes: In seeking to explain our imperfect universe, philosophers have explored the idea that before He came up with this universe--with all its limitations--God worked hard to build the best possible world. David Hume is quoted from his *Dialogue Concerning Natural Religion*, in which he speculated that numerous universes might have been "botched and bungled throughout an eternity ere this system was struck out; much labour lost, many fruitless trials made, and a slow but continual improvement carried out during infinite ages in the art of world-making."

Of course one still asks why an omniscient God couldn't perceive the results of alternate models without a trial-and-error process. The answer may lie, as suggested earlier, in the uncertainty aspects of quantum theory.

The Human Race Suffering to Make a Better Universe? But, asks Edward Harrison, in his *Masks Of The Universe*, why stop? Might not other universes thereafter have been struck by God, more splendid than the one we inhabit? This idea has captivated some who feel that the lessons of this universe are being applied, as by physicist Freeman Dyson's Socinian God or computer scientist Edward Fredkin's somebody or something who or which is running the universe program to see what will happen. And so, perhaps, we suffer for a noble purpose--a better universe-- and so our universe--and its life forms, serves some purpose and has meaning.

The Universe as a Computer: In *Three Scientists and Their Gods*, Science writer Robert Wright speaks of his interviews with computer scientist Edward Fredkin. Wright notes that Fredkin, who functions unconventionally in his own special field of computational science, "digital physics" and algorithmic (recursive) mathematics, "thinks that the universe is a computer;" that within the class of computer programs, called cellular automata, rests the ultimate principle that runs the universe.

Limits on God's Powers--Not Even God Can Take Some Shortcuts! In the interview discussion Fredkin makes a quantum leap into pure metaphysics and asks why the universe was created--why something exists rather than nothing? He reverts to his own "idiosyncratic" brand of science and comes up with the blockbuster. He says that using ordinary physics one can say God thinks of creating the universe and yes, being all-powerful he can probably go on and create it.

But Fredkin notes the existence of physical limits on omnipotence. No matter how powerful God may be, says Fredkin, there are no shortcuts--even for God. God may have various ways of creating the universe but He must go through every step of the process or, he declares, God won't get the "right answer."

And this, in effect, is Fredkin's answer to why the existence of a computer universe--there just is no way to know the "answer" but to run the program! Einstein had wondered whether God had other alternatives or was perhaps constrained by the logic of mathematics.

The Universe Set up to See What Would Happen? Fredkin thinks, says Wright, that space and time are discrete, that the laws of the universe are algorithmic, that the universe literally is a computer; that it is being used by "something or someone" to solve a problem. But then he explains that there is no way to know the future other than running the universe program; that the natural phenomena of the universe are not accidental. He doesn't know what there is or might be, but it seems to Fredkin that this universe is a result of something intelligent, set up "to see what would happen."

Augustine's Difficulties with God, Evil, and Free Will: We have noted St. Augustine's difficulties reconciling the predetermined universe (the dominating view before quantum uncertainty theory) with free will. In theism a given is God's omnipotence--and therefore His theoretical power to prevent evil. Another given is God's omniscience--and therefore His advance knowledge of human life's misguided or sinful choices. Assuming these powers (despite quantum uncertainty), the question arises why God chose not to 'fine-tune' this arrangement--either by somehow eliminating evil or by predisposing human life's disposition toward evil.

And so there are those who have said that St. Augustine can't have it both ways. That is, true free will where God predetermines and knows everything to come, abstains from preventing misdeeds, and then brings damnation and eternal torment on all. All, that is, except, in John Calvert's view, the elect--those preselected for salvation without regard to conduct or good works.

Free Will and Perfection As Mutually Exclusive: There are many views of St. Augustine's persistence in arguing free will in a predetermined universe. One is that St. Augustine saw that God had evaluated free will as being so central to meaning in this universe that everything else had to give way, even though God knew both that the price of free will was the existence of evil and that human life's free choices would be tarnished with error.

Liebniz's Best Possible World: Often cited as explanation for God-permitted imperfection in the world is the "best-possible-world" concept of philosopher-theologian-logician and mathematician Gottfried Wilhelm von Liebniz, inventor, contemporaneously with Sir Isaac Newton, of differential calculus, and founder of mathematics-based symbolic logic. While his work in mathematics is now but a footnote in history, his philosophical work still looms large in any discussions of the problem of evil in a world created by a caring God viewed as both all-powerful and all-knowing.

A Total Explanation Is Precluded Says Liebniz: It is sufficient, he observes, to have confidence in God that he has done everything for the best. To know the reasons which moved him to chose this order of the universe, to permit sin, is beyond the capacity of a finite mind. According to Leibniz, the "felicity of the spirits" is the principal aim of God and he executes this as far as "the general harmony will permit."

Liebniz explains that God has only to ordain a real world into being; and ex-planations should avoid multiplicity in hypotheses or principles. God's creation would always have been regular and in a certain order. God, he says, chose the most perfect, the simplest in hypotheses and the richest in phenomena.

He Does Not Pretend to Fully Explain the Great Mystery: But Liebniz antic-ipates criticism and recognizes the imperfect nature of his explanation of the divine wisdom. He suggests that *he seeks simply to raise our minds to conceive what cannot otherwise be expressed.* In so doing, he says, he does not pretend to explain the great mystery on which the whole universe depends.

Imperfection As Required: Leibniz goes through a rather elaborate process of describing why God must exist and the various alternative possibilities God faced in undertaking to create the universe, including the presence of evil. Having adopted St. Augustine's determinism, Leibniz of course, has tough going. Boiled down, he is saying that for human life to have meaning, requires the balance of good and bad aspects of the world, there being no other way to deal with the required complexities.

Some suggest other reasons which are not altogether unpersuasive. If there were no dangers, limitations, and if all was provided to humanity so that humans need not strive nor suffer, nor overcome, then life would have little if any meaning; life would be just a another form of vegetation.

Human elevation to a thinking, striving consciousness in a universe beset with difficulty, must be viewed as requiring the imperfect world. Imperfection is a prerequisite for achievement--fulfillment of human goals.

Good As Meaningless Without Bad: So imperfections are unavoidable characteristics of a workable universe of meaning. Thus, bad is the indispensable cor-relative to good--good has no meaning without bad. Dissatisfaction and deprivation are necessary to satisfaction and fulfillment. Human life could not be goal-oriented if all wants were providentially bestowed. No wants, no striving, equals no meaning.

In *The Fingerprint of God,* astronomer-minister Hugh Ross discusses the persistent conflict between God's omniscience and presence of suffering and evil. He suggests that the biblical plan to develop perfect people without sacrificing free will can only be achieved through the process of God's mercy and justice through pain, suffering and death and the exposure to temptations.

So, in the Liebniz view, a meaningful reality requires this world's otherwise unexplainable formulation by an omniscient and omnipotent Deity. The evil and imperfections we see in our universe, are the necessary consequence of the limitations inherent in construction of a workable universe.

Theodicy and Candide: But many questioned Leibniz's 'best-possible-world' explanation of why an omnipotent and compassionate God tolerates the kinds of evil and imperfection in the world He created, and over which, presumably, He has total control. For one thing, it implies a God with finite limitations, one not omnipotent. Distinguished theologians have not hesitated to question the idea and Thomas Aquinas, for example, argued that a better world was possible.

Perhaps the most famous challenge to Liebniz's best possible world found expression in satirist Voltaire's classic, *Candide*: "If this is the best of possible worlds, what then are the rest?"

An Overdose of Imperfection? Millions of God's favored subjects on the planet, human beings, live in pestilence, poverty, drought, natural disasters of every variety, and in subhuman conditions--no real home, no access to medical help, and without even the survival instincts of early human life. Human beings die by the tens or hundreds of thousands, some from natural disaster of flood, earthquake, famine, and others by acts of unspeakable human perversity. Could not a provident, compassionate God, omnipotent and omniscient, have avoided this huge overdose of imperfection?

Alternative Universes? Einstein mused whether God could have come up with another kind of universe, one with different parameters, different governing laws of physics. Physicist George Greenstein in *The Symbiotic Universe,* asserts the importance of exploring alternate universe configurations--to see if they permit life. If most models contrived did, then woe betide us, our supposition of a unique, intended-life universe fails, and the Anthropic Principle turns out to be egocentricity--or sheer arrogance.

A God We Don't Have to Blame: Interestingly there are, of course, theological models of God as having certain finite characteristics--limitations on his power, epitomized in physicist Freeman Dyson's concept of the Socinian God who is evolving as the universe and human life evolve. But these interesting concepts remain largely outside the mainstream of contemporary western theology.

A More Exonerating Explanation for Imperfection? Built into Leibniz's "best possible world" approach, is the proposition that the laws of physics themselves are somehow a limiting constraint on God. Are these limitations the price of life itself? If so, perhaps God's acceptance of the limitations is recognition of our importance in His scheme of things.

The Imperfection of Death--Order and Time: One of the most noted negatives about the world is that of the occurrence of death. Death is perceived as the product of the working of *time*, one of the dimensions of our spacetime universe. It seems that time and its effect to bring death, are intrinsic to a universe governed by the mechanics of order and chaos, thermodynamics, relativity and quantum theories.

In the nineteenth century, the Austrian physicist Ludwig Boltzman, pioneer in thermodynamics, thought of the universe as an eternal state of chaos or disorder. In such a thermal equilibrium state, incidentally, time does not exist, that is, it has no meaning because time is a geometric dimension of a spacetime universe which is wound up, in a state of high order, with the heat energy to produce change.

It is the thermodynamic process by which the universe winds down from high order or low entropy to a state of low order or high entropy that produces time. So time is that part of the spacetime geometry which represents change. It seems that you just can't construct a life-supporting universe without thermodynamics, order, time--and death. So, somehow, death is related to the physics of this universe structured to provide us life.

Even God cannot change the past. --Agathon

CHAPTER 31
COMMONLY-HELD NOTIONS OF GOD

Identifying Some Commonly-held Views: Despite the notably different outlooks between and within religions, between theology and science, and even within science for that matter, perhaps we can identify commonly-held "perceptions"of the nature or attributes of God, useful at least as a framework for discussions of God.

In *Now, That's a Good Question*, theologian R.C. Sproul asks *What is the average Christian's understanding of God?* He notes the widespread view he encounters in which God is reduced from the biblical portrait to a *benevolent celestial grandfather* type, sometimes with the sole attribute of love.

Sproul is concerned that the fullness of God's bible-displayed character is being abbreviated; that it should be a priority for the Christian community to increase broadened understanding of the character of God. *With respect to miracles*, Sproul expresses the view that God is not doing the kind of conspicuous miracles in the Bible, like raising the dead, because such miraculous sources of revelation are no longer needed in contemporary society.

God as a Being--or as Being: A suggested model would accept God's *existence*--the functioning or efficacy of a higher power. The first characteristic would be then that God is in a state of actual being, He exists. Whether this 'being' is 'supernatural' or 'natural,' can be left open for the moment. While questioning the existence of a theistic God, even atheism does not question the existence of the forces of nature which provide dimensionality, structure, and the panoply of remarkable physical coincidences representing the natural felicities leading to our very existence.

We might finesse theologian Paul Tillich's admonition that God doesn't 'exist' but *is* existence itself, by suggesting that God may be said to exist even if he represents existence and 'existence' applies whether one's frame of reference is theistic, deistic, or pantheistic-- and whether God is describable in supernatural or natural terms.

God Is Self-Existent: In *Philosophy of Religion,* John Hicks notes that the "Judaic-Christian Concept of God" includes the concept that God is self-existent--with two characteristics. One is: that God *is,* has "ontological independence," an "infinite richness and plenitude of being" as the unconditioned reality, and not dependent on other causes.

The Second Component of Self Existence--Eternal: It follows from uncaused or independent existence, notes Hicks, that God is eternal, "without beginning or end." We have noted the observations of physicists like Paul Davies who address the problems in physics of God, time, and eternity.

The problems of God and time, of course, don't originate with contemporary physics. Eleventh-century St. Anselm, early Christian theologian and Archbishop of Canterbury, described God as more than eternal, as being "outside all time." Even earlier, fifth-century St Augustine, suggested in effect, that God was timeless, existing apart from time since time, he noted, was created with the creation of the universe.

A Beginning or No Beginning? We have noted Stephen Hawking's quantum cosmology hypothesis in *A Brief History of Time,* in which, because of quantum effects, the universe has no technical actual boundary or beginning in time. However, physicists like Frank Tipler, in *The Physics of Immortality*, suggest that Hawking is invoking a *quantum boundary* rather than a relativistic boundary.

So we may go on to assume the relatively recent creation of time, as confirmed by the mathematics and thermodynamics of Einstein's general relativity theory and reflected in Big-Bang theory which involves creation of space and time, matter, dimensionality, and the fields and forces.

Shared Immortality? God's divinity and timeless nature is manifested throughout the universe and some physicists suggest that we are existential expressions of God, imperfect extensions of God's mind, as physicist Freeman Dyson observes in *Infinite In All Directions*. This concept of human life as derived from God also finds expression in some earlier philosophy.

The 19th century philosopher Schopenhauer viewed individual life as individuation and differentiation from unified cosmic mind, and, in his view, upon physical death the human mind rejoins the cosmic unity. Jesuit paleontologist Pierre Teilhard de Chardin went further, suggesting that upon death, the human mind rejoins the ocean of cosmic mind but retaining its developed individuality.

God as Infinite: Interestingly, St. Thomas Aquinas thought that the universe, though created, might still be eternal. As we noted earlier, physicist Stephen Hawking's quantum-gravity model of the universe posits the possibility of the universe without a time boundary--a universe that would just *be*. In any event, being transcendent, apart from the universe, God is seen as eternal, unconfined by space and time.

God As the Creative Force or Author of the Creative Force: In reconciling earlier religious ideas of God as "thinking" the universe into existence we must deal with the new physics which, as we have discussed, suggest the creation of the universe as originating in the random quantum effects of a primordial void, a quantum state of mathematics and pregeometry.

It is a condition without space, time, matter or dimensionality, from which the vitality of mathematics and quantum uncertainty generates a particle of unified forces representing perfect symmetry. And it is that randomly-erupting particle or point from which our universe developed through departures from symmetry represented by the separation of the originally unified forces.

But, asks Stephen Hawking in *A Brief History of Time,* and *Black Holes And Baby Universes*, what is it that "breathes fire" into the equations to generate the physical universe we inhabit? P.W. Atkins, in *Creation Revisited,* suggests that the universe is self-generated from mathematics, without "external intervention," implying that no independent agency is needed to "breath fire" into the equations. But Hawking leaves this open to an intervenor, whether God or perhaps "logical necessity."

This again raises questions of the nature of God as natural or supernatural and there are those who suggest that the 'supernatural' is an attenuated degree of the natural--a difference of degree rather than kind.

God As the Cosmic Controlling Intelligence: In this context, God may be referred to as the controlling intelligence of the universe. God may *be* intelligence or, as some have suggested, Cosmic Mind. Everything that exists may, in some way, be the quantum constituents of this intelligence or mind. As physicist Freeman Dyson observes in *Infinite In All Directions*, the human mind is an *extension of God's mental apparatus* and any differences are those of degree rather than kind.

God As Purposeful: Some physicists see God as purposeful. As physicists like Freeman Dyson in *Infinite In All Directions* and Paul Davies in *God And The New Physics*, point out, the incredibly complicated, inchoate but unfolding universe is an expression of that purpose. It is human life's innate imperative to seek the full expression of this purpose--and its likely destiny ever to fall short of finding it.

God, Human life and Purpose As Evolving: As physicist Freeman Dyson observes, purpose unfolds and changes as the Socinian God, human life, and the universe evolve. The purpose is largely subjective, there being no defining description valid for all sentient beings, each human being left to construct his/her own meaning, although scripture is thought by many to codify this meaning.

God As a Personal God: The very evolution of the universe and the workings of the laws of physics are such that God can be said to be personal, embrace personhood. This need not invoke an anthropomorphic figure but does represent personhood, and the derivative source for our own personal status. Apart from some Biblical references to physical parts of God--which can be interpreted allegorically, according to the weight of Western theology, *God is mental, spirit.*

In this context, and as we discuss elsewhere here, some scientists, like Freeman Dyson and Frank Tipler, speak of an evolutionary process in which human life, developing in sophistication, finds a *spiritual dimension or field* and finally achieves direct contact with or merges with Deity.

Which Are Closer to God, Atoms or Abstractions? As physicist Richard Feyn-man noted, like the universe, space, time and dimensionality, atoms, complex matter forms, mind, consciousness, subjectivity, all evolved from the beginning universe, through processes of increasing order and complexity, eventuating in highly personal abstractions.

Derivative Personhood: And which are closer to God, atoms or abstractions? asks physicist Richard Feynman. *Immaterial abstractions and physical atoms are equidistant from God*, ventures Feynman. What all this appears to demonstrate is that the source of the universe gives rise both to the physical universe and to phenomena having characteristics of personhood. This implies and invokes a source of the universe, whether we call it God or not, which can be characterized as personal. Certainly the laws of physics generate personhood.

And if we call that originating source 'God,' our human characteristics, with their profound strengths and abysmal weaknesses, are, somehow an imperfect reflection of that God's characteristics-- human beings, made in God's image but with less than God's perfection.

God As Morality: Another critical characteristic of God invokes morality. Western theology equates God with love and morality, each subsumed in the other. One might select morality as the all-encompassing descriptor for God, incorporating love, justice, compassion, mercy--and the responsibility for one's acts. Philosopher-cosmologist

Immanuel Kant, after rejecting the Aquinas proofs of God, found the necessary existence of God in morality, sense of duty, also reflected in the categorical imperative.

God As Both Transcendent and Immanent in Human Life: The preferred theological view is that God is both transcendent, existing apart from the universe and, at the same time, immanent in the universe and therefore immanent in life. Theologian Karl Barth thought God's transcendence precluded his immanence in humans. But most of Western theology is comfortable with a transcendent God who is also immanent in mankind. Mathematician Frank Tipler, in *The Physics of Immortality,* invokes a remarkable physics-mathematics process to *prove transcendence, immanence, omnipotence and omniscience* in his Omega Point God.

And, it is urged, when humanity reaches up, noble in purpose and infused with the intuitional sense of the sanctity of its divine origins, humankind shares God's being, because, at that moment, they are one--humankind in God and God in humankind. Immanence has powerful appeal and foundation for hope, salvation and redemption.

Sir Bernard Lovell's God of Being: British physicist and radio astronomer Sir Bernard Lovell, developer of the Jodrell Bank radio telescope at Cheshire, England, a sort of philosopher at heart; gave expression, in his *Astronomer By Chance,* to religious observations. Referring to theological differences between theologians Karl Barth and Paul Tillich, Lovell notes: "Perhaps as I age, I shall agree more and more with Paul Tillich that *God is Being itself*--not the supreme transcendent Being of Barth's theology who is known only to man through self-disclosure."

Paul Tillich and God As Existence: As Bernard Lovell noted, theologian Paul Tillich, who noted the freedom with which physicists bring in God in their references, rejected the reference to God as a supernatural *being*, rejected as demeaning. For him, God was *'being' or 'existence' itself.* In one fell swoop this all-encompassing definition of God as "being itself," renders *nonacceptance of God's existence as virtually impossible.* Tillich asserted that God is invoked even in the atheist's very denial of God.

But not every theologian is comfortable with God as "being" or "existence" itself, because of the pantheistic implications--reminiscent of Spinoza's God as the totality of everything, including the universe and human life. In a sense, it may be thought of as embracing philosopher-mathematician Leibniz's 'best possible world' concept, criticized as a concession to the finiteness of God, a limitation on His power and an incursion into his omnipotence.

Existence and Personhood: Tillich's definition of God as existence might better explain the limitations of the world without denigrating God or trying to rationalize His failure to prevent evil or restrain the life-destructive natural forces. The totality of existence would understandably include everything needed for a rational world, the good and the bad, the primitive and the advanced, the mundane and the sublime. Perfection is understandable only in the context of imperfection. God as existence puts in jeopardy the working of miracles unless miracles have *natural* explanations.

Revisiting the Supernatural: On the question of the supernatural vs. the natural, in *The Creator and the Cosmos* astronomer-cosmologist and minister Hugh Ross points out that some of the New Testament acts of Jesus, spoken of as miracles, are explainable in the *extra dimensions* whose existence we now recognize and, of course, available to God. It has been suggested that the term 'supernatural' refers not so much to acts beyond or in defiance of nature but to those aspects of nature beyond our knowledge or with which we happen to be unfamiliar.

God Shares His Mind and Creative Power with Humankind: It has been pointed out that God shares his mind and even his power of creation with humanity. In a sense, it may be God's presence or immanence in human life that gives human life creative power, such as the recently-acquired ability to alter genetic codes and alter the very course of evolution in the universe by creating new life forms through genetic engineering. Both Freeman Dyson and Frank Tipler speak of life eventually taking control of the physics of the universe to guide its future course and permit the eternal existence of life.

This is another manifestation of the changing universe in which human life appears to have a *significant role in modifying the universe*--in influencing its later form. Human life appears to be one of the very determinants of the nature of matter through God's sharing of divine mind. As physicist Freeman Dyson suggests, human mind is an "extension of God's mental apparatus." Of course, extraterrestrial intelligence, if it exists, may be ahead of us in this role.

The Ultimate Symbiosis--God's Need for Humanity: Scientists, philosophers, and even theologians have observed, from time to time, that in a sense, God's own existence may be related to human life. Arthur C. Clark observed that "It may be that our role on this planet is not to worship God--but to create him." Even the early theologians opined that God cannot be alone--any more than the humans can be alone. And so we are here, through the long creative process that has led to human life and its symbiotic relationship with God. Because of the coexistence of all time in relativity and quantum theory it may be appropriate to note that, in a sense, both God and humanity are eternal and *coterminous*, lending support, interestingly, to the notion of the preexistence of Jesus in the New Testament. .

We sometimes dwell on our 'material' rather than spiritual nature. But in a sense, matter may be viewed as the tangible mechanism by which God has manifested himself. This concept reaches its zenith, perhaps, in the philosophy of Jesuit paleontologist Pierre Teilhard de Chardin whose sanctification of the material world has so challenged some in the ecclesiastical community.

God Dependent on Human Life? To say God 'exists,' at least in Copenhagen interpretation quantum terms, requires an observer, a sentient being capable of observership of God's existence. If we are the only creatures in the universe that exist to contemplate the universe, and especially theorize and contemplate its Maker, as discussed elsewhere here, then human existence can be perceived as physically essential to God's very 'existence.'

Perhaps what favors human life's continued existence in the universe is that without the human mind to provide the awareness of God, to vouchsafe His existence, the universe's existence and, in a sense, God's existence is in question. Quantum theory suggests that the unknown universe cannot be said to exist because existence implies observership.

Earlier, this concept was considered purely philosophic and often disputed by those who asserted that something exists apart from knowledge of its existence. But, as we discuss in the context of quantum theory, the laws of physics themselves seemingly support the views expressed by quantum physicists like John Wheeler, that existence may have no objective reality apart from an observer.

Human Beings Looked up and Sensed God into Existence? In this context, it may be said that it took some fifteen to eighteen billion years for both human life and God to 'come to fruition.' Before that there simply existed the potential in the form of mathematics and laws of physics representing the 'process' for translating and 'geometricizing' abstractions in physics into some form of tangible reality.

So, in the quantum context at least, God and human life may be reciprocally indispensable. Perhaps it was the first moment that humans looked up and sensed an awareness of God that in the quantum theory context, they both came to 'exist.'

God and the Universe Involve Purpose: God works toward perfection by invoking purpose. Thus the Leibniz'a unavoidable imperfections of the world are wrought into reality. The universe, like humanity, is evolving, striving for perfection, say physicists like Freeman Dyson. A perfect world without striving, struggle, adversity would have no meaning-- an absurdity--or so the argument goes.

As we discuss in the review of the Anthropic Principle, the physical structure of the universe evidences exponentially-precise design, God represents purpose, the universe is purposeful, and human life as a manifestation of God, must represent derivative purpose. In *The Creator And The Cosmos*, astronomer-minister Hugh Ross argues persuasively for the anthropic principle and the relativistic big bang as proof of God and therefore of purpose.

Religion As Recognition of God's Authority: God is perceived as father or in some newer biblical text, as parent to all the human race. Religion is the intuitional recognition by humanity of God's authenticity and authority. As writers like Karen Armstrong in *A History of God*, point out, perceptions of God evolve, and God must use different mechanisms to manifest Himself at different times and places, and to different civilizations and societies.

And so all the great tried and tested religions are generally viewed as valid vehicles for dealing with God-human relationships in the context of particular societal and cultural histories. Each has something to teach us--the Eastern religions show us the abstract oneness of everything in the universe; the Judeo-Christian faiths emphasize personal aspects of the God-human relationship. American Indian religious ideas emphasize human relationship to nature.

God's Selected Instruments for Revelation: And so figures like Jesus, Moses, Mohammed, Gautama, and many other human exemplifiers of God have made critically important contributions of noble thought throughout history. Men like Gandhi are not mere accidental occurrences in history but seen as instruments for revelation and guidance.

Jesus, totally apart from doctrinal concepts of divinity, represents a particularly intense personification of divine personhood and personality, and in consequence,his influence in molding human mind and character is seen, in the Western world at least, as unparalleled in human history, reducing divine abstractions to comprehensible human values, goals, and imperatives expressed in the New Testament.

God's Other Manifestations--Other Universes--Other Dimensions: Ours may be but one of an infinite numbers of universes, some "botched and bungled" as philosopher David Hume argued, deformed and barren of life or meaning, but perhaps serving some quantum theory purpose, even governed by different laws, with different dimensions. Also, human life's perception of our own universe may be confined by human life's present relatively primitive state, and what some now regard as 'supernatural' phenomena may eventually be understood as natural. As discussed, many events once viewed as miracles now have accepted physical explanations.

We have noted possible extraterrestrial forms of intelligence the nature of which may be difficult to comprehend. These civilizations may deal with or enjoy abstract or theological concepts totally outside our experience.Suppose, for example, such civilizations have totally different biological and material forms, with none of our pestilence or sickness. While our morality concepts are often sex-related, other forms of intelligence may have totally-different reproduction mechanics, with completely different morality implications.

An advanced society may have conquered altered the aging process, making any further reproduction a threat to the the existence of their society, reproduction being punishable. Our concepts, like sin, forgiveness, compassion, love, friendship, retribution and salvation may be irrelevant to such a society.

A Threat to Earth's Religions? Physicists and cosmologists have noted, the possibility that our accustomed avenues of human thought may not be universal, and may be foreign to other intelligence in the universe. This may prove destabilizing to our societies. Their "religion" might well be beyond our comprehension, and our theology viewed by them to be primitive. As physicist Paul Davies observes in *God And The New Physics,* a religion of such a society or even an entire advanced-civilization-galaxy, might totally upset our complacency about the validity or efficacy of our own religions.

Especially vulnerable, notes Davies would be religions, such as Christianity, based on the assumed universality of a particular dominating central figure representing God. If advanced civilizations exist, which is considered more than likely, it is reasonable to assume that God has manifested himself to each civilization in appropriately different but meaningful form and concept. By contrast our own religions, the product of earlier impressionistic societies, may seem parochial or provincial.

Biblical and Big Bang Creations in Conflict? Eastern theology speaks of the unity of the universe as cosmic mind. In conventional cosmology, Western theology speaks of God as spiritual and personal, Creator of the universe and human life, all powerful, and all knowing. In quantum cosmology, physicists like Stephen Hawking, Heinz Pagels and Paul Davies speak of the universe as self-generated from the quantum-uncertainty-dominated, pre-big-bang, mega-universe quantum void, essentially a mathematical, pregeometric, quantum energy state of the potential of the laws of physics. The two approaches have been characterized by many as being in irreconcilable conflict, a view which contributes to promoting an atmosphere of hostility between science and religion.

Some Reconciliation: As we have noted, some religiously-oriented scientists, like astronomer-minister Hugh Ross, sponsor of 'Reasons To Believe,' an organization of religious scientists, speak of the latest proofs of the Big Bang as evidencing the existence of the God of the Bible, citing in particular certain passages of the New Testament.

They question the quantum cosmology of quantum void self-generation of the universe as essentially metaphysical speculation while, for the most part, accepting relativistic Big Bang cosmology as compatible and corroborative of the Biblical Creation scenario. The Creationists, of course, see the Big Bang physics as in-compatible with the biblical creation scenario based on insistence on literal biblical inerrrancy, such as the interpretation of the days of creation as 24-hour earth-days rather than geological or cosmic time days.

Reconcilability of God and Quantum Void Creation: Interestingly, though, there appears to be no necessary inherent incompatibility between quantum cosmology creation and the biblical account. Many are willing to accept the notion that the biblical account is a metaphor, *not intended as a handbook on physics* but simply an allegorical account of divine involvement in terms understandable to early society.

The primeval mega-universe quantum nothingness state, devoid of time, matter, forces, and dimensionality, can be equated with the biblical creation *ex nihilo* by a God who exists outside of time and space. Obviously that primeval condition represents a powerful *creative vitalism*, which generates a physical particle or mini-universe of unified force, space and time which unfolds into our dimensional universe with space, time, and matter. That universe further unfolds into highly personal cognitive life and mind capable of exquisite and highly personal abstractions.

Biblical Creation as a Metaphor for Quantum Cosmology: Physicist Freeman Dyson characterizes this highly personal human mind *an extension of God's mental apparatus* so that *man's personal* nature is a derivative of God's personal nature. The individual human mind is growing in cognitive capacity as a part of group mind, mankind and society as a whole, in its power of comprehension and God-derived creative power. Mathematician Frank Tipler asserts that this growth of societal mind

will spread throughout the universe to become the *Omega Point* omnipotent and computer- information omniscient God who exists *now* too because of the relativistic coexistence of all time.

The point is that in the same way that the relativistic Big Bang creation is a metaphor for the Biblical Creation, so also may the quantum creation be interpreted as a metaphor for the Divine Creation. Tipler's mathematical proofs, incidentally, also extend to the nature of God as personal.

Is There an Alternative to God? In Chapter 3 of his *The Quest for God*, English historian and journalist Paul Johnson asks: Is there an alternative to God? He discusses the *Promethean Spirit,* based on belief in man's unassisted progress and potential, the spirit of those believing they can dispense with the notion of God altogether, a view, he says, growing with "dramatic speed" over the past two and one-half centuries, but a failed concept.

He cites the examples of Prometheans as Voltaire, Richard Dawkings, author of *The Selfish Gene*, historian and writer H.G. Wells, Bertrand Russell and his disciple Sir Alfred Ayer (better known as A.J. Ayer), and Jean-Paul Sartre, all of whom essentially failed in demonstrating an alternative.

He speaks of humanism as also as a declining and failed alternative. He goes on to describe the totalitarian alternatives to God, like the economics of Marxism-Communism, and other "isms" based keyed to race politics, sexual politics, environmental politics, all involving agendas leading to extremism.

He concludes that he can see *no alternative to God* and so goes on to discuss what is the "nature of this enormously important and essential being, asking, "Can we in fact know God, describe him, define him?"

> It may be that our role on this planet is not to worship God but to create him.
>
> Arthur C. Clarke

PART XI--PROVING THE EXISTENCE OF GOD

CHAPTER 32
PROOFS OF GOD--INTRODUCTION

Source of Knowledge of God: If the physical senses do not give us access to God, how is his existence confirmed? Philosophers and theologians down through the centuries have argued over the source of knowledge of God, some arguing for and some against the authentic knowability of God.

The Role of Revelation: Western theology, including Judaism, Christianity, and Islam, base 'knowledge' of God primarily on *revelation.* The Old and New Testaments and the Koran invoke knowledge of God through revelation. Vedanta Hinduism presents the problem in reverse, asserting that *the only reality is God* and the universe is an illusion, the reality of which is subject to proof.

As many have pointed out, truth through revelation is essentially subjective, as shown by the many divergencies in so-called revealed truth as reflected in differences in holy scriptures of the various religions of the world and differences within each. Such limitations naturally generate interest in proofs which appear to have scientific foundation, so-called scientific proofs.

Apart from uncertain implications of psychic manifestations, 'mountaintop experiences,' and unexplained dream 'visitations,' we have not had demonstrable proof of God's existence, at least in physical or other verifiable terms, and so, as philosopher Immanuel Kant asserted, the question of God's existence is a metaphysical question--beyond the potential of man's knowledge. Yet philosophers, theologians, and even scientists, such as astronomer-minister Hugh Ross in *The Fingerprint of God,* discuss various arguments for and 'proofs' of God.

Two examples are earlier deductively-reasoned proofs. One is thirteenth century St. Thomas's 'cosmological argument' based on required causality, and another, eleventh century St. Anselm's 'ontological argument' that actual existence is required by definition of a most perfect being. Often cited as classical arguments for God are St. Anselm's *Five Ways* which we will discuss.

Philosophers David Hume and Immanuel Kant challenged these classical arguments for God's existence. While rejecting the validity of such arguments, Kant advanced his own argument, namely, that the existence of the inherent morality mandate implied a higher power source or *primum mobile* of the mandate.

Impact of Emerging Cosmology Physics: New interest in so-called 'scientific proofs of God' has arisen from developments in certain aspects of the physics of cosmology, especially information relating to the anthropic principle involving the remarkable life-promoting constants of physics and new proof of the validity of Big Bang theory. We explore both of these developments.

> It may seem bizarre, but in my opinion, science offers a surer path to God than religion. Right or wrong, the fact that science has actually advanced to the point where what were formerly religious questions can be seriously tackled, itself indicates the far-reaching consequences of the new physics... --Paul Davies

CHAPTER 33
SCIENTIFIC PROOFS OF GOD AND OMEGA POINT GOD THEORY

NASA COBE Findings: While some physicists, like Paul Davies, suggest these remarkable precision coincidences in the constants of physics involved in the anthropic principle as evidence of design and purpose, others, like Heinz Pagels and Steven Weinberg, question its efficacy or relevance as evidence of design. But astronomer-minister Hugh Ross insists that this principle, together with the 1992 NASA COBE findings of primordial temperature variations in the cosmic back-ground radiation from the Big Bang, represent *scientific evidence* of God's existence and role in the Creation.

Astro Observatory 2 Findings of Primordial Helium: As additional proof of the Big Bang Creation, Ross also cites the June 1995 Johns Hopkins University Astro 2 Observatory findings of primordial intergalactic clouds containing spectral lines evidencing heavier-than-hydrogen elements of helium, and of trace quantities of deuterium, lithium, boron and beryllium.

These elements, whose creation predates the galaxy formation, had to have been made within the first four minutes after the Big Bang, and exist in amounts consistent with calculations of the quantities generated during the first four minutes following the Big Bang. The findings, incidentally, also support the conclusion that most matter in the universe must exist in intergalactic space rather than within the more visible galaxies.

Abstractions As Derived from Increasing Order and Complexity: In *The Character of Physical Law*, physicist and mathematician Richard Feynman analyzes the process by which the physicist goes about depicting how physics is used in describing how everything works in our universe--its composition, how it gets its structure and operating mechanisms, the evolutionary processes which produce pockets of increasing complexity and order in a universe winding down toward total chaos. Noted, are the 'hierarchies' of increasing complexity of ideas which emerge with evolving mind.

In this view, the embodied human mind, selfawareness, consciousness, subjectivity, and highly personal abstractions like love, compassion, reverence and holiness, are viewed as *informational phenomena emerging in a holistic process* from the quantum via thermodynamic and complexity processes. At the same time that the ordered, 'wound-up' universe unwinds through a process of *decreasing* order toward chaos, this process also mysteriously generates pockets of *increasing order* and complexity.

Quantum Theory, Abstractions, Personhood, and God: Thus, emerging from the quantum, comes the physical universe, life, consciousness, self-awareness, and mind, equipped to comprehend abstractions that signify personality and personhood, elements that we normally associate with God and the Divine Creation.

So, one may ask, is this *creative unfolding process a metaphor* for God? Abstractions and enobling spiritual ideas represent ideas which, to a large extent, define the human dimension. Most people are accustomed to think of these felicities as emanating from God rather than from some mechanism of physics.

Which End Is Nearer to God, Abstractions or Atoms?: But, asks Feynman, using the religious metaphor, "Which end is nearer to God...beauty and hope, or the fundamental laws?" Before answering this central rhetorical question he digresses to observe that "we have to look at the whole structural interconnection" of the sciences, the intellectual, the connections of the hierarchies, and so forth, to connect beauty, human life's psychology, brain, neural impulse, chemisty, and so forth, "up and down, both ways."

And today, he notes, we "cannot...draw carefully a line all the way from one end...to the other, because we have only just begun to see that there is this relative hierarchy." And, he declares, "*I do not think either end is nearer to God.*"

The Unigod Who Is Both God and Physics: We see here elements involving the unity of theology and science, something physicist-astronomer Edward Harrison calls, in his *Masks Of The Universe*, the Unigod, *the God who is both God and physics*.

Feynman's well-articulated and oft-cited "hierachy of complexity" leads from the fundamentals of physics to abstractions and spiritual ideas. And all this natural sequence flows from a self-starting universe in which physics, quantum forces, are hypothesized as triggering this sequence.

Reasoning God into Existence? But we're back then to Paul Davies' question in *The Mind of God*, as to where this causation-regression stops. What is the source of the quantum and unified physics powerful enough to generate a physical universe and produce life, consciousness, self-awareness, subjectivity, abstractions and spirituality. These things are, after all, so *personal*. If physics in its complexity-generating processes can generate human beings, personhood, and abstractions, have we somehow reasoned God away or have we reasoned Him into existence?

Divinity and Physics: Although these evolving states of the universe leading to life and exquisite abstractions like love and reverence represent physical processes of the growth of order and complexity, to many, they suggest or imply the *supernatural.* We are accustomed to thinking of events such as the appearance of life and consciousness as somehow derived from *Divinity--God* kinds of things.

One may well ask what is the distinction between a personal theistic God who creates human beings who entertain sublime thought, on the one hand, and on the other, a quantum phenomenon, a power of mathematics and unified physics, possessing the capacity for generating universes which produce biologically-housed mental states, minds, which similarly dwell on concepts like reverence and holiness?

One might refer to the "Power of Mathematics and Physics" to find expression in personhood and abstraction, while another may prefer reference to the 'personal, theistic God' who employs mathematics and physics to brings into being personhood and personal concepts. Some suggest the two approaches as synonymous--involving a distinction without a difference. But some go further suggesting scientific evidence of the existence of God.

Can Science Provide Quantifiable Evidence of God's Existence? In the August 1991 Omni Magazine article *The Mind Of God*, A.J.S. Rayle and K.T. McKinney ask whether science can *reach beyond physical reality* and prove that a divine being or force oversees the universe? Although many in science view "finding God" as outside the realm of science, there are some who believe that science can produce "quantifiable scientific evidence of God." Rayle and McKinney remind us that theologians do not assert that "divine revelations--or intuitive discoveries" prove God's existence.

A Signpost That God Exists? Astrophysicist Thomas McDonough is cited as noting that the rate of stellar fusion is precisely tuned to produce a sufficient supply of carbon forms required for life; that this and similar life-proclivities involved in the remarkable coincidences in the mathematics of the constants of physics, provides a *signpost that God does exist*, which he recognizes, of course, as falling just short of actual scientific proof of God's existence. There are religious scientists, like astronomer Hugh Ross, who suggest that this is scientific proof of God.

Proving That God Does Not Exist? Scientists, of course, cannot prove the negative--that God does not exist--notes Johns Hopkins University neurophysiologist Vernon Mountcastle, who expresses the view that any belief in "science as a tool for discovering God" would be nothing less than arrogance.

On the other hand, Tulane University's mathematician-physicist Frank Tipler observes that science is concerned, after all, "with the totality of existence" and that *if science can't reach God, then God doesn't exist.*

While for some, this sounds close to asserting God's nonexistence, the forbidden scientific negative, it appears simply to suggest that if no scientific effort to prove God's existence *can* succeed, i.e., that such an effort is *unachievable*, then God's existence remains metaphysical speculation. After all, it isn't science, in science philosopher Karl Popper's philosophy of science theory, unless it can be *falsified* by observation or test.

Mathematical Proof of God: Interestingly, in *The Physics of Immortality,* Frank Tipler, in his Appendix for Scientists, provides what he qualifies as mathematical proof of an evolving God, the vehicle being his Omega Point God Theory. In that book, Tipler presents *scientific proof that God is transcendent, personal, omnipotent, and omniscient.* While God is evolving with the spread of intelligence through the universe, that ultimately perfect God exists now also, he explains, because of the *coexistence of all time* derived from relativity and quantum theories.

Science--Whatever May Be Falsified by Observation: Science philosopher Karl Popper thought and wrote extensively on verifiability or unverifiability of knowledge in science and meta-science. He argued that science consists of whatever may be falsified by observation, that is, by test or experimentation, and is therefore demonstrably 'falsifiable.'

So, in this view, science is 'falsifiable' and nonscience is 'nonfalsifiable,' that is, not subject to disproof. In *Masks Of The Universe*, Edward Harrison, who discusses Popper's falsifiability criterion, illustrates this with the proposition that "angels exist" as being nonscientific because it is an unfalsifiable proposition--cannot be proven wrong.

A Research Project to Prove God? In the Omni magazine article cited earlier, the authors suggest that the greatest barrier to a research effort to prove God's existence may be its very acceptance as a proper subject for scientific research. They note that Stanford University psychologist Stephen LaBerge points out that even if such research were to be encouraged by an offer of financial support, you would first *need to define what is meant by God*, a subject we deal with here at some length, without, of course, an accepted resolution of that question.

As we noted earlier, each religion has its own conceptual, historical, and cultural approaches to the nature and existence of God. Some envision physical forms, like Michelangelo's famous Sistine Chapel's depiction of God as the powerful figure of the white-haired patriatrch; others as representing "spirit" or some nonphysical reality, such as a "unifying force in the universe," or simply as representing the universe or the Totality of everything there is, and so forth.

A Cosmic Interconnected Mental State: Earlier we noted the relationships of personal abstractions to the hierarchies of complexity emerging or unfolding from the quantum. We discussed the fact that underlying the physical world is the quantum world where the physical things of the macro world disappear into mysterious shifting energy states, something nonmaterial and somehow triggered into some reality by observation. Underlying these energy forms is thought to be a cosmic fabric of information so that the quantum may be viewed as a manifestation of an interconected cosmic mental state, sometimes described as cosmic mind.

The Paradigm of the Creative Universe--The Cosmic Blueprint: Physicist Heinz Pagels, in *The Cosmic Code*, speaks of a software program for the universe con-taining subroutines from which life unfolds. In *The Cosmic Blueprint,* physicist Paul Davies speaks of the "self-organizing processes in every branch of science," reflecting the propensity of physical systems spontaneously to generate entirely new states of order not previously existing in the universe. And so, Davies observes, there is the *new paradigm of the creative universe* which emphasizes the synthetic, holistic, and organizational aspects of nature.

Are God, Physics, Nature, Synonymous? If, as Davies notes, Nature, representing the laws of physics, the physical forces, becomes the model or paradigm for the creation process, just where does God fit in? As we have discussed, this line of thought suggests the need to consider carefully where God leaves off and physics begins, whether, as we have discussed, the two are perhaps coterminous, two descriptions of the same intrinsics.

If so, however, the question which follows is whether God, then, represents the *natural* rather than *supernatural*, the latter used generally to refer to events beyond, outside of, or inconsistent with nature. Philosophers have argued that all events, even 'miracles' are natural events for which we have no present explanation, that God is a natural phenomenon whose natural functioning is simply not yet fully explained.

Purpose As Beyond Science: Many of the things earlier thought to be the direct handiwork of God are now accounted for by discoveries of scientists like Copernicus, Galileo, Descartes, Newton, Heisenberg, Bohr, Einstein, Hawking and many others, with more explanations in the pipeline.

The Laws of Physics As outside Space and Time: British biologist Rupert Sheldrake is quoted in his *The Rebirth Of Nature: The Greening Of Science And God*, as saying that exploring the origin of the laws of physics must conclude that they exist "outside of space and time." Stephen Hawking asks, in *A Brief History Of Time*, what it is that " breathes fire into the equations" to create our physical uni-verse? In *Creation Revisited*, Oxford's P.W. Atkins insists it is mathematics that does it all--*without outside intervention*.

Physics As Existing by Necessity: Several physicists, Paul Davies being one, asks why, if God *exists of necessity*, then why not also the laws of physics existing of necessity, perhaps on the ground that they represent the only self-consistent rules permissible, and so, like God, are built into some underlying eternal reality.

From Idea to Physical Manifestation: If this universe, and the unified superforce from which it is theorized to have sprung, originated as a fluctuation in the quantum void of the primordial mega-universe, then, of course, one can say that, like the universe, the laws would just *be*. This still leaves unanswered Hawking's question of how mathematical abstraction is somehow metamorphosed or actuated into physical form. What triggers this transition from concept to physical manifestation?

The Engineering God--God As Spiritual Energy: As do Freeman Dyson and several other physicists, Rupert Sheldrake suggests a view of cosmic evolution in which the full panoply of the laws of physics did not always exist; that, in fact, they *unfold with the universe, evolving with time*, suggesting a "mechanistic or engineering God with a mathematical imagination," reminiscent of Einstein's perceptions of God. This approach requires, Sheldrake feels, a different perspective of the evolution of the cosmos, of nature, and of the *role of God.*

If, as Sheldrake believes, God is a *spiritual energy*, then yes, Sheldrake believes that it is theoretically possible for science to prove God's existence. But this requires that we understand *this other realm of spiritual energy,* which science has not yet addressed, although he notes that as being the direction we're headed.

In *Mindsteps To The Cosmos*, astronomer Gerald S. Hawkings organizes the evolutionary development and experience of human life into "mindsteps," In Mindstep

6, he speaks of the discovery of new spiritual forces, a "nonphysical cosmos," an "age of the spirit," of telepathy, the existence of a vital force, a soul, continuation of life.

But Omni cites Dartmouth physicist and former NASA Goddard space scientist Robert Jastrow, as feeling that "integrating spiritual energy and physical reality" would *destroy the body of science and the structure of cause and effect*. Jastrow refers to the "larger questions" of the start, end, and purpose of the universe, as having a "certain narrowness," in that if you try to break the "structure of cause and effect," you have "nothing, not good philosophy, not metaphysics, and not good science."

God and the Nonlocal Nature of the Human Mind: In the exploration of Quantum Reality the human mind is discussed as 'nonlocal,' existing somewhere "out there," part of the interconnected cosmos of information and mind. The Omni writers cite physician Larry Dossey's observations about this 'nonlocal' nature of the human mind in its bodily framework of informational molecules, molecules which are "omnipresent," interconnected with the cosmos, infinite in space and time, and therefore immortal and eternal.

But these happen to be the very characteristics we ascribe to God. And so it is this "nonlocality," they feel, that will provide science with the proof of God's existence.

God As a Concept Residing in the Human Brain: Omni cites neuroscientist Michael Persinger of Ontario's Laurentian University as believing that the answers lie not in the cosmos but in the human brain, the product of human experience with the anxieties about death, which *invokes God and induces his existence*.

But this notion of the God whose existence is found in the human brain, suggests a God whose existence is the metaphor for human life's divinity imperative, his imperious needs for meaning, a conceptual God whose *actual verifiable* existence is not required. This reminds us of the oft-cited philosophic observation that if God did not exist human life would create Him and, as Arthur C. Clarke observed, "It may be that our role on this planet is not to worship God--but to create him."

Ideas of God As Arising From Human Needs: In *A History of God*, Karen Armstrong, a former Catholic nun, gives expression to this idea that down through the ages, deity forms emerged as the output of *human practical need for meaning* and for a stabilizing set of values. So new ideas of God evolve from human emotional or psychological needs. "Each generation has to create its own imaginative conception of God....All religions change and develop." And, Armstrong declares, "if they do not, they will become obsolete."

FRANK TIPLER--SCIENTIFIC PROOF OF THE OMEGA POINT GOD

Mathematical Proof of God? As we have discussed, most of the so-called "scientific proofs of God" relate to the anthropic principle, highlighting the extraordinary coincidences in the constants of physics allowing for life. However, in *The Physics of Immortality*, Tulane University mathematician Frank Tipler unabashedly offers so-called *mathematical* proof of the existence of a developing personal God, as part of his *Omega Point Theory.*

Frank J. Tipler, a full Professor of Mathematics at Tulane University, is also a theoretician in "global general relativity," a relatively new phase of physics generated by the work of Cambridge University Stephen Hawking and Oxford University physicist Roger Penrose. Tipler was coauthor with John D. Barrow, of the popular *Anthropic Cosmological Principle* and author of many professional journal articles.

Omega Point Theory: The Omega Point, in effect, is the equivalent of God who, like physicist Freeman Dyson's Socinian God, is evolving over time to the Omega Point, the end point of the universe.

Tipler reduces his Omega Point theory to a single statement: "God, who is a personal being who created the universe out of nothing, exists, loves us, and will at the end or Omega point resurrect us to live in heaven forever." 'Omega Point' refers to the mystical content of the New Testament Book of Revelation dealing, among other things, with "time" and the future, and in which we find the reference in 1:8 to: *I am Alpha and Omega, the beginning and the ending.*

"The Physics of Immortality"--Solving Metaphysical Problems as Physics: Tipler cites philosopher Immanuel Kant as posing three fundamental metaphysical problems: Does God exist? Do we have free will? Is there life after death? In *The Physics of Immortality*, Frank J. Tipler uses his *Omega Point Theory* to answer 'yes' to all three questions, offering *mathematical proof* of the physics involved. Tipler points out that the history of science has typically turned insoluble metaphysical problems into problems of physics and then solved them.

Omega Point Theory--Mathematical Model of the End of the Universe: Tipler observes that the fundamental computer is the ultimate reality, and resurrection in Omega Point Theory is accomplished by future computer simulation sweeping in all possible earlier human existences. The Omega Point Theory assumes that, based on the necessity of continued life, a closed universe is required and therefore must exist. The theory provides a *mathematical model of the end of the universe* in which the contraction phase reaches the Omega Point or boundary of spacetime, where space-time curvature becomes infinite and ceases, and quantum cosmology governs.

The book, and its supporting mathematical appendices, raises many challenging questions. In the October 1994 issue of *Omni* magazine on "Science and Religion," Tipler, answers many of the questions that have been raised.

The Theory Assumes That Life Must Spread through the Universe: Paralleling physicist Freeman Dyson's *Infinite in All Directions*, the Omega Point Theory assumes that 'life' in various forms, must spread throughout the universe, becoming infinite intelligence with infinite knowledge by the end or Omega Point, in a million-trillion years or so when computer capacity will permit.

And the intelligence thus spread-through-the-universe, representing God, will, by computer simulation, recreate all prior beings by simulating all beings who did exist and, citing many-worlds quantum interpretation, beings who could have existed. Tipler explains that he is merely revising the boundary conditions from the "beginning of time to the end of time."

Natural Theology: Tipler describes himself as a natural theologian, declaring that the only thing you'll learn about God "derives from nature itself, rather than from what He chooses to reveal to his prophets." The notion that the basic tenets of religion can be explained by physics, says Tipler, has been accepted by all great Christian theologians, quoting St. Paul to the effect that "the basic attributes of God can be seen by the natural light of reason" and citing St. Thomas Aquinas' proofs of God based on "Aristotelian physics."

Anchoring Your Frame of Reference to the Future: To understand the physics of the past and present, says Tipler, you *must use the future* as your frame of reference. You can only understand what's going on now if you "impose boundary conditions at the end of time." He notes that the reference to *point* in his Omega Point theory--in contrast to the "standard model" in which the closed universe ends in a three-dimensional sphere rather than a point--has significance because it signifies "unlimited communication at the end of time," necessary for continuation of life.

We should be interested in the future, notes Tipler, *because most of reality is there*. He makes a quick calculation of the relative physical size of the future and past. Employing the past light cone's spacetime volume, he figures the volume of our future is at least 30,000 times larger than our past. If you try to understand the future by the past, he observes, "you'll get a mess you can't possibly interpret."

Unpredictabilities As Canceling Each Other: Tipler goes on to say that the unpredictability associated with chaos theory and the unpredictability associated with human free will "cancel each other out," leaving you predictability. With respect to our possible destruction, that is now unlikely. Instead, he says, interstellar colonization will commence next century, so destruction of the earth "won't affect the postulate." (Good news!)

Does God Exist Now in Omega Point Theory? Omega Point Theory, about the God of the Future, raises an important question. The theory conceives of life and intelligence intensifying until the end of time when God representing total intell-igence will exist. This raises the important question: Does the theory imply that God does not presently exist? That question, he replies, while relevant from our point of view, is not relevant from the viewpoint of space-time [general relativity] where you deal with the entire universe at the same time--the coexistence of all time.

So while, from our perspective, God is evolving, from "God's perspective, God is "drawing the totality of reality into Himself/Herself as time goes forward." Physicist Freeman Dyson, incidentally, in *Infinite in All Directions*, also speaks of the Socinian, evolving God.

Life Reconstructed at the End of Time: While we die now, notes Tipler, we'll be revived again near the final state, and continue to live forever--in "subjective time." This is important, he notes, because the future encompasses most of reality. Tipler talks of everything, including human beings, as represented by a *pattern of information* which can be coded in a computer.

So you can write all the information needed to reproduce human beings at another time and place and a sufficiently powerful computer can emulate the entire universe. The human being represents 10^{45} bits of information, he notes, and the entire universe, 10^{123} bits. So in the far future, he says, descendants of present life forms will emulate the whole present universe. We would come into existence again, the present universe, but at a "higher level of implementation."

Is This Event an 'Information Emulation' or an Actual Physical One? Tipler's response to this question is that the event will be the "present reality... at a higher level of implementation." There will be no difference between the emulation and "the real thing."

How Do We Know We Are Not Now an Emulation? We don't, counters Tipler. We could be an emulation in [from] the far future. Beings that are *perfect copies* are no longer "copies," they are the "actual beings."

What about Souls? Tipler defines a soul as the essence of the human being. The difference between a living being and a corpse, he suggests, is "a particular program being run on the body," the brain. What about a robot and souls? Yes, so a robot could have a soul in this context. A person, says Tipler, is a program you can talk to.

And How Would You Emulate All Human Beings? You'd make emulations of "all those possible histories, all variants," and the real person would be included as one of the emulations. He cites the possibility that, courtesy of the Many-Worlds quantum interpretation, all such possibilities might actually exist.

How Will Life Control Galaxies and the Universe? It won't be Homo sapiens, he reminds us, that will be faced with this problem since present type of life will exist perhaps "only a few more million years." It is more likely to be intelligence in the form of intelligent robots which would have spread throughout the universe. As an aside, physicist Freeman Dyson sees possible evolving forms of biological life more compatible with the rigors of space as one phase of the spread of life to engulf the whole universe. Where advanced biological life ends and robot life takes over is left open.

These "beings," says Tipler, will use "the universe's chaos to force it into patterns," a necessity for survival. Tipler notes that chaos allows "a little nudge here to amplify, after a while, to an enormous change there--like a domino effect." These changes, he notes, can result in the universe collapsing at "different speeds and directions" into certain patterns. As the size of the collapsing universe goes to zero, he observes, the ultimate source of energy, gravitational energy, "goes to infinity."

All Information to Be Encoded As You Go into That Final Singularity? The singularity, says Tipler, is on the boundary of space-time and you never reach it as long as you remain in space-time, but the energy is going to infinity. Information is always encoded as occupied or unoccupied energy levels. There are discrete levels of energy--a gap between one level and the next. As you approach the singularity, he notes, just make sure that "the energy levels that encode information are at higher levels than the temperature of the environment."

How Do You Prove the Existence of God? Considering the total reality, says Tipler, "if you do a consistent physical analysis, God just falls out...He is there in an intrinsic, essential way," not just a cover for our ignorance. And, he declares, *any cosmology with unlimited progress will end in God.* He cites Exodus as God saying to Moses, "I will be what I will be" [another interpretation of the "I am what I am" language]. So, Tipler suggests, the Bible can be used to define God as the *ultimate future*--the Omega Point.

Tipler's asserts his mathematical theory as telling us that "the ultimate theory is personal," warranting the use of the term 'God.' All personalities acting together, he says, "will drive the universe into the ultimate future." Furthermore, he declares, "it will be these future 'persons' who will resurrect us."

But what about the *classical philosophic* proofs of God?

Considering the total reality, if you do a consistent physical analysis, God just falls out...He is there in an intrinsic, essential way, not just a cover for our ignorance. And, *any cosmology with unlimited progress will end in God.* The Bible can be used to define God as the *ultimate future*--the Omega Point --Frank Tipler.

CHAPTER 34
CLASSICAL-PHILOSOPHIC PROOFS OF GOD

We Assume There Is a God; Some Argue That There Is None: James Gray's *Johnson's Sermons* records the sermon in which famous lexicographer-philosopher Dr. Samuel Johnson, discussing God and the universe, observes that "It is astonishing that any man can forbear enquiring seriously whether there is a God; whether God is just; whether this life is the only state of existence." These are the questions, he says, which every reasonable being ought to consider with "attention suitable to their importance...."

For the most part in our earlier material we start with the premise of God's existence and discuss some differences in religious beliefs and theological appproaches. We have also explored some aspects of so-called scientific proof of God, revolving mostly around the anthropic principle--that the remarkable coincidences in the constants of physics are such as to permit and produce life in this universe--and, to some at least, to imply a Designer.

But as Dr Samuel Johnson observed, there are serious arguments, advanced by intelligent minds, that the existence of our universe or at least our world here on Earth, with all its teeming wretchedness, militates against the existence of God, or at least against theism's ubiquitous, personal, omnipotent, omniscient, compassionate, and perfect God dedicated to a close and protective relationship with human life.

A Hostile and Haphazard Universe of Disasters--Natural and Human-Made: The universe, judged by our experience on Earth, appears hostile and haphazard, generating not only an unending sequence of catastrophic natural disasters but of human-induced destruction by war and environmental despoliation.

Destruction of our chance-produced life on planet Earth is thought by some to be a strong possibility, perhaps inevitable, as by collision with an asteroid, comet, black hole, antimatter aggregate, proximity to a supernova stellar event, or any of the other cosmic mechanisms derived from the same random forces which account for our present precarious existence.

The Universe and Life--A Chance Event? In this pessimistic view, the world is hardly the product of perfection, of an all-powerful, all-knowing, compassionate, benevolent God, full of grace and dispensing equal justice, punishments and rewards commensurate with human conduct or behavior.

So the universe may not unreasonably be viewed as a wayward, chance event, an expression of quantum uncertainty--something that just happened, and happened also to produce life, but a life dominated by the imperfections, perversities and tragedies of a hostile and often ruthless environment.

Philosophical Proofs of God: So what proofs or arguments, then, do we have for the existence of God? Here again we face the question of defining this God. Suffice it for this purpose that the 'God' we refer to is not just a vague creative force like physics, the existence of which is reasonably a given, but the theistic personal God of the Bible with whose image we are familiar and have previously sought to define.

Earlier we discussed so-called "scientific proof" of God, primarily derived from the anthropic principle. But what about nonscientific arguments, derived from reason, logic, common sense, philosophy, and revelation? Down through the years, philosophers and thelogians have advanced so-called 'arguments' representing supposed 'proofs' of God.

But Theologians and Philosophers Have Also Been Physicists: As has been pointed out by scientists such as physicist Frank Tipler, most philosophers and theologians were physicists or mathematicians first, examples being Aristotle, Aquinas, Kant, and Descartes. Paul Tillich studied the physics of his day, including relativity theory and relationships between space and time. Thomas Aquinas was a student of Aristotle's physics, required for his famous 'five proofs' of God. Medieval theologians were required to study physics and, as physicist Frank Tipler suggests in *The Physics of Immortality*, physics must be a requirement for the study of theology.

ST. THOMAS AQUINAS'S FIVE WAYS

Thirteenth century St.Thomas Aquinas offered his famous *Five Ways* of demonstrating God's existence: The first three are sometimes treated together as the *Cosmological Argument.*

(1) The argument from *motion*. Motion or change presupposes an existent mover or causal agent--the Prime Mover;

(2) The argument from *causation*. There cannot be an infinite regression of causes-there must ultimately exist a First or Prime Mover or First Cause;

(3) The argument from *contingency*. God is logically 'necessary.' Existence comes not

from the nonexistent but from logical necessity, a necessary existent, a necessary, non-dependent, and non-contingent being.

As John Hick describes it in *Philosophy of Religion,* "Only a self-existent reality, containing in itself its own being, can constitute an ultimate ground of existence of anything else which is therefore contingent. Consequently, if there is an ultimate ground of anything, there must be a necessary being." The argument is also expressed as a function of a 'prime mover' and 'first cause.'

(4) The *ontological argument* or argument from degrees of goodness or perfection. The highest degree of perfection requires that God must not only exist in theory but in actuality. If God existed only in the abstract, He would not be the perfect being requiring actual rather than theoretical existence.

(5) The *teleological or argument from design*. Order cannot exist without an orderer, God. Everything moves toward goals so God exists to propel things toward these goals. The universe shows evidence of purpose and design, requiring a designer and promoter of this purpose. The scientific aspects of this argument are reflected in the so-called *anthropic principle*, that the precision of the constants of physics reflect extraordinary coincidence of life promotion, for some at least, implying intelligent design.

Lasting Influence on Theology, Though Challenged: Aquinas's Five Ways have continued to have lasting influence on theology and even echo in scientific literature, despite the rather devastating marshalling of counter arguments by many philosophers, especially Immanuel Kant and David Hume. Kant, after announcing his arguments against the Aquinas proofs, proceeded to propose his own proof, based on moral necessity. Kant's arguments, in turn, underwent considerable analytical attack, not so much as a defense of the Five Ways, but as a challenge to Kant's analytical logic.

Quantum Theory and the Aquinas Five Ways: As we have seen, much of the Aquinas proofs' content is undermined by quantum theory, interpreted by many physicists as positing a universe created by quantum uncertainty and the forces and dimensionality as traceable to mathematics, leaving God's role, perhaps, as the authoring mathematician. But some, like P.W. Atkins in *Creation Revisited*, suggest that mathematics itself exists by logical necesssity, a description applied earlier to God in the 'argument from contingency.'

Classifying Arguments for or Proofs of God: There are many groupings for so-called proofs of God. For convenience and for the most part, we will use the Five Ways list as modified by John H. Hick in his highly analytical and instructive *Philosophy of*

Religion. We must bear in mind, as he notes, that 'philosophy of religion' is not a branch of *theology*--the systematic formulation of religious belief, but a branch of *philosophy*--one step removed from theology. It is, says Hicks, *philosophic thinking about religion*. As he points out, the atheists, the agnostics, and persons of faith all philosophize about religion.

THE COSMOLOGICAL--FIRST CAUSE ARGUMENT

An Infinite Regression of Causes As Unlikely: Since an infinite regression of causes is either pointless or inconceivable, it is suggested there must therefore be a first or uncaused cause, that being God, thought Thomas Aquinas. God requires no cause because he is self-caused and exists of necessity. Of course, an infinite regression is not only conceivable but an acknowledged characteristic of mathematical infinity concepts. As some suggest, the cosmological argument is a way of saying that since the cosmos does not account for itself, we must assume the existence of an 'absolute' to account for it.

But an eternal universe, not caused, just there, is scientifically defensible, as we see in some of the quantum cosmology concepts such as that of physicist Stephen Hawking. And the mathematics of a universe erupting into actuality from a primordial quantum void is now at least seriously posited by science, as we have discussed. Several philosophers, including Imannuel Kant have convincingly argued against the cosmological argument.

Did God Create the Universe? We observe in several contexts that Thomas Aquinas and Liebniz built on the thoughts of Plato and Aristotle to construct aspects of the *cosmological argument*, one of the classical 'proofs of God': that the universe must have a cause, a prime mover, and the regression of causes must end in a First Cause, God.

In *God and the New Physics*, physicist Paul Davies notes these theological and philosophical arguments and asks, "Did God create the universe? Does everything have a cause?" He continues and intensifies this questioning in his *The Mind of God,* discussed elsewhere in this material.

Uncaused Events in Quantum Theory: Davies points out that the nearest thing to *uncaused* events is quantum mechanics where particles appear randomly, "with no rhyme or reason," without identifiable cause, as it were. The whole universe, he points out, is subject to *quantum principles involving absence of direct cause*. Interestingly, it was this "uncaused" feature of quantum theory, Heisenberg's uncertainty principle, that so bothered Einstein who was so hung up on the necessity for "cause and effect" in what science had previously hypothesized as a determined universe.

If God Is Uncaused Then Why Not the Universe? Davies then picks up on the idea of God as a necessary being, not requiring an outside cause. Well, in that case, if you're going to accept an absence of cause for one purpose, namely God, then, he asks, why can't the universe exist without an external cause, or be self-caused?

Others raise this same question. Philosopher and author Sidney Hook, winner of the Presidential Medal of Freedom, who taught at Harvard, Columbia, and other universities, notes in his *Living Philosophies* essay, "It Didn't Have To Be," that one of his religious hangups was this very idea.

"If God always was, and didn't require the existence of another creature or God to create Him, why did the world have to have a Creator? Why couldn't the world always have been?" Indeed, many scientific theories of an eternal universe have been advanced in the past, and more recently in the light of Stephen Hawking's quantum cosmology.

Hawking's 'No Boundary' Universe: We have explored how Einstein's general relativity theory, Hubble's finding of the galactic redshift evidence of the universe's expansion, the finding of the cosmic background radiation still present from the Big Bang, and NASA's 1992 COBE Satellite findings of primordial radiation variations, all demonstrate the universe's beginning or 'time boundary' from the Big Bang.

Big Bang Further Confirmed: The Big Bang event was further reinforced by the June 1995 Johns Hopkins University announcement of the space shuttle's Astro 2 Observatory, Hopkins Ultraviolet Telescope findings of primordial helium in very ancient intergalactic clouds. The helium, along with trace elements deuterium, lithium, boron, and beryllium, were created, not by nucleosynthesis from later- formed galaxies and stars, but by the fusion temperature conditions within the first four minutes after the Big Bang, and in amounts consistent with the 24 percent conversion of hydrogen to helium and trace elements predicted by Big Bang thermodynamics.

But recent quantum-gravity theories, like those of Stephen Hawking, still suggest that there need not have been a time boundary, any beginning of time; that our universe could have emerged from and in a sense is an adjunct to or extension of an eternal mega-universe quantum void.

So while we conclude there *was* a creation event--the Big Bang, we must consider the possibility that it needn't have been the product of volitional intervention by a "non-contingent necessary being," but could be a chance quantum uncertainty event, and the universe might simply be part of the *superspace* of an infinite and eternal mega-universe in which universes like ours periodically erupt as random quantum fluctuations.

Outside of Time: But speaking of causes and the cosmological argument, in *God And The New Physics,* Davies goes on to suggest that causal matters presuppose the existence of time, yet time first appeared as one of the dimensions, the geometry, of the spacetime universe. Since there was no 'before,' prior to the advent of time, there was no cause, no strictly temperal event.

So, he notes, it is meaningless to speak of God creating the universe in the normal causal context. Theologians Paul Tillich and Karl Barth, speak of God as being *outside* of time, just as 4th-5th century St Augustine did.

THE TELEOLOGICAL (DESIGN) ARGUMENT

The Universe As Replete with Evidence of a Designer: The design or teleological argument (evidencing purpose or goals), has found support in the existence of extraordinary coincidences in the nature of this universe. These include the remarkable balance in the constants of physics, the fortuitous blending of the exponential values of the strengths of the forces, the universe's felicitous dim-ensionality allowing for our structures, the precision-balanced ecosystem on planet earth, the development of the incredible complexity of the human body. These remarkable felicities, according to some physicists, imply purposeful design, and infer the existence of a Designer.

Paley's Watch Maker and the Universe Maker: This argument classically cites early 19th century British theologian William Paley's *Natural Theology on Evidence and Attributes of Deity*, in which Paley, also a naturalist, argues that while finding a stone on the heath allows for its always having been there, finding a watch is clear evidence of design and the existence of a Maker.

We infer from its intricate design that a watch is the product of purposeful design, implying a "watchmaker," rather than a happenstance product of chance self-construction. So, Paley's argument goes, the infinitely more complex design for a living organism, so much more evidences an extraordinarily more intelligent "universe-maker."

The example of the human body, however, may be explained by the operation of laws, for example, the Darwinian theory of natural selection and the natural process of pockets of increasing order and complexity as compatible with the general thermodynamic drive toward entropy and chaos.

God As a Mathematician: A related explanation of the design argument is the fact that since there are exquisite mathematics underlying the physics of the universe, its designer must be a mathematician, as Sir James Jeans suggested; that God actually is a name for or represents mathematics or that mathematics is a name for or represents God. In his *Creation Revisited*, P.W. Atkins insists that it is mathematics alone which triggers the universe into existence, "without outside intervention." But if God is mathematics as James Jeans suggests, then we may argue that it is God who creates the universe, via mathematics.

God and Quantum Theory: Although there appear to be equally defensible counterarguments, random chance built into quantum theory, Heisenberg's uncertainty principle, has been used to argue against human life as a product of design. But many argue that unusual properties in evolution, such as the creation of highly ordered systems in a universe otherwise moving toward entropy or total disorder, are indications of built-in *directionality*, or design.

Order As Implying a Designer: In *God and the New Physics*, phyicist Paul Davies argues that the fact that the universe started out with a considerable stock of order (negative entropy), demonstrates that the universe was created in an "orderly condition," evidence at least of a "designer-by-proxy" who at least winds things up to crank out structures, an exercise requiring "supernatural dexterity."

Order As Exponentially Sensitive to Rearrangements: Davies observes that "Ordered (low entropy) states are highly improbable and unstable...requiring the cooperation of vast numbers of individual molecules." To buttress the argument, he points out that mathematical investigation shows that *order is exponentially sensitive to rearrangements*. He cites the example of a litre of air rushing to one end of a box having a probability of a number represented by one followed by one hundred quintillion zeros!

Probability Favors a Disordered Universe: If the universe is simply an accident, he says, the odds against it containing any appreciable order are "ludicrously small." It appears hard to escape the conclusion, he notes, that "the actual state of the uni-verse has been chosen or selected from the huge number of available states,...that there surely had to be a selector or designer....

The universe (he notes) went bang in a remarkably orderly way....an accidental creation would, with a probability that is a virtual certainty, produce a totally disordered universe." He does go on, however to consider counterarguments that have been made to this argument that a highly ordered starting universe implies a designer, a "fundamental paradox of cosmology." Stephen Hawking also noted, in *A Brief History of Time*, that the "initial conditions" of this universe of high order is extraordinary in view of the far greater liklihood of chaotic starting conditions.

Freeman Dyson's View of the Design Argument: In his *Infinity In All Directions*, physicist Freeman Dyson says that the design argument was excluded from science because it dealt with metaphysics, teleology, purpose. But, he says, it has some "meta-science" merit, like the anthropic principle, and is acceptable as philosophy.

Human life As a Small Piece Of 'God's Mental Apparatus": As we have discussed, Dyson speaks of the human mind as an extension of the mind of God, a "mental component of the universe." It represents an extension of the anthropic principle to a "universal scale"--the peculiar features of the laws of nature as being hospitable to the growth of mind. If we treat this "mental aspect" as "God," he suggests, then "*we are small pieces of God's mental apparatus*."

The Question of Ultimate Purpose of the Universe: Dyson says the problem is to read God's mind. Einstein and Hawking have both spoken in a similar vein--probing the Mind of God, and theologians pursue the same goal from their perspectives. In a philosophic mood, Dyson says he must repeat Job's persistent question of God--Why do we suffer? Why is there injustice, pain and tragedy?

The Principle of Maximum Diversity: His answer, he says, is based on a hypothesis that is an extension of the anthropic principle and the design argument. The universe is constructed, he says, according to a *principle of maximum diversity*, the principle, operating at both the physical and mental level. The physical laws and "initial conditions" are to make the universe "as interesting as possible," life being "possible but not too easy"--comet strikes, ice ages, nuclear weapons, plagues, "computers, sex, sin and death." Tragedy arises because "not all challenges can be overcome."

He cites Dante's *Purgatorio*:

> Can you not see that we are worms, each one
> Born to become the angelic butterfly
> That flies defenseless to the Judgment Throne?

Philosopher David Hume Evaluates Argument from Design: 18th century Scotch philosopher David Hume, critically evaluated the design argument in such an incisive way that the argument lost much of its earlier appeal. He pointed out that life presupposes order and can be viewed as the result of design but he questioned whether this requires a deliberate designer, as distinguished from other explanations, such as the natural result of random combinations of matter.

But, says Hume, even if one infers a designer, it doesn't thereby suggest: a necessary Christian theistic designer; or even one designer in view of the variations in nature; or even a benevolent God, since there is evil; or a perfectly wise or powerful God, given the imperfect nature of our world.

The Break from Perfect Symmetry: Physicists like Heinz Pagels and Steven Weinberg refer to the intriguing role of the initial unification of time, dimensionality, matter and the fundamental forces, representing the primordial supersymmetry or "perfect symmetry." The unification inherent in this starting symmetry did not allow for structure, or time, or galaxies, or human life, until this primordial supersymmetry could break down in the phased separation of the forces, providing for structure, matter, and the processes of thermodynamics.

So human life is possible, then, only from a process involving departure from initial perfect symmetry--no imperfection, no human life. Philosopher Leibniz noted that God created the "best possible world," that imperfections of the world are beyond God's power to prevent, at least in this life. Philosopher Kant used the present imperfect world to rationalize proof of continuation of life so that justice and morality are ultimately vindicated.

God Not Omnipotent? Of course this line of argument and explanation for imperfection suggests a God not omnipotent but one with certain finite limitations, something which many theists find unpalatable but with which physicist Freeman Dyson feels quite comfortable, his Socinian God that is evolving along with man and the universe.

St. Paul and God's Use of the Design Argument: In Romans 1:20, St. Paul employs the design argument. "God's eternal power and divine nature," he says, "have been clearly seen, being understood from what has been made." Some cite God's response to Job citing his startling creations as God's own use of the design argument although others say that God's references to his astronomic formations was simply to denigrate Job's pretentions of entitlement to an explanation of God's ways.

Job's Ignorance and Our Sophistication: With human life's present knowledge of so much of the laws of physics and the macrocosmic and subatomic or quantum mechanics--the workings of the universe, many argue that God must have had second thoughts about his denigration of Job since God has since enabled human life to reach his present sophisticated grasp of knowledge of the laws of physics and the underlying mathematics of the universe.

The Fitness of the Universe for Life: Back in 1913, Harvard Biochemist Lawrence Henderson had written on the peculiar "fitness of the environment" for biological organisms, using conceptual material suggestive of what became known as the "anthropic principle," discussed earlier.

Similarly, a deep connection between astrophysics and quantum mechanics--between cosmology and atomic theory, was highlighted back in the thirties by physicist Paul Dirac who noted the strange similarity in numbers between certain phenomena related to the fundamental forces, ratios between electromagnetic and gravitational force and between certain basic time measurements.

The Anthropic Principle: In the nineteen seventies British physicist Brandon Carter invoked the 'anthropic principle,' dealing with our universe's proclivity for life and the fact that the laws of physics seemingly *required* life. His was the first real effort to catalog those laws of physics which represented the critical coincidences needed for life.

Greenstein's Symbiotic Universe: A philosophically searching treatment of the anthropic principle is contained in George Greenstein's *The Symbiotic Universe* which deals with the remarkable coincidences in the constants of physics, suggesting that the universe is the way it is because human life is here to contemplate the universe and ask the question why the universe is the way it is.

Design Argument: Natural Versus Supernatural Explanations: British Physicist Paul Davies in *God and the New Physics*, describes the creation by purely natural laws of physics, suggesting the existence of an ancient superintelligence, one *not supernatural*but built into the natural laws. But he suggests the need also to consider supernatural causes, that is, acts of God, as alternative explanations, noting the remarkable coincidences in the constants of physics which, he feels, represents the most persuasive *evidence of an element of cosmic design.*

He observes that this is a preferable explanation to the so-called 'weak anthropic principle which explains our universe as a chance, life-compatible universe--one of an infinite number of universes having all possible sets of laws of physics, ours by chance just happening to be life-hospitable.

But physicist Alan Wolf, in *Parallel Universes*, takes the opposite tack, that the infinite number of universes posited by the many-worlds quantum theory, demonstrates the extreme lengths to which God has gone to create at least one, this universe, hospitable to life.

THE ONTOLOGICAL (EXISTENCE) ARGUMENT

Existence As Needed for Divine Perfection: The ontological argument (relating to being or existence), while asserted by Aquinas as one of the Five Ways or proofs of God, was earlier asserted by St. Anselm, Archbishop of Canterbury, who proposed that a perfect God must have existence or he would lack that one necessary characteristic of perfection, actuality. So God cannot be thought not to exist.

A French monk, Gaulinon, proceeded to demonstrate that the existential reasoning, the necessary existence principle, if applied to other situations would lead to patent absurdities. Thomas Aquinas also rejected the ontological argument on similar reasoning.

The French philosopher René Descartes questioned the argument on the ground that the assumption that existence is a predicate is challengeable. The German philosopher Immanuel Kant, among others, pointed out that while existence may be a predicate in a grammatical context, it *does not follow that the subject and its predicate exist*. There is no necessary existential state to qualify for perfection.

THE PROBABILITY ARGUMENT

The two principal aspects of the 'probability' theory are the 'frequency of occurrence' and the 'reasonableness of belief,' which, in *Philosophy of Religion*, John Hick refers to as the statistical and inductive senses of probability.

Frequency of Occurrence Argument : This argument, generally advanced for a theistic God, suggests that one proof of God's existence lies in a broader scope of occurrences such as the design inherent in evolution, aesthetic and religious experience, moral experience; and that the cumulative effect, the 'statistical effect' of multiplicity of applications, represents an increase in probability.

The Reasonableness of Belief Theory: The other probability argument, reasonableness of belief, is based on the increased reasonableness of belief derived from cumulative effect, that is, a proposition is more likely to be valid if it conforms to a larger body of prior experiences.

A suggested weakness in this argument, notes Hick, is the fact that there is nothing outside the universe that might evidence its nature; that this universe may be interpreted "both theistically and nontheistically." "Nor," he says, "is there any valid sense in which it can be said that a religious interpretation of life is antecedently more probable than a naturalistic interpretation, or vice versa."

However, Hick suggests that the general perception of these probability explanations is "weighted in the negative." Many suggest that acceptance of existence of God, somehow defined, does not lead necessarily to a theistic God. The "probabilities," it is argued, don't fall any more readily on a theological explanation than a naturalistic one.

THE MORAL ARGUMENT

The Inherent Sense of Duty: This argument asserts that human ethical experience, the sense of obligation to 'fellow human beings' requires the existence of God, who represents the *supreme moral reality*, as the source of this ingrained duty. The argument has strong emotional appeal, apart from its strictly 'probative aspects.' For one thing, everyone recognizes a fundamental sense of duty, something deepseated and innate, and about which there is general acceptance.

Philosopher Immanuel Kant, after decimating the arguments in The Five Ways Of Thomas Aquinas, proceeded to advance his own proof of God. This took the form of moral necessity with which human life is equipped at birth--his sense of duty, his "intention for Good, at the heart of conscience."

The Highest Prison Walls--Conscience: Escape from the law in our workaday world, or even from the social consequence of wrong doing, is commonplace. But the highest prison walls are those of one's conscience. Many suicides are the product of self-imposed death sentences, the product of self-judgment inbred in the human mind. This phenomenon is sometimes interpreted as the act of God, impinging upon human conscience, to vindicate moral duty.

Conscience--A Divine Instrumentality? Human conscience--an inbred sense of duty, suggests itself as something fundamental, easy to think of as an instrumentality of Divinity. The very notion of conscience somehow conjures up the image of the God within us, something within us, yet independent of us and judging us. One can argue psychiatric or even genetic or other natural physical sources of conscience, but such attributions are somehow less persuasive.

Several Approaches to the Moral Argument: There are several approaches to defining the 'moral argument.' One is that God, the divine source of law, is to be inferred from the very existence of Moral Law. As Cardinal Newman observed, if we are ashamed or frightened at "transgressing the voice of conscience, this implies that there is One to whom we are responsible" and if the source of these feelings is unfathomable, then "the Object to which [one's] perception is directed must be Supernatural and Divine."

On the other hand, in his *Philosophy of Religion*, John Hick points to the body of scepticism of the validity of inferring divine rather than naturalistic explanations of the existence of Moral Law. Unfathomability can be attributed to natural causes--we have seen how abstractions are the result of the growth of complexity and order, natural manifestations.

Recognition of a Paramount Sense of Duty As Inferring God: A second form of the moral argument is the one advanced by Kant, that the supremacy of moral values, involving a transcendental sense of duty, implies the existence of a supernatural source--God. Hick suggests that this particular form of the argument has at least a qualified form of validity as moving in the direction of God but that it cannot be said to "point all the way with unerring aim."

While this may sound 'damned by faint praise,' one senses Hick's desire to lend credence to the proposition. Hick's views did not have the advantage of the more recent Hawking quantum cosmology ideas or of Feynman's tracking of abstractions as natural products of the growth of complexity and order.

THE ARGUMENT FROM EXPERIENCE OR SPECIAL EVENTS

The Limited Effect of Individual Cases: Putting aside for the moment, Scottish Philosopher David Hume's famous arguments against the objective acceptance of miracles, the argument is advanced that we have a right to rely on human experience and personally-validated events such as miracles, answers to prayers, instances of parapsychological phenomena, spiritual encounters through dreams, visions, and so-called 'mountaintop' experiences, all seemingly involving God in some way.

But, Hick cites 17th century English philosopher Thomas Hobbes' comment that God speaking to someone in a dream is simply a dream of God speaking to someone. One account speaks of a person who happened to have endocrinological problems, was deeply religious, and had visions of communication with the Divine.

When the medical problems were treated, the religious behavior, together with the religious experiences, disappeared altogether. This suggested, of course, that it was the illness that induced religiosity. Of course it may be argued that physical impairments may produce authentic special insights, as in the case of unusual mathematical dexterity attributable to an autistic savant.

The difficulty with this use of individual cases, anecdotal evidence, to argue the generalization of God's existence, however, is the absence of assurance of the truth of the individual's interpretation. It is not unlike the use of anecdotal evidence to prove the effectiveness of a medical cure only to find that a broad-based full-scale scientific test produces a contrary conclusion.

But What of the Many Transcendental Experiences? But these statements of the 'argument from experience' don't do justice to the transcendental or 'mountaintop' experiences such as those of St. Paul and Martin Luther. The experience of separating oneself from one's essentially physical self and reaching into one's upper consciousness, one's awareness of unity with God and the universe at large, happens with sufficient frequency not to be lightly dismissed simply as unexplained aberration.

These experiences, involving exhilarating flashes of intuition and insight, and the remarkable ability to express the mystery of this transcendence, have a startling universality; and they are not confined, as some suppose, to those of intense religious piety.

The search for God may not even have been involved in the experiences, although introspection born of the search for meaning, or of human trauma, is often the environment of occurrence. Of course, introspective experiences are an integral part of the religious experience of many in the Eastern religions.

God is more plausible than we are. --Abraham Joshua Heschel

CHAPTER 35--GOD MORE PLAUSIBLE THAN HUMAN BEINGS? GROUNDS FOR DISBELIEF IN GOD

GOD MORE PLAUSIBLE THAN HUMAN BEINGS

God As More Plausible: One of the great contemporary exemplifications of deep religious faith and piety, steeped in the tradition of Jewish theology, is Abraham Joshua Heschel, author of *Man Is Not Alone*, A Philosophy Of Religion, and *God In Search Of Man*, A Philosophy Of Judaism. Heschel's approach to speculations on proofs of God does not neatly fit the pattern of conventional philosophy. He sees the source of knowledge of God as intuitive--revelation. He is not preoccupied with any constructed set of rationalizations.

Resort to Intuition: His trusting resort to intuition, is admired by theology at large, as expressed by Reinhold Niebuhr, and other distinguished Protestant theologians, as well as many in Catholic theology, including the Pope who sought out Heschel for consultation in connection with Catholic examination of its relationships with Judaic thought. His views find expression in many other theological writings because of their intuitive persuasive power that gently brushes aside the very notion of a 'need' to 'prove' God.

Unconcern with Proofs: His unconcern with 'proofs' of God may be summed up by his proposition that *God is more plausible than we are*. One senses in the sublimity of his writings the *a priori* (known) or *a fortiori* (by greater necessity) existence of God. Immersed in Heschel's uplifting work, one is forced to devise a labored hypothesis to argue the nonexistence of God--and then to strain at marshaling grounds for disbelief.

The following are some of Heschel's thoughts:

Human Life's Encounter with the Sublime: We exploit reality in one realm but sense the inexpressible, the ineffable, in another realm. The sublime, the cosmic, he observes, though lying "within our reach" is "beyond our grasp." A glimpse of God? To seek the ineffable--an awareness of grandeur--is perhaps human life's imperative, something he does, not through reason but through insight, something meaningful but to which we are unable to give descriptive expression.

Reverence and awe of the sublime is a norm of human consciousness with a "transcendent preciousness" surpassing our understanding. It is the thing of which religion, poetry and philosophy are made, he says. To consider it as "self deception" (as some would suggest) is too "sophisticated to be reasonable."

Who Is the "I" ? God is *persistently aware of the universe*, he says. The fact that we exist, as independent awarenesses of the universe, is something of which we are hardly conscious. Thus the mystery of our existence is almost hidden to us. But who is the "I'? To exist may, at first, imply ownership. But, says Heschel, existence is not a property but a trust. Life visits my body--a loan. I neither initiate it nor construct its value. *I am, in effect, what is not mine.* There is, therefore, no "self" to ask the Question. I may be temporary--but there is a "timelessness to my temporality."

God of Philosophers and the God of Religion: The God of philosophers is one of Order, First Cause, Grand Architect, Greatest Conceivable, notes Heschel, citing the classical proofs of God. But, he observes, we are more concerned with a personal God with whom we identify, a God of Justice and Love, who honors the saintly. Religion is more than a creed or abstraction. It comes to life, he says, when the soul, "starved for an inkling of reality beyond rational methodology," seeks the ultimate mystery, the ineffable.

The Ultimate Question: The ultimate question--about God, begins with the ineffable, notes Heschel. The mind does not know how to phrase it, yet the soul pleads it. "We ask because we sense a spirit surpassing our comprehension." Heschel suggests that the question about God inquires what all things stand for, the transcendental; it is meaningful without being reasonable--or definable--the mystery shared by all men and a dimension of all existence, as if an object of divine thought, as if the human mind is not alone in thinking it.

The Divine Presence: It is in contemplating the sublime, says Heschel, that we achieve the intuitive awareness of the Divine "Presence"--before we have a grasp of the Divine "Essence." Religion begins when we recognize that something is asked of us; that we *owe* all that seems ours. We yearn to know what to live for. "Ultimate wonder" is the thought of God in search of a soul. *God is more plausible than we ourselves are.*

Heschel observes that *the enigma is not God but us*. Faith is not the product of our will, neither is it a conclusion. It is a mind-change induced by a power from beyond. It comes not by degree, not by decision or from search; it is an insight which comes with a flash--the ineffable has "shuddered itself" into the soul. There is no man, he says, who is not shaken for an instant by the grasp of the eternal.

The Role of Doubt in Faith: To seek God simply to appease skepticism or satisfy curiosity is not the road to faith, observes Heschel. Abstractions of logical evidence seldom prevail over intellectual misgivings. Immanuel Kant's refutation of the classical proofs of God did not result in modern human life's loss of faith, he points out.

The Divine, as a first insight, transcends both the power of the mind and the order of the world. Heschel is oft-quoted for his assertion that *the certainty of God's existence comes not from the realm of logic but from intuitional apprehension.* It comes from sensing the spiritual dimension, from "questioning the self and all its cognitive pretensions."

No Short Cut to Faith: Faith, he suggests, is not a gift received at random, or a treasure found by accident, but the "fruit of constant vigilance." Faith comes to him who passionately yearns for ultimate meaning with all of his mind, heart and soul. Faith is found in passionate care for marvel and wonder, witnessing the holy within the world's mundane affairs and by constant effort to grasp the indefinable. *Achievement of faith, says Heschel, is "the moment in which the human soul communes with the glory of God."*

What Do We Mean by the Divine? Although the intuition of God is universal, *neither an image nor definition of God is at our disposal*, just his name--and even that is wrapped in mystery. *God begins where words end.* But his will is expressed through His Word. Inquiries about God generally start with an initial assumption: the attribute of perfection. Divine perfection as an attribute, he notes, is derived not so much from the Bible but from Greek philosophy and has echoed down through the centuries. The Bible emphasizes His unity. This unity, of all that exists, inspires human life with a sense of cosmic brotherhood.

The Moment an Image of Eternity: How, he asks, do we identify the Divine? It is, he notes, unity where there is diversity; peace where discord. It means that human life is never alone; that *temporality is eternal and the moment, an image of eternity.* We sense "all men in one man, the world in a grain of sand, and eternity in a moment." "Here" is everywhere and "now" is forever. The goal of all effort is to restore the unity of God and world.

We Cannot Think of God with Usual Logic: *We cannot think of God with usual logic,* and, he declares, We cannot find him as object; we can only find ourselves in Him. He is within and beyond all things--and all ideas. He is not away from us and we are not beyond Him. We must seek not to know but to be known to Him.

God As the Subject of Revelation Not the Object of Discovery: The Bible is God's anthropology, dealing with man and what He asks of man rather than with the nature of God. *God is not the object of discovery but the subject of revelation.*

This ends our exploration of Heschel's transcendental experience in his expressions of thoughts about God. One might sum up his attitude towards philosophic proofs of God by stating that the subject of God's existence is simply beyond methodology and rationalization, and human existence is less plausible than God's.

In saying this he likely speaks for most of the great believers who would not demean God by attempting to argue His existence. But Heschel had faith and so was comfortable with the dismissal of probative philosophical speculations and so-called proofs of God.

Some Theologians and Philosophers Dismiss Proof Arguments: It is interesting to note that there are many figures in religion and theology that will have little to do with the so-called proofs of God. In fact the stronger their views on God the more insistent they are to disavow resort to so-called proofs.

Revealed Rather Than Reasoned Knowledge: The primary basis for Heschel's outlook is apparently the view that God, the supernatural, can only become apparent through intuition and revelation; that God can not be 'reasoned' into existence; that knowledge of God's existence is inspired, not reasoned, and inspired knowledge is directly-conveyed knowledge, superior to indirect knowledge derived through reasoning or empirical data. If you *have* a belief in God, intuitively arrived at, then the 'proofs' of God are may be an accompaniment--tolerable but superfluous.

Unfortunately, human life is a reasoning animal, and many are denied inspiration and left to reasoning and empericism. Intuitions are subjective, varying from person to person, some being relatively impervious to intuitional knowledge, basically skeptical of the nonprovable. And this brings us to several other arguments which have common currency.

Proofs of God--Further Study: This effort to capture the essence of philosophic and theological thought on the subject of proof of the existence of God is necessarily limited by its context in this material. There is a wealth of fascinating material available for those interested in further exploring this area.

One might, for instance, examine John H. Hick's *Philosophy Of Religion,* and Chapter V, "The Existence Of God," of "A Modern Introduction To Philosophy," edited by

Paul Edwards and Arthur Pap. Included, is the reference to the famous debate-encounter between the articulate stalwart, Father Copleston and the legendary iconoclast, Bertrand Russell. The battle of these gladiators in logic, linguistics and rhetoric, is now a classic.

GROUNDS FOR DISBELIEF IN GOD

Scientific Proof of God? We have explored earlier the proposition that God's existence, like that of angels, are metaphysical questions, not science according to Karl Popper's definition, because not susceptible to disproof. But we discussed earlier some of the new developments in physics, especially quantum physics, which lead physicists to conclude that the new physics requires a reevaluation of the role and nature of God, especially in relationship to the creation of the universe.

The stronger versions of the 'anthropic principle,' invoking the remarkable coincidences evidenced by exponential precision balance of the constants of physics, suggest an incredibly fine-tuned universe not atributable to chance and therefore implying design and, consequently, purpose.

Some go so far as to suggest these coincidences as amounting to *mathematical proof* of God. Other theories, such as Frank Tipler's Omega Point Theory, go even further to provide actual supporting mathematics, as contained in an appendix to Frank Tipler's *The Physics of Immortality.*

But There Are Also Arguments for Disbelief in God: But just as some of the arguments for God's existence are now trending toward the scientific or what physicist Freeman Dyson describes as "meta-science," other scientific arguments go in the other direction, suggesting proof of the absence of need for a Divine Creator, that natural forces self-generate the universe.

One illustration, which we discuss earlier, is *Creation Revisited,* in which Oxford Chemistry Professor P.W. Atkins constructs a model of the universe in which the primordial quantum vacuum generates the geometric precursors of present forms of energy, time, and dimensionality, all conjured from the mathematics which underlie the universe and which, alone, account for and generate the physical universe.

Recall that while physicist Stephen Hawking asked what it is that "breathes fire" into the equations to generate the physical universe and whether the laws of physics and their mathematics themselves represent, like God, the First Cause, Atkins, unequivocally declares that it is mathematics alone which self-generates the universe.

Philosophic and Sociological Theories of Disbelief: Many of the grounds for disbelief in God represent *naturalistic* interpretations of phenomena which religion takes to be *supernatural.* In *Philosophy of Religion*, John H. Hick cites two illustrations. One is the Sociological Theory of Religion in which the religious thinking and behavior of members of a society, its concepts of God, are the subconscious product of that society's accepted belief system.

God, in this theory, is the *symbol for that society's controlling group perception*, reinforced by individual dependence for security and wellbeing on the group. Here, human beings are viewed as 'creating' God to preserve their 'social existence.' Writer Arthur C. Clarke is often quoted in this context: "It may be that our role on this planet is not to worship God--but to create him."

The Freudian Theory of the Primal Protective Father: Another and related theory of religion is the Freudian theory of God as essentially a "universal obsessional neurosis" in which God is constructed from the protecting father concept, traceable to the "primal horde" of father, mother and offspring, the early tribal-family period of human life, "magnified to infinity," and adopted as a "mental defense against the more threatening aspects of nature." And the memorialized father's prohibitions take on "new moral authority."

Philosopher Immanuel Kant referred to this moral authority as the intuitional, built-in sense of duty, the categorical imperative. While Kant's imperative still has standing, both the sociological and related Freudian theories have lost their earlier appeal and popularity.

Disbelief Derived from the Imperfect World--The Existence of Evil: A strong argument against belief in God derives from evil in the world. In *Philosophy of Religion,* John Hick describes the dilemma: "If God is perfectly loving, he must wish to abolish evil. and if he is all-powerful, he must be able to abolish evil. But evil exists; therefore God cannot be both omnipotent and perfectly loving."

He refers to some solutions that might preserve omnipotence. One is that evil is an illusion or simply the absence of good. A second and more accomodating to God, is Leibniz's "best possible world" argument. But again, true omnipotence implies no such limits. A third solution, advanced by St. Augustine, is that everything created by God is good but God's original pure good has degrees and is degradable by the influence of perversion independent of God, and evil represents a low or negative level of good.

A fourth solution is that persons are not automatons, the allowance of bad choices is essential to the allowance of human beings, the very definition of which requires human freedom. A correlative solution is to assert that even God is subject to logic and it is simply logically impossible to create persons with free will who could not or would not exercise that freedom. Another expression is that being human means not having God's perfection. Free will requires that human beings must have the capacity for error and room to improve.

A fifth solution, compatible with certain aspects of Western theology, is that God's purpose was not to construct a paradise of pleasure but a place with the pain, suffering and adversity, necessary for honing souls toward the more perfect likeness of God. Only the existence of imperfections, adversity, pain, suffering, permits striving, goals, courage, fortitude, achievement, generosity, compassion, kindness forgiveness, and so on.

In a word, life free from imperfection, a perfect creation, would be meaningless. So the purposefully-created imperfect world is for human life's good, to permit progress, improvement, worthy goals--elevation. As has been pointed out by philosophers like Immanuel Kant, this argument has the advantage of logically requiring life continuing beyond our physical death here to right wrongs here.

The Scope of Imperfection Makes Any Real Explanation Difficult: Whatever other kinds of life may exist elsewhere in the universe, and in a universe so vast, extraterrestrial civilizations are possible, if not inevitable, our particular planetary world is fraught with imperfections and hostilities--of nature and species.

Every year we have climatic and geological natural disasters--earthquakes, hurricanes, typhoons, and draughts. Past cataclysmic destructive forces, like cosmic objects such as meteorites or comets, have wiped out most of existing life on this planet, illustrated by the hypothesized destruction of the dinosaur life some sixty-five million years ago. Other more subtle and troubling imperfections in the world human manifest themselves as frailties of the human species--sinfulness and the existence of evil.

Paradoxically, other earthly creatures are, it appears, essentially incapable of "intentional" wrongdoing, sin. Their levels of intelligence do not provide for the kind of abstract thought involved in making of human moral judgments. In a sense, they are not endowed with human life's range of moral decision making, freedom of will. So by nature and definition they cannot be said to be truly sinful. Is man, then, an aberration?

> Although the intuition of God is universal, *neither an image nor definition of God is at our disposal*, just his name--and even that is wrapped in mystery. *God begins where words end*.--Abraham Joshua Heschel

PART XII--LIFE

RANDOM THOUGHTS ON LIFE

Thy life...is all thou hast to front eternity with. -- Thomas Carlyle

Lives of great men all remind us We can make our lives sublime, And, departing, leave behind us Footprints on the sands of time. --H.W. Longfellow

Man is a piece of the universe made alive. Emerson

Life is the childhood of our mortality. --Goethe

There is no wealth but life. --John Ruskin

Oh what a glory doth this world put on, for him who with a fervent heart goes forth under the bright and glorious sky, and looks on duties well performed, and days well spent. --H. W. Longfellow

...all that a man hath will he give for his life. --Job

I simply believe that some part of the human Self or Soul is not subject to the laws of space and time. --Carl Jung

The man who becomes a thinking being feels a compulsion to give to every will-to-live the same reverence for life that he gives to his own. --Albert Schweitzer

The meaningless absurdity of life is the only incontestible knowledge accessible to man. --Leo Tolstoy

Your right [neocortex] brain is your...intuitive brain....You might call your right brain your *mystic*, since its wisdom comes from some hidden source, *perhaps a cosmic source*. --Richard Bergland

One life; a little gleam of time between two eternities; No second chance for us for evermore. --Thomas Carlyle

The best use of life is to spend it for something that will outlast it.-- William James

The gift of God is eternal life.--Rom. 6:23

We are survival machines--robot vehicles blindly programmed to preserve the selfish molecules known to us as genes. --Richard Dawkins

The incomprehensible suffering of individuals, the hope of religion, the insights and explorations of art and science, the madness, the glory and joy of our species are destined to an oblivion so complete that even the act of destruction will not be remembered....this is...the only rational inference that emerges from our scientific view of the cosmos.--Heinz Pagels

Between matter as we observe it in the laboratory and mind as we observe it in our own consciousness, there seems to be only a difference in degree but not in kind. If God exists and is accessible to us, then his mind and ours may likewise differ from each other only in degree and not in kind.--Freeman Dyson

As we acquire more knowledge, things do not become more comprehensible but only more mysterious.--Albert Schweitzer

CHAPTER 36
MEANING, THE UNIVERSE AND LIFE

A child asks, "Where do I come from"? A little child asks where he comes from. "From God," parent replies. Child asks how God gets child here. Parent replies that He uses love between human beings and, well, also a little bit of physics, chemistry, biology, and genetics. Child asks, "What am I made of"? Response: What space scientist and astronomer Carl Sagan describes as "star-stuff," in the form of debris of chemical elements created by long-exploded stars, material that collected to become matter, drawn by gravity to heat up and become stars and attract other debris which become planets from which life develops.

Where Did the Stars Come from? But, asks child, where did the stars come from? Response: From gravity effects in the primordial cosmic radiation. Child: So where did that cosmic radiation come from? Response: Well, it came from an incredible burst of radiant energy called the Big Bang, erupting from a cosmic particle of unified physics, sometimes called the cosmic egg.

So What Was There before the Big Bang? Where did the cosmic egg come from, what was before the Big Bang? Asks the child. Well, says parent, the poets and the Bible say that God thought it into existence. Others say that he used science, that the cosmic-particle starting universe erupted from a timeless void or emptiness, a quantum state of nothingness fraught with all the potential of the laws of physics. Child: So where does the void come from? Response: Some say from God, others that it is just there, unrelated to time because the void is not in time.

How Did Something Come from Nothing? Child: I don't understand. How did *something*come from *nothing*? Response: Well, this is a bit difficult to explain to you right now but actually, the void is thought to represent a primordial mega-universe quantum state of pregeomtery with no time, space, matter or dimensions but seething with all the potential of the laws of physics. This is a tensed up energy 'uncertainty state' from which energized universe particles may randomly erupt. This sounds confusing but you'll understand all this later from your science studies.

But if God created the universe, asks child, *Why* did He do it? Parent: Well I'm not tuned in to all the answers--and its getting late--but I'll take a quick shot at that. Some say that God, being the essence of love, needed a loving relationship and so, being all powerful, He thought the universe into existence so human beings would emerge with whom he could have a caring relationship.

God's thought process was expressed through an energy sequence, of cosmic radiation, then matter, then stars and planets, and then people and everything around us. Of course, His thought process employed the laws of physics derived from mathematics. Which is why some scientists call God a mathematician.

Why Time? Child: Why is there time? Why do I have to grow up and you grow old? I like things as they are now. Why can't 'now' last forever? Parent: Well, to have a workable universe requires "time" so that we can have and plan for the future and things may change.

Child: But why does the universe need to have this kind of time that causes trouble, like death? Parent: Response: Well, because otherwise nothing would happen--there would be no cause and effect, no now-and-then, past or future, earlier or later. Nothing could happen, you couldn't plan to get up to go to school or play ball tomorrow because there would be no tomorrow.

Child: Well why.... Parent: "It's your bedtime, so we'll continue this another time."

Not Just "Child" Questions: But these are not just 'child' questions; they are questions adults ponder when they stop to wonder about the universe and about their own *existence*. Human preoccupation with the fact of *existence*--that of the universe and human life--has pervaded philosophy down through the ages, long before even the Greek philosophers like Socrates, Plato, and Aristotle.

The Root Question of All Metaphysics: Like other 'existential' philosophers, Martin Heidegger viewed this preoccupation with existence as representing *the root question in all of metaphysics: Why is there something rather than nothing, existence rather than nonexistence.* But scientists too, like Stephen Hawking, Heinz Pagels, Paul Davies, are drawn to this and similar questions.

What Is Our Place in the Universe? In his *A Brief History of Time*, Stephen Hawking asks: Where did the universe come from, and where is it going? How and why did it begin? Did the universe have a beginning, and if so, what happened *before* then? He cites St. Augustine in his *The City of God* and Immanuel Kant, in his *Critique of Pure Reason*, as dwelling on such questions.

In his 'Conclusions,' Hawking notes that "We want to make sense of what we see around us and to ask, What is the nature of the universe? What is our place in it and where did it and we come from?"

The Original Nothingness--An Aside: People are mystified by the concept of the nothingness of the primordial void, a highly unstable, seething quantum state without time, space, or matter, but fraught with potential. From the standpoint of physics there may be no such thing as the true nothingness envisioned in the biblical account of Creation *ex nihilo* although many still argue for that concept on the ground that what existed before the Big Bang is viewed by many as metaphysics rather than science and therefore the answer to the question of what existed may be 'nothing.'

But it is not just philosophers who wonder about 'existence' versus 'nothingness.' Physicist Heinz Pagels, in his *Perfect Symmetry*, describes the nature of this highly charged pre-creation nothingness. In doing so, he re-expresses Heidegger's question on existence, in terms of the concept of symmetry in physics. He asks why there should be anything beyond the ultimate simplicity and perfect symmetry of the early universe of unified forces--or of the primordial 'nothingness' which preceded it.

Who Needs the Complexity of the World? He asks, "who needs the complexity of the world in the face of the simplicity of absolute nothingness....I feel anxiety as I imagine the unfathomable, silent abyss of nothingness which might have been....That odd feeling provoked by our recognition of the reality of existence rejuvenates our wonder and curiosity about the universe...And that wonder is the beginning of science." And perhaps, one might add, the beginning of theology.

The Central Question in All of Metaphysics: At the core of things, then, is this central question of why the universe exists and in particular, the more personal and relevant question of why *we* exist. At the heart of it all is this abiding concern over existence, built into our very awareness--our search for meaning.

It is this concern, Pagels implies, which fires the pursuit of knowledge of the nature of the universe, of the meaning of life, the ultimate destiny of the universe, and the relationship of life and the universe to Divinity.

Are Questions About Life and Our Universe Ultimately Theological? Astrophysicist Robert Jastro, in *God And The Astronomers*, describes the ascent of scientists to the mountain peak of scientific knowledge, only to find the theologists there, patiently waiting for the scientists for centuries.

Some assert that ultimate meaning of the universe, the origin and possible purpose of life, even the governing laws of physics, are all facets of the grand theological mosaic.

In this view, in its highest dimension, questions about God, Life, Death, Time and the nature of the universe are interrelated and interwoven in the theological fabric of ultimate meaning.

Life, Death, God and Meaning: Questions of the nature and purpose of the uni-verse quickly translate into questions about: the existence and meaning of deity and its relationship to human life--whether the universe is explainable by natural processes; and the ultimate confrontational question, whether some knowing and thinking part of us survives physical death.

Natural Versus Supernatural Explanations: Physicist Paul Davies in *God and the New Physics*, describes the creation by purely natural laws of physics, suggesting the existence of an ancient superintelligence, one *not supernatural* but built into natural law. But he notes the need also to consider supernatural causes, that is, acts of God, as alternative explanations. For instance, he speaks of the *anthropic principle*, discussed elsewhere here, which deals with the startling coincidences in the various constants in physics, every one of which is absolutely critical to the presence of life in our universe.

He notes that these remarkable coincidences in the physical constants represent the most persuasive *evidence of an element of cosmic design* and observes that this is a preferable explanation to the so-called 'weak anthropic principle,' standing for the proposition that since we are here the universe must have the precise configuration of physical laws and constants which favor the emergence of life.

Dyson Associates Mind and God: Princeton physicist Freeman Dyson associates human mind and God-- "what mind becomes when it has passed beyond the scale of our comprehension." Dyson speaks of human mind as "a piece of God's mental apparatus," a part of the interconnected cosmos at the quantum level. The Bible speaks of human life created in God's image. Dyson links us directly to God, likening God to "world-soul or a collection of world-souls," and describing us as "inlets of God on this planet." We may later grow with him as he grows, he says, "or we may be left behind."

Dyson likes the idea of "many world-souls" for mind, "growing to fill the universe," coming as a diversifier as well as a unifier. He refers to St. Paul's concern with "unity," in First Corinthians, but says he cares more deeply for the *diversity* of God's working than for God's singleness.

Interestingly, this view of everything evolving--life, God, the laws of physics, mind, the entire universe--has a certain compatibility with evolving concepts in quantum theory that favor the indeterminate universe, and supposedly the intrinsic vitality of free will.

Dyson's Universe As Mind at Three Levels: Dyson speaks of the universe evidencing operation of mind at three levels: *First*, matter in quantum mechanics making probabilistic choices, something inherent "in every electron." *Second*, the level of human mind--the extension of the mental component of the quantum choices made by the brain molecules. *Third*, a "mental component of the universe," represented by the extension of the anthropic principle to a "universal scale"--the peculiar features of the laws of nature as being hospitable to the growth of mind.

If we treat this mental aspect as God, he says, then, he says, *we are small pieces of God's mental apparatus*, another way, of course, of expressing the mind of human life as an extension of the Mind of God.

Cosmology and Biology: Most scientific views of the early universe are based on cosmological theories such as the Big Bang, without any particular reference to the later emergence of biological phenomena or life. Similarly, most reviews of the possible end of the universe are, for the most part, narrowed to the question of whether we have an open or closed universe and, if closed, the likely events of the eventual contraction of the universe to the final Big Crunch, perhaps into a black-hole singularity state from which it may have come.

However, as we will later discuss, Princeton physicist Freeman Dyson and Harvard astrophysicist David Layzer explore relationships between cosmology and biology by examining aspects of mind and matter. Freeman Dyson goes on to review how mind, as presently incident to earth life forms, may adapt to conditions elsewhere in the universe and become a driving force in future cosmic evolution, departing from its present biological format.

Human Existence Required to Authenticate the Universe? These laws of physics, especially quantum theory, seem to imply importance of human awareness of the universe and its role to influence and even modify the universe. We discuss the proposition that objects like people, planets, even universes, are quantum systems the existence of which, in the Copenhagen quantum interpretation, are 'observer dependent.' Their very existence is dependent on their observation or measurement by another interacting quantum system, a person being one such system.

If this view of quantum physics is valid, then, seemingly, the very existence and credibility of the universe can be vouchsafed only by the awareness of it by human life or some other intelligence in the universe. If there is no other intelligence in the universe, a possibility although not a probability, then human mind may be said to be essential to the very existence of the universe.

In any event, intelligent life becomes a critical event in the continuing evolutionary processes which promote increasing diversity, organization and order in the universe at the same time that the universe is otherwise 'running down,' toward total disorder.

A Deteriorating World Heading for Extinction: If planet Earth is the only life-supporting planet in the universe, the health of the planet, with its life-supporting mechanisms, becomes super critical and its fate has cosmic significance. In the August 1992 Omni Magazine *Interview*, Rachel Rosenthal, well known West coast actress and devotee of the performing arts, is interviewed.

In her own words, she tries "to take people away from the sense of separateness, loneliness, isolation, and alienation that is killing us" and she speaks of a world deteriorating into one "without nature or art, where extinction is a certainty."

A Calling to Understand the New Scientific Paradigms: Through a series of questions and answers she gives expression to various thoughts about the role of art and science in the universe, suggesting that artists have a calling to "understand these new scientific paradigms (like quantum physics and virtual reality) and create work that reflects that understanding, that disseminates these paradigms at the human level."

She believes everything has a "mind, consciousness, and soul" in some degree; that because of the nature of the human brain, an organ which she suggests human mind "does not know how to use," there is a "total imbalance between object making and ethical sense," leading us to "muddle on in our evolution."

On Planetary Extinction: This leads naturally to the subject of "planetary extinction," a situation in which, she says, the human species takes the "entire world as our ecological niche." This, she feels, is "an outrage" and she "can't bear to see this extraordinary planet lose its wildness and become a human cybernetic machine. There won't be anything "remotely resembling" what we call 'nature' and "everything will be managed by humans."

We Want to Do Things Better Than the Gods: And why do we continue to go down this path, she is asked. "Because," she answers, "there are so many of us, and because we're always tinkering. We always want to do things better than the gods." The "concept of progress is one of the biggest lies we've concocted," she notes, giving us "a plundered and degraded planet."

Not only are we “emotionally bankrupt, sick in our souls.” she observes, but we “don't understand our place in the universe, that we're not exempt from the laws of nature.” She declares that “We cannot continue to see ourselves as having license to eradicate anything that does not directly feed, entertain, or make us richer.” Rather than being “imbedded in nature,” she remarks, “we've created an adversary in nature.”

What Is Our Place in the Cosmos? In response to the question of what is our place in the cosmos, she observes that we are capable of working with all of nature to exist in a world with room enough for all to live in harmony; that evolution doesn't stop just because we arrived on the scene.

Our place in the cosmos? She doesn't think we really have a 'place' in the cosmos. By which she means that we have the same place everybody else has on this planet--to compete without exterminating everything else. It's not that the whole human race has "gone bad" but that we are in ecological, nature-destroying “free fall.”

Asked about Consciousness: Asked what is consciousness, she responds that consciousness is a gift, entrusted only to us among the planet's species. A universal and very sudden shift in consciousness, she feels, could save everything if everyone realized the big picture of global *interconnectedness of all life*. She makes reference to the Eastern traditions which address universal mind, the "pervasive consciousness of all creation."

Dyson's View of Purpose and Human life's Role to Share: In his Aberdeen, Scotland *Gifford Lectures,* updated in his *Infinite In All Directions*, Princeton Physicist Freeman Dyson gives expression to the thought that there is "some purpose" for our presence here, something beyond our comprehension, something "having to do with the future."

Dyson's God Who Shares in Evolution: If you like, he says, you can "call the transcendent purpose God," but, he muses, the God of the Socinian Heresy, the God who, though not omniscient or omnipotent, is "growing in power and knowledge as the universe unfolds"--an evolutionary process in which we share.

Socinianism, a religious movement organized in Poland, was founded by Faustus Socinus, the Italian Catholic reformer, based on the teachings of his uncle Laelius Socinus. Not only did Socinus have trouble with perfection aspects of God but he had trouble with the doctrine of the Trinity. Church persecution brought the Socinianism movement to a close. There is a nostalgic analogy between Dyson's evolving and impliedly not yet perfect God, and Liebniz's God who gave us his best--the "best possible world."

Liebnitz, of course, thought to avoid the heresy of the less than omnipotent God by discussing the *unavoidability* of the world's characteristics rather than dealing with Divine imperfection. But critics of Liebnitz atttacked Liebnitz's best possible world on the ground that it indeed implied that God was less than omnipotent, essentially the Socinian God--with which Freeman Dyson is so comfortable.

Dyson on Origin of Life: Life in its earliest forms, says Dyson, was "little removed from ordinary chemistry." But what then, do we make of the later development of consciousness, and of esoteric or abstract concepts such as justice, meaning, and Wigner's "joy of being alive"--and intuitive awareness of Divinity? Having asked this question, Dyson chooses not to confront the question directly but to respond by first redefining the question of origin of life as the question of how you reconcile life as just "chance" with the idea of life as God's "plan."

Dyson Suggests Three Alternatives: He narrows things down to *three alternatives:* First, say everything is chance, (nondeterminism, God's role not stated). Second, deny chance and say that God, as the predesigning cause, "knows events to come" (determinism). Third, say that chance exists because "God shares our ignorance," the Socinian Heresy of the God who, though not omniscient and omnipotent, is growing with the universe and "chance is part of his plan" (God's will employing chance, modified determinism). There are other alternatives, of course.

The Evolving God and Random Processes: For those who want both God and free will, he prefers the idea of the evolving Socinian God. Free will, he says, is the coupling of the human mind to random processes inside the brain and *God's will is the coupling of the universal mind to random processes in the universe*. Here, Dyson combines design, classical physics, with chance, probabilistic quantum theory.

He describes the human brain's perception and cognitive functions as governed by random processes--not predetermined, and God or universal mind as also operating on random processes. However one may read this interrelationship between causal classical physics and quantum uncertainty, Dyson's bottom line on the existence of free will, in the perennial determinism vs. indeterminism debate, is unmistakably free will.

But getting back to Dyson's question of the source of consciousness and subjectivity, which Dyson chose to redefine rather than answer directly, read as a whole, his materials track mind, consciousness and subjectivity to the growing organization and complexity of biological matter. In fact, as we shall see in his discussion of artificial intelligence, he does not even exclude the possibility of consciousness and subjectivity being resident in nonbiological, solid-state matter. For him, God, human life and mind, are facets of the evolving quantum universe.

Life and the Cosmic Code: In *The Cosmic Code,* physicist Heinz Pagels says that *Life is written into the Cosmic Code*--part of the grand program. He feels that science has something to contribute by bringing renewed appreciation of the human condition and the limitations of its existence in the universe. The human being, he feels, embodies *the principle of life*, a principle that cannot be destroyed because, he says, it is *written into the cosmic code.*

In a sense, this proposition seemingly endorses the vitality of the anthropic principle--even though Pagels, in directly confronting the principle, conspicuously distances himself from it. Pagel's thought of life as being an ingrained part of the cosmic code seems a cognate of Freeman Dyson's vision of Mind destined to spread throughout the universe. Both of these ideas are seemingly related to the existence of a quantum state of information and the mental interconnectedness inherent in quantum reality.

Primordial Nothingness As Inconsistent with Divinity and Meaning: We have discussed the idea of the 'nothingness' state of the pre-big-bang primordial meta-universe void, and the idea that 'nothingness' seems inconsistent with the concept of theistic Divinity in which idleness--implied in nothingness--is unacceptable, pur-pose is pervasive, and meaning is central. In that context, life--preferably intelligent life--is essential and ultimately crucial to the destiny, meaning, and perhaps very existence, of the universe.

The Crucialness of Life in Universe Models: This line of thought--the crucialness of life--seems to imply that *only* "life-involved" universes--universes with observers, may exist--or at least that life in such universes must be in some gestation process. The currently popular if not prevailing view has been that our universe started with a quantum fluctuation in the primordial mega-universe quantum void.

The Universe without Life Proclivity Exists Only As Potentiality: But whatever model best represents truth, *only a universe with life-proclivity would seem to have any meaning--any rationale for existence.* For all practical purposes, universes without life don't exist since existence implies knowledge or awareness of the fact, unless we say that God's knowledge is sufficient without the existence of any other observer or 'knower.'

Thus human obligation for preservation by assuring continuity of the life-supporting ecology of the planet assumes paramount importance, a necessity for further unfolding of the universe. Human life's criticality, of course, may be somewhat less central if there happens to exist in the universe other intelligences capable of subjective awareness.

The Universe Itself a Living Organism? It is a fact, as we have discussed, that everything in the universe consists of different forms of a single substance, energy, the principle of Einstein's famous $E = mc^2$. Similarly, it is a fact that, under the laws of conservation, while the *net* direction of the universe is toward entropy-disorder, that another concurrent process of organization-order reaches toward increasing order in the form of higher life forms--toward consciousness, subjectivity and mind.

Physicist Paul Davies speaks of an ancient intelligence pervading the universe and directing the evolution of the laws of physics toward life and mind. Some have equated this all-pervading energy substance as 'cosmic mind,' 'world soul,' or God--as in Spinoza's universe, or 'impulse traversing matter' as in philosopher Bergson's view, or a living being, as contemplated by the Jesuit paleontologist philosopher Teilhard de Chardin.

As we note elsewhere, Teilhard de Chardin viewed *the universe as a living being in transformation, involving the convergence of life and matter* into the ultimate tranquil ocean in which each drop of water, each soul, retains consciousness of its separate existence.

The *Gaia Hypothesis*: The idea of life as a primary purpose and characteristic of the universe has had many suggested applications. Through the years, scientists have addressed the life-habitability feature of the universe and, like Scottish geologist James Hutton, in 1785, pointed to life-balancing characteristics of our planet earth.

In *Gaia: A New Look at Life on Earth*, British chemist and inventor James Lovelock describes his hypothesis--called the *Gaia Hypothesis* after the Greek earth goddess--

that the earth is a "living organism," a biosphere, *a global system of collaborating organic and inorganic interacting forces governed by some sort of integrating 'intelligence.'*

Boston University microbiologist Lynn Margulis expresses similar views. While scientists cautiously distanced themselves from the *gaia* theory for some years, critical environmental warning signs have attracted renewed interdisclipinary interest by geophysicists, geochemists and marine biologists, highlighted in the 1988 American Geophysical Union conference.

Existence of Some Life-related Mediating Mechanism: Although the precise feedback and regulating mechanisms remain blurred, there is recognition that some still *undefined life-related integrating mechanism is functioning to collaborate, and mediate,* treating the whole as a unified organic-inorganic system which makes adjustments to preserve the integrity of the earth biosphere. It has, for example, kept oxygen levels and earth surface temperatures at a uniform life-sustaining level through the centuries despite developments which theoretically would produce severe life-threatening imbalances.

While, for some, this remarkably-coordinated, self-balancing-self-adjusting bio-system implies, to some, purposeful design, others feel the same results are explainable or reconcilable as simply the product of a system with a feedback mechanism that needs no necessary resort to 'purposeful design.' Interestingly, both religious and scientific groups see in this remarkable global and extraordinarily complex, coordinated *life-safeguarding system* built into the blue planet, a sign that life in our universe, whatever it may signify, whatever its purpose, somehow is not pure accident.

Is There a Nonmaterial Life Force or Spirit? Recognizing that everything in the universe representing the various present energy configurations is a derivative of the original radiant energy, the view of the cosmos as a single substance with built-in life frequencies is not difficult to rationalize. However, for some years we have witnessed a philosophic schism on the nature of life--whether an immaterial or nonmaterial 'life force,' spirit or super consciousness is imparted to the physical human body at birth or whether that life force emerges as a part of biological materiality.

Relationship to Divinity: This immaterial life force is frequently viewed as somehow related to *divinity*--part of or an extension of God--or perhaps an expression of 'universal soul.' Involved here also, is the discomfort of some in treating life simply as a step in the energy unfolding process, implying that life is just one aspect of materiality--without spiritual or theological significance.

Here is raised the conflict between so-called scientific materialism on the one hand--that everything that exists is explainable in terms of the energy, matter and biology, a single material substance--and on the other hand, religious transcendentalism--that there also exists an independent, immaterial, mystical and spiritual reality, a theory sometimes termed *dualism.*

Scientific vs. Theological in the Context of Biology: Science writer Boyce Rensberger's *Discovering New Worlds In The Realm Of The Cell*, prominently displayed in several issues of the Dec. 1988 Washington Post, suggested that while early philosophers and biologists once invoked a "mystical vital force" that breathed life into ordinary matter (Genesis 2:7--God breathes life into Adam), that present biological physics and chemistry serve to explain phenomena of life in "purely materialistic terms."

Rensberger was, of course, simply characterizing the currently accepted scientific approach to the origin of life in terms of conventional scientific materialism--that all human phenomena are ultimately explainable in objective biological terms, the product of the growth of organization and order, *without resort to immaterial, ethereal or mystical causes.* But that view, scientific materialism, is not always popular with persons of deep religious convictions.

Denial of Mystical Causes Brings 'Letters to the Editor': Rensberger's reference to a "mystical vital force breathing life into matter" as being explainable in the more natural scientific terms of developing biology, increasing order, brought highly-spirited and critical editorial responses.

A "Letters to the Editor" reaction by Christine Hanson Mercer said: "I hope he didn't mean to suggest that the God-concept is unscientific. My memory banks yield numerous examples of deeply religious scientists (citing her devout Catholic father, a scientist at the Johns Hopkins APL; Nobel prize-winning physicist Werner Heisenberg, author of the Uncertainty Principle in physics; and Albert Einstein: "... in this materialistic cage of ours the serious scientific workers are the only profoundly religious people."

She goes on to suggest that "Arguments that try to explain away God by using science, do a disservice to both." *She suggests that people who engage in such "sophistry" instead spend a little time "examining the private agonies of their repressed souls."* The very subject triggers emotions.

The How and Why of the Universe As Philosophical and Theological: In another "Letters to the Editor" response, Robert A. Boisture asserts: "...how and why our universe came to be....has always been, and remains, a philosophical and theological inquiry, not a scientific one. As we grapple with this most important of all questions, the marvelous order and complexity of life which science describes becomes a strong circumstantial clue pointing to a conscious creator rather than random chance as the final answer." Implied here, of course, is an endorsement of the anthropic principle that there is a direct connection between the laws of physics and the universe's proclivity toward life.

The Need for Change: In *Real Philosophy, An Anthology of the Universal Search for Meaning*, professors Jacob Needleman and David Applebaum, in Part Five, "The Unfinished Animal," address the problem of the future evolution of society and *need for human change* and improvement, given the human condition--conflict, cruelty, anguish, guilt, self-doubt, and the "blight of negative traits."

Are Our Deficiencies the Price of Being Human? Nietzsche Responds: About our deficiencies, the authors ask, can they be corrected, or are they inherent--the price of being human? And what are the deficiencies? Can we correct a weak will or the shackles of ignorance? Nietzsche, who, they suggest, offers one of the clearest responses, says that we have allowed ourselves to be burdened by a false notion of morality which opposes proper change.

To Be More Than What We Are Born to: The implication, say the authors, is that it is up to us to to make ourselves complete. To be more than what we are born to, it is imperative to know more, not so much factual knowledge, but "wisdom arising from within ourselves." To do this, they suggest, we need to take a direction different from that which the ego pursues.

Knowledge only feeds our competitiveness, our hunger for more territory. For *wisdom* to enter "the center of one's being," one must, they say, turn from self-love and vanity to realize that the new-found knowledge brings with it a responsibility to communicate it to others. Our tendency is to live a life according to a private intelligence which fantasizes, and schemes based on partial evidence for disharmonies in our behavior. We should, they say, let ourselves be guided by what is public or common to all.

Plato on Knowledge, Wisdom and Happiness: Needleman and Applebaum turn frequently to Plato for meaningful insights into human thinking. Plato's philosophy addresses many subjects, as we have seen. But one emphasis is on knowledge and wisdom, acknowledging that the *knowing* what is right must be effectuated by *doing* what is right. The learned, educated soul represents the good soul, says Plato, and wisdom is the greatest virtue. Thus the *best pleasures* are the intellectual, and the *highest happiness* is achieved in contemplation of the highest ideas.

The Republic: In his great work *The Republic*, Plato writes of his theory of the ideal state describing a city where social justice could be achieved. There would be three groupings of men, the small philosopher-kings class to rule through wisdom, the soldier guardians, and the worker class. Social justice and fairness would be achieved in part through a semi-communal state in which private property ownership and family life would become communal and even art would be addressed to that which benefitted rather than denigrated the state.

The city would be ruled by philosophers, whose education would assure their greater insight for daily decision-making and guidance of the public. Which brings us to Plato's *Allegory of the Cave*, expressed as a dialog.

The Allegory of the Cave: Plato's "Allegory of the Cave," from *The Republic,* is often cited to in support of the need for education, knowledge and wisdom of those who are to govern and influence affairs of the state. Most people go about life disadvantaged by limited perspectives of the world at large. In the allegorical account, prisoners are confined in a cave, their perceptions of the world limited to the shadows cast in the artificial light of the cave.

One prisoner is released to see the outside sun, the light and understanding of outside things. Upon his enlightenment he returns to his former fellow prisoners but has difficulty in overcoming their shadow-delusions of the truth. Plato uses this analogy, of course, to draw conclusions about education and sound government.

The dialog continues: "But, whether true or false, my opinion is that in the world of knowledge the idea of good appears last of all, and is seen only with an effort; and, when seen, is also inferred to be the universal author of all things beautiful and right, parent of light and of the lord of light in this visible world, and the immediate source of reason and truth in the intellectual; and that this is the power upon which he who would act rationally either in public or private life must have his eye fixed."

The Book of Proverbs: The authors turn to the Book of Proverbs for statements relating to *knowledge and wisdom:*

Happy is the man that hath found wisdom, and the man that acquireth understanding. Proverbs 3:13

My son, let them not be removed from thine eyes: keep [before thee] sound wisdom and discretion. Proverbs 3:21

The wise shall inherit glory; but disgrace shall be the portion of fools. Proverbs 3:35

Acquire wisdom, acquire understanding; forget it not, and depart not from the sayings of my mouth. Proverbs 4:5

Wisdom is the principal thing; therefore, acquire wisdom: and with all thy acquisition, acquire understanding. Proverbs 4:7

> If you expect science to give all the answers to the wonderful questions about what we are, where we are going, what the meaning of the universe is, and so on, then I think you could easily become disillusioned and look for some mystic answer. --Richard Feynman

CHAPTER 37
LIFE AND NONLIFE

The Growth of Order in the Universe: We have noted how physicists, cosmologists, astronomers and space scientists give written expression to their interest in certain aspects of biology. Examples are Freeman Dyson in *Infinite In All Directions*, Eric Chaisson in *Cosmic Dawn*, Carl Sagan in *The Dragons Of Eden, Speculations On The Origin Of Human Intelligence*, and David Layzer in *Cosmogenesis*.

In his book, *Cosmogenesis, The Growth of Order in the Universe*, Harvard University astrophysicist David Layzer, with a flair for the biological sciences, seeks to present a coherent scientific view of the universe, quantum mechanics, macroscopic physics, cosmology and biology. The reader is taken on a grand tour of cosmic evolution.

Layzer describes the paradoxical buildup of chemical and structural increasing order in a universe generally characterized by entropy, decreasing order. And along the way, he discusses biology, the development of life, genetic codes, the human brain and its relationship to mind, consciousness, subjectivity, perception, and thinking.

Dealing with Causal Relationships and Quantum Indeterminacy: His cosmic history serves nicely as the backdrop for reaching a scientifically-oriented description of the nature of our mental processes, and discussion of the question of the existence of human life's free will in a universe that is governed both by causal macroscopic physics and, quantum uncertainty.

Qualitative Changes in the Human Brain: He describes the qualitative changes in the architecture of the human brain, the transition in the neocortex of the human brain from the stereotyped function of "visual feature detection," common to lower life forms, to the higher function of "visual perception," a "far richer" visual world with a capacity to "construct and modify neural circuits" and thus to "construct, test, and store new perceptual and cognitive hypotheses,"--the stuff of human creative activity.

Instantaneous Versus Evolved Life: According to Genesis and Western theological thought, life came into existence instantaneously as an expression of divine will. According to scientific theory it came into existence as the product of evolving natural forces over vast epochs of time. Opinion is divided over whether these natural processes, the laws of physics, were themselves supernaturally originated or were generated through some natural process. But apart from so-called 'biblical inerrancy' and 'creationist' point of view, discussed in the chapter on Religion, Denominations, and Meaning, the generally accepted view is that life evolved as the product of organizational transitions in the basic component of the universe, energy.

Divine vs. Natural Creation: Quantum cosmology theories suggest the possibility of a universe that erupted from the quantum void of the eternal mega-universe as a random quantum fluctuation. Erupting from the quantum void was a particle or point, exponentially smaller than a proton, a mini-universe of space, time, matter and forces, governed by a 'superforce' representing space, time, dimensionality, and the unified forces and fields of physics.

Through a 'phase transition,' this mini-universe point went through a momentary exponential, faster than the speed-of-light inflation stage, which subsided into the linear reduced expansion rate. This expansion continued in the Big Bang sequence in which the original 'perfect symmetry' of the unified forces broke down in a phased separation of the forces to provide the space, time, energy and matter representing the present structure of the universe.

A part of this scenario is the concept that the total energy of the universe may equal zero because the universe's positive energy such as the galaxies, is derived from and equal to the negative gravitational energy of the system.

So in this view, the net energy of the entire universe is zero, the positive energy in the universe and its negative gravitational energy simply canceling each other. This explanation reminds us of Oxford professor P.W. Atkins' proposition in *Creation Revisited*, that the universe's chance eruption from quantum chaos is an application of mathematical theory that everything is built from nothing.

Reductionism--Life Built into Atoms: Assuming an evolutionary process, as physicists and biologists point out, there are wide differences in views over what happened to produce life in that evolutionary process. The *reductionist* view, sometimes referred to as the scientific or radical-reductionist view, is that life proclivity is built into atoms. So left alone, atoms will, on their own and by virtue of their intrinsic nature, combine in molecular forms, winding up as life and mind. Sometimes, it is said that quantum energy forms, particles and fields, are information units with built-in life proclivity.

Holism: The so-called *holistic* view is that through chemistry, atoms and molecules combine to self-organize into increasingly ordered forms representing entirely new *emergent* stages of complexity, a *collective* phenomenon, but something *not inherent or implicit in the original components.*

The example sometimes cited is that of Douglas Hofstadter in *Gödel, Escher, Bach,* in which individual ants exhibit very limited behavior in task performance, but at the higher group level of the ant colony, demonstrate high intelligence, the capacity, in this collective format, to perform remarkable feats of colony engineering. In the same way, says physicist Paul Davies, the collective level of the human body may be viewed as the holistic level of the constituent individual cells.

Vitalism, Dualism, Holism: But the so-called *vitalists* believe that you just can't bring inanimate materials to *life* without a vitalizing *nonmaterial* component, a *life force*, being injected. The *dualists* believe that mind and body are two distinct substances, one physical, one nonmaterial--mental or spiritual.

The predominant view is seen as favoring the holistic view of a single substance that self-organizes into emergent new states of increasing complexity taking on life characteristics. In the holistic view there is no longer a need for *insertion* of a life force, the force *emerges* or ensues at the higher holistic level.

The holistic view is analogized to the software and hardware levels of the computer, or an electric sign of electrical circuits, traversed by the flashed message. Other examples are thousands of letters or words which have no meaning until they emerge at a higher level to become the gripping plot in a story. Similarly, the multiplicity of non-significant musical notes, artfully combined, suddenly cross a threshold of individuality to become, in the aggregate, a beautiful symphony or sonata.

Four Approaches to Distinguishing the Living from Nonliving: Layzer discusses theories of entropy and developing order, concentrating considerable effort on the matter of how life evolved as a part of increasing pockets of order, and what life is, in biological and scientific terms. He describes *four approaches* to distinguishing the active, animate or living, from the passive and inanimate: panpsychism--the life force, life as physics or chemistry, dualism, and life as biological order.

The *first* approach, embracing several ideas which he subsumes in the term *panpsychism,* traceable to metaphysical speculations in 'prescientific' times, is that everything in the universe, including the inanimate, is impressed with different levels of 'inner being,' a quality suggestive of the 'vital life force' mentioned earlier.

Philosophic and Scientific Support for a Life Force: The existence of such a life-force has had its support from philosophers and scientists like Liebnitz, Goethe, and Schopenhauer, Whitehead, Jesuit paleontologist philosopher Teilhard de Chardin, and biologists Waddington and Rensch. Actually, some contemporary physicists, like Freeman Dyson, David Bohm, Paul Davies, consider life and mind as implicit in all matter, the *reductionist* view mentioned earlier.

In *Perfect Symmetry,* physicist Heinz Pagels speaks of the confrontation between Newton and Goethe, a vitalist--believer in *the existence of a 'life-force'* in living things, one that is not subject to the laws of physics. Quantum theory, notes Pagels, has been cited as an example of the abstract explanations of science that illustrate the modern humanist critique of science as being at odds with what he terms "the vital core of human experience."

Neurobiologists: Particles Do Not Have an 'Inner Life': While panpsychism, says Layzer, has the merit of a "unified view of reality," he points to the absence of scientific support of a "postulate that subatomic particles have an inner life." Most Neurobiologists, he says, reject the separate mind and body theory, and regard mental processes as adjunctive to the body, part of the natural physical world.

The Difference Between Life and Non-Life as Degrees of Order: The difference between animate and inanimate, and between the conscious and nonconscious nervous processes, says Layzer, are not produced by the *injection* somehow of some external phenomenon of spirit, soul, mind, or life-force, but results from the "presence or absence of certain kinds of order." This conclusion, he feels, is the unavoidable consequence of accepting the hypothesis of biological evolution: In the evolutionary process, he says, there was no "insertion point" for the introduction of "nonmaterial [life] substances."

Physics and Information Science: But there is not, in science, an absolute conceptual barrier to the existence of an 'inner life' or 'vital life force.' There is growing interest by elements of the scientific community in looking at quantum energy phenomena, particles, not only from the standpoint of the physics of *structure and forces*--particles, forces, fields--but also from the standpoint of their *functional* characteristics as the *information* units in an information-system universe.

Computer scientist Edward Fredkin--some of whose thoughts are reviewed here in our discussion of the universe as a computer, suggests that particles should be explored more as *information* bits than energy forms. Heinz Pagels, in *Perfect Symmetry*, also sees the possibilities in the blending of physics with information science.

The nature of energy particles, fields, and forces, as vehicles for the flow of information--a form of order as part of the workings of thermodynamics--is evidenced by the continuous 'unfolding' process involved in the evolution of increasingly complex and diverse forms of matter, life, and mind.

Whitehead's Other Physical Laws Governing Molecular Aggregates: Alfred North Whitehead speaks of these emerging higher forms of order as proceeding from and *governed by other physical laws*. He observes that the particles in living molecular aggregates seemingly reflect knowledge of how to function in the increasing complexity of their organization, a view close to that of physicist Freeman Dyson, Sir John Eccles and physicist David Bohm but at variance from that of David Layzer.

Human Beings As Aggregates of Information-Bearing Particles: Biological or living molecular aggregates, as represented in the human brain, have reached the stage where they are able to fathom and modify genetic patterns, not only through animal cross--breeding, but through human-cell-modification. Thus, particles somehow bear the information components for life creation; and human beings are complex aggregates of information-bearing particles that carry within themselves the potential for knowing and even altering the evolution of the universe.

In the discussion of quantum reality we explore the concept that underlying quantum energy forms is a universally-interlaced nonmaterial fabric of information. As computer scientist Edward Fredkin points out, science, preoccupied with energy as particles, fields, or other quantum states, has devoted comparatively little study and resources to the ultimate constituents of the universe, the "informational" aspects of quantum physics.

Extrapolating Inchoate Particle 'Life-Vitality': Returning to the so-called life-force, and physicist David Layzer's reference to the nonexistence of any "insertion point," one might not unreasonably construct from this functional-information nature of particles, the existence of *particle potentiality for organization into forms of life and awareness.*

Life Wrought into Matter: In this sense at least, one may view man's 'nonmaterial' or 'spiritual' side--the 'vital life-force,' or in Layzer's terminology, nonmaterial [life] substance--as derived from this 'information' and 'potential awareness' characteristic implicit in particles, fields or other quantum forms. In this context, life is literally *wrought into* matter and 'emergence' is an implementing *fulfillment* function, an idea supported by the views of physicists Freeman Dyson and David Bohm.

Some Philosophic Views Reinforcing a Life Force: The concept of the existence in matter of a life force, specified in the cosmic code, as Pagels says, derives a unifying coherence and reinforcement from science, philosophy and theology, as exemplified, for example, by Georg Hegel's universe as Mind, Anglican Bishop George Berkeley's universe as mental substance, physicist Freeman Dyson's view of mind as dominating matter, philosopher Henri Bergson's life as "impulse traversing matter" and related views of others.

Refuting Reductionist Philosophy--The Mind As Apart from the Physical World: Repeatedly raised is the question whether this subjective consciousness or awareness can survive death, the dissolution of its organic framework. In *God and the New Physics,* Davies concludes that while the 'new physics' requires a re-examination of conventional ideas of the role of God in the creation, that the human mind is something apart from our physical world--that we are something more than just a co-location of atoms, as philosopher-mathematician Bertrand Russell had so bleakly declared.

Mind As Capable of Disembodiment--More Than Mounds of Atoms: Davies observes that "It would be foolish to deny that many of the traditional religious ideas about God, human life, and the nature of the universe have been swept away by the new physics." But, he says, our search has turned up many positive signs too. He notes that the existence of mind, for example, as an "abstract, holistic organizational pattern, capable even of disembodiment, refutes the reductionist philosophy that we are all nothing but moving mounds of atoms."

Energy As Information, Mind and Life: As we see in our discussion of physicist Freeman Dyson's views of mind in the universe, there are concepts entertained by highly-regarded scientists that suggest a greater role for mind, "independent of flesh and blood," a sort of *scientific transcendentalism* in which mind is literally structured into energy, an extension of the idea of particles as information carriers in the *universe as an information system.*

In the context of Einstein's special relativity theory, we have described the universe as consisting of a single substance, energy, underlying which is information, which, in the aggregate, represents mind at the cosmic level. In this context, energy represents the physical mechanism for the storage and communication of information, denoting the functioning of cosmic mind, of which the human mind is an extension. This is essentially the analogy drawn by physicist Freeman Dyson in *Infinite in All Directions*, describing human mind as an extension of the mind of God.

'Mind' has been thought of by some, as 'spirit,' 'soul,' 'higher consciousness,' and so on, all terms implying intelligence, information and non-materiality. Thus quantum energy forms may be considered the transfer mechanism or medium for information and intelligence. The brain, itself an elaborate configuration of solid energy is a reservoir of intelligence-- structured information.

Mind and the Quantum: Frequent in literature is use of terms such as 'world mind,' and 'world soul,' 'cosmic mind,' 'the force,' all involving the concept of the universe as some form of living mentalness, mind, or the mind of God, reminiscent of Pierre Teilhard de Chardin's universe as living being and the *Gaia Hypothesis* of the planet Earth as a living organism.

The concept of mind as a durable, configured energy state, not dependent on the physical body for continuance, as physicist Paul Davies suggests, has appeal for those that seek some kind of reassurance that life has projection and meaning beyond its perishable physical form. This view of mind as having independent nonphysical quantum existence may signify more commonality or interdependence between science and theology than envisioned in the past-- but something not necessarily attractive to either.

Leery of Materiality: Science seeks independence from theology and theology, in turn is leery of science seen as suggesting everything reducible to materiality. The idea of quantum commonality of body, mind and soul may represent an amalgam which, for theologists, may be one step too close to scientific materialism.

Reassimilation of Mind upon Death: This independence of the mind, as we have noted in the discussion of death, might also imply that physical death involves a reassimilation, a transference to or reassimilation with the underlying originating cosmic mental source, as envisioned by philosophers Schopenhauer, Bergson, Feuerbach, Simmel, and paleontologist Teilhard de Chardin.

The various approaches to the nature of mind and matter--dualism, monism, pluralism--coalesce in the vitality of a single ultimate substance--mind--having both physical and nonphysical quantum manifestations. It is not surprising that increasingly, quantum theory is explored by theologists.

Energy, Matter, Mind, and Love: Relating ideas on energy and mind suggests interesting patterns of relationships. The universe consists of a single thing, original radiant energy, evolving to take various forms (as specified in Einstein's $E=MC^2$ energy-matter equivalency equation): motion energy: radiation--fields and particles; frozen or structured energy (matter or rest energy forms like planets and people) interlaced with that which is mental. As Freeman Dyson notes, energy is weird enough that God can do anything he wants with it.

In this analogy, energy, including life as *thinking energy forms*, capable of generating abstract concepts of love, justice, compassion, nostalgia, contrition, all part of the underlying cosmic information bank, invoked through the mental processes of the higher forms of biological consciousness, God reaching out through the quantum connection to the human mind.

Bodies: Temporary Biological Structures for the Mind: Bodies, in this view, are nature's biological structures to house and nurture aggregates of mind, mind-supporting mechanisms. Interestingly, as physicist Freeman Dyson points out, *body forms are likely to change* by human efforts in the future, both in the near space of our solar system and in the far reaches of the galaxy, in order to spread mind through the galaxy and the universe.

Freeman Dyson's Two Books: Physicist Freeman Dyson speaks of God having given us *two books*, one we call the Bible, dominated by religious and theological ideas, the other, The 'Book of Nature,' representing so-called 'natural theology,' describing reality in terms of the laws of physics. But in the 'new physics,' quantum uncertainty implies that underlying our material and objective reality, there exists a subjective component, dependent on mind.

This unifying formularization, it would seem, makes easier the theological prop-osition that the origin and nature of the universe, the laws of physics, man's presence and meaning, have their ultimate source in pure information, perhaps representing divine idea.

Hegel's Universe As Mind; Berkeley's Universe As Mental Substance: Interestingly, this expression of Energy-Mind-Life "natural theology" is also reminiscent of philosopher Georg Hegel's universe of pervasive Mind and unfolding thought, and Irish Anglican Bishop Berkeley's idea of the universe as a single 'mental' substance. One may or may not make the final extrapolation to Spinoza's universe as the single substance, God, but may leave God in the exclusively transcendental role of theologian Barth, or theologian Tillich's God as existence itself.

Related Theological And Philosophic Ideas: Some theological and philosophic ideas related to these questions are the immanence of God in human life; human life as an extension of God; the human mind as an extension of the mind of God--as described by physicist Freeman Dyson, and man as part of Spinoza's concept of God as the single universal substance.

Schopenhauer's *Individualization* of the Unified Mind: In the chapters on Death, we note that philosopher Schopenhauer perceived birth as an "individualization" into our temporary, physical, more limited and finite form, from an earlier undifferentiated form. On death, our spirit--our "determined infinite" form or spirit, to use philosopher Feuerbach's concept of the meaning of death, is reassimilated into its undifferentiated form again. Death, here, in the sense of our status as individuals, is real, but it represents only the end of our "separateness."

On the other hand, as we have noted above, the French Jesuit philosopher-paleontologist Pierre Teilhard de Chardin, stops short of Feuerbach's total "assimilation" and asserts that in the physical reassimilation, *the individual consciousness remains.*

Georg Simmel: Only the Individualized Self Dies: As we discuss in the chapters on death, the philosopher Georg Simmel, as pointed out by Joachim Wach in his essay on “The Problem of Death” in *Understanding and Believing*, is that *death is immanent in life and a formative agent of life.* Here, one's awareness is enhanced by death and the transcendental self survives the death of one's ‘individualized’ self.

Layzer's Second Category of Life as Physics or Chemistry: But getting back now to physicist Layzer's description of the four alternative views of life, he goes on to describe a *second* view of life with explanations that shift as science has evolved--the actions of atoms, particles governed by Newtonian physics--that life is chemistry; that distinctions between animate and inanimate, between mind and matter, are illusory.

The Third View--Dualism: Matter and Life As Two Separate Realities: The *third* view, in Layzer's enumeration, is *dualism.* In this view, he says, while matter and energy represent the physical reality, mind or soul is a *second reality*--one beyond the grasp or purview of science. Socrates, Plato, and Aristotle, as he points out, distinguished body and soul. And, says Layzer, the philosopher Henri Bergson advocated this dualism--life as "impulse traversing matter"--and Bergson's view, in turn, says Layzer, influenced James, Shaw, Proust and Piaget. Dualism has a distinguished background in philosophy, as we shall see.

The Dualism of Mind and Body--Descartes, Locke, Kant: French philosopher mathematician René Descartes spoke of dualism, mind and body as two distinct substances. Although both are traceable to God as the absolute substance (an echo of Spinoza's concept of the universe as a single substance--God), mind and body are otherwise separate substances, according to this view. So how can the mind, separated from matter, know the real external material world? We note in passing that British philosopher John Locke and German philosopher-astronomer Immanuel Kant were also dualists.

Descartes' mental substance is not just a historical philosophical notion but has adherents in modern science. Sir John Eccles, the Nobel-prize-winning neurophysiologist asserts dualism in his view of consciousness as a nonmaterial force separate from the body biochemistry, which enters the body during embryonic development.

Layzer's Fourth View--Life As Biological Order: Layzer's *fourth* view (which coincides with his own), he says, is that of modern biology. *Living matter, in modern biology, is different because of the way it is organized--not because it has a nonphysical component.* The critical element is *biological order*.

But as we have seen, the opposition by Neurobiologists, and Layzer, to panpsychism, the existence of a life-force 'insertion point,'is side-stepped either by viewing matter, particles, as inchoate mind, or as an emergent characteristic in evolution, a new form of order, as we have noted earlier, a view referred to as *holism*.

Freeman Dyson and Life As Aspects of Energy: Along this line, are treatments of life forms as aspects of energy in thermodynamic terms. As we discuss a little later here, Princeton physicist Freeman Dyson proposes that with sufficient time, life can adapt itself to *any* environment, based on his *hypotheses of abstraction and adaptability.* His "primary theorem of cosmic ecology" is that for a fixed-complexity life form adapting itself to various environments, *the rate of metabolism of energy varies with the square of the temperature.*

Material Reality As Interacting Quantum Fields: The point is that quantum theory, organizational hierarchies of energy, matter, thermodynamics and entropy, information and order, all these are interrelated aspects of the same thing. Heinz Pagels, in *Perfect Symmetry*, with his unique facility for constructing classic descriptors, says that all material reality thus consists of "interacting quantum fields," the "transformation and organization of field quanta....the ultimate triumph of the field concept in the human attempt to comprehend reality."

He is implying, at least, that all evolved forms of order, organizational developments, including life, are part of the universe of interacting quantum fields. It would appear that the entire universe with all its phenomena, galactic structures, fields, forces, quantum states, are needed for the mental construct of even a single human being.

Life and Nonlife Based on Encoded Information: In *Life Beyond Earth*, biochemist Robert Shapiro and physicist Gerald Feinberg suggest that the distinctions between life and nonlife on earth blur into the earth's single identity as a "biosphere." In describing what is life and nonlife, their emphasis is not on chemical content such as nucleic acids but on the degree of order, the encoded information in a system.

Matter to Life Transition Leads to Discussion of Subjectivity: We have been discussing Nonlife and Life, that is, Matter and Life, Body and Mind, Life Force, using Layzer's categorization of the alternative views of the nature of Life. The central theme running through the discussion, however, is expressed in *The Cosmic Code,* where physicist Heinz Pagels declares that *Life is written into the Cosmic Code*--part of the grand program.

He feels that science has something to contribute by bringing renewed appreciation of the human condition and the limitations of its existence in the universe. The human being, he feels, embodies the principle of life--a principle that cannot be destroyed because, he says, *it is written into the cosmic code.*

But perhaps the subject of life and nonlife and of the life force is better delineated through discussion of the evolution of the brain, nature of 'mind,' the advent of something mystical called consciousness, subjectivity, and human capacity for reasoning, contemplation, and abstract thought. And so we continue with the nature of life versus nonlife via discussion of Consciousness.

> Whence? whither? why? how?--these questions cover all philosophy.--Joubert

CHAPTER 38
CONSCIOUSNESS

The Mystery of Consciousness: Is consciousness simply a manifestation of the physical brain or does it derive from some other source or have some other explanation? Is there a 'mental substance' as asserted by Descartes and Kant? Are individual consciousnesses interconnected, part of a single cosmic mental fabric at a quantum level? Must consciousness have a biological framework, or can non-biological material be used to create or sustain it? Is consciousness peculiar to human life or do other life forms possess it?

Can artificial intelligence be developed to achieve "thinking" and consciousness? Can a robot intelligence be created to contemplate the universe and deal with abstract thought? Can a silicon brain simulate the subjectivity built into the human biological brain or is there some impediment to non-biological devices crossing over into consciousness or self-awareness?

Intelligence-Consciousness in Non-Biological Form: In *Infinite In All Directions*, physicist Freeman Dyson argues that life is a holistic phenomenon that *resides in organization of matter*; that consciousness is inherent in the way the molecules are organized, not in the biological configuration of the molecules themselves. It therefore makes sense, he suggests, to imagine *life detached from flesh and blood and embodied in networks of superconducting circuitry.*

He points to the work currently going on to create computer advances in artificial intelligence, 'artificial life,' by simulating the organization of biological matter using non-biological materials. Contemplated, is a *non-biological evolutionary process* that mimics the phenomena that drive biological changes. Such an evolution is thought by many to be currently in process.

In a similar vein, the idea has been advanced that extraterrestrial visitations are more likely to be by highly-intelligent robot-creatures, sent perhaps by some form of biological intelligence forms from elsewhere in the universe.

Where Does Consciousness Come from? In our earlier discussion we ask where consciousness, awareness and subjectivity comes from; whether there is a mind or 'mental' substance, as Descartes' dualism asserted, apart form the physical substance of the body. Several views are considered, whether the mind, consciousness, is simply characteristic of and identical with the physical brain, whether it is implicit in matter, or whether it is something separate--an emergent property, and whether, as an emergent property it is something *nonphysical*.

Earlier, in our discussion of life and nonlife, we discussed one approach, Cartesian dualism, mind and body as two separate substances. We also discussed another view, life as a vital substance, life force (vitalism) injected at birth. But there is a growing consensus that life is a holistic phenomenon, the result of the growth of order and information, an 'emergent' quality.

Consciousness as Emergent Order: In considering the source of consciousness we noted earlier that many scientists, like physicists Freeman Dyson in *Infinite in All Directions*, and David Layzer, in *Cosmogenesis*, suggest life as the product of increasing order--the amount of information encoded in a system.

Something *qualitatively new* appears, Layzer says, at each level--molecules, organelles, cells, tissues, organs, organisms, and communities. It is, he says, a "new form of order," just as much a part of the external world as energy and matter.

Something Mysterious Emerges: But the nature of *consciousness* and subjectivity is more abstract and uncertain than that of bodily life. We know all kinds of life that do not possess the faculty of self-awareness and consciousness. In his *Cosmogenesis*, Harvard astrophysicist David Layzer describes this as a *metaphysical emergent property* of living organisms, a part of the cosmic evolutionary process in which elements *that did not previously exist,* emerge in a *hierarchy* which includes all previous elements. This is an expression of the so-called holistic view.

Layzer goes on to suggest that emergence of consciousness could not have been predicted; that it is *radically contingent*; that, *scientifically*, it has "no rhyme or reason;" and consequently that it is emergent in the *metaphysical* but not scientific sense. The eruption of consciousness is viewed as a holistic phenomenon, the emergence of a higher level in the hierarchy of development.

He cites Samuel Alexander's five hierarchal levels: spacetime, matter, life, mind, and deity. Physicist Gerald Feinberg and biologist Robert Shapiro also suggest life and consciousness as primarily the amount of organization, order, or information that is present in a system.

Molecular Complexities As Inherent in Components: But Layzer then goes on to describe--*and disagree with*--the 'counter theory,' that is, *reductionism* (sometimes called radical reductionism or scientific reductionism)--that all the characteristics of a complex structure are implicit, that is, built-in as components. His view is that while the properties of molecules *depend on* those of the component atoms, they are not '*implicit*' in them, a point worth further consideration.

Consciousness As Derivative, Implicit in Matter: But, as we discuss later, other physicists suggest the reductionist point of view that life, consciousness and mentalness are indeed, physical--implicit in matter, in particles. Physicists Freeman Dyson and David Bohm, view particles themselves as possessing an elementary form of intelligence and consciousness, suggestive of mind as being implicit in matter, part of the hierarchy of intelligence represented by the universe.

Consciousness and the Inchoate Potential of Organized Matter: Of course, it can be argued that the very term 'emergent' means emerging from an earlier state, that is, imbued with a derivative quality, implying at least that consciousness has some form of earlier inchoate presence. Perhaps this is another example of pseudo differences traceable to linguistics. Physicist David Bohm and others look on consciousness as nonmaterial, pure information, a form of the universal subquantum software, part of the underlying cosmic mental fabric of the quantum universe.

Is Consciousness Implicit in Matter? And so we are led naturally to discuss whether mind is *implicit* in matter and quantum energy phenomena such as particles and fields or whether it is rather an emergent property, not one previously implicit in its components.

While some, like Gregory Bateson, David Layzer and Paul Davies, see consciousness as an emergent property, that is, not implicit in matter, others, like physicists David Bohm and Freeman Dyson, view particles themselves as possessing an elementary form of intelligence and consciousness, suggesting that mind is implicit, built into, matter. These differences may be linguistic refinements in one sense, because if life and consciousness are 'emergent' qualities this implies some form of derivation from an earlier state.

In *Disturbing the Universe*, Princeton physicist Freeman Dyson takes the so-called reductionist point of view that mind or consciousness is indeed implicit in matter--in every electron.

Consciousness As Holistic Rather Than an Inserted Substance: Paul Davies, in *God and the New Physics*, says that mind and body are not dual components (Descartes' dualism of physical body substance and mental mind substance) but "different concepts drawn from different levels in a hierarchy of description," and without any mysterious extra substances (a reference to an inserted "life force" of vitalism, or the mental or soul substance of dualism).

You are not just a "moving mound of atoms," he says, but a self that "transcends any particular living body." And so for Davies, the mind, consciousness, invokes holism. It is part of the two-level description--a philosophy that is "gaining ground with the emergence of...the cognitive sciences." In this new philosophy, sometimes referred to as *functionalism*, the essential ingredient of mind is not the hardware, the brain, but the software or program.

Alfred North Whitehead--Molecular Behavior in Living Bodies: A somewhat different approach was taken by Alfred North Whitehead. In his lectures at the University of Edinburgh (published as *Process and Reality*, and reprinted in part in *The World of Physics,* Vol III, an anthology by Warren Weaver), Whitehead, the distinguished mathematician and philosopher, coauthor with Bertrand Russell of the monumental *Principia Mathematica,* discusses philosophy of the organic theory of nature in his article on "Science and Philosophy."

After a sweeping review of early and modern philosophy, Whitehead reviews the Meditations of Descartes and the contrasting work of William James in *Does Consciousness Exist*, involving a discussion of the *subjective and objective nature of cognition.*

In this process, he asks *whether molecules exhibit, in living bodies, properties not found in 'inorganic surroundings.'* He anticipates more recent research in neurobiology and neurophysiology, referring to the self-preservation actions of living bodies and body responses to "determinations of will," as suggesting the "modification of molecules in the body as a result of the total pattern."

Additional Laws of Physics for Molecules of Higher Organisms? Whitehead goes on, in his discussion of "philosophy of organism," and of cognitive experience and consciousness, to say that there may be "physical laws" governing the basic organisms when they form part of higher organisms, speaking of modification of the cell and altering the molecule "or some subtler entity." His reference to "some subtler entity," is interesting in that it is suggestive of a line of thought related to the nature of particles as being part of the *universal information system.*

Survivability of the Consciousness: There are different opinions on whether the individual human consciousness and self-awareness can survive the dissolution of its organic framework--death. Both reductionism and scientific materialism suggest that consciousness is dependant on its originating organic format. The concepts of holism, quantum theory, spiritualism and religious transcendentalism, on the other hand, allow for disembodiment of the consciousness, an independent existence and a reassimilation of individual consciousness with the cosmic consciousness.

Teilhard de Chardin and Consciousness: French Jesuit philosopher and paleontologist Pierre Teilhard de Chardin specifically posits individual consciousness not only as surviving death, but as retaining its uniqueness. Whether this is an intuitional insight, born somehow of faith or insight, or derived from scientific rationale, is not clear. It is probably the result of both these powerful influences in his life.

Physicist Paul Davies, as we've noted, speaks of the survivability of the con-sciousness, based on an assumption that it is not a physical part of our spacetime universe. Certain versions of quantum theory suggest that the consciousness is part of an underlying cosmic mental fabric, consisting of nonphysical information and software and so it survives physical death.

How Much of Our Self Relates to Our Bodies? But physicists who speak of survivability of consciousness also ask what it is that might survive death of the body because so much of what we are is physically related, part of or directly related to our body. We know that sickness alters our personality. Strong bodies generate sports-mindedness and promote ideals of Olympic competition, while body weakness or infirmity predisposes toward intellectual interests and pursuits. Pain and suffering and its effects are physical. Love, hate, disgust, and all the other emotions are physical and are altered by physical conditions and events.

Personalities Are Altered: Personalities are altered by endocrinological changes. The account is described of a person's deeply religious outlook as being totally altered when hormone imbalances were corrected. Drugs can produce or alter mental states, take you from depression to euphoria, and brain surgery can alter personality. Many of our thoughts, hopes, prayers, and plans for the future are related to our physical needs and condition. We also know that much of our personality, disposition, and sets of inclinations are inherited characteristics, differing from body to body and family to family.

So What's Left of Us When We Die? So Davies asks, what, then, is left of us when the physical part of us, around which our personalities are built, is gone? If all emotional qualities and personality characteristics are wiped out at death, what remains to describe us? Davies asks whether we would retain a sense of humor, of reverence, of boredom. It is debatable, he says. But what of memory? We are a product of experience which gives us our sense of identity.

And What of Memory? As we discuss here, scientists disagree on what and where is memory, some associating it with various locations in the brain, others suggesting that it is somewhere else, like consciousness and self-awareness. Davies feels that it is unclear whether someone with no memory can really be thought of "as a person at all."

Mind Enduring by Transfer of Information: Davies goes on to discuss a number of principles and concepts relating to individuality and says that some of these ideas hold out the hope that we can make "scientific sense of immortality," for they emphasize that the essential ingredient of mind is *information*. He cites the analogy that Beethoven's Fifth Symphony does not cease to exist when an orchestra finishes playing it. So, he concludes, "the mind may endure by transfer of the information elsewhere."

While research leaves little room for the traditional idea of the soul, he observes, "they leave open the possibility of survival of the personality." It is interesting to note here the oft-repeated religious proposition that our physical life is given us to refine and hone our birth spirit by the wealth of experience that our physical lives produce.

Pre-existence and Post-existence: The question of where our consciousness comes from sometimes becomes the question of where did our 'existence' come from and is that existence eternal? If we had some kind of 'pre-existence' in some inchoate form, as Plato asserts, then this will influence our view of its continued eternal character, the post-death continuation.

According to St. Augustine's view all time was laid out at the creation, a concept also reflected, as we have noted, in Einstein's relativity theory. In a sense, at least, we exist from the beginning and throughout eternity. This seems implicit in the dimensionality of spacetime geometry, differences in past, present and future, as Einstein asserted, being illusion.

In his discussion on whether the soul and consciousness transcends time, Paul Davies has difficulties with "timeless souls." He cites the Catholic Encyclopedia: "The notion that God has a supply of souls that are not any body's in particular until he infuses them into human embryos is entirely unwarranted by the evidence....The soul is created by God at the time it is infused into matter." He notes that this same 'temporal dilemma' is common to all immortality dis-cussions. He notes that Jesus spoke of "life everlasting," implying unending time but he refers to the difficulties of reconciling this with God's laws of physics, suggesting that "there may be no 'everlasting' anyway."

Part of a Unified Cosmic Consciousness? Do we originate as part of a primordial and infinite, unified consciousness? This is sometimes expressed as whether we originate as an undifferentiated part of the universal mind, perhaps part of the unity of God. We may be viewed, in a sense, as being as old (and young) as the universe, composed of 'rest-energy' forms fused in the stars from energy derived from the radiant energy of creation.

Despite our cultural and personal diversity, many think that quantum theory implies that in a sense we all share a common mental source, the unity of the primordial cosmic mind. As the Eastern religions highlight, even in our diversity and conflict, our enmity and wars, our cultural tensions, we are One.

Consciousness, Space, and Time: We have explored, in several chapters, certain aspects of physics. We have also dealt with our physical existence in our spacetime universe of four dimensions, and the possible existence of other dimensions, and even of adjoining but unreachable and unverifiable universes. Having noted that mind and consciousness are nonmaterial or metaphysical things that exist *somewhere,* Davies goes on to discuss the question of *where* the soul (consciousness) might be located, and whether it exists in time (our spacetime dimensionality).

He discusses the idea of spacetime being divisible into abstract quantum subunits, of the universe extending beyond our presently conceived spacetime, leaving an "elsewhere" of quantum spacetime bits, and of mind as possibly imbedded in a higher-dimensional space. So, the *soul, consciousness, may not be in the space of our universe at all.*

But what about time? He asks whether consciousness is in our time dimensionality. Elsewhere we discuss whether God exists 'in time,' and the problems of how he may be regarded as performing temporal functions like planning, if He does not exist in time. Well, Davies feels that both human mental processes that are time-dependent, and our very perception of time, require our consciousness to exist in time.

Relationship of Information to Meaning: In *Three Scientists and Their Gods*, science writer Robert Wright, in constant search for "evidence that life is not devoid of meaning," profiles several scientists and notes their views about relationships of information to meaning in a universe which consists ultimately of something nonphysical, information.

The Knowability of Everything: Relevant here is the dialogue between Wright and economist Kenneth Boulding, whose scientific background is chemistry. They get to discuss the "knowability of everything." Boulding muses that with our particular "biological equipment", it would be presumptuous to assume we can know everything. [Wright notes that Boulding is a man of some intellectual humility.]

On the other hand, Einstein's intuitional perception of the human mind viewed it as capable of comprehending the universe through the power of the mind to grasp mathematical concepts and other fundamental abstractions that account for our universe and its natural laws.

Religious Experience as the Adoration of the Unknowable: Well, Boulding goes on to refer to the "sense of mystery in the unknowable," noting then, the importance of *religious experience*, in which worship is "almost the adoration of the unknowable." He takes the occasion to acknowledge his "community" with persons in the "mystical tradition," like George Fox, founder of the Society of Friends, and especially the Jesuit paleontologist Pierre Teilhard de Chardin, citing his *The Phenomenon of Man*.

Wright, Boulding and the Weirdness of Consciousness: Well, they finally get onto the subject of consciousness and Wright goes right for the big one--the *weirdness* of consciousness. You recall physicist David Layzer's characterization of consciousness as "radically contingent" and like life, also an "emergent" property, but emergent in a *metaphysical* sense, something that could *not have been predicted*. So Wright, Layzer, and as it turns out, Boulding, agree on the essentially weird or mystical character of consciousness.

Consciousness As Giving Meaning to Life: Wright notes that this mysterious thing, consciousness, was not a necessary ingredient in the evolutionary development of human life; all the biological functions could go on without it--you could *be* alive without the intense subjectivity and the mysterious *feeling of being alive*.

Perhaps other intelligent creatures in other planetary systems could have evolved without this subjectivity ever emerging. *This mysterious quality of consciousness is very significant to Wright, a man in search of evidence of meaning in the universe.*

It is important, says Wright, because this mysterious component of our being, consciousness, sentience, knowing *what it is like to be human,* to *be alive*, is “precisely what gives life at least a modicum of meaning.” In a world of strictly scientific materialism there is no need for this mysterious subjectivity. Evolution explains about everything--but it doesn't begin to explain consciousness, subjectivity or feeling human. It explains *being alive,* he says, *but not feeling alive*.

Introducing Consciousness into a Computer? Wright goes on like this, as he says, seeking to apply his rhetorical skills to get Boulding further into the subject. Well, Boulding picks up on it and gets involved. He notes that we really don't yet understand our access to the mind, that is, we don't have a “model of consciousness” and, he feels, we don’t have the slightest idea of how to *introduce consciousness into a computer*. David Layzer too, doesn't quite see achieving computer consciousness, but then, he notes, look what we've been able to do with energy.

Actually, as we discuss later on artificial intelligence, there are those, physicist Freeman Dyson and others, who do hypothesize our ability to introduce subjectivity into artificial intelligence. But Boulding’s proposition is that we really don't yet have a model of consciousness, just some theories revolving around organization and complexity.

Evolution Tilts toward Consciousness--Notions of Divinity: But Boulding goes on, noting that matter and energy are "coders and transmitters" of information and we could provide the information needed "to get my genes transmitted" without referring to consciousness--"another order of reality."

They then get into DNA, noting that each person's DNA is "a record of the last couple of billion years of the earth's history." Boulding is amazed that the universe's "awareness of itself" seemingly evolves from its randomness; that evolution "has a prejudice toward complexity and consciousness...[with a note of humor] *almost enough to make a case for creationism.*"

Boulding Picks up on Genesis and Evolution: Wright gets worried at this near "reconciliation of science and religion" by Boulding and gently leads Boulding toward "a less radical destructive" accommodation of science with "some notion of divinity." This is Boulding's signal to pick up on the Genesis and evolution battle noting that, in his view at least, the evolution process is a more impressive creation technique than direct creation from zero.

An Intelligence that Can Produce the Powerful Natural Selection Process: His own statement of the point as taken by Wright, is far more expressive. He is far more impressed, he observes, with an intelligence that would create "a process...beyond our comprehension... as subtly powerful as natural selection--a process that, given a few self-replicating molecules and a few billion years, leads to a theory of itself," than he is with creating the human species "out of thin air."

But this point, Wright observes, has also been made by many others, long ago, for instance, in a letter to Charles Darwin by Anglican clergyman--a naturalist--Charles Kingsley after he had read a courtesy copy of Darwin's *Origin of the Species*.

Wright Relates Consciousness to Determinism and Free Will: This mystery of consciousness, knowing what it is like to be alive and to be human, reminds Wright of the question of determinism. He points to men like computer scientist Fredkin and Einstein as believers in determinism. This is the view that all events, including human behavior are the theoretically-predictable and inevitable consequences of the state of the universe and the operation of its physical laws.

Subjective thought would just be some physical neural transactions. Which leaves free will, the product of supposedly subjective human thought, in the realm of myth. You recall the quandary St. Augustine got himself into trying to explain free will in the face of his description of creation as the predestiny of everything that was to happen for all eternity.

Wright and Boulding Get into the God Role: Here Wright gets into the God role and says that he is more comfortable with accepting that God is immanent in everything rather than functioning as a bearded, robed, score-keeper. But people, he muses, seem to want to play for high stakes--an evaluator God, along with punishments and rewards, heaven and hell, and to "equate faith and certainty." He puts this to Boulding: whether there's "anything, you know, *out there*."

The 'God Question' Turns into the 'Existence of Potential': Boulding reflects an element of uncertainty and responds that what he *is* certain of is *the existence of potential*, that this "enormous potential" out there is "the real core of religious experience." Boulding says we can *ally ourselves with this potential*, "go with the flow," as Wright paraphrases it. Wright sees this as "evolution's directionality," the growth of complexity and consciousness, and we "go with it" by "gentling" and subordination of selfishness--becoming part of the whole.

Bolding, devoted to human betterment, notes that all the great Western and Eastern religions emphasize control of the ego and extending fraternal bonds--in the direction of *unity*, sensing the totality. He notes the world as on the brink of apocalypse. He sees the world as an organism, sees some sort of international ecosystem, the reconciling of stability with autonomy, turning “enemies into opponents.” But he worries some about *purpose*, teleology, as implying determinism, preferring his idea of "potentiality" which better implies the existence of freedom.

Our Physical World As a Subjective Holographic Projection: In the ‘new physics’ of physicists like David Bohm, our so-called objective, physical reality may be thought of as a multidimensional holographic projection into the spacetime of our four-dimensional universe from a subquantum underlying order which interconnects and joins everything in a single multidimensional cosmic consciousness not limited by our macro universe's dimensions of spacetime. Einstein may have sensed this timeless consciousness when he spoke of distinctions between the past, present and future as being an illusion

Consciousness as Tuned into Pribam's Frequency Domain: Stanford University neurophysiologist Karl Pribam, who deals with consciousness in his *Consciousness And The Brain*, thinks of this holographically projected reality as a derived from a "frequency domain" to which each human brain is tuned and, like a TV, the brain converts these consciousness frequencies to construct physical images that "resonate" with what may be out there.

Long before the ‘new physics’ or even whispers of quantum theory, the great German philosopher Immanuel Kant came remarkably close to Priban's idea when he suggested that human beings can never know what’s “out there,” the "thing in itself," they can know only what the human mental construct delivers to comprehension, acknowledging that this construct may give human beings some approximation to physical reality.

The Dynamics of Mind--What Criteria Do We Use to Define a Mind? When does a system become a mental state describable as mind and having some form of consciousness? In his *Mind And Matter*, Gregory Bateson thought of mind as pure information, neither a substance nor an energy form, but flowing within and between minds and between minds and the environment organized in patterns of regularity and interpenetrating to form all aspects of our human behavior. *The system must be able to receive, process and utilize information both for its external responses and for its self-maintenance*, as does the human mind.

Extension of Cybernetics to Natural and Biological Phenomena: Although these features were earlier recognized by MIT mathematician Norbert Weiner's cybernetics, the study of self-regulating systems, Bateson was able to extrapolate their application not only to natural phenomena like oceans, atmosphere and rain forests, but to various social organizations, cultures, and systems of belief used by societies for self-maintenance.

Also interesting is the significant contribution of the study of self-regulating systems to the understanding of the complex biological dynamics of self-organizing, detrimental, mental states such as addiction, a self-organizing system employing addict rationalizations as a part of its reinforcement and self-aggrandizement process. The point here is that the demarcation between life and nonlife, as suggested by scientists like Lazer, Dyson, Margulis, Shapiro, Feinberg, Lovelock, and others, is based on *degree of order, the encoded information in a system.*

How Does Increasing Order, Life, Occur in the Entropic Universe: But in a universe governed by the second law of thermodynamics, in which systems, like the universe itself, tend toward entropy, states of *decreasing order* or increasing disorder and chaos, how do we explain the phenomena of pockets of increasing complexity, *increasing order*, increasing concentrations of information, like the emergence of life and then consciousness?

Cases of increasing order, such as evolution of life, were considered an anomaly or deviation from the entropy law and were the subject of great speculation. In his *What is Life* papers, Erwin Schrödinger, father of the quantum uncertainty principle, suggested that there must be some more refined explanation for the drive toward life in the universe than conventional thermodynamics.

One Set of Laws--Two Forms of Behavior and Spontaneous Self-Organization: In *God and the New Physics*, Paul Davies discusses how a single set of laws produce two different forms of behavior--"two pathways," one living--progressing toward states of increased order, and the other, the "lifeless," moving toward increased disorder under the second law of thermodynamics.

He notes that the mystery of life can be made less mysterious by the study of inanimate systems that achieve spontaneous self-organization, citing the example of the appearance of vortices in fluid flow, atmospheric circulation patterns, storms. He cites the work of Nobel-prize recipient Ilya Pregogine at the University of Brussels, who describes the mechanisms and mathematics of self-organization which turn out to be the same in biological systems as in inorganic chemical reactions.

Existence of a Bifurcation Point--Dissipative Structures: The common link between organic and inorganic systems in Pregogines work is that when both systems exceed some critical threshold value, a *bifurcation point*, in which they are "driven far from thermodynamic equilibrium," their instability leads to *dissipative structures*, spontaneous organization to greater complexity, involving "long-range order" in which the system "acts as a whole."

Other Types of Ordered Systems: As Carl Sagan and Freeman Dyson observe, in this view also, "life" elsewhere in the universe may consist of all sorts of ordered systems such as organized forms of radiant energy, plasmas, or patterns of the fundamental forces and may be based on ammonia or other molecular aggregates occurring naturally throughout the universe.

In the *Gaia hypothesis*, for example, the earth is a living, organized, and self-regulating system. Astronomers and cosmologists report on the great gravitation-dictated selforganizing structures represented by galactic organization into galactic clusters, filaments and voids.

Prigognine's Universe of Self-organizing Systems: In *Order Out Of Chaos*, Ilya Prigognine and Isabelle Stengers, recount how Prigogine, a student of self-organizing phenomena, observed that while increasing disorder characterized closed systems, open systems exchange information with their environment and "fluctuate." When the fluctuation reaches a critical threshold, a "bifurcation point," it either deteriorates into a lower order, or surges into a higher level of order or organization, a "dissipative structure," into which is absorbed the energy and information arising from the fluctuation.

Biological Life Structured into the Laws of Matter: The tendency of matter toward self-organization represented a new property of matter which implied a revolutionary departure from the earlier concept of self-organizing systems as an anomaly. Such open systems were now seen as a primary characteristic of the universe. Self-organizing systems, known to exist in living systems, with a key role in life evolution, were now known to exist in pre-life nonliving systems.

Biological life, rather than being a random exception to the rule of increasing disorder, was structured into the very laws of matter. Matter was no longer viewed as totally inanimate, but part of the *life process continuum*. In this context, matter is an earlier stage of life, and in that sense, matter is alive, that is, aliveness is a matter of degree. For this work, Prigognine received the Nobel Prize.

The Poltergeist as a Self-Organizing System: The subject of self-organizing systems surfaces in the context of parapsychological phenomena. There are numerous accounts in the literature of poltergeist ghost-like hauntings, phenomena normally associated with children.

In Michael Talbot's *Beyond The Quantum*, he describes a poltergeist haunting of his childhood home, evidenced first by gravel showering on his home and then, inside his house it stomped and rattled and pelted him with small objects, made crashing sounds, moved objects and threw the vacuum cleaner around, engaged in acts of levitation and teleporting, that is, moving objects some distance.

Most peculiar was the manner in which its mischief was more or less benevolent in parallel with his own positive or negative mental and emotional states. Talbot felt that the poltergeist phenomenon was somehow a psychic manifestation of its "human focus," understandable in terms of the self-organizing properties of psychological phenomena.

Physicist Fred Allen Wolf, he said, noted that quantum theory contemplates the existence of certain life-like phenomena in nature. We have noted how systems self-organize and take on characteristics of intelligent self-regulating behavior, like addictions, like a rain forest, a tornado, the whole planet earth--the Gaia hypothesis. So there are degrees and types of aliveness, all self-organizing systems.

Creating New Awarenesses--Holographically Projected into Physical Reality : And so Talbot announces his "boldest heresy," that the poltergeist was a self-organizing system entity, a state of order or a piece of software, alive and existing in the realm of information, holographically projected into his physical reality. He goes on to cite the case, documented by the Canadian Broadcasting Corporation, in which a Canadian group periodically assembled, came up with a fictional character, "Philip," giving it character detail of likes, dislikes, and attempting to make contact with their own created ghost entity.

Contact was finally established and, through repeated encounters, the entity developed a distinct personality, evolved with moods, and a detailed past. The entity moved objects around and engaged in give and take discussions (communicating via knocking and scratching noises).

Self-organized out of Collective Consciousnesses: Who was "Philip"? Well, says Talbot, he was an amalgam, a collection of pieces of the minds of those present at the sessions. His personality changed with changes in the group. *He had self-organized out of the collective consciousnesses.* What did he consist of? Pure information, says Talbot. *He was a part of the universe that had crossed from nonexistence into existence* achieving the status of awareness, an awareness which characterizes the planet and the universe.

The Power of Human life to Bring Awarenesses into Existence: Thus, it is thought, human beings can trigger into existence self-organizing systems which become alive and aware. The implications for artificial intelligence are enormous. In the subquantum reality of unified mind, consciousness and information, Philip existed only as a potential, to be constructed from a selected collective of consciousnesses.

When does a system become a mental state describable as mind and having some form of consciousness? The system must be able to receive, process and utilize information both for its external responses and for its self-maintenance, as does the human mind. --Gregory Bateson

Perhaps only the present instant 'exists' for me. Perhaps the 'me' of tomorrow, or of yesterday, is really *a quite different person with an independent consciousness.* Perhaps I am actually living *backwards in time*, my stream of consciousness flowing into the past, my memory really telling me what *will* happen to me rather than what *has* happened. Does the distinction between *that* picture and the normally-experienced time-progression actually 'mean' something, that one is 'wrong,' the other 'right'? To be able to resolve such questions, a theory of consciousness... is needed. --Roger Penrose

CHAPTER 39
ARTIFICIAL INTELLIGENCE

What Is Artificial Intelligence? Living organisms demonstrate various levels of intelligent behavior. Despite erratic and sometimes destructive behavior, human beings are credited with the most advanced level of intelligence presently known to us. Human beings have produced increasingly more sophisticated mechanical devices to perform functions previously associated with human effort, like calculators, robots, and computers.

For some time now, artificial intelligence (AI) computer systems have performed speech recognition and language processing, face recognition, medical diagnostics, and other 'intelligent' functions. Such systems perform the functions of receiving information, configuring a response, and producing some intelligent result or product such as a geometric proof, playing a chess game like IBM's *Big Blue*, responding to some test, or evaluating sickness symptoms to identify a medical cause.

This mechanical functioning mimics human function performance and is said to reflect an artificial equivalent to aspects of human intelligence. As such it generates questions about human intelligence, whether the mind or brain is essentially a biological machine, whether artificial intelligence (AI) can reach a level of sophistication which activates that which we call consciousness, and a bit further, some form of self-awareness. Would this nonbiological intelligence be an artificial 'mind?'

Can We Introduce Consciousness into Artificial Intelligence? In *Cosmogenesis*, physicist David Layzer addresses the matter of simulating biological intelligence. He asks whether it is possible for an artificial-nonbiological mechanism (such as a silicon brain), *functionally identical* with a biological brain, equipped with non-biological sensory equivalents, to possess consciousness. Consistent with his view of consciousness as a metaphysical reality, he thinks consciousness is not duplicatable in nonbiological form. But then, in a moment of self doubt, he cites energy and electric charge as realities we still can't explain either.

Interestingly, we have David Layzer and Freeman Dyson, both physicists with abiding interests in cosmology and biology, coming to somewhat different conclusions, Layzer, that consciousness is not likely in nonbiological form and Dyson, that it is possible--and likely. Freeman Dyson's view of mind spreading across the universe speaks of "flesh and blood" as not being required as the vehicle for mind--or consciousness.

Rapid Progress in Developing Silicon Intelligence: In *Disappearing Through the Skylight*, O.B. Hardison Jr. discusses the evolution of machine intelligence into the silicon brain and the surprising level of intelligent response programmers have already been able to construct. It is startling to realize that this *non-biological silicon intelligence has made more progress in the last few decades than biological intelligence in the last two billion years*. His discussion of the significance of human life's self-created reality and the "transformation of the human spirit," is reminiscent of Dyson Freeman's view of mind spreading throughout the universe.

Meaning of the Turing Test: In his Washington Post March 5, 1990 article on *Thinking Like Humans*, Staff Writer Malcolm Gladwell describes the test devised by British mathematician Alan Turing to compare human with machine responses to inquiries and the view of University of California at Berkeley philosopher John Searle who argues that such programmed mechanical response is *symbol manipulation* and not the equivalent of human thought. But is human thought itself a form of biologically-induced symbol manipulation?

Human & Mechanical Information Processes--A Theological Divide? Gladwell cites the rote learning by children which permits correct responses without really understanding the questions or the answers (reminding us of our own school days). He also cites Indiana University computer scientist Douglas Hofstadter as suggesting that the brain is, after all, governed by the laws of physics; that Searle's differentiations between human and mechanical information processes represents a conceptual gulf, "almost a religious divide."

Evolution of Nonorganic Intelligence Forms: A little later here we discuss physicist Freeman Dyson's thoughts on genetic engineering of the human mind to permit its spread to other places in the universe. Included in that discussion is the likely nonbiological evolutionary process that would mimic the phenomena that drive biological changes. The process involves the competition between carbon-based life forms and silicon-based or other artificial life forms more readily adaptable to space travel than genetically engineered changes in human life as discussed by Freeman Dyson.

The advent of highly-intelligent-silicon robot-creatures, without all the esoterics of human values, is probably as inevitable as it is scary to many. One nightmare links these ceramic-plastic artificial 'people' to programs for perfecting genetic engineering in microorganisms, handling nuclear fusion assignments, and perhaps even filling the shortages in judges in our court systems!

Can Artificial Life Provide Consciousness and Subjectivity? We recall astrophysicist Layzer's view, described earlier, that consciousness and subjectivity are unlikely characteristics of artificial intelligence, and physicist Freeman Dyson's view, that all mental states are likely to be simply products of organizational order rather than dependent on "flesh and blood." This, of course, raises the question of what it means to be *alive.*

The Distinction between Natural and Artificial Life to Disappear? It also raises the question of whether differences between slow-evolution-driven biological life and the rapid human life-driven evolution of 'artificial life' are essentially the same. Will the distinction between natural and artificial life disappear? Many fast-moving developments suggest that both evolutions may simply be part of a single phenomenon. It has been suggested that *we may be on the threshold of the greatest evolutionary event in the history of the universe.*

Co-Evolution of Carbon-Based Species and Silicon-Based 'Life': The next century is seen as witnessing an evolution of silicon-based artificial life, side-by-side with further evolution of carbon-based life forms. Taking a page from Freeman Dyson's book, some artificial life enthusiasts suggest that 'genetic takeover' is under way, that we may be the first species "to create its own successors," that human life is a mere evolutionary stepping stone on the road to a more intelligent silicon-based species--the ultimate put-down of the human species.

J. Doyne Farmer of the Los Alamos Complex Systems Group suggests that artificial life may be human life's ultimate accomplishment, humanity's "most beautiful creation." He muses that "to shun artificial life without deeper consideration reflects a shallow anthropocentrism." On the other hand, there are those who see artificial life as a malevolence threatening to humanity, reminding us of the 'takeovers' described in science fiction.

Dualism, Life Force, and Artificial Intelligence: Earlier we discussed questions about dualism--separateness of mind and body, and the possible 'life-force' 'inserted' at some stage in the organization of biological materials. Such questions can be hypothesized as arising also in the human-controlled evolutionary development of nonbiological artificial life forms.

Consciousness as *Structure or Software*--Functionalism: In our review of the cognitive sciences, we have discussed theories of monism, dualism, materialism, holistic and reductionist and computer-information system approaches to mind and intelligence.

Unhappy with trying to conceptualize consciousness as a simple emergent of the evolving brain--materialism, or as a separate mental substance--dualism, another approach is proposed, called *functionalism*, in which consciousness is *structure*, the *software*. In this view, consciousness, representing pure information--*software*, would be an emergent, nonphysical holistic phenomenon--more than the sum of its component parts.

In *The Mind-Body Problem*, Scientific American, Jan. 1981, MIT psychologist Jerry Foder discusses the cognitive sciences, concluding that information processing systems relating to computational theory, cybernetics, artificial intelligence, and the human consciousness are all capable of being or possessing mental states. Included, are disembodied spirits. A more elaborate treatment of this subject is contained in Michael Talbot's discussion of "What and Where Is Consciousness?" in *Beyond The Quantum.*

Workshops on Artificial Life: The Los Alamos National Laboratory Center For Nonlinear Studies and the Santa Fe Institute recently sponsored the AL II workshop on Artificial Life, interestingly reported by Curt Suplee in the May 6, 1990 Washington Post.

Abstracting Life from Its Present Material Basis: Suplee notes biochemist Stuart Kauffman's observation that if being alive is self-organization rather than organic reaction, then we could "abstract the logical form of life from its material basis" to apply it to artificial life, abandoning organic, carbon-based systems, along the lines suggested in physicist Dyson Freeman's *Infinite in all Directions*.

Artificial Life partisans, says Suplee, suggest there may be *degrees of life* or, in the words of Norman Packard of the Santa Fe Institute, "discard matter altogether and have an organizational [dynamic] form entirely independent of matter."

Computer Created 'Genes' that Self-Improve: Suplee discusses other computer evolution modeling systems such as Genesys of UCLA's David Jefferson, creating strands of computer code with a set of *genes*. These 'genes' are bits of binary digits, that determine its propensities and go through an evolutionary selection of the fittest process, 'mate,' pass on their characteristics, and raise their success scores in running a hazards and rewards maze. MIT's Artificial Intelligence Lab, says Rodney Brooks of MIT, is "one prototype away" from a robot to live and perform intelligent functions on the moon.

Learning Algorithms and Systems That Modify Their Own Genomes: Computer programs will be made to be more like animals, adaptable with learning algorithms that permit programs to 'profit from experience.' Would these be alive? We have to cease distinguishing between what's made directly by nature and what's made by nature through human life, notes Langson of Los Alamos.

Los Alamos physicist J. Doyne Farmer notes that Artificial Intelligence could provide a "symbiotic Lamarckian evolution" *permitting our systems to modify their own genomes*, representing a "major change in the global rate of evolution comparable to the acceleration produced by the advent of culture....a landmark event in the history of the earth, and possibly the entire universe....*We may be the first species to create its own successors*."

Development of Neurochips--Silicon Neurons: Efforts to create artificial simulation of human brain activity has been constrained by reliance on computer chips which must deal with information in the digital mode, broken into encoded bits. But neurons in the human brain are data processors which operate in a continuously flowing and smoothly-varying analog mode.

In the Dec. 19, 1991 Nature magazine, Misha Mahowald, of Caltech Pasadena, and Rodney Douglas, University of Oxford, report their creation of a human brain-like silicon chip with transistor wiring which duplicates the electrical charge processes that go on in the neurons. The chip thus mimics the analog characteristics of neurons in the cerebral cortex.

The Drive toward Artificial Intelligence in High Gear: The new silicon, analog-mode neurons are expected not only to provide a new threshold for study of the way the brain functions, but also to trigger exciting new technical product applications, like facial recognition for security devices, voice recognition for vocal input to computers, and still other applications. But the point here is that the drive toward true artificial intelligence is in high gear, with all the implications that this has for possible crossing the threshold into a form of artificial consciousness and self-awareness.

The Nature of Mind --Can Machines Think? Science writer Robert Wright, in the March 25, 1996 issue of Time, refers to Garry Kasparov's chess match with IBM's Deep Blue computer. He cites Alan Turing's test of whether computers can think, described in Turing's 1950 essay in *Mind.*

Turing suggested that to the extent that a human interrogator of concealed responding human and computer entities is fooled by the computer's responses, the successful responding computer may be said to think. Wright's response to the question, whether machines can think, is that perhaps computers *can* think. And this staggering proposition is triggering new debate on the nature of mind.

Machine Versions of Mind? The better these machines become, Wright says, the greater likelihood they represent machine versions of mind and consciousness. The brainier computers get, he notes, the more feasible the notion of some "ethereal existence" of mind as a separate phenomenon--*and the more leeway for theological speculation.*

Wright cites Marvin Minsky of MIT to the effect that the biggest challenge is giving machines 'common sense.' While Big Blue's calculation manipulations are impressive, the question remains whether its response is 'thought' versus rote calculation. The computer is programmed to analyze the play and select programmed conclusions. But people look beyond data flow to inner feelings, involving the question of consciousness. Does the computer have some form of consciousness? Can enough information be programmed in to produce "subjective experience"?

Different Approaches of Chalmers and Dennett: Wright cites the different approaches between David Chalmers in *The Conscious Mind*, Oxford University Press--that consciousness is a deeper puzzle than realized, and Daniel Dennett of Tufts University, in *Consciousness Explained*--that consciousness is no longer a mystery, having been reduced to a set of manageable problems.

In *The Concept of Mind*, Gilbert Ryle, Dennett's mentor at Oxford, dismissed the idea of a separate human soul, everything being explainable in materialist terms, the mind *being* the brain, a view, Wright notes, that soon became 'mainstream'--for awhile.

But Wright observes that the idea that consciousness is readily explained in material terms, has faded and, observes Rutgers University philosopher Colin McGinn in *The Problem of Consciousness*, many now take the problem more seriously. For human beings to try to grasp how subjective experience arises from matter, McGinn says, is "like slugs trying to do Freudian psychoanalysis."

Why the 'Extraness' of Consciousness: In addition to physical data flow, Chalmers reminds, human beings have feelings of events, subjective experience, and asks why that is--why the "extraness" of consciousness. He goes on to ask why God (meaning whatever is responsible for the nature of the universe) inserted conscious experience of pleasure and pain, love and grief-- "a fairly central source of life's meaning."

As an aside, Wright goes on to suggest that consciousness may not be "extra." That it actually does something in the physical world, like influence behavior. The extraness problem is what Chalmers sees as one of the "hard" questions of consciousness. According to one model, our brain subconsciously generates competing theories about the world, but only the "winning" theory becomes part of consciousness, "wresting control" of our neurons and thus of our "perceptual field."

Convert the Consciousness Mystery to a Theological Question? Presuming the existence of some physical explanation for subjective experience, Chalmers doubts the ability of the human mind ever to grasp it. And McGinn, says Wright, thinks it tenable to convert the mystery of consciousness to a theological question, something more rational than strict materialism, because, he says, it respects the data, the "yawning and perhaps eternal gap in scientific understanding." Materialist dismissals of the mind-body problem may seem persuasive on paper, says McGinn, but you start to see the limits of a concept once it gets "realized."

Differences So Deep that Communication Fails? Dennett responds that for the extraness problem--the question of what function consciousness serves--if you are a strict materialist, believing the mind is the brain, then consciousness must have a function. After all, the brain has function and, consciousness is the brain.

Consciousness As a Nonphysical Property of the Universe: Consciousness is one of those questions so deep, he says, that people with different views not only fail to convince one another, they "fail even to communicate." Chalmers thinks that a solution to the consciousness puzzle is possible. But, he feels, it will require recognizing that consciousness is something beyond the physical and then building a theory which, though some might call metaphysical, he prefers calling *psychophysical*.

In *The Conscious Mind,* Chalmers sets out a psychophysical theory. Perhaps, he says, consciousness is a "nonphysical" property of the universe, vaguely comparable to physical properties like mass, space, or time. And maybe, by some other law of the universe, consciousness accompanies certain configurations of information such as brains. Maybe information is a special incarnation of matter, he says, with two sides, the physical and the experiencial.

Chalmers thinks it possible that AI research may some day generate, may even now be generating, new spheres of consciousness unsensed by the rest of us. Wright cites Alan

Turing's statement that in attempting to construct such machines we may not be irreverently usurping God's power of creating souls, any more than in the procreation of children. Rather we are, in either case, he says, instruments of his will.

Theological and Other Implications of Artificial Life: We have been discussing the growing sophistication of human life's efforts in creating artificial intelligence, even subjectivity and consciousness. Its only a small jump then to questions of the meaning of that kind of life. Does Albert Schweitzer's "reverence for life" take on a new realm of application to artificial life forms?

If subjectivity, consciousness, thinking and contemplative cognitive functions need not be "flesh and blood" as physicist Freeman Dyson suggests, and can be built into nonbiological life forms, will they experience compassion, love, be capable of and liable for their acts of evil, be subject to punishment, experience repentence and remorse? Will they acquire a 'life force,' spirituality, a soul? Will they become entitled to repair as human beings become entitled, in enlightened societies, to 'medical repair?' Will they be tried for crimes 'knowingly' committed? Will they be 'alive?

Will they have the right to reproduce, as do human beings, by building vast production lines? Will they be subject to the rights and obligations of citizenship, become a part of our political process, including the right to vote and run for office? Would you vote for a highly qualified artificially-produced "person?" Will they make war--and peace, sign and renege on treaties? In other words will they become nonbiological human-type beings?

Will we send these artificially produced 'people' to explore other planets? And might other civilizations send their artificial people to visit ours? And how will we treat them when they arrive?

In *The Emperor's New Mind*, Oxford University physicist Roger Penrose comments that *consciousness* is so important that he cannot believe, as the strong AI do, that it emerges just through a computation process; that consciousness is the only phenomenon that can conjure a putative 'theoretical' universe into actual existence!

> Artificial Intelligence could provide a symbiotic Lamarckian evolution *permitting our systems to modify their own genomes*, representing a major change in the global rate of evolution...a landmark event in the history of the earth, and possibly the entire universe....*We may be the first species to create its own successors*.--J. Doyne Farmer

CHAPTER 40
THE HUMAN BRAIN

Information Transfer--The Human Brain at Work: In *his Masks Of The Universe,* Edward Harrison insists on the "unknowability" of the outside Universe, as distinguished from those models of the universe, "cosmic belief systems," which each society constructs in each age. But *individuals* are not all confined to their societal universes; they have their own *comprehension systems* which form subjective "world pictures" of the world in which they live.

These pictures are the result of the *physiological and neurological processes* by which light rays, sounds and other signals are processed by the sense organs and transmitted to the brain via the neurological pathways utilized to govern the flow of information to the brain.

The Nervous System, the Brain and the Network of Neurons: So far as we now know, the most advanced and complicated structured matter in the universe is the human brain, far more advanced and complicated than any computer or mechanism created by human life or, so far as we know, by nature. The brain, still little understood in its subjectivity aspects, is viewed, says Edward Harrison, as "a hierarchy of horizontally and vertically coordinated structures."

Arrays of interconnected neurons in the cerebral cortex, characterize human behavior, providing short-term memory and thinking processes. Both here and "at lower levels of the brain," are the "soft-wired structures" programmed by previous personal experience, language, culture, and cosmic "inheritance."

The Brain Collects, Processes, Creates and Acts on Impressions: Harvard physicist David Layzer points out that *any* nervous system collects and processes information to produce muscular responses. But the *human nervous system*, the brain, is more, providing consciousness, subjectivity, and creativity.

It analyzes collected information, identifies problems, remembers the past, creates alternative hypotheses, solutions, and evaluates them, to initiate the selected solution. It comprehends, wonders, contemplates, triggers emotional states, constructs abstractions and invents. It *creates something new,* he notes.

A Network of Coordinated Structures of Interconnected Neurons: These functions are moderated by a network of specialized cells arrayed in a hierarchy of horizontally and vertically coordinated structures--interconnected neurons in the cerebral cortex which "generate and transmit electrical impulses." Carl Sagan says the brain has about ten billion of these switching elements, neurons.

The Switchboard: The *thalamus* is the central switchboard which receives and relays information to the cerebral cortex for analysis and decisions. It coordinates sense information, memory, movement speech and thought. It is the *prefrontal region of the cerebral cortex* that accounts for human as contrasted with primate behavior--such as creativity, planning and complex skills.

Various Brain Functions: The *cerebellum* coordinates muscle and skeletal movement. The brain's emotional center is the *Limbic system*; the *brain stem* triggers to action and regulates body functions like respiration and blood circulation. *Frontal lobes* control certain aspects of thinking such as planning. The *parietal lobes* are the sense and motor receptors registering touch. *Temporal lobes* deal with memory, time, and hearing and the *occipital lobes* analyze vision-derived information from the retina..

Interdependency and Redundancy: In *The Dragons Of Eden* Carl Sagan discusses the interdependency and redundancy aspects of the various parts of the brain--how redundancy between portions sometimes substitutes for brain-damaged portions, but in other situations, fails to substitute. Olfactory information is maintained between left and right hemispheres via the anterior commissure, for example, when the corpus callosum is cut.

After the first two years of life, the brain, unlike other body organs such as the liver, kidney, or pancreas, is apparently unable to self-repair by brain cell regeneration or by relocating functions between left and right hemispheres of the neocortex, although the journal Science, on May 4, 1990, reported a *major breakthrough* in brain study by the Johns Hopkins School of Medicine neuroscience department, opening up new horizons of experimental research.

Neuron Cell Breakthrough by Johns Hopkins Medical School: In the Johns Hopkins breakthrough, a cultured gram of hormone-treated neuron cells, derived from brain surgery on an 18-month old, survived, divided and grew in the laboratory. Molecular biology techniques permit the production by these cells of chemical substances like the dopamine hormone needed for Parkington's disease and acetylcholine needed for Alzheimer's disease.

These cells could be used in place of the controversial use of aborted fetal-cell-transplants now used because of fetal-cell adaptability and non-rejection. Insertion into brain cells of genes that produce these chemicals would permit implantation of such hormone-producing cells to supply hormone deficiencies.

Knowledge Requires Both Left and Right Brain Hemispheres: In the dim past, thinks Sagan, our early ancestors may have had comparable and redundant spheres which became specialized. He suggests the right hemisphere proposes patterns for analysis by the left hemisphere, so that "pursuit of knowledge requires both hemispheres." Thus Einstein's intuitive grasp of relativity by the right hemisphere required left hemisphere analytical evaluation.

Evaluations of Einstein's behavior have lead some experts on autistic behavior to believe that he may have had high-performance, so-called savant autistic brain characteristics which tend to augment right brain performance and diminish left brain capabilities although Einstein gave no indication of diminished verbal skills except delayed speech acquisition symptom as a child.

Left and Right Hemispheres of the Neocortex: The left hemisphere is the site of verbal skills, suppressed in the dream state, and the 'intuitive' and creative right hemisphere with nonverbal visual patterns, emotion, melody, is 'functioning well.' Sagan points out that the left works 'in series' and processes information 'sequentially' and the right works 'in parallel' and processes information 'simultaneously.' His analogy is the left working like a digital computer and the right like an analog computer.

Dream Problem-Solving: During the dream state, perhaps the left is busy doing 'maintenance,' data-dumping from short to long-term memory. He speaks of pattern-recognition types of *problem-solving during sleep* as characteristic right hemisphere creative activity, citing psychoanalyst Erich Fromm's reference to dream-state insights.

Some people have learned to "assign" conceptual problems to night-time dream activity for recapture on awakening, if the brain is so disciplined. But unless you awaken with the captured insight, you may lose it even though you recall experiencing the insight.

The Triune Brain: Carl Sagan's description of the brain, in *The Dragons Of Eden,* is geared more to its origins in the distant past. We look at each other, says Carl Sagan, through *three different mentalities*, the "triune brain....three interconnected biological computers," corresponding to distinct *evolutionary steps*, with notably "different distributions of the neurochemical dopamine and cholinesterase."

The Reptilian or R-Complex: The *first* and most ancient brain, he describes as the *reptilian or R-complex*, perhaps several hundred million years old, known as the *neural chassis* (representing almost all the brain of fish or amphibian), is referred to as the brainstem, the hindbrain and midbrain, consisting of the spinal chord-medulla oblongata and pons. This is sometimes described as the "basic neural machinery for reproduction and self-preservation," essentially involuntary or automatic activity, such as heart regulation, blood circulation and breathing.

Sagan says that the R-complex, with an important role in aggressive and bureaucratic behavior, and social hierarchy, is still sort of performing its reptilian and "dinosaur" functions.

The Limbic System: Surrounding and bordering on the R-complex or neural "chassis" is the *second evolutionary step, the limbic system*, about one-hundred-fifty million years old, including the thalamus, hypothalamus, pituitary (controlling the endocranial system and various states of mind, including even religious feelings), amygdala and hippocampus, and olfactory cortex. It is related to emotions, memory, sexuality and smell.

The Hypothalamus and Shifting the Body Clock: The hypothalamus contains the body clock, a sort of biological metronome. Resetting the body clock for night-shift workers is important. The work of the team of Dr. Charles A. Czeisler of Boston's Brigham and Women's Hospital as described in the New England Journal of Medicine, May, 1990, shows how biological rhythms can be readjusted in just two nights, shifting the daily cycle low point from early morning to late afternoon, by sitting before a bank of lights while at their night work and using blackout shades during the daytime sleep time.

Limbic Stimulation--Cocaine and Crack: Sagan points out that electrical stimulation of this area imitates psychedelic and hallucinogenic drugs. Limbic malfunction can produce aggravated "rage, fear or sentimentality." Sagan muses that altruism may come from the limbic system but sex, in contemporary human life, apparently now derives from all three systems.

Drugs and Blocking the Chemical Messengers: Studies at the Addiction Research Center of the National Institute on Drug Abuse, show that *snorting cocaine and smoking crack to produce euphoria, interferes with the brain's chemical messenger movement*. It blocks the chemical messenger, dopamine, from being recycled back to the despatching nerve transmitting cell neuron, confining it in the synaptic space where it keeps stimulating the receiving cell neuron receptors in the nucleus accumbens of the limbic system.

Thus its retention in the synaptic space overstimulates the receiving neuron in the limbic system's nucleus accumbens which communicates with the part of the brain, the motor cortex, governing movement, that is, influencing movement toward pleasure sources. This *continuing pleasure response is addictive, driving the user to increasingly repetitive use of the drug*, the overdose leading eventually to death.

As a result of recent medical advances, new compounds have been developed which have the effect of interdicting and dulling the uncontrollable cocaine urge. The effect of the medication is essentially to restore the efficacy of the patient's power of avoidance.

Manufacturing Morphine and Codeine in the Human Body: Studies in the past several years, according to Nature Magazine, show that mammal tissue manufactures molecules, peptide proteins, called endorphins and enkephalins, which are chemically similar to morphine, codeine, and heroin. Brain cell receptors are "lock and key shaped" to fit the morphine molecule shape. Morphine precursors, ring-like "skeletons" of morphine molecules, have been found in various animals such as cattle, rabbits, and rodents. Mammalian brains thus seem to produce their own pain-suppressing chemicals.

Alcohol Addiction May Result from a Defective Gene: Speaking of addictive behavior, new approaches to dealing with alcoholism may be indicated. The alcohol impulse may result from suppression of dopomine activity resulting from defective directions in the construction of portals on neurons supplied by the defective dopamine D_2 receptor gene. Drugs that adjust the dopamine metabolism might be a fallout of the studies. Also interesting is the possibility that such genes may be involved in other addictions or forms of abnormal behavior.

Addiction Anatomy: Loss of control, and the inability to adhere to limits, is what distinguishes the makeup of those who have *controllable behavior preferences versus addictive compulsions*. While low self-esteem is characteristic of the addicted, it is not clear that there is a so-called "addiction personality." Depressive tendencies may trigger use of drug stimulants that produce a high. Compulsive addiction behavior may manifest itself by *crossover from one addiction to another*, as alcohol and gambling, drugs and sex, alcohol, or overeating and compulsive spending.

Allergies and Addiction: Allergies or addictions may mask each other. Addictions to chocolate, tobacco, coffee, alcohol, and foods may be allergy- related. Many chocoholics are allergic to chocolate, smokers to tobacco, alcoholics to alcohol. Treating the allergy may eliminate or diminish the addiction.

Food Addiction and Eating Disorders: A propensity to food addiction, like tendency toward alcohol addiction, may have genetic origin. Reports in The May 24, 1990 New England Journal of Medicine indicate that a person's "vulnerability" to becoming fat is significantly determined by the genes and resulting metabolic rates rather than being overfed as a child or as the consequence of psychiatric factors.

Sets of identical twins at Laval University in Quebec were fed extra calories with limited exercise. The weight gain in each set, with identical genes, was essentially the same but wide variations resulted from one set of twins to another set, the different twin sets involving, of course, different genes.

Food As a Tranquilizer: Food, like other addictive substances may produce a tranquilizing effect that is stress-relieving. Sugar consumption stimulates the production of certain proteins, opiod peptides. Interdicting these peptide proteins reduces the desire.

Persons with food disorders, such as anorexia or bulimia, may have started their condition with alcohol. Compulsive eating and compulsive spending or gambling and even compulsive working, workaholicism, may be related or crossover addictions although career-intense activity has many other explanations, such as insecurity, desire to excel or be competitive, the aggression or survival instinct, perfectionism. Work may become a tranquilizing diversion that consumes stress, a coping mechanism, and then an addictive necessity.

Addiction Treatment: Addiction treatments fall generally into four somewhat overlapping categories:

(1) Addiction-fighting self-help organizations, such as Alcoholics Anonymous, Cocaine Anonymous, Overeaters Anonymous, are built around certain therapeutic self-help principles, mostly adopted from those in the Alcoholics Anonymous program. *The primary ingredients are acknowledging the addiction dependency and resort to a 'higher power,' a spiritual resource.*

(2) Medically-oriented treatment are used by eating-disorder clinics like Philadelphia's Renfrew Center, and addictive disease programs run generally by hospi-tals, like Atlanta's Charter Peachfield Hospital program of Medical Services for Addictive Diseases. Some hospital programs may include treatments involving alteration or interdiction of brain-protein activity and new addiction-suppression medications like those for cocaine addiction.

(3) Use of Psychoneuro-immunology (PNI). This involves use of the mind to control disease which asserts a biological relationship between emotional, brain, and immunological systems in which the three systems communicate. The patient's knowledge of the actual neurophysiological processes by which the brain functions to produce certain chemical effects, helps the person to mobilize his resistive response.

Psychotherapists and mental-imaging specialists, so-called "visualization therapists," believe that people can control their illnesses through mental imaging. Included here are *biofeedback* techniques in which, for instance, using the proper thinking sequence, the mind is employed to modify body functions. An easily-induced and common example of a biofeedback response results from a thinking sequence which raises the temperature of cold hands or feet.

(4) Positive Outlook, Optimism, And Happiness Increase Immunity. Cancer patients in a UCLA study who participated in an *interactive mind-stimulating group therapy*, fared better than another group that did not, despite the same medical routines.

Optimists test better in disease fighting than the pessimists. Happiness also tests out as a positive factor in disease control. Test groups engaged in physical exercising have fewer colds and flu, and recover faster than a nonexercising sedentary group. Their NK (natural killer) cell activity increases in the exercising group.

Walking and simple exercise seems to promote self-esteem and foster good feelings, apparently *influenced by endorphins called neurotransmitters*, hormones produced by the brain, nervous and endocrine systems.

The Power of Faith: Recovery through faith, so-called faith-healing and the "laying-on-of-hands" procedures may be mechanisms which influence the immune system, an aspect of mental imaging, power of suggestion, or psychoneuro-immunology. Recovery from illness via faith or power of the will, is thought, in medical terms at least, to result from the power of the mind to mobilize the body's biological defense processes like the immune system. Religious feelings are considered to have a powerful effect on brain "willing" mechanisms and are viewed as legitimate reinforcing instruments in psychotherapy.

Whether there is some form of external *spiritual intervention*, as some have suggested, cannot be known. The theological view of divine intervention may, it is suggested, coincide with PNI theory based on theories of "immanence," God's omnipresence and his presence within us--manifested, for instance, through the immune system.

Resort to the Spiritual and a 'Higher Power': The "resort to a higher power" concept embedded in addiction programs like Alcoholics Anonymous, is thought, in medical terms at least, to utilize these PNI and positive outlook principles in a support group environment.

Psychotherapists acknowledge the role of faith and religious conviction in support of psychotherapy measures. The lives of many are favorably influenced by these addiction-coping organizations. However, the biomedical problem is only in remission and resurfaces if the association with the program is discontinued. "Cure" can only come from eliminating or altering the underlying physiological deficiencies.

Religious persons may look upon faith healing, resort to God, as being healing by divine intervention. But the power of the mind to influence body chemistry and immune reactions is increasingly evident. If faith is a reinforcing mechanism for this process, as it seems to be, then so much the better.

If God intervenes in some miraculous way, even He must use some chemistry-altering physical process. In this same way, the addiction organization resort to a "higher power" is similarly necessarily invoking a physiological biochemistry-altering process.

Altering Habit Patterns and Pleasure Responses in the Brain: Hospitals and other organizations have rather successful programs to deal with the cigarette-smoking addiction. The key to the high success of these programs involves an interesting combination of influences involving destruction of habit patterns and changing mental attitudes to diminish "acceptability" of the habit. Participants are taught gradually to develop an actual dislike for cigarettes, a loathing of dirty lungs and smelly cigarette butts. The program seeks to interdict and modify the brain's pleasure responses.

Decreasingly-addictive and less-pleasing cigarettes are introduced. Habit patterns involving lighting up after a meal or in tension situations are "interrupted" by new sequences which alter and modify the smoker's normal addictive behavior pattern. The group functions as a support group and each person has a support sponsor to contact when the going is tough. The program results in both physiological changes, changes in brain biochemistry pleasure responses, in mental attitudes and in self-esteem which plays a significant role in regaining control.

Addiction as a Self-Organizing System in Chaos Physics: Elsewhere in these materials we discuss the study of *chaos,* the physics of the two opposing thermodynamic forces of increasing disorder, or entropy and the countervailing force of so-called self-organizing centers of increasing order and complexity.

Neurophysiologists and others have noted the curious parallel between increasing addictive behavior and so-called self-organizing systems in nature. Such systems include weather phenomena like tornadoes and hurricanes, life forms, planetary evolution and even galactic organization into clusters, filaments and voids.

The tendency in nature toward self-organizing centers seems to play a role in addiction where the addiction becomes a goal-oriented, self-organizing center within the brain chemistry, one seeking absolute control over the other biochemical processes. This powerful force generates a stubborn rationalization process in which any action to interfere with or in any way ameliorate the addiction is interdicted by rationalizations and actions to defend against interference.

The addict will carefully prepare for and nurture the actions needed to assure opportunity for addiction satisfaction and fulfillment. Anything that might arrest or interfere with the onset of the craving or its fulfillment is carefully eliminated or controlled. Hours of work and sleep are rearranged to facilitate addiction satisfaction. The addiction self-organizing center acquires a mind of its own that dictates the addict's total behavior.

Coping with the Addiction 'Mind' by Confrontation: Jane Meredith Adams writes about commencing addiction control in her "Confronting The Addict," in the July 17, 1990 Washington Post Health Section. Part of the addict behavior is self convincing and convincing others to avoid interfering, that the addict needs to *'hit bottom,'* to deal with the addiction on his own.

What 'Hitting Bottom' Means: This, of course, is part of the arsenal of defensive rationalization devices employed by the addict to protect and nurture the addiction. Unfortunately, as she points out, 'hitting bottom' usually means *severe consequences,* which can translate into losing one's family, divorce, losing one's job, and becoming a derelict in society, a street person, a public dependent and whatever may be worse. Too often, hitting bottom simply means 'over and out.'

The Rehearsed Confrontation: Well, one therapeutic approach to interrupting this plunge to the bottom is the *rehearsed confrontation* in which the family and friends of the addict meet with the addict and "delineate behaviors they will no longer tolerate"--because they love him too much to allow him to continue in his delusion that he could control the addiction, a delusion entertained by most addicts.

The 'intervention' is about 'presenting reality' to a person by those human beings representing his own personal world, in a way that demonstrates caring. "I know I'm not responsible for my disease," said one addict with a family history of alcoholism, "but I'm responsible for my recovery," he acknowledged.

The goal of the intervention is to force redirection of correction efforts into some more drastic therapy for eliminating the addiction, such as resort to focussed therapy and medical assistance such as available through specialty treatment centers. The point is, of course, that while the problem is related to mental attitudes, it is also directly related to brain biochemistry, neurotransmissions.

The Neocortex and Human Cognitive Functions: But getting back to the buildup of the human brain, the *third* and most recent brain addition (perhaps less than one hundred million years old) is the *neocortex*, most highly developed in the advanced mammals--the human being, the dolphins and whales--and is the site, says Sagan, of "many of the *characteristic human cognitive functions*": in the frontal lobes, deliberation; parietal lobes, human symbolic language like reading and writing, and exchange of information between brain and body; the temporal lobes, perceptual tasks, perhaps including sound and sight stimuli coordination; the occipital lobes, vision.

Sagan says that damage to the parietal lobe produces the greatest decline in intelligence related to "activities in everyday life." Human symbolic languages, particularly reading and writing and mathematics, seemingly require cooperation between the temporal, parietal, and frontal lobes, and perhaps the occipital.

Joan of Arc--Brain Irregularity, And Religious Insights: Engagingly described by staff writer William Booth in the May 2d, 1990 Washington Post, is a new view of Saint Joan of Arc which was presented at the May 1990 annual meeting of the American Academy of Neurology. In "Saint in a New Light, Joan of Arc May Have

Had Epilepsy," he discusses the work of neurologist Lydia Bayne of the University of California at San Francisco, and author Elizabeth Foote-Smith that suggests Joan's visions as an "altered consciousness."

Temporal Lobe Overstimulation Producing Hallucinations: This takes the form of auditory and visual hallucinations from "complex partial seizures," a less known form of epilepsy resulting from *temporal lobe nerve cell overstimulation*, sometimes taking the form of visions and receiving "messages." Her visions gave her euphoric pleasure, a "voice from God" and "the voice of an angel," preceded by church bells and accompanied by a "great light."

Bayne says that this story of an adolescent woman with stereotyped dreamlike spells and "visions of angels and voices with a messianic message," is typical of complex partial seizures.

Mystical Personality Traits of Reflex Epileptics: Particularly interesting is Baynes description of the personality traits frequently associated with this rare form of epilepsy: "a sense of personal destiny, a deepening of all emotions, feelings of grandiosity and exhilaration, *hyper-moralism*, a loss of libido, anger and aggression, and *deep religious beliefs*, often idiosyncratic and mystical." She refers to musicogenic epilepsy, the action of the relex epileptic "to self-stimulate her own pleasurable seizures," and reports from contemporary associates of Joan that she would have the church bells rung at Vesper-time, presumably to trigger the visions and voices.

Epilepsy and Great Figures in History: Alexander the Great, Napoleon, St. Paul, and other great figures of the past are said to have suffered mind-affecting events. Could epilepsy be some form of instrumentality for divine communication? The prophets speak of visions and voices. Many unusual religious experiences are reported by persons with various forms of brain or related abnormality, such as endocrinological or autistic illnesses.

The Triune Brain and the Human Psyche: Sagan goes on to develop views of the human psyche as reflecting the three evolutionary brain systems, citing Freud's view that the "capacity for neurosis [is] the obverse of...the capacity for cultural development." He also cites Freud's two divisions of the mind, first into *the conscious, the preconscious and the unconscious*, and second, into the *id, ego and superego*, the id representing the beast-like R-complex, and emotions, including religious feeling, as the ego function, part of the limbic system.

Tree-Life Enhances Human Life's Agility: Sagan describes behavioral changes in human life-like life forms, the shift from tree-life to ground-life, pointing out how human intelligence benefitted first from the tree-life requiring agility, manipulation, balance, and binocular vision, and then from the ground-life, requiring thinking processes for self-preservation, using tools, protecting living space and producing food by hunting and primitive agriculture. Early technology required manipulative skills well suited to human life's skills derived from his tree-life ancestry.

Life on other planets that missed this tree-phase might well lack our manipulative skills, just as other highly intelligent life on this planet--the dolphins and whales--never progressed beyond the skills related to their water environment, even though they have faculties like sonar, beyond ours.

Walking on Two Legs: As reported in Omni magazine of June 1992, Director of biological and earth sciences Peter Wheeler, at Liverpool Polytechnic University, discusses how human beings wound up walking on two legs rather than all four limbs. Using a model of an apelike creature, the Australopithecine, that evolved into the early hominids, he suggests the environmental stresses of strong solar radiation and absence of sufficient drinking water, the upright position for walking would be cooler, require less water, and permit longer forays for food; it would also permit them to become taller, thinner, and their heat-sensitive brains to grow larger.

Development of Humans and Increased Brain Size: Discussing the pre-human hominids, Sagan describes the two kinds of Australopithecines, a different form, *homo habilis*, "the first true man" (appearing coincidentally at the time of the first true horse), and *homo erectus* with increased human -like brain size.

Sagan notes that it is this evolutionary, increased brain size that accounts for childbirth pain in *homo sapiens* and cites Genesis 3:16, "In pain shall thou bring forth children." He cites the other evolutionary forms like *Proconsul, Neanderthal,* and highly intelligent *Cro-Magnon*, with actually greater brain size than our own.

Might Primate Intelligence Overlap Human Intelligence? He discusses whether some of the contemporary primates, like chimpanzees, might have intelligence resembling human characteristics, as seems likely from recent primate research based on communication via sign language and computer manipulation. Sagan notes human inability to communicate with the animals, traceable to an evolutionary course of events which parallel the metaphor of the Biblical Fall.

The Eden Metaphor: "The Fall from Eden seems to be an appropriate metaphor for some of the major biological events in recent human evolution," the repository for which, he says is the genetic code. There is a definite nostalgic-romantic twist to Sagan's book, evident even from the very title of *The Dragons of Eden, Speculations On The Origin Of Human Intelligence*, speaking of the portions of our brain, the most incredible complex mechanism known in the universe and traceable to primordial reptilian, animal and remote cosmic sources.

The Dinosaurs Still Thunder in Brain Phenomena: In his discussion of the development of the human brain, Sagan sees interesting interrelationships of certain brain phenomena, such as mysterious dream activity, as somehow traceable to human 's early hostile involvements with the once dominating reptilian life on the planet, and in our dreams, he notes, from our reptilian brain the dinosaurs still thunder out of the primordial past.

The Reptile and Human Exile from Eden: He points out that God's proscription on knowledge was of an abstraction--the difference between good and evil. Human exile from Eden depicted in Genesis, he points out, was the work of the reptile-serpent, offering the fruit of knowledge of good and evil, which he terms "abstract and moral neocortical functions," a more recent brain evolutionary event than the earlier reptilian and limbic systems. Sagan cites God's commanding of the eternal hostility between humans and the reptiles, Genesis chapter 3.

He speaks of the original niche of the mammals as being in the night to escape the day-active reptilian predators which were immobilized at night; and how the disappearance of the biggest reptilian threat, the dinosaurs, then created the safer mammalian daytime environment of today; how human 's very intelligence was honed by reptilian-encounter-survival experiences.

Sleep and Awake States Governed by Different Sets of Neurons: Layzer points out that two different sets of neurons in the brain-stem mediate the waking and sleeping activities. In sleep, one set suppresses the activity of the second set and disables the wakeful responses. In the dream state, skeletal muscles lose their tension to become useless; dream-state brain stem impulses *suppress the motoneurons*, interdicting any higher brain center commands directing use of skeletal muscles.

If the suppression mechanism is impaired, however, physical dream acts may become real, as in sleep-walking. Dream actions may be totally unrelated to reality, as the human being who, in a dream state, leans over to pet his cat--one that he doesn't have. Medical treatment may, however, restore the suppression process.

Dreams and Dragons: Speaking of the R-Complex, dream states, and reptilian life forms such as serpents and dragons, the term dragon has different meanings according to usage--evil and Satan in the Book of Revelation, or benevolence, as in ancient China. But these and related reptilian life forms are by no means gone from the planet. Apart from reptilian forms like the American Iguana, the chameleon, the crocodile and the alligator, it is thought that all birds and fowl are descendants of the dinosaurs.

As Sagan points out, The Komodo dragons of remote parts of Indonesia, represent one of the few remaining large predatory dragons in existence. They look just like the dragon in Donatello's painting of St. George slaying the dragon, in the Chiesa di Or San Michele, in Florence. There are also smaller dragon versions; some climb trees to eat the eggs of birds, creatures descended from the dinosaurs. Thegeneral pervasiveness of dragon myths, says Sagan, probably reflects their likely greater prevalence as recent as a thousand or so years ago.

Childhood Dreams of Monsters: Sagan wonders whether children's dreams of monsters are somehow vestigial remnants of early neural patterns, instinctive defenses against the dragons recalled as a part of childhood brain development to wake us up "to see if anyone is about to eat us."

Events in the Dream State: What's going on in the dream state? Well neuro-scientists speculate that a number of therapeutic functions are performed such as brain "maintenance" programs, like computer hard-disk-maintenance programs--cluster rearrangement, defragmentation, deletion of useless information, cross-linking and the like. Sagan uses terms like “buffer-dumping” and “memory storage functions” to describe some of the "maintenance" activity. Neurologists have succeeded in inducing vivid "dream states" by electrical stimulation.

It has been suggested that in dream states the brain does *disease analysis*, initiating corrective processes involving the immune system by a process involving neuro-transmitters and neuromodulators. Reduced sleep, common in the aging process, may result in reduced "maintenance," thus playing a role in mental deterioration.

Replaying the "Warfare between the Reptiles and the Mammals": Discussing the brain's primordial R-complex, Sagan cites Freud's work on aggressive, hierarchal, ritualistic and sexual brain functions, the "inhibition center" developed in the temporal lobe to "turn off" much of the reptilian brain function, and the "activation center" in the pons, harmlessly to turn on the R-complex during sleep.

Descended from the reptiles and early mammals, muses Sagan in his customary fascinating way, we may, "in the daytime repression of the R-complex and in the nighttime stirring of the dream dragons...be replaying the hundred-million-year-old warfare between the reptiles and the mammals."

Knowledge of the Future in the Dream State: Several examples have been reported of asserted cases of *dreams of the future*. The question has been raised about theoretical "memory of the future," associated, perhaps, with the general relativity's geometry of spacetime. Einstein had noted [without fully explaining] that distinctions between the past present and future were illusion.

Can dream states somehow conjure or "retrieve" future events into present recall. Interestingly, long before Relativity theory, St. Augustine spoke of the universe, including all time, as now existing, having been laid out at the creation. One interesting dream account described a dream of detailed technical instrumentation of an atomic submarine. So what? Well, the dream occurred years before there was an actual atomic submarine. Of course anecdotal proof has its limitations.

Sagan's Dream of Future 'History': Sagan describes the occasion of his own dream of thumbing through a book. Not so unusual, but this book *was a history book which included the future*. He could make out individual words but not combine them in sentences; he was "alexic," he declares, and muses that perhaps this future-reading disability is somehow a "*metaphor of the unpredictability of the future*." Well, there are many reported cases of paranormal or parapsychological prescience, "seeing" future events--with precision accuracy of detail.

Quantum Uncertainty and the Dream State: There are those who suggest that conscious recall, willed cognition, may be less responsive than dream-recall mechanisms, the nature of which are not yet fully understood. The question *whether Heisenberg's quantum uncertainty forecloses causal predictability*, including predicting the future in the macro universe, is still the subject of debate.

The Hierarchy of Neuron Circuits: Getting back to brain circuitry, the organizational hierarchy of neural circuits, Layzer notes, include five levels: The *first* or bottom level consists of local circuits which carry either inhibiting or exciting signals into and out of the circuit and contain interneurons whose connections are within the circuit. The *second* level connects local circuits within the same region. The *third* level are circuits connecting regions; the *fourth* set of circuits are components of broader systems such as sensory and motor pathways that transmit information to the central nervous system. At the top is the *fifth* level consisting of systems that control the organism's overall behavior.

The Brain's Functional Organization: Layzer cites A.R. Luria's three interacting functional units, strongly connected by neural pathways, which are described in his *The Working Brain.* The *first unit* controls arousal states of wakefulness, sleep, dreams, and attention. The *second,* controls sensory information receipt, processing and storage. The *third unit,* which controls conscious activity, occupies the front or *anteria cortex* consisting of *three zones*.

The *first zone* consists of the *motor cortex* (neurons that sends signals through the spinal column) and the somatosensory cortex (which receives muscle, joint and skin receptor information), performing essentially as s single functional system.

The *second zone,* the *premotor cortex*, assists in "integrating" individual muscle movement into coordinated actions such as stepping. The *third zone,* the highest, is in the *prefrontal region of the cortex* that makes the human brain the most distinctive from the primates and *accounts for what we describe as human behavior*--such as creativity, planning and complex skills.

Short-Term and Long-Term Memory: Sophisticated long-term human memory is traceable to the neocortex. As we age, we tend to forget "recent" or short-term memory items but vividly retain long-term events, like early childhood memories. Short-term memory is badly impaired by hippocampal lesions, Sagan notes.

Short term and long-term memory seem to reside in different regions of the brain. The problem in aging recall of short-term memory items seems traceable to difficulties in "accessing new material into the long-term memory." This may be due to old-age insufficient blood supply to the hippocampus resulting from physical disabilities such as from arteriosclerosis.

A Biological Electrical Signaling System: The brain is a biological electric signaling system employing *three processes*, notes Layzer. The first is a conductor--body fluids; the second, a voltage potential or voltage difference--a voltage drop at the cell membrane produced by greater cell membrane permeability to positive than negative ions, mediated by the cell-membrane sodium-potassium pump which transfers potassium and sodium ions through the membrane.

The third process is a device for initiating and signaling changes in electric potential--*neuron junctions or synapses*, some direct welds and some chemical synapses involving neurotransmitters synthesized in the cell body which escape to bind to matching receptors to trigger opening of ion channels.

Incredible Number of Synapses: Carl Sagan says that the brain has an absolutely incredible number (some 10^{13}) of synapses, functionally different configurations of the brain. He wonders whether this might account for the unpredictability of human behavior and suggests that there must be an enormous number of unused synapses available for new learning. The sheer magnitude of human brain potential is almost beyond comprehension. Apparently you can't overload the human brain.

The construction of new synapses, requires the synthesis of protein and RNA molecules. There is considerable evidence, he says, that these molecules are produced by the brain during the learning process and that the *learning is contained within the protein or RNA*. Layzer attributes learning to the chemical process in neuro-modulation. Sagan notes that to process the information in the human brain, a computer would need be some *ten-thousand times larger than the human brain.*

Neuromodulators Possibly Account for Learning and Memory: At some synapses, says David Layzer, the binding of chemical messengers, neuromodulators, to the postsynaptic membrane, instead of triggering a voltage impulse in the postsynaptic neuron, trigger chemical reactions that alter the response to later signals. Layzer notes that this finding suggests that neuromodulation may explain the learning and memory processes.

The Neocortex As Producing Perception: Higher brain functions, says Layzer depend on neurons in the outer "bark-like" (highly convoluted) layer of the brain, the *neocortex,* covering both hemispheres. Layzer suggests that it was the *expansion of the neocortex* that accounts for the transition from visual feature-detection to visual perception--the construction, testing and modification of hypotheses.

How Does the Nervous System Hypothesize? He cites immunobiologist-neurobiologist Gerald Edelman's model for dealing with perceptual hypotheses as having been derived from analogous phenomena between the nervous and the immune systems. Involved in the immune system is the recognition of foreign particles by antibodies and an interactive process by which genetic variation produces a "repertoire of genotypes" which are acted on by natural selection to produce *immunological memory*.

Candidate Neural Modules for Hypothesizing: So a repertoire of neural modules of perceptual hypotheses is created in the course of embryogenesis and development and the motor cortex contains these "units of perceptual function." Thus Layzer says, perceptual stimuli leave "enduring traces in the form of facilitated neural pathways" and a repertoire of cortical neural modules function as candidates for selection (perhaps one million modules of as much as ten-thousand neurons), plus a huge number of "modifiable neural pathways" connecting these units. By modifying the strength of existing neural connections, hypothesis construction defines new neural configurations from a vast collection of candidates.

Can We Describe Mental States Scientifically? Layzer discusses consciousness, left and right brain cognition functions, whether we can give a complete scientific account of the objective aspect of mental states, and how they are related. The view, he says, that experience gives us direct, though partial, access to what is going on in the brain seems well supported by available evidence. He cites specific mental tasks associated with specific brain activity patterns as recorded from PET or positron-emission tomography techniques.

Thus, reality seems to consist, he says, of two parts; a part we can describe but can't directly experience, and a part we experience directly but can't fully describe. The part we can describe, the external world, transcends individual human experience expressed in the universal language of mathematics. The part we experience directly isn't wholly indescribable, but, he says, it does contain an indescribable residue--a residue that makes one's inner life distinct from another's.

The Thinking Process Mimics Biological Evolution: In the *Cerebral Symphony: Reflections on the Structure of Consciousness*, neurobiologist William H. Calvin, meditating about the operation of the human brain, likens the problem-solving character of the human thinking process to the evolutionary process of life. In thinking, he notes, we test alternative possibilities, discard the unworkable, improving conclusions through *an accelerated miniature refinement process that mimics biological evolution*. Thus, he muses, we see that the Darwinian life-shaping principles have "again blossomed in human heads on an even more accelerated time scale."

Brain Size and Intelligence: In his *The Dragons of Eden*, Sagan discusses the relationship of brain size to intelligence and points out that the highest brain mass to body weight is found in mammals, first in humans, then in the dolphins. He goes on to discuss the extraordinary human-like dolphin intelligence, including their complex communication. Being mammals, they will drown if not able to breath air and dolphins will hold up an immobilized dolphin to permit it to breath.

More on Mammal Intelligence: The National Science Foundation, the U.S. Navy and the U.S. Air Force all supported the dolphin research of the brain researcher John C. Lilly (perhaps best known from the film *Day of the Dolphin*). Lilly's work demonstrated high dolphin intelligence, vocal communication skills, learning and teaching ability and concern about the needs of other dolphins. Lilly's experiences have been chronicled in *The Man Who Talked To Dolphins*, *John Lilly, So Far*, by Francis Jeffrey and John C. Lilly.

The brain of the sperm whale, closely related to the dolphin, is more than six times human brain mass and Sagan wonders what the whale does with all that brain. "Are there thoughts, insights, arts, sciences and legends of the sperm whale?" he wonders.

Abstract Reasoning and Human Behavior: Sagan speaks of abstract association, reasoning, curiosity, problem-solving as the most significant human behavioral characteristics, specifically, "mathematics, science, technology, music and the arts," and one might add, religion and philosophy. But, he notes, higher mammals like whales, dolphins, elephants, may be as "humane" as human beings, having the brain features to deal with certain abstractions.

Is abstract thought a matter of degree in thinking rather than a distinct mental capacity so that the distinction between human and other animal abstract thinking is simply one of degree, depth, scope or subject matter? Alfred Russell Wallace, like Darwin, associated with work in natural selection, noted that a baby orangutan behaved like a human child in the same situation, and chimpanzees cooperated to perform intelligent acts.

Primates and Symbolic Behavior: Chimpanzees prefer movies about chimps, jazz over rock music; they learned to make modestly complex requests via a learned animal-oriented computer language and to construct sentences. So certain primates can learn to use symbolic language, previously considered an attribute exclusively of *human* behavior.

Human Arrogance about Other Life Forms: This discussion of course, raises questions about human disregard for other intelligent species on this planet, something which Sagan notes as representing a form of human arrogance. Sea mammals, dolphins, are not only regularly drowned by the thousands when caught in tuna nets, but imprisoned in aquatic parks where their lives are of shortened duration. The purpose? Human entertainment--one step from the Roman Coliseum gladiator-show mentality.

Peter Singer Inveighs against Human Cruelty and Speciesism: In his popular and scholarly book on animal rights, *Animal Liberation*, Australia's animal rights activist and Monash University professor, Peter Singer, articulates the human abuses of the earth's other "beings," members of the various animal species coinhabiting our planet.

He inveighs against human cruelty toward the earth's animal species, including of course, killing and eating them. The so-called Lethal Dose 50 tests illustrates the 'expendability' of animals philosophy by indiscriminately exposing large numbers of animals to agonizing destruction to determine toxicity levels for consumer products deleterious to human life.

Speciesism and the Bible: To racism and sexism, Singer adds 'speciesism,' a belief in superiority or primacy of the human species, as one of the human chauvinisms. He does not, incidentally, condone the violence practiced by some animal rights activists, for example, such as the groups which have aggressively attacked biomedical facilities which use animals in developing cures for human diseases.

Although animal rights discussions deal with needless slaughter and cruelty to animal species they also highlight the tension between those who would equalize the value of all life--save animals even over babies whose survival is tied to animal biomedical testing--and those who assert the view of human life as privileged over animal life, a view thought by many to be traceable to the Biblical rights of human 's dominion over the animal world.

Kant, Darwin, and Human Ethical Obligations toward Animals: Some argue human rights to use ("dominion over") animals is based not on the Bible but on the Kantian philosophic argument of reciprocity; that since animals possess only instincts and are incapable of moral standards (many animals are carnivorous) and they have no reciprocal rights to have moral values applied to them.

Barbarous infliction of pain and destruction on other species by one specious that has been elevated to moral capacity would seem to violate even minimum standards of stewardship and trusteeship inherent even in Biblical "dominion," to say nothing of love, compassion, justice, mercy, the golden rule, and general concepts of moral conduct.

The other argument is not based on ideas of superiority or moral reciprocity but simply on the Darwinian argument that the laws of physics, chemistry and biology, the laws of nature, dictate that the fundamental drive in all life forms is survival of the species and, since Eden at least, Nature fiats or at least tolerates a predatory life survival mechanism. One wonders, though, whether our philosophy toward other animal species would change if we were suddenly confronted with a predatory extraterrestrial perception of human beings as just another primitive predatory life form.

Reverence for Life: Fortunately, we are reminded that human depravity is not universal; the saintly philosopher-medical missionary, Dr. Albert Schweitzer, of course, came down on the side of reverence for life, life in any form. There is a growing awareness of the preciousness of the planet and its life forms, the dangers to the planetary life-supporting ecology. Suppose it is the only such life-supporting ecology in the entire galaxy? The Universe?

Science and Religion Combining to Save Earth's Environment: Science and religion are often at odds, as we have discussed, but Catholic theologian Donald B. Conroy, president of the North American Conference on Religion and Ecology (NACRE) says that his organization hopes to influence American public opinion toward more prudent treatment of the environment, emphasizing the moral dimension of the ecological crisis.

A May, 1990 conference at the Washington National Cathedral, "Caring for the Creation," partly supported by the Humane Society of the United States and the World Wildlife Fund, attracted large numbers of religious and scientific leaders, Carl Sagan, EPA Administrator William Reilly, and British Prince Philip, president of the World Wide Fund for Nature.

Emergence of an Environmental Theology: "Self-inflicted" ecological damage, such as ozone depletion and species destruction, threaten the planet, it was noted. The Genesis grant to human life of dominion over the other living creatures of the planet has been opportunistically rationalized.

But a kind of new environmental theology is developing to cope with the dilemma by enlightening human life's stewardship of planetary life. That which is looked upon as sacred is more likely to be respected, it was noted. Church opposition to prudent population self-control has been mentioned as an environmental negative.

Washington Post Staff Writer Laura Sessions Stepp, reporting on the conference, cites Old Testament scholar Theodor Hiebert's research pointing out that the scenario of Adam and Eve tending the Garden of Eden was written much earlier than the seven-day creation sequence, and "is closer to how the Divine intended people to live."

A Punishable Offense against Galactic Law? Our own primitive behavior with respect to other Earth life forms, viewed by advanced extraterrestrial intelligent life exploring our planet's habitability potential--or perhaps inspecting our observance of galactic life-preservation criteria, may indeed classify us as dangerously primitive--technologically and socially deficient.

The Zoo Hypothesis: So far and despite our efforts, we have been unsuccessful in receiving communication from extraterrestrial intelligence in the universe and speculate on why, considering our own broadcast transmissions, if there are so many life-potential planets, that we haven't received a single message. Well, the so-called 'zoo hypothesis' explanation suggests that we may be regarded as a just-emerging primitive life form not yet ready for interstellar dialogue.

In the zoo hypothesis, we are being observed, and treated, as we treat our own earth animals in a zoo. We earthlings don't have many sophisticated efforts to communicate with our own zoo inhabitants--except for the dolphin research and some primate communication experiments.

The Only Species Endangering Other Life in The Galaxy: But being the ecology-damaging creatures we are, our continued existence may be judged by some galactic oversight court as unwarranted, especially since the human species represents the only one knowingly endangering planetary ecology and the existence of precious other earth life forms. This may represent a heinous offense against Milky Way galactic life-survival obligations.

Hopefully, though, enlightened extraterrestrials may see some redeeming features in our species, and seek simply to rehabilitate our species into conformance with galactic or perhaps universal life-promoting standards of behavior.

Future Brain Evolution: In *The Dragons of Eden, Speculations On The Origin Of Human Intelligence*, Carl Sagan discusses future brain evolution. Larger brains with larger skulls, sometimes depicted in pictures of extraterrestrials, could not evolve, given the size limitations in pelvic and birth canal physiology. But availability of Ceasarian section and possible advances in outside-the-uterus fetal development, he notes, offer possible chance for increased brain and head size.

Brain Surgery and Implantable Knowledge Modules: Developments in brain surgery is another area of potential. We are learning about new nutritional approaches to brain development that may affect future intelligence on an evolutionary level. Sagan also discusses the growing possibilities for use of cognitive and intellectual prosthetic devices. Perhaps such devices could be implanted in the formative early brain as an extension of the birth sequence.

As miniaturization progresses, suggested possibilities include education modules for knowledge of languages, or of mathematics, medicine, or law. You want to be a doctor, lawyer or scientist, get a suitable module installed. They'll likely be priced according to their income-producing potential, at least if the free-enterprise system continues along present lines. This might free schools for major breakthroughs--if they survive the economics of such competition. Or perhaps they will concentrate on developing course content for more advanced modules.

Machine Intelligence to Augment Human Intelligence: But some say that more promise lies in the development of machine intelligence to augment human intelligence rather than trying to re-engineer or modify the brain itself. Later here we discuss physicist Freeman Dyson's views on evolution and re-engineering of the human mind to facilitate its spread throughout the universe. As we have seen, he does not foreclose the development of solid state brains, even with consciousness and subjectivity.

The Evolution of Intelligence Elsewhere in the Universe: In his *The Dragons Of Eden*, Sagan can't help extending his speculations to the evolution of intelligence elsewhere. He points to the universal application of the laws of physics and what he perceives as the consequential universal trend toward the growth of intelligence, however "stunningly exotic" may be the biological, psychological, sociological and political differences on other planets. Biological evolution on other planets, as Sagan and other space scientists point out, may take strikingly different forms. Even here on our own planet we have amazingly diverse intelligent life forms, like humans and the dolphins.

Life Forms Based on Sulphur Chemistry? One of the earliest forms of life on this planet, *Sulfolobus* bacteria, for example, traceable back some 3 billion years during the earlier more active volcanic period, live in hot, acid environments, and use energy sources differently from other bacteria. French scientist Louis Bec, has computer-modeled a veritable zoo of *sulfanogrades*, sulphur-based life forms which we might find elsewhere in the universe, or for that matter, perhaps genetically engineered here.

Creatures with Totally Different Personality Traits: The point is that possible kinds of life forms defy even science fiction. Brain and mind development may have taken totally different paths than even the diverse forms on Earth. Maybe some day we'll encounter advanced benign life forms, perhaps like E.T., creatures in whom evil either never evolved or was bio-engineered out as a survival necessity.

Because of the confined nature of the small biosphere, the inhabitants of a small moon or earth-orbiting space station, or a space vehicle bound for Mars or the moons of the outer planets, will need be carefully selected for their benign and nonaggressive personalities. In any event, there is no doubt that the first encounter with an advanced civilization will have the most profound effect on Earth life.

Consciousness and Separateness of Brain and Mind: Perceiving, knowing and thinking, involve both physical-biological and psychological mental processes. But, asks David Layzer in *Cosmogenesis*, how are the two related--are there two part of a single physical process or are there two distinct phenomena, one physical and one an immaterial something else?

The question, says Layzer, divides into two problems, the first one scientific--the nature of biological structure in which perceptual and cognitive processes occur. The second is viewed as philosophical, referred to as the 'mind-body problem.' Here, Layzer goes back to Western philosophic roots. Long before René Descartes' dualism, Socrates held mind to be separate from body. Democritus, although agreeing with Socrates on free will, opted for mind-body as a single substance.

What Do We Mean by 'Mind?' What do we mean by the term 'mind?' Does the mind 'tell' the brain and body what to do? How can a mind, a mental and nonphysical thing, make the body run, or eat, or do anything physical? What triggers the muscles and limbs to action? How can a thought, having no physical properties, initiate the physical acts of a body? Nobody knows the answers for sure. Neurobiologist Richard Restak says we "have to admit that there is little there to grasp onto."

The monist, and the scientific materialist, generally looks on the brain and mind as the same, a single mechanism. The dualist, the most famous being the French philosopher and scientist René Descartes, as we have discussed, spoke of two separate substances, mind and body, joined only in God. A third 'new mentalist' and 'split brain' view, exemplified by Nobel prize-winning Roger Sperry, suggests upward and downward control by 'atomic and molecular actions,' neurons determining cognitive events exerting 'downward control' on the brain, triggering neural reactions affecting the body, software activating hardware.

Something in Personal Experience that Defies Science: After seeing science successfully uncovering the secrets of time, space, the composition of matter, the secrets of life and human thought, many feel, says Layzer, that science can ultimately explain all reality. Yet, he says, *something in subjective personal experience doesn't seem to belong in the objective world of science,* citing the 'life force' Henri Bergson saw in the 'irreducible subjectivity of experience,' and the 'nonphysical reality' recognized by neurobiologists Sherrington, Eccles, and Sperry.

Subjectivity Is Not within the Purview of Science: But Layzer suggests that subjectivity is not a physical phenomenon in the context of scientific understanding; that *objective or scientific knowledge must exclude the subjective aspect of experience.* This conclusion of his--that subjective experience of consciousness, one's personal gloss on things, is 'irreducible' (not translatable into a biological occurrence), comes somewhat as a surprise to the reader who is struck by his documented buildup toward ultimate knowability of biological phenomena.

How Are Objective and Subjective Aspects Related? Layzer shows how specific mental states are associated with patterns of brain activity derived from EEG (electroencephalograms) and Pet (positron emission tomography) scans, for instance, which show two aspects of reality, one that can be "described but not directly experienced" and one that can be "experienced but not fully described." The part experienced, he says, contains an "indescribable residue" that makes one's "inner life" unique from others.

The Brain as Anatomy and Physiology: In The *Dragons Of Eden*, physicist-philosopher Carl Sagan describes mind as another descriptor for the brain, and its workings are but the result of "its anatomy and physiology." On the other hand, Canadian neurosurgeon Wilder Penfield, echoing Descartes, thought it "more reasonable" that the mind may be a "different essence" from the brain which can perform complex tasks without conscious thinking.

British brain specialist Sir John Eccles, a dualist, feels the mind equals brain concept as demeaning. If that's so, he notes, then we're mere "passive spectators," creativity and free will are "nothing but illusions," if not superstition.

Dualism and Death of the Brain: Dualism, brain and mind as two separate entities, is also important for another reason, because, as Wilder Penfield points out, survival of the soul or mind after physical death of the brain requires "a connection with a source of energy other than that of the brain."

Of course, Wilder is here asserting that mind, soul, or spirit may represent a quantum state, part of the quantum 'nonlocal' field of cosmic mind and information underlying the universe, something timeless, spaceless, and dimensionless.

Matter Is Weird and Doesn't Limit God's Use of It: In this context of cosmic mind, we recall physicist Freeman Dyson's reference to the human mind as an extension of the mind of God and his assertion that "matter is weird enough" so that it does not limit God's freedom to do with it what he pleases. We also recall Schopenhauer's proposition that, on death, our *individualized* segment of the cosmic consciousness is reassimilated with its undifferentiated cosmic source; and French Jesuit philosopher and paleontologist, Pierre Teilhard de Chardin's reassimilation in which *the individual consciousness remains.*

The Infancy of Understanding Quantum Energy Theory: All these concepts necessarily invoke aspects of quantum theory in which all that exists originated in a single unified quantum energy state; that everything, energy, mind, matter, is related via this primal quantum connection existing at the creation.

As we have discussed above, the dualism of mind and body, the physical and ethereal, are in this view, related aspects of the same thing. As Dyson puts it, scientific materialism and religious transcendentalism are not mutually exclusive in a universe created by God of quantum states the nature of which we are just beginning to grasp.

Dualism and the Bible: As more fully discussed in the material on death, contemporary Western religion favors dualism, a separate and eternal soul, spirit or mind, although Old Testament scripture speaks of physical life and death and, in the end, resurrection.

In the religiously-oriented magazine *Truth*, May/June 1990, "The Mind--Inside The Hidden Universe," Paul Kroll discusses dualism in a religious context and cites various sources in the Bible: Genesis 2:7, God "formed man of the dust," Eccl. 9:5, "...the dead know nothing." Zech. 12:1, God "forms the spirit of man within him." I Cor. 2:1 "...the spirit of the man which is in him." I Cor. 15:44, Resurrection is promised--"raised a spiritual body."

As we discuss later, theology is uneasy with science over the general view of biologists that the mind is simply a biological manifestation, a form of organization of matter. This oversimplification, it is felt, leaves no room for spirituality, felt to be a real and significant force in human life.

Subjectivity a Part of the Physical World? But physicist Layzer with a yen for biology and neurology, notes in *Cosmogenesis*, that if the irreducible subjective component of consciousness exists, he says, it "can't interact causally" with the physical world because *it is not a part of that world.* So how does this subjectivity or truly mental state get "annexed to the external world?" This is not explained by neurobiology, he says, and neither is it explained why this subjectivity exists.

The Brain-Mind As Simply Physiology: But this view of consciousness and subjectivity as not being part of the physical world, as we have seen, is not necessarily shared by all scientists interested in biology. In his *The Dragons of Eden--Speculations On The Origin Of Human Intelligence*, astronomer-space scientist and philosophically-oriented Carl Sagan suggests that there is not a "shred of evidence" to support mind-body dualism.

His fundamental premise is that the brain, the mind, are the result of "anatomy and physiology, and nothing more," [adding] that mind may represent the action of the "components of the brain severally or collectively."

We are, he says, "the results of the interactions of an extremely complex array of molecules," citing the original mystery of human biology as now explained by the chemistry of nucleic acids, DNA and RNA, and "their operational agents, the proteins."

The Order behind and in Chaos: How does order, spontaneously *self-organize* in a universe governed by entropy, by increasing disorder? In *Chaos*, N. Y. Times science writer James Gleick discusses new developments in the science of disorder, chaos research, the scientific search for connections between different kinds of irregularity in nature. In the science of chaos, he observes, some see a major change in the direction of all of physics. Behind order lies chaos, and within it an "eerie type of order."

Chaos science deals with objects on a human scale, the macro universe "we see and touch." It speaks of "fractals and bifurcations, intermittencies and periodicities." He points out that chaos research engenders speculation about evolution, about determinism, free will and "deterministic predictability."

Life as the Product of Organizing Principles: Biochemist Stuart Kauffman of the University of Pennsylvania speaks of trying to discover some of the "organizing principles that allow complex systems to be possible, here or on other planets." Astrophysicist Layzer's description of life, consciousness and subjectivity as an emerging quality in the evolutionary process is somewhat analogously described by Kauffman as a *critical perculation property* of matter, in Kauffman's description of life as a product of the self-organization of matter.

Kauffman modelled general rules for formation of complex polymers and cutting and splicing of molecular strands, discovering the environment for initiating "autocatalytic sets," the self-reproducing reaction chains, suggesting that perhaps the primordial self-organizing reproduction process was not mere chance but what he calls a sort of "critical percolation property," part of the "directionality" or design of nature [perhaps another facet of the anthropic principle].

The Enjoyment of Discussing Anatomy of Higher Brain Functions: Because of its contact with "all areas of human endeavor," discussion of the brain-mind subject, notes Sagan, is ever a source of enjoyment and delight. He goes on to cite Rousseau's caveat, in his 1754 *Dissertation on the Origin and Foundation of the Inequity of Mankind,* that scientists had [then] made "too little progress" to warrant definitive conclusions on human life's "natural state."

But Sagan points to the enormous progress in knowledge of anatomy and human and animal behavior since that time, and the likelihood that future progress in neuroanatomy will more precisely identify the anatomy of the higher brain functions.

Mind-Body Dualism and Materialism vs. Transcendentalism : These discussions of the source of consciousness and subjectivity invariably lead, in any religious context, to the *materialism-transcendentalism dichotomy.* There has always been the tension between those who find everything in the universe ultimately explainable scientifically in terms of energy or 'matter,' and those who find 'answers' expressed primarily in theological terms.

But there is growing appreciation that so-called *scientific materialism*, and theological or *religious transcendentalism* are not mutually exclusive. We explore this mutual exclusivity, or its absence, in connection with new scientific phenomena, especially in quantum theory.

Most analyses on the evolution of life lead inevitably to a discussion of whether all of our reality, derived from energy, is explainable in terms of (1) a single material substance which includes mind and spirit (scientific materialism), or (2) a single substance--God (pantheism), or (3) two independent substances, material (body) and immaterial (mind), as we have discussed earlier and as René Descartes and other 'dualist' philosophers have supposed.

The Weirdness of Matter: Princeton physicist Freeman Dyson asserts that "matter is weird enough" so that it does not limit God's freedom to do with it what he pleases; that *scientific materialism and religious transcendentalism are neither incompatible nor mutually exclusive.*

Physics Sometimes Deals with 'Religiousness': The recurring arguments about the distinctions between the material world and something entirely immaterial--spiritual or religious--frequently surface, strangely enough, in discussions of physics, the physical phenomena of the universe.

In this vein, in his *Perfect Symmetry*, physicist Heinz Pagels could not resist philosophic asides. He quotes an observation by Albert Einstein in 1932 which captures this preoccupation with the sublime:

> "The sense that behind anything that can be experienced there is a something that our mind cannot grasp and whose beauty and sublimity reaches us only indirectly and as feeble reflection, this is religiousness. In this sense I am religious. To me it suffices to wonder at these secrets and to attempt humbly to grasp with my mind a mere image of the lofty structure of all that there is."

Is Consciousness Just Nerve Cell Activity? In *The 3-pound universe*, Judith Hooper and Dick Teresi explored with scientists of various disciplines the question of whether consciousness was merely the name for nerve-cell activity. But definitive resolution of the question escaped their efforts. We do know, however, that the thinking mind, whatever that may be, involves energy transactions in the brain.

We also know, for example, that nonbiological information-processing activity such as that in computers programmed with artificial intelligence, involve energy transactions. Somewhere there is an energy link between mind and matter. Freeman Dyson suggests, in *Infinite In All Directions*, that consciousness is the result of the organization of atoms and molecules and that "flesh and blood" are not prerequisites for such organization.

How Does the Brain Produce Subjectivity? What is the biological basis for a subjective inner life? Layzer notes that even though all organisms have a nervous system, they do not have an inner life. He points to the *autonomic nervous system* which controls our automatic functions like digestion and respiration--part of our bodies but their sequences not subjectively dictated by us. So 'consciousness' is not this kind of automatic nervous activity.

The *somatic nervous system,* on the other hand, does regulate by receiving and processing sensory information and physically reacting. A reflex action of an animal is of this type but does not involve subjective inner experience. But inner experience has more to do with internal workings like "thinking, remembering and expecting," then it does with sensing and reacting.

'Feature Detection' Versus 'Perception: Layzer distinguishes 'feature detection,' like a frog's retinal capture of a bug's presence, and the reflex muscular eating response, from human 'perception' which requires a *cyclic process* in which the brain constructs, tests, and modifies, a perceptual hypothesis.

So perception requires *hypothesis construction.* Perception= construction, where *seeing the scene triggers the perception cycle.* He points out that a desire, an internal stimulus, can also trigger the construction of cerebral representations and result in images.

From experience of meditation--I used to think of psychic phenomena as New Age flimflam...and reincarnation as myth, soul as metaphor....Now I know there is a God--my God, in here demanding not faith but experience, an inexhaustible wonder at the richness of this very moment. Now I know there is a consciousness that transcends science,..toward which our species is sputteringly evolving, a welcome development spurred ironically by our generational rendezvous with mortality. --Marty Kaplan

CHAPTER 41
MIND OVER MATTER
MIND'S INFLUENCE OVER ILLNESS

Psychoneuro-immunology (PNI)--Mind Mobilizing to Control Disease: The developing field of *psychoneuro-immunology* or PNI posits a biological relationship between emotional, brain, and immunological systems in which the three systems communicate. Psychotherapists and mental-imaging specialists, so-called "visualization therapists," believe that people can control their illnesses through mental imaging. This is described in *Getting Well Again* by O. Carl Simonton and Stephanie Simonton, citing cancer and other cures that seemingly demonstrate the mind's curative power.

Biofeedback instructional materials cite the case of a child's brain tumor reported to have disappeared over a period of time after the child's intense mental imaging, in this case, visualization of persistent atomic ray-gun attacks on the tumor.

Norman Cousins and Mental Control of Illness: In *An Anatomy of an Illness*, the New England Journal of Medicine describes the case of Editor of the Saturday Review Norman Cousins, diagnosed as having an incurable disease of the connective tissues, collagen illness. He adopted a *laughter regimen*--and recovered.

He became a biochemistry researcher at UCLA School of Medicine to study his theory of mental control over illness. He found a *substantial increase in immune cell production resulted from a cheerful frame of mind.* This confirmed his view that since negative feelings--depression and stress--suppress the immune system, positive feelings should enhance the immune system.

Positive Outlook, Optimism, and Happiness Increase Immunity: Cancer patients in a UCLA study who participated in an interactive mind-stimulating group therapy, fared better than another group that did not participate, despite the same medical routines. Of the two groups of women studied at Ohio State University, one married, the other single or divorced, the married group had stronger immune systems. *Optimists test better in disease fighting* than the less resistive pessimists. Happiness also tests out as a positive factor in disease control.

Exercise and the Immune System: California Loma Linda University researchers have found that an exercising group had fewer colds and flu and recovered faster than a sedentary, nonexercising group and the natural killer cell (NK) activity increased in the exercising group.

Walking and simple exercise seems to foster good feelings, apparently influenced by endorphins called neurotransmitters, hormones produced by the brain, nervous and endocrine systems. Exercise increases the immune-stimulating proteins such as interleukin-1 which raises body temperature and triggers increased activity of T lymphocyte white blood cell activity as well as antibody-producing B cells. Blood-warming procedures have also been used to fight cancer cells.

Exercise also helps to reduce tension and fatigue and thus increase immune system effectiveness. Stress and fatigue produce cortisol stress hormones harmful to the immune system, as well as suppressing disease-fighting killer-cell activity. The American Medical Association Journal of November, 1989 disclosed lowered mortality rates, improved heart and cardiovascular activity and even reduced likelihood of cancer, possibly through removal of carcinogens by better elimination or strengthening of the immune system.

Self-induced Recovery through Faith and Will: Recovery through faith, so-called faith-healing and the "laying-on-of-hands" procedures may be mechanisms which influence the immune system, an aspect of mental imaging, power of suggestion, or psychoneuro-immunology.

Recovery from mortal illness via faith or power of the will, is thought, in medical terms at least, to result from the power of the mind to mobilize the immune system. Religious feelings are considered to have a powerful effect on brain 'willing' mechanisms and are viewed as legitimate reinforcing instruments of psychotherapy.

Whether there is some form of 'spiritual intervention,' as some suggest, cannot be known. The theological view of divine intervention may, it is suggested, coincide with PNI theory, based on theories of 'immanence,' God's omnipresence and his presence within us--manifested, for instance, through the immune system.

Dr. Bernie Siegal's *Love, Medicine & Miracles* cites cases of persons with cancer remission by altering their attitude toward themselves. Anyone who has seen him on TV comes away impressed, at least with his own strong convictions.

A Shift Toward Spirituality in Use of Medical Alternatives: A successful shift in the use of medical alternatives was reported in Time of June 24, 1996, in which faith healing to pray for AIDS patients was an included regimen under the control of Dr. Elisabeth Targ of San Francisco's California Pacific Medical Center. A 1995 Dartmouth study of open-heart surgery reported significant improvements on patients who drew comfort from prayer, and other studies show that frequent church attendance have "half the risk of dying from coronary-artery disease" than non- or rare attenders.

Dr Herbert Benson of Harvard, in *The Relaxation Response,* and in *Timeless Healing,* addresses the advantages of meditation . He suggests that *humans are actually engineered for religious faith.*

He Got Religion: In the same Time magazine, Marty Kaplan, a self-declared "cultural Jew, an agnostic, a closet nihilist," reports reading a Deepak Chopra book which lead him to try meditation. He got more from mind-body medicine than bargained for. *He got religion.*

Some people, he says, "manage to find purpose...making from wholly earthly values a sufficient basis for moral choice and meaningful life. What attracted him to meditation was its religious neutrality...90% of meditation was about showing up." The God he found is "common to Moses, Muhammad, to Buddha and Jesus." In his religion it is known, he says, as "Tetragrammaton, what the Cabal calls Ayin, Nothingness,... Spirit, Being, the All."

I used to think of psychic phenomena as New Age flimflam," he says, "and reincarnation as myth, soul as metaphor." Now, he says, "I know there is a God--my God, in here demanding not faith but experience, an inexhaustible wonder at the richness of this very moment. Now I know there is a consciousness that transcends science, ...toward which our species is sputteringly evolving, a welcome development spurred ironically by our generational rendezvous with mortality."

Caveats: But cancer and other biomedically-oriented specialists feel that so-called 'positive thinking' cures are *misleading* because cancerous cell growth demonstrates an independence from the usual body influences and thinking processes which might have some relationship to remissions.

Even if the theory is correct, say some, not many people may be able to mobilize the brain via the will, to affect the body processes, and there may be psychological damage from a patient's ineffectiveness in marshaling mistakenly-assumed curative powers.

We do know, however, that stress, grief, discouragement have chemical consequences negative to the immune system--can make us sick, and being sick can itself trigger negative emotional responses.

Voodoo Death: Voodoo death--death from sorcery, is suggested as a phenomenon of the power of the will. Where the curse is communicated to the victim, the death of voodoo victims is thought to result from some form of subconscious acceptance of the inevitability of their death. Where the curse is not expressly communicated and death results, apart from coincidence, of course, the suggestion has been made that the information transfer may be by a form of psychic communication power, as in the case of the twin analogy.

The Twin Analogy: In the twin analogy, information is seemingly registered or felt *concurrently* by the sending and receiving mental apparatuses, perhaps involving some sort of quantum theory "informational" linking, a phenomenon sometimes referred to as the "quantum connection." In the twins case we have a single sperm and single egg which subdivided, sometimes viewed as the same person, subdivided.

As we discuss elsewhere here, all particles may be viewed as information units in the universal information system, a product of the primordial unity when all particles, fields and forces were unified. Thus, all sending, receiving, and information components have a primal linking by their very existence, needing only some mechanism to trigger and receive, to *activate the information into the consciousness.*

> Humans are actually engineered for religious faith. -- Herbert Benson

CHAPTER 42
SANCTITY OF LIFE, RIGHT TO LIFE, RIGHT TO DEATH

What Characterizes the Sanctity of Human Life? The most familiar concepts of the sanctity of human life derive from the Bible, made in the image of God, a little lower than the angels, and given dominion over other forms of earth life. But there are many who view human attitudes toward other life, whether or not based on the Bible, at least cavalier, if not arrogant.

There are many organizations throughout the world that seek to promote respect for and preserve other life on the planet. Besides, dominion over other animal life does not mean disrespectful attitude or action toward other life. There may be other, far more advanced, forms of life in the universe who might perceive human life as primitive, destructive, a threat to cosmic morality. In any event, what distinguishes human life from other life is intelligence, consciousness and subjectivity, a product of evolution of the human brain.

Attitudes toward Sanctity of Human Life: In *The Dragons of Eden, Speculations On The Origin Of Human Intelligence*, astronomer and space scientist Carl Sagan discusses brain evolution. In the process he talks of what characterizes the sanctity of human life from other life, concluding that the distinction lies in the *level of intelligence*, traceable to the "development and functioning of the neocortex."

The Human Brain and Abortion: In this view, humanity commences when neocortical activity begins, as determined by EEG signals (electroencephalograms) which measure the electrical activity of the neural network. In his book *The Dragons Of Eden* and in his article *The Question Of Abortion*, the April 22, 1990 Parade Magazine, Sagan, discusses the history of attitudes toward sanctity of life and the different views of various religions and cultures in different times.

The Slippery Slope of When Life Begins: He notes that both pro-choicers and pro-lifers are "pushed toward absolutionist positions by parallel fears of the slippery slope." "The slippery slope" is a reference to the inability to fix on a point in pregnancy when life is deemed to have emerged--when we "become human." Thinking, our blessing and our curse, says Sagan, occurs principally in the top layers of the convoluted gray matter of the cerebral cortex, the roughly 100 billion neurons in the brain which constitutes the material source of human thought.

When Do We Become Human? *Roe vs. Wade*: The "large-scale linking up of neurons" starts in the 24th to the 27th week, the sixth month. But brain waves with regular patterns characteristic of adult human brains appear in the fetus near the start of the third trimester, the 30th week. Allowing for early or precocious fetal brain development, we might draw the line, says Sagan, at six months, where the Supreme Court drew the line in *Roe* v. *Wade* in 1973, based, however, not on the inception of human thinking ability but on the concept of 'viability,' meaning by that, survivability outside the mother.

Scientific Views in The *Webster* Case: Neurobiologists do not view brain development in the fetus as capable of human thinking, 'personhood' functions, until after 28 or so weeks into gestation, according to the large group of scientists and physicians, including Nobel laureates, who presented their views to the Court in the 1989 Webster case.

Sagan's Suggested Balance between the Conflicting Claims: That case allowed abortion in the first trimester, and in the second trimester for health-related exceptions. It permitted state restriction in the third trimester except for serious life or health threat to the mother. A fair balance between "the conflicting claims of freedom and of life," lies, he feels, in the criterion of the commencement of the ability to think, a condition which is verifiable by brain-wave signals, but allowing for earlier or 'precocious' fetal brain development, at six months. This just happens by coincidence, to be where the Supreme Court put it, but based on 'viability' concept.

Early Human Life Status and the National Right to Life: Of course, in recent years, religious thinking keeps moving the human being status back to earlier points. Jack and Barbara Wilke of the National Right To Life speak of the creation of a "new being" when, "at fertilization, the 23 chromosomes from the sperm join the 23 chromosomes from the ovum," to create "a unique being, containing within itself a genetic package, 46 chromosomes, completely programmed for and actively moving toward adult human existence."

Prefertilization Sanctity of Life? Sagan asks whether the trend toward treating earlier fertilized states as constituting human life or personhood, will lead to some form of prefertilization human sanctity for sperm and ova, which represent *potential* life. Of course, atoms, quarks, leptons and other quantum states also have 'life potential' and we discuss elsewhere here the entire universe as a living organism and mind as a quantum state of matter.

Some day, Sagan points out, science will likely be able to develop a form of scientifically-induced immaculate conception, that is, clone, without fertilization, a human being from a single cell (which just happens to have the needed 46 chromosomes for life). Will it become unlawful to lose the cells in a drop of blood, he wonders.

Theology, Philosophy, Biology and the Start of Personhood: Views differ widely on the time that human personhood commences; but several schools of thought have evolved with theological, scientific-philosophical, and the right-to-life at-fertilization, sometimes called the 'biological' view.

The theological view is based on the dual body-soul theory involving the soul's 'insertion point,' the time when the soul enters the body. Views differ on when this occurs. Some see this as the point of fertilization, as viewed for instance by the National Right To Life organization, while others suggest the point when the embryo gives a human appearance. But judicial imposition of a theological view is seen as a violation both of the constitution's separation of church and state and the woman's fundamental physiological rights.

The prevailing scientific view, as discussed here and as described above by Sagan, is that consciousness, subjectivity and cognitive thinking functions associated with personhood are the product of organizational development in the brain, not the point of 'soul insertion,' the latter being a questioned concept in any event.

Denominational Views on Abortion: Religious denominations vary widely in their views, edging toward antiabortion but, for some, with certain exceptions like endangerment of mother's life, rape or incest. But opinion polls of individuals tilt toward free choice, regardless of denomination. Until the Webster case, public demonstrations and political pressures were dominated by pro-life or antiabortionists. Since that case, however, public pro-choice demonstrations have increased to become a political factor, affecting various elections. Adherents of both views quote scripture for support. Abortion issues now play a major role in the American political arena.

The Value of Life in Medical Impairment Situations: A hospital is sued for the suffering and misery in the continued life of a seriously ill patient whose wishes for discontinuance of treatment were frustrated by the hospital's continuing the treatment despite the patient's protestations and insistence on terminating treatment.

What's at stake is the tension between the patient's subjective perception of a life worth living--the quality of life, versus society's present emphasis on 'unconditional' life continuance, continuance at any cost, without reference to meaningfulness or value of the kind or quality of the life to be preserved, and without regard for the wishes of the *possessor* of that life or the wishes of the family.

President's Commission Recognizes the Patient's Views: In considering ethical and biomedical problems in life continuance, a special commission, The President's Commission for the Study of Ethical Problems in Medicine and Biomedical and Behavioral Research, considered what kind of life a person's condition offers, whether the condition permits "a life that he or she views as worth living."

Saving Lives in the Present As Destroying Lives in the Future: Omni magazine interviewed microbiologist and ecologist Garrett Hardin, lover of nature. Influenced by British economist and writer Thomas Malthus, Hardin is the author of *Living Within Limits, Exploring New Ethics For Survival,* as well as the essay, *The Tragedy of the Commons.* These works all warn against future population growth as leading to ultimate chaos.

'Progress' Leading to Chaos: Hardin deals with the repression of thought on sensitive issues relating to the dangers of economic, medical, and social so-called 'progress' in the real world. Valiant efforts to reduce infant mortality, to enhance medical facilities for backward peoples, to assure greater longevity for example, inevitably accelerate huge population increases, taxing the planet's resources and hastening ultimate ecological chaos leading to the end of the human species. Such ideas have begun to influence social concepts and have led to the formation, for example, of the ZPG, Zero Population Growth, organization which encourages reduced child-bearing.

Reproductive Freedom without Responsibility: Hardin points out that this appeal would influence reduced fertility in the more educated without achieving fertility reduction in the less educated. Reproductive freedom, without responsibility for their care, transfers the burden, financial and social, to society at large--the clear prescription for disaster.

Harbin makes the case for better planning for the future through more emphasis on assisting birth control than in reducing infanticide. Sending food to Ethiopia increases population while Ethiopian food production declines, the land's carrying capacity declining with overuse. He gives the Chinese credit for restricting childbirth and suggests that the most effective population control for the U.S. would be to end immigration, pointing out that the U.S. accepts more immigrants than the rest of the world combined.

Hardin, Abortion, and *Roe v. Wade:* His ideas of population control lead naturally to a discussion of abortion policy, and his preference for upholding Roe v. Wade. He emphasizes the dangers and inconsistencies in regarding the newly fertilized egg as entitled to all the laws and protections for human beings. Thus the loss of 10,000 acorns should not be treated as the loss of 10,000 trees.

Hardin's view is that "If a woman wants an abortion, either because she doesn't think she will be a good mother...it's not in society's interest to urge her to have the child. We have enough poorly-taken-care-of children already...." Regarding *Roe v. Wade*, he observes, "I can see no good resulting from forcing women to have babies they don't want. The cost of raising these children will sooner or later reach the general public.... He notes that Conservatives usually strive to avoid taxes and higher expenses. And yet, by opposing abortion, they ultimately ensure higher taxes...."

Living Wills: Many are now executing 'living wills' which set forth the wish of the declarant, made when the declarant is of sound mind and in full possession of faculties, that machine or other artificial methods for life continuation not be employed or continued once recovery or normal living is no longer achievable.

Although not necessarily honored in all jurisdictions, these statements of purpose and intent are being used to prevent the life-continuance practices normally followed in hospitals because of medical philosophies or fear of malpractice liability. Many documented case histories bear witness to the months or even years of agonizing caused the families and loved ones of patients whose lives have been artificially machine-prolonged when the patients have been long brain-dead.

Euthanasia Aspects of Rationing Health Care: The burgeoning cost of medical care in the United States, fueled by factors such as the inordinately high cost of new medical technology, is said to be virtually out of control. Projections show this cost as threatening the solvency of the national medical system, especially Medicare and Medicaid.

In a world in which profitability of new life-saving developments is the primary driving force, the economics of development seems pitted against understandable humanitarian efforts to make each new development, no matter how expensive, available for the saving of life. Stories are told of insurance company reluctance to authorize reimbursement for such procedures on the grounds that the particular technology, even though proven, is new--in a test stage, and therefore not yet reimbursable.

The problem is especially acute in the rising costs of care for the elderly. Another factor is creeping in where the young are perceived as being asked to pay increased taxes for the medical care of the elderly and a new old-young tension is growing.

Inevitably, reluctance to raise taxes to accommodate the costs of such new treatment has generated thoughts about some method of "rationing" health care. Some suggest that patients must bear a higher share of the cost of the medical procedures; others suggest that availability of such expensive new technology must be limited in some way, another name for rationing.

This, in turn, is viewed by many, including the National Right To Life Committee, Inc., as economics-driven euthanasia--and that's just about what it may amount to. But what's the answer to the rock-and-a hard-place problem where people will needlessly die unless other people increasingly pay for this compassion through increasingly unaffordable higher taxes?

Medically-Assisted Death: Although the practice of assisting in life termination is accepted in some European countries, without reference to the law, it is not accepted here in the United States although there are, no doubt, individual cases where long-suffering and hopeless condition of patients have influenced physicians to diminish or terminate possible treatments.

The Suicide Machine: But occasionally views of a physician sympathetic to the plight of a suffering patient with a terminal condition leads to condemnation and criminal charges. The Michigan doctor-assisted suicide cases come to mind. There patients with no medical potential have sought out Dr. Kevorkian for assistance.

Playing God? Accessory to a Crime? Some have charged that in assisting in the death inducement procedure he was playing God; others that he was guilty of homicide or as an accessory to a suicide, a criminal act in some jurisdictions. Still others charged him with violating the Hippocratic Oath. He responds that he turns away those with any recovery potential; that these desperate persons would find other more traumatic or damaging ways to self-destruct, such using a gun, whereas his assistance provides effective, painless, and comforting medical supervision.

Supreme Court: The courts are ambivalent and state law varies. In January of 1997 the Supreme Court heard arguments on the the right to doctor-assisted suicide--whether people have “a voice in their own suffering.” If the issue is decided, it will be a landmark ruling, comparable to Roe v. Wade. Justice Scalia asked why society can’t determine that it is wrong to kill yourself just as it is wrong to kill another.

The Doctor's View: Dr. Kevorkian cited the historical background of the Hippocratic Oath which seemingly did not preclude providing medical assistance for the act. He saw nothing in the Michigan law forbidding his assistance and sees no inherent interest of the state in useless life prolongation.

Turning away from someone desperately seeking to end needless suffering from a terminal debilitating disease which in any event destroys personhood, he says, could hardly be offensive to a loving God; that in his compassionate act of succor in this patient's hour of desperation, he should rather be viewed as being an instrument of God's own love and compassion. He has been acquitted in the several cases brought against him, two based on statute and one described as a 'common law' violation.

Families Who Help the Suffering in Terminal Illness: Similar questions have been raised where one family member makes a gun, or lethal dose of some medication, available for use by a suffering family member with a terminal illness. But society in general is negatively disposed to these acts and even courts are often unwilling to sanction turning off high-tech life support.

How Do Doctors Feel about Help to the Terminally Ill? The Washington Post of June 12, 1990 carried a quote from the National Capital Hemlock Society solicitation of local medical views on the role of medical assistance in cases of the haplessly ill. Fifty six percent of the respondents believed such patients should have the right to assistance ending life.

If decriminalized, forty-eight percent would assist through lethal dosing. Sixty-eight percent would disconnect artificial life support if requested by the patient. Three responses acknowledged having assisted the patient to die by making available or actually administering lethal doses. But the relatively small sampling, five hundred, is too random and limited to be considered a broad consensus.

Some Church Views on the Right to Die: As a general rule churches tend toward life-preservation along the lines of their views on abortion. But there are differences. At a Denver regional conference of the United Church of Christ, a resolution was passed reflecting concern over 'biological idolatry.' 'Abundant life' is something more than simply the 'avoidance of death.'

The resolution, suggesting support of the right of persons "in hopeless and irreversible conditions" to terminate their lives, was sent along to the church's general synod. Although suicide is not advocated, it was suggested that there should be "Christian compassion" for those who elect that route.

CHAPTER 43
WHERE DID LIFE COME FROM?

Life And Monod's "Chance and Necessity": The assumption that everything that occurs in the universe was predetermined at the creation was severely challenged with the advent of quantum theory's uncertainty principle which caused many physicists and philosophers to opt for the proposition that the universe is fundamentally indeterministic, governed by chance.

Contrary to Einstein's belief that God would not 'play dice' with the universe via quantum mechanics, Hubert Reeves, in *Atoms Of Silence*, feels, as do most physicists today, observes that God does indeed 'play dice' with the universe, at least to the extent of the functioning of the quantum uncertainty principle.

He quotes French biologist Jacques Monod in *Chance and Necessity*, as describing the innumerable combinations inherent in the thousands of atoms in the "fundamental units of life," the proteins, enzymes, and nucleic acids. "It is well accepted that life arose from inanimate matter....This "miracle" of life seemed to have been "an exceedingly improbable event...a juxtaposition of coincidences," leading, in earlier years, to the conclusion that we must be alone in the universe.

Reeves describes, however, how nature's invention of DNA has controlled or, as he puts it, "bridled chance." Monod's view was that the role of chance demonstrated the absence of *intentionality* in nature and, unlike physicists, like Paul Davies and others, who see a universe of purpose in the remarkable design of the universe, the anthropic principle, Reeves sees no symbiosis or necessary interrelationship between the universe and human life.

In Monod's view, human life was an accident in the cosmos, the product of chance. But Reeves, not quite satisfied with Monod's total abandonment to chance, would avoid total abandonment of the universe to chance. With DNA, he says, nature has "tamed chance" and "made it an admirable ally."

Chance As a Creation Mechanism: Reeves concept of God's use of chance as the mechanism for the evolution of life, a *creation mechanism,* is echoed of course by others--that the creation of human life could not have come about except for and through a chance selection and organizing mechanism. And so, in this view, the chance inherent in quantum uncertainty is part of the reality designed into the macro universe in which we live.

Randomness As Inadequate to Account for Life: The idea of life as the product of pure chance has many other questioners. In *The Intelligent Universe*, Sir Fred Hoyle suggests that life is not reasonably explainable simply as a product of random occurrences; that life implies a purposeful primordial intelligence.

The liklihood, for example, of just the right enzymes coming together to form a single living cell, he suggests, is an astounding 10 to the 40,000th power to 1. He describes the universe as run by an interlocking hierarchy of intelligences which functions purposefully to bring about life and mind as a form of order in a universe otherwise governed by the thermodynamic processes of entropy, disorder.

Organic Molecules in Space: The finding of a profusion of *organic molecules in space*, recently identified in Halley's comet, for example, suggests degrees of molecular evolution in outer space as well as on planet Earth and, of course, the possibilities if not probabilities of life forms elsewhere in the universe.

Evolution: David Hume's "Botched And Bungled" Universes: Scottish Philosopher David Hume speculated, in his *Dialogue Concerning Natural Religion*, about going beyond evolution in our universe to an evolution of universes. He speaks of numerous universes that might have been "botched and bungled throughout an eternity ere this system was struck out; much labour lost, many fruitless trials made, and a slow but continual improvement carried out during infinite ages in the art of world-making."

But Edward Harrison in *Masks Of The Universe*, commenting on Hume's bungled universes, asks, "Why stop?" Might not other universes thereafter have been struck out more splendid than the one we inhabit?

People Find It Difficult to Accept Evolution: Hubert Reeves, in *Atoms Of Silence,* speaks of the eons of time, billions of years, that it took nature to bring about life and points out that people find it difficult to accept evolution. We do not refer to the Genesis argument here but the sheer implausibility of "chance" producing our form of life, people.

The odds against our life forms approach infinity, he says. Generalized explanations of chance in a random universe are insufficient to explain how it happened that the first molecule evolved that was capable of replication. As suggested earlier, other explanations have been proposed, especially those relating to the tendency of matter toward self-organization, the dynamics of self-organizing systems. A related concept is that of matter as an earlier stage of life.

How Does Increasing Order, Life, Occur in the Entropic Universe: In a universe governed by the second law of thermodynamics, in which systems, like the universe itself, tend toward entropy, states of *decreasing order* or increasing disorder and chaos, how does one explain the phenomena of increasing complexity, *increasing order*, increasing concentrations of information, like the emergence of life and consciousness? One answer to the implausibility of life as a product of random chance lies in the discussion of the nature of systems of increasing order.

Cases of increasing order, such as evolution of life, were considered an anomaly or deviation from the entropy law and were the subject of great speculation. In his *What is Life* papers, physicist Erwin Schrödinger, one of the founders of the quantum uncertainty principle, suggested that there must be some more refined explanation for the drive toward life in the universe than conventional thermodynamcs. Even Darwin, father of the theory of evolution, noted that the development of the human eye with its incredible complexity could not possibly be explained simply by evolution.

Prigognine's Universe of Self-organizing Systems: *Order out of Chaos*, by Ilya Prigognine and Isabelle Stengers, recounts how Prigognine, a student of self-organizing phenomena, observed that while increasing disorder characterized closed systems, open systems exchange information with their environment and 'fluctuate.' When the fluctuation reaches a critical threshold, a 'bifurcation point,' it either deteriorates into a lower order, or surges into a higher level of order or organization, a 'dissipative structure,' into which is absorbed the energy and information arising from the fluctuation.

Chaos Subordinated to Order; Matter Part of the Life Continuum: The tendency of matter toward self-organization represented a new property of matter which implied a revolutionary departure from the earlier concept of self-organizing systems as an anomaly. Such open systems were now seen as a primary characteristic of the universe. Self-organizing systems, known to exist in living systems, with a key role in life evolution, was now known to exist in pre-life nonliving systems.

Biological life, rather than being a random exception to the rule of increasing disorder, was proposed as structured into the very laws of matter. So matter was no longer totally inanimate, but part of the *life process continuum.* Matter, in this view, is an earlier stage of life, and in that sense, matter is alive, and aliveness is a matter of degree. For this work, Prigognine received the Nobel Prize.

Life a Disease of Matter Or Part of the Cosmic Intelligence: Explanations for life abound. The present genetic material, DNA and RNA, are the distant progeny of that ancient replicating molecule which, says Heinz Pagels, in *Perfect Symmetry,* must have made countless billions of copies "in an orgy of reproduction" until an error occurred that produced a "*different self-replicating molecule*," starting molecular evolution. So life may be the result of molecular aberration, *a disease of matter*, citing the musings of physicist Gerald Feinberg. Evolution is too implausible to have been *imagined*, it had to be *discovered*, he says.

But, we have noted the physics of self-organizing systems giving rise to 'bifurcations' which lead to new thresholds of more advanced development. We have also noted the ability of biological systems when confronted with crisis to somehow cross time barriers and draw on information from the future, via what is sometimes described as the quantum connection, the underlying informational quantum cosmic mental interconnectedness, the underlying universe as mind.

Mutations in Nucleotide Sequences: Carl Sagan points out that the sources of evolution are mutations in the gametes, the eggs and sperm cells, representing inheritable changes in the particular nucleotide sequences which constitute the heredity instructions in the DNA molecule. Mutations are caused by cosmic rays, such as gamma rays, from outer space, or from 'spontaneous rearrangements' of the nucleotides, a 'statistical property of randomness.'

What Happens in the Oceans? Subjected to continuous electrical discharges from the lightning of the primal atmosphere, says Hubert Reeves, the oceans, as demonstrated by laboratory experiments, develop organic materials, fats, amino acids, alcohols and sugars. New molecules with 'hooks' combine repetitively in chains of polymers, carbon playing the role of catalyst, hooking molecules together, molecules that are 'predatory,' having the ability to feed on alcohols and sugars for energy and to join with fatty molecules to provide a membrane to make them watertight, like the oil in duck feathers, but semipermiable so as to admit sugars for energy--digestion.

Advent of the Special Molecule, Chlorophyll: But in the molecular mutation process, notes Reeves, there evolves a new molecule, predecessor to chlorophyll, with the ability of capturing photons from the sun and storing their energy. Thus starts the process which provides for the use of solar energy, through *photosynthesis*, to power or nourish animal and vegetable life of the planet, using water as the energy transportation system.

The Crucial Role of Water: Physicist George Greenstein's *The Symbiotic Universe* highlights the fact that life's very presence in the universe is tied to an extraordinary substance, water. In all living things, water predominates, apart from a bit of other materials for structure and functioning.

Every living thing, plant or animal is a factory with things being moved around--oxygen, nutrients, waste products, glucose produced by leaves through photo-synthesis, nutrients spread throughout the trees and plants from roots. The trans-portation system for all of this? Water--Water that transports these materials mostly by acting as a solvent, by dissolving them.

For electrical reasons, the solvent water molecules cluster to surround and isolate the solute materials as ions, in which the negative electrons have been removed, so the dissolved ions are positively charged. The negative ends of the water molecules are attracted to the ion positive charge. This electrically-induced clustering of water molecules, and its dissolving strength, is what makes it so suitable as a conveyor in the factories of all organic life, notes Reeves.

What Is the Source of Our Breathable Oxygen in the Atmosphere? But life, as we know it, requires oxygen. Where does it come from? Well, says George Greenstein in *The Symbiotic Universe*, spread over much of the planet is a magnificent *solar energy conversion system.* It traps the energy of sunlight and converts it through the process of photosynthesis, to a form in which it is biologically usable, directly by plants and, through consumption of plants [or animals], by animals.

In photosynthesis, water is combined with carbon dioxide in the atmosphere to form carbohydrates, *glucose*--the basis, finally, for all foods--and, in the process, releases oxygen back into the atmosphere as a waste product. So all life on earth is energized--fueled from sunlight, from the waste product of photosynthesis.

The human brain, notes Reeves, runs on chemical and electrical energy, as, for instance, the process by which the eye scans this material, translating the perceived images into understanding, thoughts. We know, from Einstein's equation that defines the universe, $E = MC^2$, that sunlight is matter, and about *160 tons of sunlight* fall on the earth each day. And all oxygen in the atmosphere comes from this photo-synthetically-processed water.

From the Oceans Come the Amino Acids for DNA: Reeves refers to the ocean-evolved molecules, the amino acids which form the proteins and the nucleotide bases which form DNA. The atomic arrangement of the twenty amino acids which are used in the transmission of the genetic code, is described, such as that of aspartic acid, glutamine, glycine, lysine, tryptophan and valine. The molecular structure of the nucleotides (A-adenine, T-thymine, G-guanine, C-cytosine) is outlined.

Molecular Chains--Genes: The genetic components, genes, are molecular chains of the nucleotides, the nitrogenous base subunits or links. The two-stranded chain, called nucleic acid, or DNA for Deoxyribonuclei acid, is commonly displayed as coiled in a double helix. It is the principal component of the 46 chromosomes in each human cell which transmit the characteristics of heredity.

Information in a Chromosome: How much information is in a single human chromosome? In *The Dragons of Eden, Speculations On The Origin of Human Intelligence*, Carl Sagan says that the human chromosome equals some five-hundred million words--two million pages, or some four thousand volumes of five-hundred pages. Sagan points out that the Mars Viking Lander contained "slightly more genetic information than a bacterium, but significantly less than an alga."

Amino Acids: Triplet combinations of the nucleotide bases represent an amino acid, for instance, TGC is aspartine. Needed proteins for cells are selected from the amino acids. Each type of animal--and each individual--has its or his own unique sequence. Reeves gives the example of the spatial structure of a molecule "cytochrome-C," used to capture molecules of oxygen from respiration.

The Cell As the Foundation for Life: Human life is sometimes referred to as being midway between the atom and the stars. Well, Reeves refers to the cell as midway between the atom and the animal, a biological instrumentality performing "all the great functions of life," the foundation of all living things. Animal life forms are "assemblages" of varieties of cells, the *human body containing over a hundred thousand billion cells* functioning in different roles--bones, hair, blood--all in a controlled harmony.

So how big is a cell compared with an atom? *A cell has about a thousand billion atoms,* says Reeves. What's in a Cell? The cell is the producer of proteins and in the chromosomes in its nucleus it stores the genes containing the "genetic code" for living and reproduction--encoded information in the form of an "immortal" sequence of molecules called DNA or deoxyribonucleic acid.

Cytoplasm: Organelles, Mitochondrea, and Ribosomes: The cell's material is surrounded by a membrane enclosing the gelatin-like 'cytoplasm' which envelops the genetic-code-containing nucleus. The cytoplasm contains the 'organelles' that perform human-organ-type functions like those of the lungs and liver. The 'mitochondria' receive oxygen through the cell membrane, cellular pulmonary respiration, and release energy to be stored by the ATP (adenosine triphosphate), the cell's 'rechargeable batteries,' until they deliver the energy at energy-requiring locations in the cell.

The protein-building process is performed by the 'ribosomes' from nucleotide bases in amino acids derived from digested food. The amino acids are hooked together to form proteins as specified by the genetic code.

The Vegetable Organelle 'Chloroplast' Produces Oxygen: The 'chloroplast' organelle of vegetable cells, with the help of sunlight, take water absorbed by the plant's roots and, through the leaves, carbon dioxide from the air. These combine to form glucose or sugars--stored energy for plant-eating animals. In this process, photosynthesis, *oxygen is released as waste*,--the source of the oxygen in our atmosphere.

The 'mitochondria,' the cell's respiratory and energy producing components, receive blood-delivered oxygen through the cell membrane and release energy to be stored by the ATP (adenosine triphosphate), the cell's 'rechargeable batteries,' until they deliver the energy at energy-requiring locations in the cell.

Reproduction of Plants and Animals: Animal and plant reproduction is preceded by cell reproduction. DNA molecules prepare copies of themselves for transmission by RNA, messenger molecules, to opposite sides of the cell which then divides to form two identical cells.

Sometimes though, essential information for producing the proteins has been omitted from the genes. Yet, strangely, the missing information is supplied by "RNA editing," the recent explanation for which is that the mitochondria--which release energy to be stored by the cell's batteries, called ATP-- produces "guiding RNA" pieces identical to the genetic material implanted into the RNA messenger molecules. Life for each of us began, say Reeves, with a single cell, the egg fertilized by the sperm, dividing into two, four, eight and all the way to trillions of cells in an adult.

Origin of the Cell? One theory, notes Hubert Reeves, is that cell formation results from collaboration, simple systems "pooling their talents," part of the working of the anthropic principle and physicist George Greenstein's *The Symbiotic Universe,* involving a crucial interdependency of the structure of the universe and human life.

The first oceans, says Reeves, abounded in complex molecules. Blue algae, single-celled organisms capable of producing photosynthesis, thrive even today in temperatures of a hundred degrees. They are found in hot water from geysers in Iceland. They do not have a nucleus, being a gelatin-like mass within a membrane. Also, surrounding the geysers are similarly organized single-cell bacteria.

According to the theory of collaboration, says Reeves, it is thought that these two simple cell forms associated to form the complex cells leading to human beings. The bacteria, as mitochondria in animals, provided respiration and the blue algae, as chloroplasts in plants, provided photosynthesis.

The Secret of Life: In his *Cosmogenesis* chapter on Molecules, Genes, and Evolution, physicist David Layzer describes the function of DNA, the secret of life, as encoding a program of regulated and coordinated chemical reactions for the development of an organism, describing the control role of the genes and proteins. A gene directs the synthesis of the proteins it encodes through RNA molecules and contains the same information as the protein it encodes.

At the molecular level, as he points out, the laws governing lifeless matter also govern interactions in living matter. Biological organization is attributable to nucleotide base sequencing in DNA which undergoes change as it passes through later generations. While information is transferred from DNA to its encoded proteins, so far as is known, there appears to be no reverse information flow by which the genes could communicate experience in the organism's replication process. The information in a gene is the same as that in the protein which the gene encodes, written in an "alphabet of DNA triplets" rather than in amino acids.

What Distinguished Living from Inanimate: What distinguishes a live cell from an inanimate version of the same molecules, Layzer says, is "its organization and the organization of the processes that go on within it and at its boundary," organization that reflects information encoded in the cell's DNA which directs the assembly of organisms.

In all life forms, DNA is assembled from the same four nucleotides and the proteins assembled contain the same twenty amino acids. All forms of life, he notes, use the same genetic code linking a specific amino acid to every sequence of three consecutive nucleotides.

What Encoded DNA in the First Place? But how did the information on molecular combination get encoded in the DNA? *Evolution,* Layzer answers. Did all life forms evolve from a single ancestral source through replication with chance errors in copying of DNA?

Drawing on the work of men like Lamarck and Spencer, says Layzer, Charles Darwin framed his view that evolution is the result of 'differential reproduction' (historical accidents) and undirected genetic variation (natural selection). In *his The Origin of the Species*, Darwin asserted, in 1859, his theory that descent of human life resulted from "modification through variation and natural selection."

Layzer points to Mendel's 1866 genetic studies as reinforcing Darwin's theory, pointing to inheritable traits as transmitted by genes which were later discovered to be subject to spontaneous mutations; fitter genes become more frequent.

Commonality--All Organisms Derived from a Single Genetic Source: Every living organism has chromosomes with the genetic material of inheritance, nucleic acids (DNA). The fact that every organism uses the same code to translate the information it contains into the proteins that mediate the organism's development is prima facie evidence, says David Layzer, of *descent from a common ancestral population.*

Human life and other kinds of molecular evidence point to a common origin for all existing life forms. ATP (adenosine triphosphate) molecules, for instance (the standardized medium of exchange), provides the high-grade energy needed to drive the chemical reactions that maintain the structure of a cell against entropic decay and enable it to do work in its surroundings.

In *The Dragons of Eden--Speculations On The Origin of Man's Intelligence*, Carl Sagan points out that the genetic instructions of all life forms on earth are written "in the same language, with the same code book," that this "shared genetic language" is one element of evidence that "all the organisms on Earth are descended from a single ancestor, a single instance of the origin of life some four billion years ago."

From Molecular Anthropology, New Insights on a Common Ancestor: Washington Post Staff Writer Larry Thompson recently noted the growing conviction of biologists that through some 1.6 million years, *all of earth's present five-billion or so human life population is descended from a single mother* called, you guessed it, "Eve."

Mitochondrial Lineage: Early work in this aspect of molecular biology by Allen Wilson of the University of California at Berkeley and Douglas Wallace of Atlanta's Emery School of Medicine relate to studies of the human life cell's mitochondria, shown to have DNA material different from human life chromosome DNA, reflecting the fact that mitochondria developed like bacteria, outside of and independent of human life cells. The common-lineage conclusion, based on mitochondrial inheritance, was reached some years ago based on an analysis of the *mitochondrial genes.*

Unlike *chromosomal genes* which are mixed in the reproduction process, mitochondrial genes are not mixed and only daughters pass on the mother's mitochondrial genes which can change only by random mutation. Noting that mutations in different populations permits measuring relationships, the fewer the mutations, the closer the relationships. Interestingly, Thompson notes the information that distinctions between some ethnic groups, such as Asians and American Indians are as recent as 15 to 30 thousand years ago.

The analysis disturbs earlier assumptions that hominids like the Neanderthal and Homo habilis were part of human life's family tree. It turns out, says Thompson, that those species were simply "dead- end branches," though grown from the same trunk as human life.

Opposition to Darwinian Theory: Layzer notes that Darwin's theory found both scientific and religious opposition, one ground being its seeming *implication of meaninglessness of life*. But despite its opposition, it has generated broad scientific acceptance with the increasing biological, geological and paleontological data tending to support it. New metaphysical conceptual ideas about matter, some rooted in science or supported by scientific discovery, have also contributed to mellowing of theological views.

But some feel that while Darwinism reasonably explains 'microevolution'--like changes of the type promoted by animal breeders--it falls short of explaining 'macro evolution,' quantum leaps or highly significant major advances, like the advent of the extraordinary human intelligence. Layzer agrees with biologists Medawar and Stebbins that Darwinism has not provided the basis for understanding the more esoteric occurrences illustrated by the growth of *biological complexity.*

Francis Crick--Life a Near Miracle: As we have earlier noted, the idea of life as the result of pure randomness leads many to suggest other explanations like the dynamics of self-organizing systems and the theory of matter as part of a built-in life continuum.

Darwin himself questioned randomness as ever explaining certain aspects of evolution such as the intricacy of the human eye.

Nobel-prize winning Francis Crick, co-discoverer of DNA structure, questioned life as an accidental process. Computer scientist Edward Fredkin suggests the universe is an experiment of some sort, unfolding according to a computer software program directed by an ancient intelligence, someone or something running the program to see what evolves.

The Universe a Chain of Intelligence Dictating Events in Evolution: In his *The Intelligent Universe*, British scientist Fred Hoyle questions biological life as a random occurrence, suggesting that the universe is governed by organized intelligence, an interlocking hierarchy of intelligences which *dictate the direction of evolution through organizing principles* that are not confined by limitations of time. They draw on the future to make advances at critical or 'bifurcation points.'

This intelligent cosmic mental structure, chain of intelligence and information system, operating outside the confining limits of relativity's spacetime, injects *striking and sudden changes in the evolutionary process* rather than being confined to purely gradual changes of randomness.

Pierre Teilhard de Chardin and Spiritual Energy Driving Evolution: We have discussed the work of the distinguished Jesuit paleontologist-philosopher Pierre Teilhard de Chardin who saw sanctification in materialism, thought of both the universe and society as organism, and viewed death as a reassimilation of individual consciousness into the cosmic consciousness, a "new break-through and a re-birth, this time outside Time and Space." He spoke of an "etherised universal consciousness."

Some of Teilhard's nonconforming views, of course, were naturally questioned by his Jesuit peers. But he saw *design-purpose teleological aspects in evolution* and found evolution and natural selection inadequate to explain the process of growing complexity, the "higher psychisms," and "the great drive of consciousness." He preferred explanation based on the growth of a "spiritual energy," based on some "inner principle," and "a powerful field of internal attraction."

So, in this view, growing complexity and natural selection was driven by a spiritual energy. While somewhat criticized by some, like theologian Philip Hefner, Teilhard de Chardin's views have had a tremendous influence on many, perhaps because there is a natural beauty to these thoughts and because they imply the existence of meaning. His views bear an interesting correlation, as we noted earlier, to the later-expressed ideas of physicists like Paul Davies and David Bohm who see in quantum theory an underlying, intelligence-driven, directed fabric of cosmic information and mind.

Organic Evolution: But getting back to organic evolution, Reeves highlights the "most evolved" organism at each stage in the history of the planet. In 4 billion years, says Reeves, we passed from blue algae to human beings, in three billion of which, bacteria and blue algae "seemed to have held the ascendancy," since the oldest multi-cell organism appears to be the jellyfish, appearing some seven hundred million years ago. This was followed, a hundred million years later by the skeleton-containing crustaceans, and then, in a similar further period, by the "reign of the fish."

Life Emerges on Land: Thus far, life is confined to the water regions of the planet. It is only three hundred fifty million years ago, says Reeves, that life emerged from the oceans. Reptiles and birds appear on land under an ozone-layer-protected atmosphere, the product of photosynthesis from aquatic plants. Mammals appear a short time later but are limited by the hostile environment of the dinosaurs until the disappearance of the dinosaurs about sixty five million years ago.

Mammals Appear: Appearing at this time, a small mammal, the shrew, "carries the promise," says Reeves, "of the human life brain in its genes," and from this source comes the various primates, then the hominids and early human life. Thus evolution has developed an aggregate of about 10^{29} (one hundred billion billion) "elementary particles," to form human life, illustrating nature's organizational propensity.

Life First Appeared: As to creation, cosmology and astronomy speak in 'cosmic' and 'astronomical' time, representing two-digit billions of years, and in terms of evolution of life, geology and paleontology speak of 'geological' time, representing millions and single digit billions of years ago.

A fall 1996 issue of *Science* article describes molecular research using genetic changes as molecular clocks, showing single-cell organisms dating back 3.6 billion years, and more complex species 1 to 1.2 billion years, and a figurative big bang of emerging life forms over a short span of some 20 million years. A November 1996 *Nature* issue speaks of evidence of microorganisms in rocks dating back 3.85 billion years, the time being based on the amounts of Carbon 12, a by-product of living organisms.

The Scientific Approach: The scientific approach to the time sequence involved in the appearance of human life suggests that human life evolved as a part of the evolution of life on this planet, the present *homo sapiens* species evolving from or in conjunction with earlier hominids over some several million years, present homo sapiens, appearing within the two hundred thousand or so years.

Species in Evolution of Present-day Human Life: There is a vast amount of information (based on carbon, potassium-argon, uranium-lead, and other chemical and atomic dating methods) on approaches to the sequence of species in modern man's development. The principle species commonly described are: *Australopithecus ramidus* (Ethiopia: a forest-living, chimp-sized ape-man, described by Liverpool paleontologist Bernard Wood as either the common ancestor of apes and humans "or damned close."); *Australopithecus afarensis,* living in family groups and walking upright, dubbed "Lucy" (Tanzania: 3.5 -2.9 million years ago).

A. Africanus : Also included are: *Australopthicus Africanus* (South Africa: 3.3-2.9 million years ago); *Homo Habilis* (Tanzania: 2 .5 billion years ago, reputed to be the first maker of tools); *Homo Erectus*--"Java man" (Java and Africa 1.8 million years ago); *homo neanderthalensis* (Germany, Belgium, France and other parts of Europe, toolmaker and hunter, said to be extinct some 35,000 years, though some insist that through interbreeding,"their genes survive today.") *Homo Erectus* and *Homo Sapiens* are thought to have overlapped in Java about 20,000 years.

Homo Sapiens : Finally came *homo sapiens* (France: 130,000 years ago, and China, 200,000 years ago) and still in vogue. Homo sapiens is thought to have evolved perhaps some 270,000 years ago. A study of genetic materials show that female ancestry is traceable back 200,000 years to Africa. Some scientists suggest that the *neanderthals*, the least understood of the evolving species, and sapiens, had thousands of years of overlap, the neanderthals mysteriously dead-ending or perhaps being absorbed.

Primates and Hominids: Speaking of primates and hominids, on a remote African jungle island, a German scientist, Bernhard Meier of Ruhr University, recently photographed a race of small primates long thought to be extinct. Called "lemurs," this primate evolved in isolation from other species like apes and humans, perhaps an illustration of so-called 'concurrent' as contrasted with sequential evolution.

Recently Discovered Link between Reptile and Mammal Families: In the May 21, 1990 Washington Post, William Booth notes the report, in the British Journal Nature, of a recent striking reevaluation of significance to evolution of a long-stored, cat-sized skull at the University of Toronto, originally discovered in 1897 in Texas and thought to be of the reptile pelycosaur family. The animal, now identified as *Tetraceratops* *insignis*, is an evolutionary link between the *pelycosaurs*, belonging to the reptile family, and *therapsids*, subsequent pre-mammal animals, precursors to mammals.

Gene Study More Accurately Tracks Evolutionary Relationships: In "Redrawing Life's Family Tree: Genes Offer New Insights on Evolution," Washington Post Staff writer Susan Okie reports on revisions in life's family tree resulting from the ability of molecular biologists to use the study of DNA mutations and gene characteristics of various life forms as a more accurate method to track evolution of life forms than reliance on comparative anatomy features.

Gene studies group humans more closely with chimpanzees and gorillas than with apes such as orangutans and gibbons. The gene analyses of various species of crocodiles around the world suggest common ancestry from a single location, occurring long after the reptile age some 66 million years ago.

Deciphering Genes to Study Evolution: Genetic relationships between all divisions of living organisms, including human beings, can be traced by study of differences in ribosome gene sequences, revealing common ancestry. Recent PCR (polymerase chain reaction) technology permits multiple-copying of a single copy of a gene, even from an extinct organism, to facilitate study. Molecular biology can analyze molecular sequences to show relationships between dissimilar life forms.

The physical merger of unrelated organisms to create new life forms represent shortcuts in evolution, as illustrated by mitochondria thought originally to have been bacteria which infected and merged to become part of a single-cell organisms through a process called endosymbiosis.

Finding of Alien Amino Acid Molecules: Washington Post staff writer William Booth in the June 9th, 1989 Post notes the report in the journal Nature, of the detection of alien, that is, extraterrestrial source, molecules in earth sediments. Scientist Jeffry Bada of University of California-San Diego, Scripps Institution of Oceanography describes the two kinds of *alien amino acid organic molecules*, molecules not resulting from earth's own natural evolution.

The finding of large amounts of iridium, a metal rarely found on this planet but frequently found in meteorites, tend to confirm the theory that the dinosaur age was brought to a close by a collision with earth of a large extraterrestrial object such as an asteroid or perhaps a comet, a theory proposed by father and son geologists Luis and Walter Alvarez. But other scientists suggest a ‘domestic’ source of the iridium as volcanic eruptions from the earth's own molten mantle.

Life and the Increased Organization of Matter: We discussed earlier the universe as winding down, governed by entropy, decreasing order and increasing chaos. But we also noted the countervailing tendency of matter in open systems to self-organize. Reeves speaks of this mysterious tendency of matter to organize itself--the molecular tendency in interstellar space, the proliferation of life forms in the jungle, the nucleosynthesis in the stars. He points out that the molecules which incorporate more than three atoms, include atoms of carbon, "a structure equipped with four hooks particularly well adapted for molecular combination."

Silicon-Based Life Forms? The existence of carbon is, of course, not confined to the earth but permeates the universe. Reeves asks whether there are other life forms that use silicon rather than carbon as the molecular catalyst but then points out certain limiting factors. There are few molecules that use silicon, their relative rarity in the universe, the fact that the atomic bonds of silicon atoms, like helium, are too strong to play the catalyst role of carbon, and the fact that it is not soluble in water, the transportation system for the biological processes of human life. So silicon is an unlikely molecular catalyst for life.

The potential for extraterrestrial life forms, the calculated life-elsewhere probabilities in the Greenbank or Drake-Sagan equation, are based on carbon-related life. Silicon is disregarded in the computations. If a silicon life system has evolved somewhere in the universe, well, so much the better but it hasn't been factored into the extraterrestrial-life-potential equation. One scientist suggests that silicon life forms may have evolved earlier in the universe before the formation of carbon-oxygen life forms.

Life Involves Different Levels of Reality: All the levels of the real world: past, present and future, and subatomic, atomic, chemical and molecular, come together in everyday life, says Reeves. Quarks were joined into nucleons in the first microseconds of the universe; nucleons formed nuclei in the stars. In the primal Earth ocean, increasing molecular complexity started the process of cell formation and evolution of the genetic code. But, he notes, the process of increasing organizational complexity continues today, in chemical and cellular activity.

Reeves traces the path of nature as it experiments with moving around particles to form elements. The unique role of water in promoting life, as described above, is related to the abundance of hydrogen and photosynthesis, removing carbon dioxide and providing oxygen in abundance.

The role of phosphorus, sulphur, iron, zinc, manganese, chlorine, boron, fluorine, selenium and tin are described, as nature puts these elements to use in the organization of life. He gives examples, such as zinc in digestion of alcohol and proteins, cobalt in the synthesis of the genetic code, copper in bringing iron to the blood.

No Reason to Expect Evolution to Stop at Present Life Forms: And, he asks, will this process of continually evolving complexity stop at the point of the existence of the human being? We have no reason to think so, he responds, and perhaps on other planets other stages have already begun, he muses, and what untold wonders are being prepared by cosmic gestation in each one of us? Biologists suggest future human life as being characterized by increased brain size and capacity, perhaps like science-fiction extraterrestrials.

Physicist Freeman Dyson suggests the evolution to other body forms and the existence of mind outside the body or at least our present body form, as incident to our exploration and colonization of space.

Communicating Via Brain Waves Rather Than Language: Science fiction depicts extraterrestrials that have no need for communication via language; they 'read' transmitted brain waves to understand. We already use electromagnetic transmissions from the human brain to detect certain types of medical information and so the idea is not new although we might be concerned over the loss of privacy in an open system where language is not available to edit thought. However, if such a brain evolutionary development has occurred elsewhere, it is not unlikely to evolve here. *We better start learning how to control our thoughts.*

Another View: Biological Evolution Has Been Diverted to Culture: But, in *Cosmogenesis*, David Layzer suggests that human significant control of the environment, the ability to meet the needs of the species as a whole, has arrested the normal evolutionary change process and that *branching in the human species has therefore dead-ended.* In this view, contemporary evolution has shifted from primarily biological to predominantly cultural, reflected in "memes" or ingrained cultural patterns, rather than the genes of biological evolution.

Memes, the New Replicators: Layzer's views on memes echo the work of British zoologist Richard Dawkins. In his *The Selfish Gene*, Dawkins discusses the self-protection strategy of early self-replicating molecules, surrounding themselves with protein walls to become cells, then organizing themselves in self-protecting colonies--animal and human bodies--to perpetuate themselves in these reproducing bodies, a "taxi service" for the genes. The newer replicators are information units, ideas communicated between human minds, which he calls *memes*, living structures which parasitize the brain for continued propagation.

> The genetic instructions of all life forms on earth are written in the same language, with the same code book; this shared genetic language is one indication that all the organisms on Earth are descended from a single ancestor, a single instance of the origin of life some four billion years ago. --Carl Sagan

CHAPTER 44
GENES--THE BLUEPRINTS FOR REPRODUCTION AND THE GENOME PROJECT

The Cell As the Foundation for Life: The cell is a biological foundation unit of all life and contains billions of atoms. Animal life forms are "assemblages" of different types of cells, the *human body containing trillions of these cells* performing different roles--bones, hair, blood, enzymes--all in a controlled unity.

The cell, which produces the life-sustaining and replicating proteins, houses in its nucleus the chromosomes which contains the genes, segments of the DNA molecule, *the genetic blueprint or code for living and reproduction.*

The Structure of the Cell: The cell material is contained within a membrane which houses the gelatin-like ‘cytoplasm’ containing the machinery of reproduction. These include "organelles" that perform, *at the cell level*, body organ functions like those of the lungs and liver; "mitochondria," which perform cellular pulmonary respiration and deliver energy within the cell via the ATP (adenosine triphosphate), the cell's ‘rechargeable batteries.’

The machinery of reproduction also includes ‘ribosomes’ which build proteins from *amino acids* in accordance with genetic instructions delivered by messenger RNA molecules which have encoded instructions from the DNA molecules. At the cell's center lies the *nucleus* containing 46 chromosomes which house the genetic material, molecules of deoxyribonuclei acid, better known as DNA.

Molecular Chains--Genes--The Double Helix: The genetic components of the cell's chromosomes, are genes, the molecular chains of the nucleotides, the nitrogenous base subunits or links. The two-stranded chain, called d.b. nucleic acid, or DNA for deoxyribonuclei acid, is commonly displayed unraveled as a *double helix*, each strand of which has four steps, resembling a spiral staircase, or an eight-rung ladder twisted in the center.

Information in a Chromosome: How much information is in a single human chromosome? In *The Dragons of Eden, Speculations On The Origin of Human Intelligence*, Sagan says that the human chromosome equals some five-hundred million words--two million pages, or some four thousand volumes of five-hundred pages.

Reproduction of Plants and Animals: Animal and plant reproduction is based on cell reproduction. The DNA molecule in the chromosomes of the cell's nucleus unravels into two twisted strands (consisting of sugar and phosphate groups) resembling the rails of a (twisted) ladder which forms a double helix, the structure of the gene, first identified by James Watson and Francis Crick at England's Cambridge University.

Design of the DNA Double Helix: Between the 'ladder rails,' of the DNA molecule's gene, double helix, are four DNA subunits, the steps or rungs of the ladder, consisting of the nitrogen-rich nucleotide bases identified by the first letter of each base: A (Adenine), T (Thymine), G (Guanine), and C (Cytosine).

These bases are paired off as AT and GC. Thus each strand or rail of the ladder has the bases A, T, G, and C. In the reproduction process, enzymes travel to the cytoplasm's reservoir of bases and match each A with a T and each C with a G to create two copies of the original double helix. *The number of base pairs of human life cell DNA has been estimated to equal the number of people on earth.*

Enzyme Protein Molecules human life the Production Line: It should be noted that the cell's bases--its reservoir of spare parts, would be useless, a veritable parts warehouse, without the production-line workers, the protein molecule enzymes, trained to latch on to the right production parts, the matching bases. Enzyme molecules, constructed not from DNA material in the chromosomes but DNA material in the mitochondria, function to power the cell by disassembling the glucose sugar molecule upon its arrival in the cell so that the energy thus produced energizes the cell. These enzymes consume oxygen in converting glucose into energy.

Each Gene Strand a Model for Duplication: Thus, As David Layzer points out, in *Cosmogenesis*, each strand or rail of the twisted ladder with its A, T, G, and C bases is a model or template for duplication in the protein building, that is, for assembling strands with 'complementary' bases. This protein assembly process is performed by the cell's 'ribosomes' which assemble proteins from the reservoir of bases, amino acid molecules, floating in the cell's cytoplasm.

The Protein Assembly Process: Enzymes encode the chromosome's DNA molecule genetic sequence instruction into "messenger RNA" molecules which exit the cell nucleus to reach the cytoplasm's (ribosome) protein factory. The ribosome "reads" the single-stranded messenger RNA's encoded genetic information to establish the order of the (twenty) amino acids to form the protein's shape for fulfilling the cells's "form and function." Another molecule, the 'transfer RNA' molecule, brings the messenger RNA-specified amino acids to the ribosome for assembly. The process leads to the creation of identical new cells.

Thus, in the nucleus of each cell in every living organism are chromosomes with the DNA nucleic acid blueprint molecules of genetic inheritance. Every organism uses the same code to translate the information it contains into the proteins that construct the cells for the organism's development.

THE GENOME PROJECT--Mapping 100,000 Human Genes

So we have seen that the cell's DNA encodes the entire genetic blueprint for replicating the human body; that the mechanisms for transferring the DNA's code into reproduction are the body proteins; that the sequence of bases in each protein is dictated by a segment of the DNA called the gene.

The Government-approved *genome project,* presently managed by the NIH, National Institute of General Medical Science, is a massive undertaking designed to map the location of all the DNA pieces of the human blueprint, the genes, some 100,000, which are distributed along the 46 chromosomes in each cell. The chromosomes contain the helix-shaped genes, pieces of the DNA molecule, deoxyribonucleic acid, the body's genetic blueprint.

We have noted that each gene has four DNA subunits, nitrogen-rich nucleotide bases, identified by the first letter of each base: A, T, G and C, and paired as AT, GC, TA, and CG.

A Genome Project for Plants: The Agriculture Department Agricultural Research Service plans a similar DNA gene-mapping project for plants. The random process involved in cross breeding of plants takes many years to accomplish what might be accomplished more quickly by use of genetic mapping, to permit genetic designing based on knowledge of the genes and their protein functions.

The Sequence of Letters Specifies the Proteins: The genome project is designed to identify the sequence of all the bases or letters in all the genes. It is the sequence of these bases or 'letters' that determines the shape and function of the various types of body proteins such as enzymes, antibodies, hemoglobin, bone, skin, cartilage and muscle. The letters in the resulting human body genetic alphabet numbers in the billions. Some thousands of volumes will be required to print the entire human genetic code.

A set of genes, perhaps from one person, would be fragmented and duplicated using a microorganism like yeast so that cultures of these fragments would represent the human genome.

Internationalizing the Project: One interesting aspect of the project is its international dimension--the potential for involving universities and research organizations throughout the world by assigning responsibility to different countries for mapping and sequencing of specific chromosomes, or some other organizational division.

But this pattern of division may need rethinking since particular chromosomes are already of intense interest to and research by certain institutions and countries, as illustrated by the shorter chromosome 21 which is related to Alzheimer's disease and Down syndrome.

Genes Dictate Protein Synthesis: A protein consists of chains of polypeptides made up of base units called amino acids. The amino acids are encoded by the DNA, that is, a gene's genetic letters dictate the assembly of amino acids, bases, via the messenger RNA, the ribosomes, and the transfer RNA, as previously described.

Finding and Decoding the Gene Identifies the Protein: Once a gene is isolated and decoded, that is, its sequence of bases determined, the gene's protein product can be determined accordingly. Finding and sequencing a gene which causes a disease is done through a process called 'chromosome walking,' examining segments of the DNA molecule in the chromosomes until the 'address,' the sought after gene is reached.

'Telomeres' and Chromosome Boundaries: Until 1989, biologists were hampered in the gene search for disease-related genetic defects by the absence of information on where the chromosomes ended. A major breakthrough in the mapping process was made recently when researchers in the Department of Energy Los Alamos National Laboratory in New Mexico found the chromosome boundaries, DNA structures called 'telomeres.'

The search for the Huntington's disease gene in chromosome 4 was accelerated by using the telomere as a starting point. In Staff Writer Larry Thompson's Washington Post article of May 9, 1989, he points out that instead of dealing with a region of *millions* of letters, as before telomeres gave the chromosome boundaries, researchers were last reported as being within *thousands* of letters of the Huntington disease gene.

A Living Molecular Fossil: In an interesting aside, Thompson notes that we share this DNA sequence, telomeres, with 93 living species of vertebrates, and with extinct species like the dinosaurs; that telomeres are 'living molecular fossils' representing a link between the human and prehistoric species.

There Are Thousands of Genetically-caused Diseases: Most of the time we tend to think of ourselves as 'catching' a sickness or disease by exposure to some external source. But there are some 4000 genetically inheritable diseases, that is, diseases that you can be handed simply by being born. Examples include such diseases as Cystic Fibrosis, Huntington's disease, Hemophilia, Sickle-cell anemia, Thalassemia, Tay-Sachs disease, and Down's syndrome.

Scientists have recently identified a gene that may increase the the liklihood of inheriting the mental illness schizophrenia,which may explain the chain-smoking tendency of schizophrenics where the nicotine overrides the brain defect to provide some calming relief for schizophrenics. The nicotine, like certain medicinal drugs, stimulates a brain receptor that helps filter information.

Homosexuality As Genetically Induced: A July 1992 report of Doctors Allen and Gorski of UCLA, in the *Proceedings of the National Academy of Sciences*, confirms the earlier finding of differences in sexual orientation as being not a matter of choice, or environmental or social factors, but dictated by nature. These are matters of biological design--anatomical differences--between the brain structure of homosexual and heterosexual men, probably surfacing before or slightly after birth.

In studies of rats, at least, the brain seems to be basically female and to develop male characteristics requires the presence of specific hormones at birth time which cause certain brain structures to grow larger in the male than in the female.

A core of nerve fibers in the brain--the anterior commissure--which connect the two brain halves at the back of the skull, to permit communication between the two halves, is larger (34 percent) in homosexual men than in heterosexual men and 18 percent larger than in heterosexual females.

Earlier, Dr LeVay of the Salk Institute for Biological Studies at San Diego, California had reported that in another region of the brain, the interstitial nuclei of the anterior hypothalamus, affecting sexual behavior, was smaller in homosexual men.

Genetic Aspects of Alzheimer's Disease--Early Detection: There is increasing evidence that Alzheimer's disease, which causes brain cells to die and leads to memory loss, decreased mental capacity and personality change, has genetic causes. Reported in the American Medical Association Journal in June, 1990 was the development of a definitive, quick and accurate test, by a scientific team of Abbot Laboratories headed by Hossein A. Ghonbari, to detect ADAP, Alzheimer's Disease Associated Protein, available from spinal fluid so that it needn't come from the higher accumulations found in the brain's memory storage area.

Finding the protein early, before brain damage, should permit corrective drug applications hopefully to prevent the disease. The defective gene produces amyloid, the group of proteins identified in patients with the disease.

At the University of South Florida, researchers dealing with the genetic flaw responsible for Alzheimer's disease, said, in July 1992, that the finding of the new genetic flaw will permit tracking the disease from its beginning. Anyone carrying the defective gene will later develop the disease. The discovery permits generating a laboratory model for the disease and developing chemical treatment. Prenatal amniocentesis could identify the gene, thus predetermining later occurrence of the disease.

Other Genetic Source Diseases: In addition, defective genes are the source of forms of other illnesses and diseases such as: Down's syndrome and certain other mental disease, muscular dystrophy, heart disease, Alzheimer's disease, retinal cancer, and colon cancer. Tracking the genes responsible for these diseases has been the object of considerable research for some years.

Alcohol Addiction May Result from a Defective Gene: In April of 1990, Kenneth Blum of the University of Texas, San Antonio and Ernest Noble of UCLA reported, in the American Medical Association Journal, that a flawed DNA segment, an "aberrant" gene, the a-1 allele, of the so-called dopamine D_2 receptor gene, appeared to be associated with alcoholism, in the group they studied.

New approaches to dealing with alcoholism may be indicated. The alcohol impulse may result from suppression of dopomine activity resulting from defective directions in the construction of portals on neurons supplied by the defective dopamine D_2 receptor gene. Drugs that adjust the dopamine metabolism might be a fallout of the studies. Also interesting is the possibility that *such genes may be involved in other addictions or forms of behavior*.

Finding Defective Gene Locations: Jim Merritt's Minicourse, *Design of Life*, in the June-July 1990 issue of the AARP publication, Modern Maturity, notes that the gene for Huntington's chorea, is in chromosome 4, retinal cancer, in chromosome 13 and some mental diseases, like types of schizophrenia and and manic depression, in chromosomes 5 and 11.

The gene for Alzheimer's disease is thought to be in chromosome 21. Researchers at the University of Utah and of Oulu in Finland reported, in the journal Science in June, 1990, finding the defective or mutated gene that accounts for Alport syndrome, a widespread disease that leads to progressive kidney failure. Funded by the NIH Institute of Diabetes and Digestive Diseases, the research disclosed the gene's location on the X chromosome.

Down's Syndrome and Chromosome 21: At the 1989 annual meeting of the American Society of Human Genetics in Baltimore, researchers at Johns Hopkins University Center for Medical Genetics reported on the creation of artificial chromosome 21 material involved in the mental retardation Down's syndrome. The chromosome material, duplicated in yeast cells, is a part of the DNA molecule twenty-five times longer than the DNA pieces previously reproducible.

Human Beings and Roundworms--Some Things in Common: Relative to Alzheimer's disease, Columbia University researchers report in the journal Nature, the finding of three genes in the *C. Elegans* roundworm, responsible for brain cell death. While it may seem humbling, roundworms do share with human beings, many of the same genes, proteins, and neurotransmitters, and the information may assist in the search for Alzheimer's defective genes, in the test for those at risk, and in the development of preventive techniques.

'Genome' Project, Genetic Engineering and Medical Advances: The genome project, expected to take about 15 years, would initiate a whole new phase of molecular biology, provide a new threshold in the battle against disease, and accelerate the science of genetic engineering and gene therapy.

Once the 100,000 genes are mapped and their corresponding proteins determined, the information may be used for all kinds of corrective genetic work. The genes may be modified and injected, the actions of their proteins modified, the proteins synthetically duplicated.

Genetic Engineering and Biological Therapies: Biological therapies, that is, use of the body biological products, gene manipulation, injection of correcting genes, treated cells, modified viruses or bacteria are all increasing. Biotech firms are marketing genetically modified enzymes, clot-dissolving hormones, artificially mass-produced proteins such as the sugar-metabolyzing protein insulin, interferon, and interluken-2.

Other proteins stimulate blood cell production in bone marrow to counter blood-cell-destroying diseases or chemotherapy, or to stimulate growth-hormones, like interleuken-2 for promoting growth of cancer-attacking white blood cells, and for producing human cell and tissue transplant material to supply missing chemicals. Some products are produced by bacteria which have been genetically modified.

Gene Reproduction Made Possible by PCR: Of considerable help to genetic engineering, clinical research and other applications is the development of DNA polymerase, an enzyme that permits the endless reproduction of a single gene through the use of polymerase chain reaction, PCR. Although developed several years earlier, Science, the American Association For The Advancement of Science journal, gave the discovery its Molecule Of The Year Award.

NIH's Use of TIL Cells: The NIH has been working on the use of TIL cells, tumor-infiltrating-lymphocyte white blood cells, as an alternative to injecting altered genes into bone marrow stem cells and made the first such transplant in May, 1989.

In March of 1990 NIH researchers removed pieces of melanoma skin cancer tumor from cancer patients, isolated from the removed tissue different types of immune system white blood cells. Cancer-resistant genes were then implanted into the removed cells, the modified TIL or tumor-infiltrating lymphocytes, allowed to reproduce and then reinjected into the patient. A selection process, using a toxic compound called G418, was used to kill all the tumor cells except for the modified TIL cancer-killer cells.

Such gene therapy experimental procedures require specific prior approval by an NIH subcommittee. Upon approval, they plan to put tumor necrosis factor-producing genes into TIL cells. This factor is a protein which, unaided, has the property of being able to kill cancer cells.

Cell Transplants to Aid Immune Systems: An immunologist at Johns Hopkins Medical School, Drew Pardoll, characterized as a 'landmark' advance, a cell transplant development in human immunotherapy reported by University of Washington researchers led by Stanley Riddell and Fred Hutchinson and reported in a 1992 issue of the journal *Science.*

Patients whose immune systems were destroyed by chemotherapy were protected from the deadly virus cytomegalovirus, by white blood cells, lymphocytes, transfused from laboratory-grown cells used to defend against invading viruses. The cells, grown for months in the laboratory, thrived upon transfusion. The experiment presages new possibilities for combating diseases like cancer and Aids. Plans exist for 'tinkering' with the gene structure of removed cells to enhance their virus-defense function.

Use of Modified Viruses to Transport Genes: Modified viruses such as retroviral vectors virus shells can be used to transport modified genes for incorporation in body-cell DNA. This transport system would help in providing enzymes for diseases like ADA deficiency, involving a deficiency of the ADA (adenosine deaminase) enzyme, a deficiency which diminishes reproduction of infection-fighting T cells.

Lab-produced Cells Materials and Artificial Organs: Newer methods are being developed for overcoming problems in returning treated cells to the body by 'growing' and implanting back into the patient an artificial material which will function to produce needed proteins. A possible example relates to Alzheimer's disease, associated with loss of nerve cells using acetylcholine for nerve communication. The implantation of cells genetically-engineered to produce nerve growth factor may prevent such loss.

The implantation materials are 'grown' by using a framework or scaffold consisting of a plastic polymer pad appropriately coated with cell-growth stimulating natural substances. The patient's gene-corrected cells of the proper type can be added. When implanted the material develops into a 'substitute' mechanism for creating the needed protein product, like insulin.

The material can also be impregnated with heparin-binding growth factor, HBGF, to stimulate the growth of new blood vessels and nerve tissue, producing what is essentially a new organ to generate required body substances such as produced by the liver or pancreas.

One effect of the developing technology will be to diminish the need for body-part transplants. Only during the embryonic stage do mammals have the limb or organ correction or regeneration ability, although adults, in some more primitive life forms, like frogs, can regenerate body parts like limbs.

Replacement Parts: In Jan. 27, 1997 Newsweek, Geoffrey Cowley writes about developments in creation of substitute tissues and replacement body organs. An ear built from human cartilage cells and surgically implanted on a mouse, thrived. Dr. Joseph Vacanti of Boston's Children's Hospital feels that we are learning to generate "living prosthetics for every organ system in the body," the first tissues likely to be structural, like skin , bone and cartilage.

Firms are working on skin substitutes for burns and wounds starting with cells from the foreskins of circumcised infants. Engineers, says Cowley, can now mass-produce small sheets of durable tissue that the body accepts. A Massachusetts firm has developed both dermis and epidermis for deep nonhealing skin ulcers. Researchers are learning to create fully functional bone substitutes and fabricating valves for the heart and urinary tract. Dr. Vacanti and associates are said to be striving to grow an entire liver around artificial scaffold of synthetic mesh.

Microinjection and Transgenic Animals: Microinjection and implantation of genes from one animal to the egg of another results in changes in the DNA genetic code of the recipient. Cow genes, for example, have been transplanted in pig ova to increase the proportion of the pig's meat over fat content, resulting in transgenic, that is, inheritable changes in the pig.

Similar work on plants has produced disease-resistant and cold-resistant plants and plants that accept exposure to arid or salt-water locations. Monsanto Co. researchers recently gave two new genes to the Russett Burbank potato that make it resistant to potato viruses responsible for worldwide potato damage.

Possible New Gene Delivery System: Microinjection is a complicated procedure. Recently, a more simple gene-delivery system for creating genetically-altered transgenic animals was announced by scientists at the University of Rome and the Institute of Biomedical Technology, and reported in the June 2, 1989 issue of the scientific journal, Cell. The process may disclose how genetic characteristics may meld between species as part of the evolutionary process.

The process involves using sperm, a naturally more efficient genetic delivery system, for mixing genes, DNA material. Genes from one species of mouse is added to a sperm mixture of another mouse species, the sperm latching on to the new genes. The transgenically-altered sperm is then used to fertilize eggs in vitro which are then implanted to pass along the added genes to later generations of offspring. The process is being tested by other laboratories for verification.

Mitochondria DNA the Source of Some Genetic Disorders: There are hundreds of *mitochondria*in the cell's gelatinous cytoplasm which function to convert glucose into energy to power the cell. Mitochondria, before melding with cells, were once cells which, like bacteria, lived independently of the human cell and consequently have their own genetic material, DNA, with genes which are different from the DNA genes of the chromosomes. Unlike chromosomal genes, mitochondrial genes are inherited only from a mother, the father's being non transferrable and only a daughter can pass on these genes.

Recently, the mitrochondria are becoming the focus for more study as the source of genetic diseases derived from its own DNA material. Thus mitochondrial gene source inheritable diseases come only from the mother even though the diseases are inheritable by both male and female offspring.

Mitochondria and 'RNA Editing': Research at the University of California in Los Angeles finds a cellular self-correction system which corrects incorrect gene instructions. Information missing from the DNA gene material is supplied by the mitochondria in an editing processes in which the mitochondria furnishes the information in the form of 'guiding' RNA material identical to the genetic information edited into the molecules of messenger RNA.

Energy-deficiency Genetic Diseases: Because of the mitochondrial energy production function for the cell, mitochondria gene-related diseases involve deficiencies in energy production and are more likely to damage those cells requiring the most energy such as those in the eye muscles, the nerves and the brain, thus accounting, for example, for Leber's hereditary optic neuropathy. One symptom of mitochondrial deficiency is droopy eyelids. A laboratory that tests tissue samples for such defects is the New York's Columbia-Presbyterian Medical Center.

DNA Uniqueness and Identification of Criminals: The fact that DNA material in each cell is unique to that person has opened an important phase in criminal law permitting, for instance, the matching of a accused DNA material such as blood or hair, with evidence left at the crime scene.

Entering Cells to Perform Genetic Repairs: Gene splicing was early recognized as having great disease-treatment potential. Viruses may be used to carry sound gene forms to cells with defective or missing genes to permit the gene to produce an essential protein like the blood-clotting enzyme needed by type A hemophilia.

The sugar-metabolizing protein insulin artificially produced for use by diabetics who are allergic to animal source pancreatic insulin, was one of the first proteins created by gene snipping and implantation. Bacteria which are altered by implanting genes from another bacterium are then cloned in identical multiple production quantities.

Gene Therapy and Muscular Distrophy : Recently reported by the University Of Tennessee was an experimental treatment, named myoblast, to improve muscle function and structure by taking healthy muscle cells capable of producing dystrophin needed for reinforcing and stabilizing muscle tissue, and injecting it into inherited muscular dystrophy patients whose cells cannot produce the protein. It is expected that the healthy introduced cells will assist in producing the needed dystrophin. While not a cure, it represents progress toward the time when gene therapy can achieve a cure.

Genetics, Virology, AIDS & the Human Immunodeficiency Virus: Vaccines employ the principle of the body's immune defense--the building of defenses to an invading bacterium or virus, by introducing a killed or other harmless version of all or part of a virus, for instance, which builds some form of immunity response in advance of the true microorganism invasion, such as in Smallpox, Polio, and Measles.

Until recently, not much hope has been held out for an AIDS vaccine, despite the existence of many approaches, in some sixty-seven AIDS vaccine studies conducted in at least six countries.

Discussing AIDS vaccines in his May 21, 1990 article in the Washington Post, Staff Writer Malcolm Gladwell quotes NIH Institute of Allergy and Infectious Diseases Director Anthony S. Fauci as finally being able to express confidence in the development of an AIDS vaccine in the decade of the nineties.

Rhesus monkeys were recently given protection against the SIV, the simian immunodeficiency virus, when inoculated with the killed SIV virus, closely related to the AIDS virus.

Killed, Live Or Partial Virus Vaccines: Several approaches to vaccines are candidates. One approach is using *killed* virus vaccines but since these have only short term effect, a second approach has been to consider a *live but genetically altered* virus. However, the proclivities of altered or errant live viruses are too uncertain to generate confidence.

A third approach is to use protein parts of the virus, *pieces of the viral DNA*, to stimulate the immune defense. But different protein subunits of the virus generate different defense responses in the form of killer T cells, antibodies, and macrophages, so the thought is to combine more than one subunit. Johns Hopkins University announced in June 1990, the development of a promising test vaccine using this third approach.

Blocking HIV Reproduction by Implanting Altered Genes: Several molecular biologists have taken extracted bone marrow cells from AIDS patients, injected the altered genes into the cells and replanted the cells in the bone marrow to produced new blood cells with the mutant genes which make altered and defective versions of proteins required for HIV reproduction, the defect effectively blocking reproduction of the virus in its virulent form.

Drugs to Interfere with HIV Reproduction: Using molecular maps of the chemical sequence used by the protease enzyme which breaks down larger into smaller protein molecules, the Upjohn Co. and Smith-Kline Laboratories have designed drugs that bind to the protease enzyme protein to interfere with HIV reproduction. The expectations are optimistic.

The End of Aids? Since the early 1980's, an HIV positive diagnosis was essentially an automatic "death sentence," the only drug help being AZT. In his December 2, 1997 Newsweek article, John Leland discusses the dramatic change in life potential following HIV prognosis.

The HIV virus is, in effect, a protein-coated piece of genetic code of nine genes which penetrates white blood cells to splice the nine genes into the cell's chromosomes leading to further replications.

The object is to disrupt this virus's "assembly line process" by neutralizing the integrase enzyme which splices the viral DNA into the white blood cell which then produces new viral RNA. The inhibitors fall into two categories, RT or reverse transcriptase inhibitors, like Retrovir (AZT) Videx (ddl), and Epivir (3TC), "1592", and protease inhibitors, like Norvir (Ritonivir), Invirase, Indinivir (Saquinavir), and

Crixivan (Indinavir). These function in combinations or "cocktails" to arrest the disease by attacking the virus's ability to reproduce or replicate.

The problem, of course, is the millions of dollars needed to fund these inhibitor treatments, the regimen's therapies involving precise time table administration of the 30-40 daily pills, and the complacency generated by the treatments encouraging carelessness in life style drug and sexual habits. But until successful prevention vaccines are available, the inhibtor technology will dominate the HIV world.

Chimpanzees with New Strain of Virus Resembling HIV: The scientific journal Nature of May, 1990 contains a report on a virus found in Gabonese chimpanzees which is a very close to the human HIV virus, sharing many of the same genes, and much closer than the SIV or Simian Immunodeficiency Virus affecting rhesus monkeys. It is unclear whether the chimp or human versions are derived one from another or are offshoots of a predecessor.

A New Plague--The Mutant Respiratory AIDS Virus? According to World Health Organization reports, the AIDS Virus has spread through some 149 countries, affecting perhaps as many as ten million people. Moreover, there are other diseases lurking in the tropics which may have similar deadly potential--at the very time that we are cutting back on tropical medical watch stations.

Recently, antibodies to the virulent Ebola disease, affecting monkeys, were discovered in two persons working in a Reston, Virginia laboratory which had received infected Philippine monkeys, the first indication of possible transmission of the disease to human life since a devastating African epidemic some years ago.

Rick Weiss reports in Science News of the concerns of virologists that a chance mutation of the AIDS virus itself, could make it transmittable through respiratory transfer, like the flu or the common cold. Citing the World War I 1918 flu virus that took a toll of twenty million lives, a mutant form of the AIDS virus is pictured as having the potential for starting the greatest plague in human history, capable of wiping out most of the human race.

But there are those who say that the process of AID's mutation to respiratory transmission form is likely to weaken its impact to that of the common cold. Others, more sanguine, are not so sure that we couldn't have a respiratory AIDS killer virus. Although human gene evolution might eventually counter such a virus, much of the human race could be lost before the human body immune defenses caught up or was genetically re-engineered--and the doctors manning the front-line defenses might be the first to be wiped out.

Ethical Concerns over Genetic Engineering: Gene therapy or manipulation to correct genetic defects or prevent inheritable disease sounds worthy and promising. But what about manipulation to bring about desired characteristics? Jim Merritt cites Princeton's Lee Silver description of this as a "more ominous use" although under the proper publicly regulated criteria it might prove a boon--in a sense altering the laws of chance to improving the human lot.

Misuse of Good Ideas: On the other hand, we have seen how governments can turn good ideas to evil uses and there are considerable misgivings. Forced genetic engineering to produce "submissive" populations come to mind. The vision of shifting from natural evolution to a "directed" evolution surfaces. Science fiction has already constructed some desperate scenarios.

Religious organizations, as Merritt points out, have given expression to the concern that "History has shown us that there will always be those who believe it appropriate to 'correct' our mental and social structures by genetic means, so as to fit their vision of humanity."

Guidelines for Genetic Engineering: In *The Clash Between The New Genetics And Human Values*, authors David Suzuki and Peter Knudtson, proposing safeguarding genetic principles, discuss recombinant DNA technology, analyzing the pros and cons and medical risks of genetic engineering involving gene manipulation, adding, removing, redistributing genes, and crossing lines between species.

Manipulating genes in the sperm or ovum is particularly sensitive since an error in judgment may affect generations of the patient's lineage. The variety of new biological agents is infinite and some could prove as devastating as AIDS.

Worries over Genetic Screening: But even more fundamental or serious concerns are expressed over genetic screening. It is known that most people carry genes which are defective in some way. All persons with inheritable disease potential disclosed by screening might be denied insurance protection and employers might well use screening to exclude such cases from employment eligibility and thus reduce their operating costs. One's economic destiny might be set at birth.

Even more controversial perhaps, is the possibility of increasing incentives for abortion following prenatal screening. Disclosure of diseases that take their toll in early childhood, or even in the teens or twenties, might well influence abortion philosophy.

Testing the Human Egg for Genetic Defects Prior to Fertilization: *In vitro* fertilization research suggest that one approach to avoiding an abortion situation is testing of the human egg for genetic defect before *in vitro* fertilization. The mother, a carrier in the case of human genetic disorders, has, in many cases, one good and one bad copy of a gene. If the baby acquires the mother's bad copy and also a bad from the father, the defect will be inherited. People with inheritable disease genes may well prefer testing *before fertilization* than be faced with an abortion decision.

The man who becomes a thinking being feels a compulsion to give to every will-to-live the same reverence for life that he gives to his own. --Albert Schweitzer

CHAPTER 45
HUMAN LIFE'S DESTINY IN THE COSMOS--
ROLE OF MIND

Eric Jones: Fourth Step in Human History-- Human Life into the Cosmos: In the essay *From Africa to the Stars*, anthropologist Ben Finney and space scientist Eric Jones *reduce human history to four steps:* First, from the trees to the grasslands, 4 million years ago; Second, The movement over land from the warmth of Africa to the more rigorous environments of the other continents, one million years ago; Third, the use of the seas to move and travel, using skills of shipbuilding and navigation, only three thousand years ago; and *Fourth,* the present, the move from planet Earth "to the stars," *to become creatures of the cosmos.*

Humans to Use Genetic Engineering for Venturing to the Stars: Physicist Free man Dyson, citing Finney and Jones' work, says that *step four* is the 'irreversible expansion' of human life into the reaches of the universe, requiring genetic engineering, artificial intelligence, adapting asteroidal environments to colonization, breeding programs, new propulsion methods.

Mind to Spread across the Galaxy: Mind is patient, he says, it has waited 3 billion years to compose its first string quartet. It may have to wait another 3 billion years to spread all over the galaxy. This philosophy is certainly nothing less than a faith-like conviction that spreading throughout the cosmos is human life's manifest destiny.

Dyson: Life Is a Universal Phenomenon: His more fundamental point though, discussed later here, is an expression both of science and theology, that life is a "universal phenomenon...of cosmic dimensions." He says that the human species will one day be faced with its "most momentous choice...to remain one species...or *diversify...[into] a million intelligent species exploring diverse ways of living... across the galaxy."*

He cites H. G. Wells' as a prophet of doom in *The Time Machine*, reflecting the failures of human society, and as a prophet of hope, in his lecture on *The Discovery of the Future,* in which he expresses the greatness of human life destiny and the basis for a surpassing faith.

Faith in a Future-Related Purpose Transcending Comprehension: Dyson uses Wells' approach to faith and reason to express his own faith, that we are here to some purpose that has something to do with the future, that transcends our understanding. He does not feel qualified to go further into theology but "If you like, you can call the transcendent purpose God...[but] a Socinian God, inherent in the universe and growing in power and knowledge as the universe unfolds, [and] our minds are "expressions of its purpose...and contributions to its growth."

Each Species Has Its Time--Potential of the Human Mind: Paleontology points out that each species has its time--and is gone. High intelligence species on other planets, whether or not like human life, may exist--or they may have come and gone. The age of the universe and the process of stellar evolution has been sufficiently long-enduring to permit the evolution of human or humanoid species on many planets. It may be though, that through evolutionary development, human life's departure from the scene may take the form of transition into something quite different than we can conceive.

Also, as physicist Freeman Dyson points out, with appropriate energy and mathematical computations, survival of the human mind may be possible entirely apart from its present physiological framework. Such concepts are thus not limited to science fiction speculation as physicists are caught up in the study of the potential of the human mind as an aspect of the Mind of God.

Freeman Dyson's Mind of God--Human Mind: In quantum theory, the human mind seemingly functions as an 'extension' of the Mind of God. Human life in the universe of the laws of physics becomes a sort of divine adjunct with a direct function in the quantum process.

Freeman Dyson, in *Infinite in All Directions*, speaks of the role of human life in the universe and applies the God-human life adjunct concept to the Mind-of-God and the human mind. If God exists and is accessible to us, then, he says, his mind and ours may differ from each other only in degree and not in kind (human life made in God's image).

Dyson Speculates on Tendency of Mind to Control Matter: The human mind as an extension of the Mind of God, says Freeman Dyson, plays a role in determining the nature of matter. Speculating on the purpose and future of life in the cosmos, he suggests that it is human life's destiny to influence the universe to his own purpose;

that the tendency of the mind to infiltrate and control matter is a law of nature; that if our species does not choose to lead the way, others will do so; and if our species is eliminated, others will be wiser or luckier.

Sharing God's Power of Determining Particle Characteristics? As Freeman Dyson says, the *human mind and the mind of God may differ only in degree.* His other statement that *particles are weird enough that there is no limit on what God may do with them,* implies something much further, that given enough time, human life will find no limits on what it can do with matter. Expressing this in another way, *is it not possible that human derivative mind shares this power of determining particle characteristics?*

In this context, several physicists speak of the laws of physics as being in an evolutionary process, like Dyson's Socinian God. Elsewhere here we discuss an ancient 'natural god' or intelligence which molds the laws of physics, and intelligent civilizations which modify the laws to cope with the conditions of a universe contracting toward its end in the Big Crunch.

MIND TO ALTER THE UNIVERSE

Role of Mind to Alter the Universe: Human life already alters the universe almost daily when it develops new materials, new chemical combinations, new breeds of animals, new genetic materials--an awesome responsibility, altering the Maker's handiwork. Or are we the Maker's surrogates for doing just that? In a sense, human life's presence in the universe may be the Maker's necessary instrument for accelerating the process of bringing about certain changes which the laws of physics could not otherwise achieve.

It is postulated that an advanced civilization can: travel to the farthest reaches of the galaxies; colonize the universe; create or even *become* non biological intelligence; even harness the incredible energy of black holes and bypass speed of light limitations. It is said that humans can do what they can conceive. Derived from and sharing the Mind of God, perhaps human life's only limitations are self-imposed --mind over matter.

Astrophysicist Freeman Dyson's Lectures on Cosmic Biology: Princeton physicist Freeman Dyson's Gifford Lectures, which were updated in *his Infinite in All Directions*, are an interesting source of reflection on cosmic biology, possible forms of life, especially the kinds of life which may be adapted to space from life on earth--and presumably any similar planet.

What If We Find No Extraterrestrials? If after our search for extraterrestrials we get no enlightening messages, and finally conclude that in the universe life exists only here, then, as the physicists suggest, it becomes the human role to spread life and mind in the universe. In our planet's history, species have come and gone, through natural processes. Apparent human inability to control its pattern of misuse of the planet's environment makes it even more urgent that the spread of intelligence elsewhere in the universe be undertaken.

What Genetic Changes are Needed to Spread Life? Dyson asks, what would be the "modifications" needed for earth life to spread through the cosmos? With the advent of genetic engineering he says it is not absurd to consider redesigning earth life for viability in space or on other planets, and once we have planted a variety of life forms in space, and provided for dispersal of their spores, we can expect mutation and the evolutionary processes to take over and our early mistakes will be corrected by the diversification of their offspring.

Impediments to the Spread of Life in the Universe: He notes that Zero-Gravity, Zero-Temperature and Zero-Pressure are the three major impediments to spreading earth-type life throughout the universe. He points out that fur and feathers, for instance, are better insulators in a vacuum than in air or water and low temperature is better suited to regulating an organism's internal temperature--keeping warm on Pluto is easier than keeping cool on Venus. Of course, adapting life to zero temperature would require evolution to warm-blooded plants--living greenhouses spreading over a cold planet.

The life-liberating movement from air to vacuum, requiring adaptation to zero-pressure, is as necessary and basic as the movement earth-life made from water to air. He cites the Russian engineer Tsiolkovsky's *Dreams of Earth and Sky*, in which he speaks of creatures with flexible, impermeable glassy skin which protect against material loss, with large radiant energy-converting appendages, the flow of their thoughts communicated through their transparent skin by continually-playing pictures formed by colored fluids channeled through the skin surfaces.

Envisioning New Life Forms: In a recent item in the Washington Post, Boyce Rensberger cites the work of scientists Fassbinder and Vali of Ludwig Maximilian University of Munich and Stanjek of Munich Technical University in the field of magnetic bacteria which extract iron to make their bodies magnetic, upon their death, leaving behind 'magnetofossils.' The point here is that biology, quantum theory, the fundamental forces, all have relationships to future evolutionary or genetically engineered life forms that cannot now be envisioned.

Life Can Adapt to Its Impediments in Space: He assumes that life can solve the technical problems in adapting to zero-G, T and P and "making itself at home" throughout the universe. He goes on to say that, based on his assumption that *life is based on the organization rather than substance, then one can hypothesize life "detached from flesh and blood* and embodied in networks of superconducting circuitry or in interstellar dust clouds."

Hypothesis Of Abstraction: He calls the assumption that life can be described abstractly, the 'hypothesis of abstraction.' He then extrapolates to define living creatures as characterized by complexity quotients representing the quantity of entropy (waste heat divided by temperature) produced in it's metabolism during the time consumed in performing a simple response to a stimulus. The achievable life complexity in any environment is proportional to the available matter, he says.

Hypothesis of Adaptability: Dyson then makes his stronger 'hypothesis of adaptability,' that with sufficient time, life can adapt itself to *any* environment. He calls these his *hypotheses of abstraction and adaptability.* The primary theorem of cosmic ecology, he says, is that for a fixed-complexity life form adapting itself to various environments, *the rate of metabolism of energy varies with the square of the temperature.*

This means, he says, that cold environments are more hospitable than hot ones, the colder the environment, the less energy required. Life is an ordered form of matter and low temperature favors order. If we live in a closed or finite universe, he points out, life will have just enough time to spread through the cosmos before the collapse begins. But if it is an open universe, infinite in space and time, the hypothesis of adaptability assures matching of falling metabolism with the square of the dropping temperatures.

Human Life Metamorphosing into a Minispacecraft? Speaking of human life's physical adaptations to travel throughout the universe, Dyson mentions his 'space butterfly' project. The DNA of the caterpillar, he notes, is programmed to wrap itself in a cocoon, to metamorphose and emerge three weeks later, transformed into legs, antennae, and wings.

If evolution can develop a metamorphosing creature like the caterpillar, then by figuring out the process, declares Dyson, we can learn to do the same thing. We can, he observes, convert to a biological minispacecraft cocoon, be launched riding on a laser beam into orbit, and metamorphose in space.

We would sprout wings for use as solar sails, grow long legs for landing on asteroids and telescopic eyes for sight, grow antennae for transmitting radio signals, and electricity-generating organs for the interplanetary magnetic field, to navigate and report observations back to Earth.

Although there's no guarantee of these ‘space butterflies by the year 2010, he acknowledges, we will "have something equally new and strange" if we keep our eyes open for the "opportunities ...beckoning us into the twenty-first century." With eminent visionary physicists like Freeman Dyson producing mathematical models of events in the future, there may be a diminishing future for science fiction.

Ultimate Proton Decay and Transferability of Mind: He refers to the *prediction of ultimate proton decay* (in 10^{33} years) suggested by present physics theory as the ultimate test of adaptability but, he says optimistically, if the abstraction and adaptability hypotheses are correct, “the patterns of life and consciousness should be transferrable without loss from one medium to another.” We would still have electrons, positrons and photons, and, he says, *immaterial plasma may do as well as flesh and blood as a vehicle for the patterns of our thought.*

Desmond Bernal and Transitions of Life to Ethereality: Dyson cites crystal-ographer Desmond Bernal whose vision of humankind soared into the future to explore the *transition of life from material substance into ethereality,* a new life which would conserve none of the substance and all of the spirit of the old, “consciousness becoming ethereal, and particle masses communicating by radiation, ultimately resolving into light.”

Mind Has Taken Control of Matter: Dyson then violates the scientist proscription against science crossing into theology and proceeds to combine them. Mind, he says, has infiltrated and taken control of matter on planet earth, fulfilling a law of nature that requires this and makes the process unstoppable. And, he declares, if we falter others will--or "may have already done so."

Mind, he notes, has taken three eons (billions of years) to control this planet and three more eons may pass before it covers the galaxy. "In contemplating the future of mind in the universe, we have exhausted ...our puny human life science. *This is the point at which science ends and theology begins.* Because of vast distances, while interstellar communication was earlier considered feasible, actual physical visitation was not. But more recent information substantially reduces the number of likely advanced civilizations and, of course, the potential for interstellar communication.

Intelligence-Consciousness May Be in Non-molecular Form: Freeman Dyson's argument is that *life resides in organization of matter;* that his consciousness is inherent in the way the molecules in his head are organized, not in the substance of the molecules themselves. It makes sense then to *imagine life detached from flesh and blood* and embodied in networks of superconducting circuitry or in interstellar dust clouds.

Life Forms Not Based on Carbon-Oxygen Systems: In *Are We Alone?--The Possibility Of Extraterrestrial Civilizations,* Robert T. Rood and James S. Trefil take off from the 'Greenbank Equation,' sometimes called the 'Drake Equation' or the 'Sagan-Drake Equation.' These equations were used in the NASA SETI program for projecting the possibility of detecting extraterrestrial life, using optimistic, pessimistic and conservative scenarios which express a 'carbon chauvinism.'

This is a predictive preference for carbon-oxygen systems and the likely presence of organic materials and water, for biological energy transfer systems like our own, rather than another type of system such as Bec's sulphur-based system, or Reeves' silicon-based system.

Nonbiological Artificial Life: Physicist Freeman Dyson's discussions on engin-eering and genetic changes in life forms needed for human life's travel and colonization of the galaxy bring to mind the work currently going on to create computer advances in artificial intelligence and to create 'artificial life' by simulating the organization of biological matter using other materials.

Contemplated, is a nonbiological evolutionary process that mimics the phenomena that drive biological changes. The process involves the competition between carbon-based life forms and silicon-based artificial life forms more readily adaptable to space travel than genetically-engineered changes in human life as discussed by Dyson.

The advent of highly-intelligent-silicon robot-creatures, without all the esoterics of human life values, is probably as inevitable as it is scary to many. One nightmare links these ceramic-plastic artificial 'people' to programs for perfecting genetic engineering in microorganisms, handling nuclear fusion assignments, and perhaps filling the shortages in judges in our court systems!

Metaphysical Extensions of the Energy Cycle of the Universe: A number of scientists have sought, as Dyson does, to connect cosmology and biology by speculating on the universe as a progression from radiation energy, to matter energy, to mind energy, from 'materiality' to 'ethereality.' From the physics of the three energy forms of radiation, matter, and mind there is a comfortable but metaphysical extrapolation, a picture of the universe as the progression from the ethereal, idea of God or Cosmic mind, to pure energy--radiation, to structured rest-energy--matter, to living energy life and mind and ultimately, perhaps, back to the ethereal, God or cosmic mind.

The Socinian Heresy--Dyson Wanders into Theology: Dyson cites a view he expressed which he learned was the *Socinian heresy,* one that he feels somehow to be consistent with 'scientific common sense', that *God is not omniscient and omnipotent but 'learns and grows' as the universe unfolds.* He speaks also of the possibility that the very laws of physics are still evolving.

Dyson Associates Mind And God: Dyson does not make clear distinction between mind and God--"what mind becomes when it has passed beyond the scale of our comprehension." He likens God to 'world-soul or a collection of world-souls,' and us as "inlets of God on this planet." We may later grow with him as he grows, he says, or we may be left behind.

He likes the idea of 'many world-souls' for mind, growing to fill the universe, "comes as a diversifier as well as a unifier." He refers to St. Paul's concern with "unity," in First Corinthians, but says he cares more deeply for the diversity of God's working than for God's unity.

Whispers of Immortality: Professor Dyson speaks of the play, *Spring 1600*, by Emily Williams, which ends with two actors meditating in the dark of the London playhouse, after the Queen and the crowds have gone. One asks the other, do you hear anything? The other answers. "Yes, I heard whispers of immortality." Dyson muses: "We know very little about...the destiny of life in the universe....*Letting our imagination wander among the stars, we too may hear whispers of immortality."*

MIND--BRAIN, KEY TO THE UNIVERSE

To Understand The Cosmos We Must First Understand The Mind: To understand the cosmos, one must understand the use of the thinking mechanism by which our species receives, records, and processes information. We must understand the nature of the human mind, and how it reached its present perception potential. As we discuss elsewhere, the very existence of the universe may require mind.

'Made in God's image' refers, as Freeman Dyson says, to human mind as an extension of God's. And to seek to know the mind of God, and the nature of the universe, requires exploring the nature of the human mind. One of the most romantic and descriptive accounts of the human mind and biological development, is contained in the writings of Cornell Astronomer and space scientist-philosopher Carl Sagan's *The Dragons of Eden-Speculations On The Origin of Human Intelligence*.

Brain Studying the Brain Is Society Trying to Understand Itself: In *Masks of the Universe*, Edward Harrison quotes Colin Blakemore's BBC Reith Lectures, *The Mechanics of the Mind*, citing our limited knowledge of how the brain works, asking *how can we have confidence in the brain's reconstruction of the external world.* In concluding, he says, "The brain struggling to understand the brain is society trying to understand itself."

All of this, of course, underlines the essentially *fragile mechanism by which we measure our comprehension of the universe.* One is reminded of Kant's admonition of the limits of our mental constructs to really know what's really out there.

Each Brain Map of the External World As a World Picture: The events of the external world lack direct correspondence with the processes of the mind. Each 'mapping' of the external world by the brain forms a unique configuration--world picture, says Edward Harrison, in *Masks Of The Universe*, which is peculiar to each person and those shared in common are peculiar to the sharing society. Harrison paraphrases Thomas Huxley's article in the first issue of the 'Nature' journal, that *the Universe fashions pictures of itself in the human mind.*

Understanding the Mind to Understand the Universe: It is interesting to observe how much science focuses on the need to understand the human mind in order to understand the nature of the universe and to grasp its meaning. It is the human mind, grouped both by societies and by the individual that shapes the many resulting different 'universes.' We see references to the many universes that correspond to the views of different societies on our planet as well as the different time periods in the history of the human race.

Each 'universe' a Conceptual Scheme That Organizes Human Thought: In *Masks Of The Universe*, University of Mass. Physics and Astronomer Professor Edward Harrison, speaks of the essentially 'forever unknown' (capital 'U') Universe, as contrasted with the many (small 'u') universes our societies construct. We alone, he says, create and organize the universes (he refers to them as the 'masks of the Universe') in which we live, and no such (small 'u') universe is or ever can be the (capital 'U') Universe. Each such universe or mask, he notes, presents a "conceptual scheme that organizes human thoughts and shapes human understanding."

Succeeding Cosmic Belief Systems: Like the *Mindsteps To The Cosmos* of astronomer Gerald S. Hawkins, Harrison describes the universes as "cosmic belief systems," of the earliest or 'magic' universe, and the succeeding "mythic, geometric, medieval, infinite, and mechanistic" universes. The magic universe is that of hundreds of thousands of years ago when *Homo Sapiens* first learned linguistic skills.

The mythic universe of ten to twenty thousand years ago, ruled by "remote and mighty gods" of Egypt, Mesopotamia and India, is followed by the rise of the "Ionian, Pythagorean and Aristotelian world systems," of philosophy, mathematics and logic, then the "magisterial medieval universe" of the Middle Ages, ...an age of devout wisdom, great scholarship, new science and innovative technology " which, he notes, "laid the foundation of modern Western society."

Here was formulated the ideas that led to the "Newtonian infinite universe," which led to the Age of Reason in the eighteenth century (age of evolution), ushering in the modern physical universe.

Why Do Societal Universes Change? Why do these societal universes constantly change? Is there some principle of natural selection, Harrison asks. Should not cosmologists study the pathology of universes, as psychologists study the pathology of the mind, Harrison asks, to understand why some universes are psychotic, as in the case of the "witch universe"of the Renaissance?

Evolution of the Human Mind: Evolution in our own universe has demonstrated the importance of the human mind and its apparent influence on the universe. Certain aspects of quantum mechanics, highlighting the influence of the observer and the effect of measurement of subatomic particles or fields, suggest to some an evolutionary process that transcends physical evolution. We are not speaking of theological implications, even though they may exist, but principles of physics, psychology, physiology, biophysics and biochemistry.

Natural Phenomena--Determined by Us Or by a Mechanical Universe: James Jeans, in *The Growth of Physical Science*, speaks of natural phenomena as being a product of our minds, *determined by us and our experience rather than by a mechanical universe independent of us.* But this probably minority view, seems to Martin Gardner somehow to be narcissistic, as he points out in his *The Relativity Explosion*.

By contrast, he quotes Einstein's philosophically more humble biographical observation:

> Out yonder there was this huge world, which exists independently of us human beings and which stands before us like a great, eternal riddle, at least partially accessible to our inspection and thinking.

Perhaps Jeans is not being narcissistic but simply echoing Immanuel Kant's philosophy that we cannot know the universe as the "thing in itself," only the perception of it represented by the construct of our minds; what our programming permits.

Philip Frank discusses "Metaphysical Interpretations of Relativistic Physics," in his *Philosophy of Science*. The metaphysical implications introduced by the advent of the theories of relativity are a matter of divided opinion, of course, but science seems to find little metaphysical correlation. And yet, Stephen Hawking says, in effect, that we are fathoming the Mind of God as we unlock the secrets of the forces of the universe, a thought also implicit in some of Einstein's observations.

THE GROWING POWER OF THE MIND--MINDSTEPS

Advancing 'Mindsteps': In *Mindsteps To The Cosmos*, Astronomer Gerald S. Hawkins organizes the evolutionary development and experience of human life into "mindsteps," the stages by which human life's object of thought evolved from a preoccupation with earth to the search of the universe--how, from a creature of planet earth, he became a creature of the cosmos.

In a broad sweep into the future, Hawkins speculates on Mindstep 5, establishing communication with extraterrestrial intelligence, that stage in our development when we face squarely the matter of contact with the outer universe.

Possible MindStep 6--Time Travel and Discovery of 'A New Field': Mindstep 6, involved time travel and discovery of "a new field." Possible travel in time, is involved, bypassing special relativity speeds through travel in the time warps, something predicted by general relativity. He kind of doubts this will happen. But he does look forward to a new type of field, nonphysical field related to life, perhaps related to the spirit--and possibly construable in religious terms.

The Nonphysical Cosmos? Continuing on Mindstep 6, he speaks of the discovery of a *nonphysical cosmos*, an “Age of the Spirit,” of telepathy, the existence of a vital force, a soul, continuation of life. He cites Nobel prize-recipient Sir John Eccles, renowned for his work on human brain structure, who mused that the cosmos was not running down (its life) "for no meaning." He wondered whether this present life is "all to finish in death," or whether there was to be an unknowable "transcendent experience."

Mindstep 7--The Ultimate--The Alpha and the Omega of Mindsteps: Could a Mindstep 7 follow a Mindstep 6? Mindstep 0 would be the alpha and 7 the omega. After communicating with other Intelligences (5), and the "spiritual domain" (6), the ultimate step (7) could only "involve" contact with the Omnipotent, the universal mind of Aristotle, overcoming the "transcendental barrier," finally substituting proof for belief. The search for scientific evidence of a supernatural power, divinity, is discussed elsewhere in this material.

KNOWABILITY--WHAT CAN HUMAN BEINGS KNOW?

Locke and Our Source of Knowledge: British Philosopher John Locke, a dualist, said that knowledge of the outside world is probable, coming only from sense impressions which have their source in a real world which does, therefore exist. But our knowledge of that world comes only from our ideas and we can be certain only in our knowledge of ourselves and God.

Immanuel Kant's Knowability As Limited to a Mental Construct: Philosopher and mathematician Immanuel Kant, a believer in Descartes' dualism of mind and matter, spoke of two universes, the one of experience (phenomenal)--the scientific world, and that of reason (noumenal), the practical world. Kant said that we can't know what it is that exists in the real world, only what our mental construct conveys to us--an idea brought to us by our reasoning and but permitting us to act toward it as real.

David Hume's Unknowable Universe--A Succession Of Impressions: In the view of philosopher David Hume, also a mind and body dualist like Descartes, we are shut up in the confines of our minds which simply convey a succession of impressions representing our ideas, the cause of which cannot be known. As with Bishop Berkeley, existence of objects depends upon them being perceived.

Hegel's Universe As Unfolding Thought: The philosopher Georg Hegel explained the universe as a totality, one of thought, in which Mind is everywhere, involved in an unfolding or evolution upward from nature to God. Although Hegel's universe has a reality, not merely a mental construct as Bishop Berkeley's, it is Idealism, based on the thinking process.

Knowability and Pragmatism--William James & John Dewey: In our modern world, there are many who believe that knowability, especially practical knowability, is limited; that philosophic abstractions are fruitless and that it is pointless to spend time in philosophic speculations about the origin and nature of the universe and life.

The universe, in this pragmatic view, is real, not just mental or idea, and consists of what one's experience suggests. In fact, the universe is simply what human life experiences it to be, and in this view, *human life is thus the measure of the universe.* Here, mind and body are real but duality distinctions are of no practical consequence.

> To me it suffices to wonder at these secrets and to attempt humbly to grasp with my mind a mere image of the lofty structure of all that there is. --Albert Einstein

PART XIII--SEARCH FOR MEANING IN LIFE

INTRODUCTION

Meaning In Individual Life: Perhaps of more immediate daily concern than purpose in cosmic terms, is meaning or meaningfulness in individual lives, as reflected in day-to-day thought, experience, goals and the very value of life continuance. Human life is forever in pursuit of reassurance that there is some meaning to life beyond mere continued physical existence. This Part addresses this drive for meaning, what meaning means, where and how it is found. For convenience the Part is divided into five chapters dealing with various aspects of meaning in life, with subjects as shown:

Chapter 46--Religion--Denominations and Meaning
- The Reformers
- Grace
- The Changing Denominations
- Creationism and Anti-evolutionism

Chapter 47--The Multiplicity of Meanings
Chapter 48--Other Thoughts on Meaning
Chapter 49--The Endless Search for Meaning
Chapter 50--Bad Things Happen to Good People--
Loneliness --Risking One's Life--Materialism

Living without a sense of purpose or reason for existence, without hope, love, friendship, and a relationship with God, or in persistent pain and suffering, or in painful terminal illness, raises questions of the existence of meaning in our lives, the value of life, and even moral choices over life continuance.

What is the meaning of meaning ? The source of meaning? Who measures it and sanctions its authenticity? What role is played by: Religion and denominational differences? By the Bible? By inherent moral mandates? By friendship? By belief or nonbelief in God? By different religious approaches to the nature of God? By suffering? These subjects, and much more, are here explored.

CHAPTER 46
RELIGION, DENOMINATIONS, AND MEANING

Religion's Essentiality: In this chapter we will be discussing meaning as reflected in religion and its practice and doctrinal content. Religion's essentiality is supported by the fact of its very universality in all societies from the earliest of times. Historical accounts suggest that religions evolved with the emergence of societies, to establish beliefs that deal with the great mysteries of life.

Mythology and Fear of Death: It has been suggested that most mysteries of nature are expressions of the fear of death and the fundamental drive for survival; that without death and the uncertainty and concerns that it brings, there would be no preoccupation with the existence of an omnipotent person or spirit, no compelling impulse for religion.

So religions, like those of early Egypt, Greece and Rome, are viewed as evolving from beliefs and concerns of early societies to account for awesome natural phenomena or to counter fearsome threats, and to answer fundamental questions about the creation, about the meaning of human life and death, the reason for the world's imperfections, and human concerns and obligations--to society and some higher authority.

Religion as Addressing Human Concerns: As societies matured, religions addressed human life's concerns and uncertainties, and its questions of the meaning of the universe and life, sharing common concepts: supplication and prayer, holy days and ritual to follow, buildings set aside for worship, obligations to fulfill, standards of ethics and conduct to observe, and written guidelines to standardize beliefs and provide common instruction.

Systems of Belief--Matters of Faith and Matters of Science: Early Greek philosophers and later philosophers like Kant and Hume pointed out the metaphysical nature of religious statements, about God, death, heaven, hell, and so forth. Though subjectively reinforceable by confidence or faith in 'revealed truth,' religious beliefs and concepts were essentially philosophy, centering around the hypothesized but not provable assumed existence of an omnipotent personality, power or spirit that controlled everything.

And so it is accepted by philosophy and theology that assertions relating to the existence--or nonexistence--of God, and what the term means, are declarations of belief, beyond verifiability. Unfortunately the unverifiability of beliefs has not prevented them from playing a central, indeed life-and-death role in the world's social and political structure and history. Wars have been fought, thousands killed, and governments risen and fallen, over these deeply held but unverifiable beliefs. So the very centrality of beliefs to religion represents one of the greatest causes of divisiveness in world society.

Interestingly, differences between religions deal not with concepts of what is justice, right and wrong, but to the existence and meaning of God, methods of access to God, the sinfulness of human beings, the continuance of life after death and access to and nature of immortality.

Scientific progress has demonstrated the questionable nature of certain biblical portrayals, such as the anthropocentric centrality of planet Earth in the universe, the description and order of the creation sequence. Apart from those who assert the 'creationist' biblical 'inerrancy' view of the literal correctness of biblical account, the biblical creation scenarios are now generally accepted as allegory, the 'days' of creation, for instance, representing eons of cosmic time rather than the duration of the Earth's revolving on its axis.

Religion As a System of Thought and Practice: Theologian Paul Tillich defined religion as 'ultimate concern,' embracing the ideas or concepts to which people are most dedicated and to which the greatest satisfactions relate.

For many, the major source of meaning in life derives from or is associated with religion, through sermon reflection, congregational and private prayer, biblical study, charitable, religious, and missionary outreach, and perhaps most important, putting religion into daily practice. Religion is sometimes described as a system of thought, feeling, and action toward God and other human beings, involving reverence and devotion, generally characterized by congregant or group participation.

It involves shared beliefs, and spiritual and emotional attitudes toward the mysteries of the unknown, that which is not explainable by reference to nature, described sometimes as the *supernatural*. It is personified in Western monotheism in the term *God.*

Religion in our society is conspicuous for the social environment which it generates and in which it thrives. Some congregants, confused by, or simply not that concerned with the distinctions or complexities of religious theory or dogma, rather than dwelling on doctrinal nuances, preoccupy themselves with the social, practical, and practicing aspects of religion, doing what is 'right.'

Religion in a Broader Context: Often cited are religious reformer Martin Luther's observations about the crucial nature of religion in terms of being able to have and experience in the heart something to be revered, whether this is God, a supernatural being, or some more mundane object of veneration, like career, power, affluence, pleasure.

So in its broader context, 'religion' is dedication to something which is meaningful. It may not require shrines, doctrinal complexity, or even clergy so long as it is something to which people can dedicate themselves to find meaning. It need not be confined to the worship of one God as in Western theism but may include ideas such as those in Confucianism and early Buddhism, dealing primarily with guidance for human behavior, and more contemporary concepts of 'humanism' and 'ethical culture,' although, understandably, both of these sometimes draw fire from religious conservatives who view them as somehow threatening to the Judeo-Christian belief system.

Meaning in Individual Life: Perhaps of more immediate daily concern than purpose in cosmic terms, is meaning or meaningfulness in individual lives, as reflected in day-to-day thought, actions, values, and goals. This meaning may be reflected in dedication or commitment to some important cause or mission which may have even life or death significance to particular individuals.

Idealizing God-Human and Human-Human Relationships: For many, 'practicing' religion involves not only worship and observing biblical injunctions and admonitions, but constantly living, managing and idealizing the whole spectrum of God-human and human-human relationships which involve every facet and activity of everyday life. In this context, the practice of religion supplies the values, purpose, and meaning for life as well as promoting reconciliation with death.

Amish Outlook: Nowhere is this daily 'living one's religion' more pronounced than among the Old Order Amish communities which, like the Mennonites, derive from the16th-century Switzerland Anabaptists, baptized or baptized again as adult believers.

Meaning in life for them places emphasis on self-reliance, family values, a religious life including biblical teaching and observance, community activity, peaceful living, avoidance of public contact and government involvement, including forgoing certain government benefits. Dressing in plain clothing of their own making and reminiscent of the past, they avoid use of modern mechanical conveniences such as telephones, T.V's and cars. They do, however, take advantage of modern medicine.

The Higher Level of Reality: In *Real Philosophy, An Anthology of the Universal Search for Meaning*, authors Jacob Needleman and David Applebaum discuss certain ideas under "Does God Exist?" Noted is that while earlier societies placed human life under the control or jurisdiction of some higher power or order of intelligence, best subsumed in the term 'God,' modern science has tended toward a 'single-level' universe, without necessarily invoking the existence of such a higher level-God. One example of this is the quantum genesis theory of quantum-uncertainty-self-generated universe.

Certain Proofs of God: But the authors note that through the ages this has run against the intuitive feeling of the existence of a superior intelligence, again termed 'God,' for which several arguments or 'proofs' have been advanced: The *design argument* arising from the universe's characteristics providing for life; the *cosmological argument* that the causes of things must finally reduce to a primal first cause; and the *ontological argument* that God, as the greatest perfection, must actually exist since merely the idea of God is less perfect.

Philosophers like Hume, Kant, and Descartes were considered to have essentially discounted these proofs. Kant asserted that the most exalting wonder is not God, whose existence is subject to metaphysical speculation, but something verifiable, our *commitment to moral action* derived from our own conscience--and this experience evidences the existence of God as its source.

Plotinus, the Enneads: Under this question of 'Does God Exist,' several classical thought sources are cited. In *Plotinus, the Enneads*, Plotinus, an Egyptian student of Platonic ideas, writes inspirationally of a higher reality of the Good, Beauty Supreme, the Fatherland, where the Father is, and towards which we may ascend through purification of the soul and dedication to becoming godlike.

Maimonides, *The Guide for the Perplexed*: Also cited is twelfth-century Maimonides, *The Guide for the Perplexed.* Influenced by the philosophy of Aristotle, Maimonides, perhaps the most renowned of Jewish philosophers of the Middle Ages, seeks to prove, through deductive logic, the existence of God as first cause, and the nonmateriality of God.

Kant, *Religion within the Limits of Reason Alone*--What Can We Know? Any search for *meaning* must consider the fundamental question of what can we really *know*? No philosopher is better known for his examination of this question than 18th century cosmologist-philosopher Immanuel Kant. In his *Critique of Pure Reason*

Kant addresses the nature of *reason*--What is truly knowable? By 'reason' he refers to pure or *a priori* reason, knowledge apart from that derived from sense or physical experience. Beyond the physically-experienced, he notes, lies the metaphysical, about which, he observes, we can engage only in merest speculation.

Among Kant's conclusions are that 'inner' experience is possible only on the assumption of 'outer' experience and further, that consciousness of ones own existence proves the correlative existence of exterior objects, reminiscent of René Descartes search for truth and his conclusion that the first truth is that the fact of his engaging in thought demonstrates his own existence, made famous by his declaration: *Cogito ergo sum*--I think, therefore I exist.

God and the Source of Morality: One such speculative area of metaphysics, according to Kant, has to do with the existence of God and the source or foundation of morality. Kant went on to demonstrate the flaws in the design, cosmological, and ontological so-called arguments for, or proofs of, God. In his *Pure Reason* Critique he goes on to discuss *morality*. He argues that awareness of moral law and motivation to live by moral law must originate from the innate, from sources other than experience, from the external.

The Categorical Imperative: Regarding morality, one of Kant's most renowned contributions to moral philosophy, so crucial to meaning, relates to his so-called 'categorical imperative' of obeying rules and fulfilling obligations considered universally applicable, guidelines by which everyone should abide, frequently expressed as to act only on those principles which can be applied universally.

Human Moral Predisposition: In his discussion of morality he speaks of the generally-prevailing "predisposition for good," respect for "action conforming to duty." One thing in our soul about which we must wonder, which is exalting, he declares, is the "original moral predisposition" in us by which we judge ourselves unworthy if we cater to "unlawful satisfactions."

God's Existence Demonstrated by Innate Moral Predisposition: Kant refers to this mysterious built-in predisposition for moral conduct as being so "incomprehensible," as to suggest divine origin, acting upon the human spirit "to the point of exaltation," exciting the sense of "sublimity of his moral destiny." So Kant, having decimated conventional arguments for God goes on to argue God's existence as derived from the existence of something we can witness, the innate human predisposition to morality.

Fundamental Meaning in Religious Concepts: Thousands have died for and, for that matter, killed for, deep-seated religious beliefs, as illustrated by the various religious wars like the Crusades, by inquisitions and trials for heresy. Galileo risked his life and suffered house arrest for his views endorsing the Copernican view, that the earth, like the other planets, revolved around the sun, challenging the anthropocentric view of earth as the center of the universe.

THE REFORMERS

Martin Luther and Justification by Faith: Martin Luther risked his life in his reformation crusade with his Ninety-five Theses posted on the door of Castle Church in Wittenberg, and in particular, his determined stand on justification or salvation by faith alone, now viewed as the basic doctrine of the Reformation.

"Here I Stand, I Can Do No Other": Called by the ecclesiastical authorities to recant at the Diet of Worms, he declared, "Here I stand, I can do no other. God help me. Amen." There are hundreds of Protestant denominations stemming from the Protestant Reformation, all divided in some aspects of church doctrine and organization. Baptist and Congregational churches, for instance, place ultimate authority in the congregation rather than church hierarchy, the ultimate in the Reformation concept of the priesthood of all believers.

Luther and Zwingli--Dispute over the Significance of the Eucharist: Martin Luther and his contemporary reformer, the great German speaking Swiss Bible expositor, Ulrich Zwingli, were kept from joining their efforts by their dispute over the significance of the Eucharist.

Calvin, Predestination and the Doctrine of Election: French John Calvin, younger than Luther and Zwingli, leader of the French-speaking reformation in Geneva, founder of Calvinism, was the author of the renowned *Institutes of the Christian Religion*, considered by many as the most influential work on Protestant doctrine to emerge from the Reformation.

His teachings emphasized grace over works, the sovereignty of God over human affairs, particularly in the doctrine of salvation by election, in which only those selected or predestined for salvation by God are saved, regardless of conduct or good works. His theology has been seen as influenced by St. Augustine's views on predestination, all time and events being laid out by God at the creation in an unalterable cause-and-effect sequence.

'Divine Right' as Prevailing over State Right: Despite his doctrinal tenacity, Calvin's theology was a major influence in achievement of political liberty, and his church successfully claimed independence, by 'divine right,' from state control. Unfortunately, his uncompromising doctrinal outlook gave religious support to tyrannical, theocratic, and inquisitorial state enforcement of church doctrine, in which exile, imprisonment, and, more likely, capital punishment, were meted out to the nonconforming deemed 'heretics.'

The Fate of Heretics: For his departure from the orthodox version of the Trinity, for example, Spanish physician-theologian Michael Servetus was seized in Geneva at the order of Calvin and burned at the stake. The fanaticism of the church is illustrated by a then-contemporary commentator who expressed his preference for having to face an armed regiment than "a single Calvinist doing the will of God."

Reformer John Wesley and Meaning: Meaning in religious outlook is illustrated by the evangelical revival in England led by John Wesley, a signal event in English religious history. A strong believer and promoter of the Reformation doctrine of justification by faith, his emphasis was on personal evangelism with the message that Christ had paid the price for our redemption from original sin, restoring our fellowship with God whose power can infuse and transform us.

Wesley and Free Will: Unlike the Calvinists, who espoused predestination and the exclusionary doctrine of predestined election for salvation, Wesley was an Arminian, a believer in free will, that predestination represent's God's omniscient awareness of how each one exercises free will to accept or reject Christ.

The world of that time, of course, followed classical cause-and-effect physics in which everything was laid out at the creation, as St. Augustine had said, and as Lutheran and Calvinistic theology had accepted. In that period, of course, there was no inkling of the subatomic universe as being governed by quantum uncertainty, and its role in undermining predestination and, for that matter, its effect on free will, discussed later.

"Be Not Weary of Well-doing": Wesley proclaimed that "The Gospel of Christ knows no religion but social, no holiness but social holiness.... Be not weary of well-doing. Go on, in the name of God and in the power of His might, till even American slavery shall vanish away before it."

Although he and his followers, called by various names such as Methodists, were not exactly welcomed in the Anglican churches, Wesley thought of himself as a high-Episcopal Anglican to the very end. He preached often to the poor in whom he found "pure genuine grace unmixed with paint, folly, and affectation." And so in his own reform movement, we see how meaning and ideas play a critical role in religious life.

The Heresy of Arianism and the Nicene Creed: Battles and wars have been fought over differences in religious concepts and dogma, over apostasy and heresy. The great battle over the heresy of Arianism, that Jesus was not of the same essence as God, led to the Council of Nicea and the Nicene Creed articulating Christian orthodoxy and confessed in churches today.

GRACE

The Matter of Grace --The Most Crucial Concept in Christian Theology: The meaning and function of 'grace' in Christianity has diverse explanations. In *A Handbook of Theological Terms*, Van A. Harvey characterizes the term as representing *perhaps the most crucial concept in Christian theology* because it refers to the 'free and unmerited act' through which God restores to himself his creatures who were 'estranged' (the alienation from Garden-of-Eden disobedience--original sin). Harvey suggests that the concept is more easily understood in the context of the major differences in the use of the term between Roman Catholicism and Protestantism.

While there are differences in theological interpretation regarding the exact role of original sin, as well as on predestined limits on qualification for salvation (doctrine of election), since the Reformation, the administration of grace through the Catholic sacraments understandably does not apply in Protestant theology.

Distinctions Traceable to Martin Luther's Justification by Faith: Compton's Encyclopedia refers to grace as the enjoyment of God's favor; the spiritual gift of God, by which people are able to choose the right and receive salvation, and, in Roman Catholicism, a similar state of grace achieved through the sacraments. This primary distinctions is often traced to Martin Luther's conception of justification as by faith alone, in which God, from his deepest being, acts to be present for human life regardless of demonstrated merit or any qualifying process.

Is 'Belief' Alone Sufficient to Qualify for Grace? But what act, event, or expression of will, if any, leads to entitlement to the receipt of grace? Generally the response is reference to New Testament scripture in which salvation comes to those who acknowledge belief in and accept Jesus as the Messiah, sent by the Father to be the Savior of the world.

What Does the Bible Say about Access to Grace and Salvation? In John 5:24, Jesus is quoted, that "whoever hears my word and believes him who sent me has eternal life and will not be condemned." In John 6:44, "No one can come to me unless the Father who sent me draws him." And in John 14:6, "No one can come to the Father except through me." What do these quotations mean? Is simple acknowledgment of 'belief' in Jesus as Savior sufficient? Some answer yes, if belief is genuinely expressed.

What Does 'Belief' *Mean*? The March-April 1996 Worldwide Church of God Plain Truth, *What is Christianity--Coming to Know God,* elaborates on the meaning of 'belief,' speaking of the invitation of Jesus to come into relationship with God, affirming, with the Apostle John, "that we live in him and he in us," returning God's love "through worship and obedience."

Evangelist Billy Graham speaks of 'belief' as *acceptance, surrender* and *commit-ment.* All these explanatory annotative adjectival additions imply *living* that belief, which means reflecting that belief in confirmatory conduct, which implies acts and deeds. So for most, 'believing,' in this context, is 'living' in God, living the role which grace implies.

Grace and Forgiveness--Does God forgive Murder or Adultery? In *The Plain Truth* magazine for February 1996, a delightful publication of the Worldwide Church of God, Paul Kroll says that under the concept of 'free and unmerited grace,' God forgives all sinners 'under grace,' including *even for murder, rape, adultery,* and other 'mortal' or conventionally-serious sins.

Excessive Forgiveness? Kroll notes that this concept of grace is a source of persistent controversy; that God is perceived by many as excessively forgiving of sin; that the rules of grace are not "the rules of fairness as commonly understood and practiced in our society." Nevertheless, says Kroll, human beings, no matter how despicable their actions, are forgiven upon belief in and acceptance of Jesus Christ as Savior, entitling them to eternal life.

The 'Scandal of Grace': What he calls this *scandal of Grace,* he says, bothers, confuses, and offends many. Of course, in Catholicism, while less serious (venial) sins are forgiven to those under grace, serious or grave (mortal) sins, such as murder or adultery, are not forgiven under grace and result in eternal damnation.

The Protestant departure from this distinction between mortal and venial sin in the context of salvation has been said to have been based largely on the view of sin as opposite to 'faith' rather than to 'virtue,' and the view that since all sins or transgressions are damnable in the sight of God, grace embraces all sin. Of course, in practice, Protestantism has sought to make certain distinctions vaguely reminiscent of the Catholic distinctions. Blasphemy of the Holy Spirit, for instance, is said to be unforgivable, although even this is not free from dispute.

Faith and Works: The Kroll article on the Scandal of Grace draws attention to the tendency of Christians, despite being recipients of grace through Christ's atonement, to try to *merit* God's grace through good works, to *earn* and demonstrate worthiness for personal relationship with God based on *performance.* Also, there are those who insist that we are not saved simply by the expedient of expression of belief in and acceptance of Jesus, or even by an emotional commitment which, although deep, has no counterpart in deeds or works. So, in this view, we are not saved without what is termed "a conversion that reveals itself in a new life."

Contrasting Statements from James, John and Paul: Often quoted are New Testament statements differently interpretable: James 2:17, that "Faith by itself, if it is not accompanied by action, is dead." The words of Jesus in John 5:24 that "whoever hears my word and believes him who sent me has eternal life and will not be condemned." And the words of Paul in Ephesians 2:8-9, that "It is by grace you have been saved, through faith--and this not from yourselves, it is the gift of God--not by works, so that no one can boast."

If there is a consensus of interpretation it seems to revolve around the proposition that when we live the new life in Christ, belief and deeds reinforce each other, work together. And so we are *not passive observers of grace,* and living in God means we do good works even though 'saved by grace alone.' As in Ephesians 2:10, "We are God's workmanship, created in Christ Jesus to do good works, which God prepared in advance for us to do."

As theologian R.C. Sproul puts it, true faith will "yield the fruits of obedience and works of righteousness." Luther's preoccupation with Paul's concept of justification by faith alone and omission of James' caveat of the meaninglessness of faith without works, have been rationalized by some commentators as not being in conflict on the basis that good works go not to qualification for grace but to genuineness of assertions of faith. Protestations or claims to faith, alone, without living that faith, are meaningless.

Faith and Works Bound Together? So in this view, faith and works, then, are "organically bound together." The further point is made that it is the gift of grace that credits the righteousness of Jesus to sinners and justifies them before God and *not even* *faith is an act of merit* qualifying for grace. We are not justified because we have faith, justification comes by grace. According to Paul, faith is the conduit that makes receipt of God's grace possible.

THE CHANGING DENOMINATIONS

Meaning in Denominational Distinctives--The Baptist Movement: Many look for meaning and are influenced by philosophical and theological perspectives from religious and denominational outlook and belief systems. We might use the Baptist denomination to illustrate.

Denominational Origins: The Baptist movement, with its relatively simple worship style, had its origins in a 17th century reform movement continuing the Reformation. Part of the Puritan movement in England, which spread over Europe, the Baptist movement sought separation from the Anglican Church worship formalities derived from the Anglican Church's Roman Catholic origin.

Baptists, who played a key role in the adoption of the First Constitutional Amendment on freedom of religion, believe that their simple worship service, placing less emphasis on creed, dogma, and the sacraments, is closer to the early Christian environment, with emphasis on living faith and practice.

Baptist Views: Despite differences between the 27 Baptist denominational variations with over 100,000 churches, there is agreement on certain principles. A derivative of the Reformation, Baptist churches take the Bible as final authority in belief, believe in equality of all members, practice adult baptism by immersion for believers, believe in the priesthood of all believers, and the independence of each congregation. Baptists endorse separation of church and state, although currently there is a strong undercurrent of support for school prayer and strong support for other government reforms.

Baptist Organization: Baptists, who presently have the highest membership of the U.S. Protestant denominations, take particular pride in the independence of congregations. They do show collective unity, however, through affiliation with conventions or associations, of which the Southern Baptist Convention is the most prominent. Baptists promote fellowship within the conventions and the churches join in state and local conventions or associations.

The largest convention is the Southern Baptist Convention in which membership is not limited to Southern states, and is made up of some 37 state conventions. Apart from the 'moderates,' the Southern Baptist Convention membership tends toward the more conservative or 'fundamental' theology, including holding to biblical inerrancy, than do the Northern conventions.

Biblical Inerrancy--The Growing Rift: Serious conflicts have arisen within the Southern Baptist Convention because of measures taken by those of the fundamentalist persuasion to realign the convention organizations and supported seminaries. These measures, considered by 'moderates' to be draconian and violative of religious freedom, are viewed as designed to diminish the influence of the 'moderates.'

This adversarial atmosphere has led to the emergence, in 1991, within the SBC of the 'centrist' or 'moderate' Cooperative Baptist Fellowship (CBF) movement which is concerned with the "current direction" of the SBC. However, in June of 1996, the CBF voted to refrain, at that time, from becoming a separate convention.

Similar schisms are developing between moderates and conservatives in the state conventions as in Virginia where the Southern Baptist Conservatives organization, disenchanted with the Baptist General Assembly of Virginia, voted to separate from the Baptist General Association of Virginia, on the grounds of liberal theology and absence of loyalty to the SBC, and to form a separate conservative state convention. This leaves Virginia Baptist churches with the dilemma of deciding on their respective affiliations.

The problem has many financial and organizational aspects relating to administration and to the managing of funds for churches, domestic and foreign ministries, retirement programs, and for Baptist-supported theological seminaries, colleges, and religious institutions.

The Interpretation Schism: The schism revolves, among other things, around what the moderates see as over-literal interpretation of the Bible by the conservatives who, through commitment to biblical 'inerrancy,' seek to avoid the slippery slope of possible unrestricted and unauthentic Biblical interpretation. In Catholicism, through apostolic succession, the Papal interpretation of scripture carries authenticity or 'infallibility.' While there is no similar biblical-interpretation mechanism for Baptist churches, which prize their independence, the 'conservative consensus,' in practical terms, does serve as a constraining influence on divergent thought.

Meaning and the Different Perceptions of the Holy Books: Martin Luther and the Reformation eliminated the Papal control for the Protestant world, highlighting the inherent right of all individuals--the priesthood of all believers--to biblical interpretation. Baptist conservatives, currently the dominant voice in the SBC, feel that the assurance of sound and consistent belief requires reliance on 'inerrancy' of literal biblical text. The problem illustrates how everyday life for so many is significantly influenced by varying human perceptions relating to the role of the world's holy books.

Some point out that the books selected for inclusion in the bible, and the versions adopted, especially of the New Testament, for which it is said there were innumerable scriptural candidates, like the letters of Bishop Clement of Rome, were the product of human rather than divine effort, which is said to militate against inerrancy of particular words in any one book.

The contrary argument is that the decision-making process had the *imprimatur* of apostolic authority. Although he was not an apostle, Mark's work derived apostolic authenticity through the apostle Peter and some interpret the authority of Paul's letters to Peter and to acceptance by the prevailing early church community.

Divisive Issues--Collaborative Responses: At the same time that there are divisive issues fueling separatist movements within established denominations, there are counter efforts by church organizations to seek constructive accord. The General Board of the American Baptist Churches, for example, established a Commission on Denominational Unity to seek reconciliation on issue such as homosexuality. Another major divisive issue is over a *U.S. constitutional amendment* asserted as needed to preserve religious liberty as related to *school prayer*. The ordination of women surfaces here as in other denominations.

Still another issue relates to ethical aspects of genetic technology, involving the use and the patenting of human and animal genes and genetic engineering. By new cell-duplication technology, absolutely identical sheep have been produced in Britain, giving rise to nightmares about similar human genetic engineering.

And of course the issue of *abortion* continues as the most heated of the controversial issues, involving society's interest in protecting the right to life for the conceived unborn versus the asserted inherent right of women to the rights of privacy and control over their own lives. There continues, of course, the issues of the theory of evolution and 'creationism,' relating to so-called science of the Genesis Creation which is discussed later here.

Changes in Worship Practices and Denominational Overlap: Earlier, we discussed religious polls on various beliefs, on denominational strengths, and the striking shift in the balance between (decreasing) mainline Protestant (Episcopal, Methodist, Presbyterian, Lutheran) and (increasing) Evangelical (Fundamentalists, Pentacostals, Charismatics) denominations.

The 'Evangelicals' are reported as more inclined to modify worship practices such as use of soft-rock bands, song sheets instead of hymnals, and Saturday worship. Services offered include counseling and fitness centers. But perhaps even more significant may be the changes in worship practices which are *crossing the lines between denominations*.

Worship, Ministry & Outreach More Significant Than Denominations?
Increasingly, notes church history professor Leon McBeth of Fort Worth Southwestern Baptist Theological Seminary, emphasis is shifting from affiliation based on *denomination* to affiliation based on the *type of ministry and worship practices* followed. On such a basis, new church categories or groupings may be described as charismatics, evangelicals, progressive evangelicals, moderates, liberals, conservatives, independents, or fundamentalists-- and some others.

Many will select churches based *less* on their denomination than on their worship-style, type of ministry, quality of caring, particular outreach program adopted (such as Habitat for Humanity), or the social or political issue supported, such as pro-life (or pro-choice), prayer (or not) in public schools, and gun control (or not). Another development is the emergence of 'independent' churches, some being 'mega-churches' which are completely independent of, or only loosely affiliated with, an established denomination, the affiliation being primarily for some logistical or pension arrangements.

Protestant Religious Grouping Descriptions--Cross Currants of Affiliation: An Associated Baptist Press listing included the following frequently-identified 'groups' and their (paraphrased) descriptions, reflecting the trend toward 'crosscurrents' of affiliation based primarily on worship style:

Fundamentalists are often viewed as ultra-conservative with a set of very specific beliefs from which little variation is permitted. It embraces five fundamentals of doctrine: Scripture as inspired word of God; virgin birth of Christ; substitutionary atonement for sin through Christ's death on the cross; bodily resurrection of Christ; second coming of Christ. Fundamentalists usually embrace a pre-millennial coming and dispensational view of the end times.

Charismatics, emphasize personal communication from the Holy Spirit, miracles, healing, and spiritual gifts such as speaking in tongues.

Evangelicals, share theological beliefs generally similar to those of the fundamentalists but more open to dialogue on other beliefs. While all Christian churches are evangelistic by virtue of New Testament exhortation, the term 'evangelicals' refers to a broad grouping, with numerous subgroups, of churches whose central focus is evangelical.

Evangelicals seek acceptance in the mainstream of American culture while holding their distinct doctrines. They embrace mainstream accredited educational efforts and are more likely to view the Bible as more 'infallible' than 'inerrant,' the former allowing for some degree of allegorical context, the latter relying primarily on the literal or verbatim wording although even here there is some interpretive room to avoid patent absurdity.

The distinction between infallibility and inerrancy probably surfaces most often perhaps in the context of interpreting such biblical text as related to the Creation and to the eligibility for forgiveness and immortality through grace (salvation-redemption and justification). Evangelicals place a high priority on evangelism and foreign missions.

Progressive evangelicals are a strongly emerging subgroup of the evangelicals. With the same 'doctrinal base' as evangelicals, they include elements of a more expansive or more liberal social agenda. Thus, progressive evangelicals usually oppose abortion and the death penalty, and favor gun control.

Moderates, a term applied in a broad sense not fully related to its common usage among southern Baptists, emphasize "actions more than belief, and freedom more than absolute truth." Although not placing a primary emphasis on doctrine, moderates do embrace most "traditional doctrinal boundaries that separate them from liberals." Most mainline Protestant denominations, Lutheran, Methodist, Presbyterian, Episcopal, fall in this category.

Liberals emphasize God's "benevolence over his judgment," usually to the point of asserting that all people will be saved in the end or that all religions ultimately "lead to the same place." Many liberals do not hold critical certain basic tenets embraced by fundamentalists and evangelicals, such as the virgin birth, the bodily resurrection of Christ, and the infallibility of the Bible. Examples are the Unitarian-Universalists.

The Problem of 'Power Evangelism': The term 'power evangelism,' is used to describe such manifestations as healings, speaking in tongues, and deliverance from demons, as tools to attract unbelievers to Christianity by persuading them of the power of Jesus.

These practices, traditional with certain Pentecostal and charismatic sects, have gained the attention of a growing number of mainstream foreign mission groups engaged in active evangelism in the uncommitted world. Power evangelism, it is noted, has gained a following in places like India, Malaysia, the Philippines, and parts of South America and Africa. In power evangelism the gospel message is reinforced by 'acts of power' designed to show the power of the Christian God.

The concern over power evangelism is with demonstrative excesses. At issue is the phenomenon of 'falling,' also known as being 'slain in the Spirit,' viewed by many as "outside of generally accepted practices of Southern Baptists." 'Falling,' when a worshiper falls to the floor apparently in an unconscious or semi-conscious state, and other such manifestations, is being used to influence and convert non-Christians.

In many parts of the world it is said that evangelism can't be done without showing miraculous intervention which may include signs and wonders not a customary part of normal Southern Baptist church life in the U.S. Although charismatic practices, such as speaking in tongues, prayer for healing, and exorcism, are part of worship in a few Southern Baptist churches, most shun such practices. And so do most missionaries.

Power evangelism is 'not mainstream' among the 4,000 or so Southern Baptist missionaries but, it is conceded, many of them do 'pray in tongues.' While Baptists run the risk of 'missing out' on the movement of God by closing the door to power evangelism and charismatic gifts, Southern Baptists, for the most part, lean toward a more conventional approach. Pointed out is the fact that the Gospel was spread through part of the world in the first century by healing the sick, casting out demons, miracles, by dealing with spiritual forces, and by intercessory prayer.

Where Do You Draw the Line? In the Jan 11, 1996 Religious Herald, Virginia Baptists Weekly, Greg Warne and Bob Allen note that whether in pristine American suburbs, or in the far-flung mission fields, the charismatic question confronts Baptists with a dilemma. *Where do you draw the line?* Most Baptists are reluctant to dismiss outright, 'speaking in tongues,' healing, and other charismatic practices which may enhance conversion potential. But Baptists are uneasy with "mysterious or esoteric" spiritual experiences which tend to "polarize believers and divide churches."

Absence of Clear Biblical Foundation for Extreme Worship: Christians overseas often adopt practices or traditions which, it is said, seem unusual or troublesome to the American Baptist. The Baptist Faith and Message, the denomination's confession of faith, is silent on the issues. Southern Baptist Convention Foreign Mission Board trustees suggest the absence of clear biblical foundation for adoption of extreme religious practice such as represented by 'falling down,' also called being 'slain in the Spirit,' in which a worshiper falls to the ground in a trance-like state attributed to the power of the Holy Spirit.

But other charismatic practices are harder to exclude on biblical grounds, like speaking in tongues, prayer for healing, casting out demons, and other phenomena attested in Scripture. Jesus is quoted, in Matthew 10, as instructing his disciples to heal the sick, raise the dead, cleanse those who have leprosy, and drive out demons.

Some suggest that Southern Baptists should, perhaps, allow what the Bible does not *expressly* prohibit. Others point out that there are certainly practices in the Bible that are offensive or bizarre in the contemporary environment. We have not seen the last of this problem.

A Post-Christian Culture? The January 1996 Worldwide Church of God 'Plain Truth' article, *The Church of the Future*, declares that many observers and critics believe that America is a "Post-Christian culture," in which traditionally Christian values and beliefs no longer prevail. As much as one third of Americans, it notes, do not believe in the personal God described in the Bible. A majority of Americans believe it is irrelevant to whom or in whose name you pray. Four out of 10 Americans believe that Christ was human rather than divine, imperfect, and either sinned, or could have.

Most Americans, it notes, cannot name half of the Ten Commandments or identify the preacher of the Sermon on the Mount. The article cites a Canadian religion professor as claiming the Bible is "largely irrelevant to Canadians." The article goes on to describe the current Western Europe *mindset of paganism* the result of "rejection of Christianity." Are these evidences of the advent of a post-Christian culture?

Evangelical activist Charles (Chuck) Colson is cited as answering yes to this question, observing that the environment for evangelism has become "more like Athens than Jerusalem."

What Do These Emerging Conditions Mean for the Church? The article continues, pointing out that the Christian divorce rate, and both unworthy T.V. subject viewing and reading habits, are virtually on a par with the rest of society. What, the article asks, does this mean for the church, and living faith in Christ and, under these conditions of religious anarchy, *who is to tell us what to believe*?

The author cites the quandary of divisive issues: the abortion minefield of different views, issues of ethics and technology, such as cloning, in-vitro fertilization, surrogate motherhood, gene surgery, assisted suicide, euthanasia, and 'organ farming.' The article speaks of the 'aberrant' new cults and religions, ESP classes, and self-declared neo-pagan cults.

Robert P. Meye, professor emeritus of New Testament studies, observes that every new generation of Christians has to be "born again" and the challenge is to *reinvent the church without jeopardizing the gospel.* How can Christianity remake itself to serve today's 'spiritually thirsty' ?

The image and credibility of Christians, the article notes, took a beating in the 1990's when televangelist scandals rocked the nation and took their toll. There is deep spiritual hunger and profound need for healing, it is suggested. Congregations need to be "homes, hospices and havens." Christians must be open-minded without compromising the basics--a tough act.

Need Innovation and Experiment: Pastor Leith Anderson comments that it is not a compromise to make the gospel message "convenient and accessible." Churches, the article suggests, must have the courage to experiment--try new models of worship and fellowship. It takes planning , teamwork and trust, but mostly initiative and courage. The article acknowledges risks but, it notes, the church that does not set out consciously to remake itself in the next decade runs the greater risk of being perceived as irrelevant in a "post-Christian culture."

Baptist State Conventions or Assemblies are recognizing the changing religious scene, and the fact that the American church is growing among ethnics, is signaling the need for ministerial students to be trained to deal with our multi-cultural world. Yet there is a reluctance to move in any direction which would sacrifice Baptist distinctiveness and congregational independence.

The New Age Movement: In the May-June 1996 issue of the Worldwide Church of God's "The Plain Truth," Keith Stump notes that "Turned off by the do's and don'ts often attached to the gospel of Jesus Christ, many New Agers reject Christianity and look elsewhere for answers to life's questions."

Stump notes that many New Age devotees once attended traditional, main-stream Christian churches but wanted to find concrete ideas for leading a more spiritually fulfilling life and so turned to the New Age movement which took shape in the early 1970s, offering "a dazzling array of alternatives to conventional religion." It now numbers in the millions around the world. What unites New Agers and Christians, he notes, is recognition of the need for more spirituality and less materialism.

New Ager Beliefs and Practices: The New Agers are a loosely-knit trend or movement with no central authority or organizational hierarchy. Their concepts differ from Christianity in a number of ways.

God: In the New Age view, God is an impersonal energy or force permeating the universe in which all is God, including human beings, who are therefore also divine, so that God is sought from within the self.

Jesus, in the New Age view, was a great spiritual master inspired by a divine energy which is available to all and the higher self is the Christ within.

The Bible, according to New Age outlook, cannot be trusted to convey God's message since it has unauthentic messages and, has been tampered with through the centuries.

Sin: Since there is no transcendent God there are no absolute moral imperatives to rebel against, sin representing ignorance of one's "inner divinity."

Jesus, for the New Agers, did not die for our sins, and salvation is irrelevant since the soul is part of the universe and never dies but is reborn in a succession of future lives. Good or bad 'karma' in one's life determines one's later incarnations. We should aim to progress to higher states of consciousness and there are many paths to spiritual perfection.

The salvation of the world depends on human beings harmonizing their energies to achieve an era of spiritual enlightenment. There are as many criticisms of the movement as 'pagan' as the movement has criticisms of conventional religions as outdated mythology.

Meeting the Challenge--Dealing with the New Age Movement: Keith Stump points out that much of the search for the deeper spiritual dimension is occurring outside the framework of Christianity. And, he asks, what is needed to meet this challenge? He points to Saint Paul's willingness to be influenced by elements of value and truth in then contemporary non-Christian cultures.

Certain things, of course, he classes as non-negotiable, such as the centrality of Christ. Included in the nonacceptables are pantheistic ideas, and use of "advice from channeled entities of uncertain identity and motive." God, he notes, does communicate personally with Christians, through prayer, meditation and Bible study.

Must Do a Better Job of Answering Questions: But, he suggests, if Christianity is to meet the challenge of the New Age, "churches will have to do a better job of answering questions about God's existence, about the nature of the mind and consciousness, the concept of the spirit, the reality of space and time, the reconciliation of science and spirituality, and the compatibility of psychology with Christianity."

Meaning and Collective Repentance: But getting back to conventional religion, history is replete with examples of sinful conduct on a mass scale, by various segments of society, various social or religious groups, even whole countries and more. Examples are Slavery, the Crusades, the Holocaust, ecclesiastical repression as in the Counter Reformation. We need only recall the burning at the stake of Spanish theologian Michael Servetus at the urging of Reformer John Calvin. We are familiar with individual repentance in everyday church life. But group or collective contrition and repentance?

In the search for meaning, *individual* repentance has always been a significant aspect of life. But society as a whole, or large segments of society have seemed incapable of collective repentance. We have seen how colonies of ants perform extraordinary feats of engineering, a sign of the holistic effect of *collective intelligence*. We have also seen how collective human intelligence achieves enormous technical and scientific conceptual breakthroughs.

We have seen collective anger, leading to wars, but collective repentance by countries, societies or segments of societies has not exactly stood out or attracted our attention. But there is something distinctively new evolving--collective contrition, by entire denominations.

A Phenomenon Sweeping the Christian World: In the March-April 1996 "Plain Truth" *Healing the Wounds,* Keith W. Stump notes that as Christianity nears its third millennium, an important phenomenon is sweeping the Christian world, a wave of *contrition for past collective sins*. Many credit this *collective repentance* to the "working of the Holy Spirit in the body of Christ worldwide," a "surging tide of God's grace," to wash Christianity clean of its historic sins and achieve long-overdue reconciliation and restoration of harmony with the wronged of the past.

SBC Resolution of Apology for Slavery Endorsement: The most widely publicized was the 1995 formal resolution of apology to African Americans by the Southern Baptist Convention (SBC) for earlier endorsement of slavery and the condoning of racism. Also cited, is the declaration of Pope John Paul II, that the Roman Catholic Church cannot enter the third Christian millennium without purifying itself through repentance for historical transgressions. These included complicity in the African slave trade, in Catholic exploitation of Latin Native Americans, and the Catholic-Protestant wars in Europe during the Counter Reformation.

Lutheran Apology: The Evangelical Lutheran Church in America expressed contrition in April, 1994 for the anti-Semitism of its founder, Martin Luther. European Christians gather in France to express remorse for the slaying of thousands of Muslims and Jews by European knights, in the name of Christ during the medieval Crusades.

A New Threshold Advance in the Human Species? In *Healing America's Wounds*, Baptist John Dawson addresses this phenomenon and the practical steps to cope with past mass wrongs, observing that "The kingdom of God will be vilified publicly as long as self-righteousness marks our relationship with the World and with each other." What we seem to be witnessing is nothing less than a major breakthrough, a new threshold of moral responsibility, a quantum leap in human perception of the human role in the world, giving new meaning to morality.

The Changing Religious Landscape: In *The Wisdom of Faith with Huston Smith: a Bill Moyers Special*, a five-part TV series on world faiths, Moyers notes that the purpose of the series is the review of the values which religions teach, not the divisiveness or supposed superiority which some religious movements assert for themselves. And so the values of Christianity, Judaism, Islam, Buddhism, Hinduism, Confucianism, and Taoism are portrayed without implications of superiority or ranking. He cites the changing religious climate evidenced by the appearance of Mosques, Hindu Temples, and other recent places of worship reflecting the growing diversity of religion in America.

He notes the changing patterns of religious predominance and, he notes, the "white Protestant culture that used to predominate in this country no longer does." Citing new religious and cultural interrelationships, he refers to his "good Jewish and Protestant friends who go off to Buddhist retreats." America, he says, is now a "constantly evolving structure of colors, religions, and creeds," with Islam as the fastest-growing.

There is a profound realization in this country, he says, that "more is not enough." That "none of this getting to the underlying disease of the late 20th century, a period profoundly bountiful and profoundly cruel." It is clear, he declares, that we need "more to live" than a Super Bowl scorecard or bank credit card.

CREATIONISM AND ANTI-EVOLUTIONISM

Age of the Universe: Interpreted literally the Judeo-Christian Bible suggests the creation of the universe, and the creation of human life, some 6,000 years ago in a six, presumably earth-day, sequence. The Big Bang start of the universe, derived from general relativity theory and the laws of thermodynamics, on the other hand, puts the age of the universe, the beginning of time, at some 15 to 18 billion years, although recent Hubble Space Telescope findings suggest an age of some 8 to 12 billion years, less than the age of some stars which, according to stellar physics, are fourteen billion years.

Appearance of Life: So, as to creation, cosmology and astronomy speak in 'cosmic' and 'astronomical' time, representing two-digit billions of years, and in terms of evolution of life, geology and paleontology speak of 'geological' time, representing millions and single digit billions of years ago.

A fall issue of *Science* article describes molecular research using genetic changes as molecular clocks , showing single-cell organisms dating back 3.6 billion years and more complex species 1 to 1.2 billion years, and a figurative big bang of emerging life forms over a short span of some 20 million years. A November, 1996 *Nature* issue speaks of evidence of micro-organisms in rocks dating back 3.85 billion years, based on the amounts of Carbon 12, a by-product of living organisms.

The Scientific Approach: The scientific approach to the time sequence involved in the appearance of human life suggests that human life evolved as a part of the evolution of life on this planet, the present *homo sapiens* species evolving from or in conjunction with earlier hominids over some several million years, present homo sapiens, first appearing within the last two hundred thousand or so years. Advocates of biblical inerrancy continue to challenge the idea of the evolutionary development of man from primates ("man from monkey"), suggesting that no one has identified the so-called evolutionary "missing link."

However, the continuing discoveries of various species of human prototypes demonstrating walking upright, increasing brain-size, increasing use of tools, decorative and art capabilities, and other manifestations of evolving intelligence and of "human" characteristics are all suggestive of a collective meeting of the "missing link" criterion.

Species in Evolution of Present-day Human Life: There is a vast amount of information (based on carbon, potassium-argon, uranium-lead, and other chemical and atomic dating methods) on approaches to the sequence of species in modern man's (*Homo Sapiens*) development.

This is more fully described in chapter 43, "Where Did Life Come From?" but the various species include, among others, *Australopithecus Ramidus*, *A. Afarensis*, *A. Africanus*, *Homo Habilis*, *H. Erectus*, *H. Neanderthalensis*, and finally, *Homo Sapiens*, currently in vogue. The *neanderthals*, the least understood of the evolving species, and *sapiens*, had thousands of years of overlap, the neanderthals mysteriously dead-ending or perhaps being absorbed.

The Sun and Moon: Geology, paleontology, and atomic-dating techniques show the sun, earth and moon to be some four to five billion years old, the dinosaurs to have perished some sixty-five million years ago, and human 'predecessor' hominid forms to have originated within the last four or five million years or so. With the exception of some religious schools, most educational systems use this 'scientific' version, but some note the existence of the biblical and 'creation-science' approaches.

Creationism: As to the start of the universe and the appearance of human life, some religious conservatives associate themselves with 'creationism,' applying 'biblical inerrancy' to the Genesis creation scenario, both as to the recency of creation and the time of appearance of human life in its present form.

As to the creation event, the Vatican, and many Protestant denominations, accept the Big Bang Theory and cosmic or astronomical time, looking upon the Genesis creation scenario as allegorical--Genesis not intended by God as a *physics handbook.* As scientist-clergy point out, 'days' which represent planetary rotation time, vary enormously between heavenly bodies. And so biblical 'days' represent, in this view, vast epochs of time. While there is at least partial religious acceptance of the Big Bang theory since it does confirm a Creation event, the situation on the appearance evolution of human life is more controversial.

Evolution and the Scopes 'Monkey Trial': We see many examples of the continued objection to the teaching of the theory of evolution as being both contrary to the Genesis Creation account and as being only one of possible theories, and therefore not entitled to treatment as 'fact.'

In *Creation or Evolution?*, a neat 1995 Worldwide Church of God book, discussing the "historic standoff between science and religion," the author reviews the famous Scopes 'Monkey Trial' in which teacher John T. Scopes is prosecuted by former Secretary of State and presidential candidate Williams Jennings Bryan for violation of Tennessee law against teaching evolution.

A Moral Victory? Defended by famous Chicago lawyer Clarence Darrow, Scopes, though found guilty and fined a nominal $100 (case later overturned by Tennessee State Supreme Court) was considered by many to have won a moral victory based on questioning which revealed technical deficiencies in the Genesis account of Creation.

The author cites *God: Religion and American Politics*, in which historian Gary Wills points out that the story of how Clarence Darrow "made a monkey" out of Bryan and the biblical passage, has perpetuated an inaccurate impression in several ways. For one thing, the State itself was only 'halfhearted' about the anti-evolution law and Bryan himself was not personally opposed to the teaching of evolution.

A Newer View of Genesis: *Creation or Evolution?* describes a newer approach by those who share in belief of the biblical truth of Divine Creation. This view is that the Genesis account was never intended as a physics handbook which could only have been a mystery at the time, but was tailored for the then unsophisticated level of readership, to communicate *theology*, not cosmology.

It was to tell who brought the universe into being, to announce God's existence and role as Creator, to announce that because God created the universe there was *purpose and meaning to human life.* Genesis was, then, a call to worship the one God, the Creator, who is "not subject to space and time."

The Factual Nature of Biblical Content As Religiously Irrelevant: In this view, science and religion are not in conflict but share in the cooperative search for truth. The December 1995 issue of Time addressed the question of whether events in the

Bible represent historical fact. In a letter response o
1996 issue, entitled *'The Bible, Fact or Fiction,'*
Lendner describes the question of the Bible's factua

The Bible As a Spiritual Communication Not Su
if "every event portrayed in the Bible is absolut
false," he declares, "my faith would remain unal
eternal spiritual and moral truth." The Bible,
communication...not subject to historical analysis."
as that cited earlier from the Worldwide Church of
discussing the "historic standoff" between scienc

Creation and Biblical Confirmations: Hugh Ross
a Congregational pastor, and lecturer on apolog
scientific Big Bang description of creation as full
ularly passages in the New Testament.

Committed to Inerrancy But Not Creationism
Believe," an organization he started which reco
scientific view of creation on the one hand, and rel
the National Association of Evangelicals and the
Inerrancy.

Though committed to Biblical inerrancy, Ross is
creationist commitment to the literal six Earth-
inconsistent with scientific discovery and, as well
which imply that these 'days' represent long epo
God, and *Creation and Time*, he cites portions of t
days as long periods of time, and elaborately det
sequence, as contrasted with the creationist 24-h

The Big Bang As Confirming the Bible: In his
has no problem reconciling the laws of physics an
recent scientific Big Bang confirmations, like the 1
of primordial temperature fluctuation remnants o
findings.

Other serious believers, of course, see evolution a
for us and to spread intelligence throughout the un
the Genesis depiction as allegorical rather than liter

adherents are creationists and deny the Big Bang as the source of creation, Ross finds the Big Bang cosmology as completely consistent with the Bible, the 'days' of creation being vast epochs of cosmic time, as St. Augustine opined.

Various Non-Evolutionist 'Science' Theories Advanced: There have been many efforts to create a 'fundamentalist front' against evolution by which fundamentalist and evangelical Christianity might convert world scientific thought to creationism or at least to the non-evolutionist viewpoint. These efforts were diluted, however, because of factional disagreements in interpreting Genesis or in constructing a suitable scientific alternative.

Some of the various 'anti-evolution,' 'non-evolution' and 'anti-false science' proposed 'scientific' alternatives include: the 'day-age' theory, the 'gap' or 'ruin and restoration' theory, the 'new catastrophism' theory, the 'new geology' or 'flood geology' theory, the 'multiple' or 'sequenced' creation,' sometimes called the 'creation-on-the installment-plan' theory. These non-evolution theories range from the concept of God creating the universe with the "the appearance of age," to fairly sophisticated explanations which allow acceptance of biblical scenario language in the context of certain events such as a phased-creation process.

Ross and Evolution: Among Ross's points on evolution is that while micro changes in species regularly occur, so-called 'macro' changes are not found; that significant changes appear suddenly after eons of time--not through 'gradualism.' But recent fossil findings do suggest the existence of 'missing link' hominid species, evidencing large brain size, speech potential, and bone structure changes emphasizing walking potential and ground-living over climbing and tree life.

Darwinian Evolution Theory a Continued Object of Attack: The continued opposition to Darwinian evolution as contrary to Genesis and as a theory or hypothesis rather than "fact," recurringly brings resort to litigation over those school text materials which generally fail to deal with 'religiously-oriented science.' The advocates of such 'religious science' point to the different explanations of the same phenomena as recurringly surfacing; that so-called 'proven' theories, like Newton's theory of gravity as attraction is replaced with Einstein's gravity as geometry.

In some states, there are, from time to time, agendas calling for the teaching of so-called 'scientific creationism' along with evolution, or the inclusion in text books of *caveats* that evolution is a controversial theory.

Darwinism and the Catholic Church: In 1950, Pope Pius XII, cautioned (encyclical *Humani Generis)* though evolution theory promoted the argument against God as creation source, he regarded it as a serious study-worthy hypothesis.

Evolution Theory "More Than a Hypothesis": In his Oct. 1996 message to the Pontifical Academy of Sciences session on the origin of life, Pope John Paul II observed that "Consideration of the method used in diverse orders of knowledge allows for the concordance of two points of view which seem irreconcilable."

Representing an historic change in the church's position on the theory of evolution, he announced that "Today, ...new knowledge leads us to recognize that the theory of evolution is ***more than a hypothesis***." However, he cautioned that the soul comes from God, even if "the human body is sought in living material which existed before it." This view has some foundation in earlier Catholic theology, going back to St. Augustine, that Genesis is "poetic and theological" rather than technical or literal.

Seeking God on the Web: In Dec. 16 *Time* magazine, Joshua Cooper discusses how believers & nonbelievers are exploring religion, faith and spirituality on the Internet. He cites E-mail of the remote Arizona Monastery of *Christ in the Desert* (solar panel electricity), c/o *porter@christ desert.org*, designing Websites (& helping Vatican).

Thousands of "faithful" & "faithless" on Internet discuss religious beliefs and there are almost 20,000 discussion groups on Usenet. With hundreds of religion electronic bulletin boards on America Online and CompuServe, the church has got to be on the "exploding" Internet, says Sister Judith Zoebelein. Barna Research in Glendale, CA says that "the faithful" are moving on line & churches that don't establish a *presence in cyberspace* will lose touch with their parishioners. Can [the Net] create "cyber-churches" with "spiritual consensus... in virtual space?"

The kingdom of God will be vilified publicly as long as self-righteousness marks our relationship with the world and with each other.--John Dawson

CHAPTER 47
THE MULTIPLICITY OF MEANINGS

The Diversity in Outlook on Meaningfulness: In everyday life we see various expressions purporting to convey meaning. "Life is a Beach" says one bumper sticker. "Love is all there is," says another. " Man is God's noblest creation," says one essay. One scientist expresses the ultimate put-down of human life, describing its origin derived from "a disease of matter."

"Man is made in God's image," says Genesis. " Man is an accidental co-location of atoms." "The mind of man is an extension of the Mind of God." "Human mind is but an extension of the nervous system of the insect." "Man has yet to find a justification for his existence or a set of values for him to live by." "Spirituality is our eternity, materiality the bane of our temporary worldly existence."

Dedication to Causes: We see meaning reflected in the pursuit of meaning by dedication to some cause. This is illustrated, for example, by Mother Teresa's dedication to helping the poor, Van Gogh's pursuit of artistic creations, Albert Schweitzer's medical missionary effort and reverence for life, Martin Luther's mission to reform the church and preoccupation with grace, Einstein's drive to redefine the universe, George Washington's drive to free his country.

Self-transcendence: In his world-renowned *Man's Search for Meaning,* Dr Viktor Frankl speaks of meaning as reflected in self-transcendence--doing for others or permanently changing something for the better, implying an element of immortality. It is in individual lives that we find the deep existential concerns, life's driven search for meaning.

An Endless Set of Meanings: There is a virtual endless set of expressions of meaning, values, and purposes in life, some sublime, some seemingly absurd. Where do we start in the search for meaning? How are we to decide what really is and is not important? We know from our experience in daily life and from the literature and wisdom of the ages that there are innumerable differences in outlook and meaning. Each person's perception of the universe, and of meaning, is influenced by genetics, the intellectual and practical influences of his environment, the society in which that person lives, and even state of health.

Random Thoughts on the Meaning of Life: Great classical literature and philosophy of the past address meaning in general and meaning of life, as do many recent publications, some of which we touch on. The Little, Brown book, *Meaning Of Life*, by David Friend and the Editors of LIFE, quotes interesting thoughts of certain individuals. For Austrian physician Helen Caldicott, the meaning of life is the "wonder of evolution that produces the extraordinary mix of species...which ... *human life...an aberrant ...[is] now rapidly destroying [and in the process] destroying...the meaning of life*."

Although the product of chance, we are part of a "collective consciousness," says architect Mayo Lin, interconnected through our works; we project our presence into the future by our accomplishments which "extend beyond our own time." Pastor Norman Vincent Peale spoke of each being born "for a purpose," with a lifetime of "insatiable curiosity," carrying our childhood "into old age," helping others through our "God-given potential."

Russian emigre Yakov Smirnoff quotes his mother's story about Tolstoy. When, as an old man planting apple trees, Tolstoy was questioned by a neighbor as to his purpose, he replied that when he's gone, people will eat the apples and think of him. Thus Tolstoy sees meaning in extending one's influence beyond this life.

Can We Read the Meaning of Life in Nature? Paleontologist and essayist Stephan Jay Gould notes that if you wind back the tape of life on earth and replay it, "you will never get man a second time." "We may yearn for a higher answer than the...quirky, improbable, unrepeatable and utterly unpredictable pathway of our evolution...but none exists." We cannot read the meaning of life in nature; We must, he observes, construct it ourselves "from our own wisdom and ethical sense."

Differing Perceptions of the World: In his thought-provoking *Masks of the Universe,* Edward Harrison, with great erudition, describes the subjective universes of different societies in succeeding time periods, and the distinctive 'world picture' constructed by each human being. As others have also suggested, of course, Harrison suggests that *each person constructs his own universe*.

And if, as we discuss elsewhere, there are extraterrestrials in more advanced civilizations which have, perhaps, learned to travel in the dimension of time, or utilize the incredible power of black holes, then there are countless other subjective personal universes beyond our power of perception.

Multiple Meanings: We discuss the term meaning, both in the sense of meaning of the universe and in the sense of the meaning of existence with the emphasis on meaning of life. In *Three Scientists and Their Gods*, science writer Robert Wright, senior editor of New Republic, engaged in the constant search for "evidence that life is not devoid of meaning," notes the views of scientists about relationships of information to meaning in a universe which consists ultimately of information.

Meaning of Life in General: Wright cites the *Encyclopedia of Philosophy* as speaking of two contexts of meaning. The *first* sense is related to divine purpose, whether a supernatural intelligence created us "to serve some end," a reference to *purpose of life in general*.

Noted, are the views of scientists like astronomer Fred Hoyle that we really have "scarcely a clue" as to our significance. Of course there are philosophers and scientists, such as physicists Freeman Dyson and Paul Davies, who suggest evidence of design and purpose, implying a *guiding intelligence* of some sort.

A 'Guiding Intelligence': Some physicists are more comfortable in describing this 'guiding intelligence' without commitment to whether this represents a *natural* or a *supernatural* phenomenon. Some convey this idea in terms of the universe as having *directionality*. Some wonder whether these are intrinsic or just linguistic differences.

One might observe here that when you get down to what is meant by this 'guiding intelligence,' whether this is a veiled reference to 'God,' many scientists seem suddenly to dematerialize into protective ambiguities of one sort or another. In our discussion on the anthropic principle (that the constants of physics are incredibly correlated to produce life), we note the work of physicist George Greenstein, in *The Symbiotic Universe*.

This is a delightful adventure into the physics of the anthropic principlein which Greenstein comes *razor-edge close* to demonstrating what some refer to as 'mathematical proof' of the existence of God. And because he realizes full well how uneasy science is about such matters, he then proceeds, painstakingly, to extricate himself from the implications of his all-too persuasive efforts. He proves God, or so it would seem, and then dutifully excuses himself, of course, for doing the unpardonable.

Back to Meaning in Individual Life: But getting back to the meaning of meaning, the *second* meaning of life relates to meaning or meaningfulness from the standpoint of *individual lives*, reflected in dedication or commitment to some important cause or mission: Mother Teresa's helping the poor, Van Gogh's artistic creations, Albert

Schweitzer's medical missionary effort and his reverence for life, Dr Viktor Frankl's self-transcendence--doing for others or permanently changing something for the better, implying an element of immortality.

Wright's Profiles: In *Three Scientists And Their Gods*, Robert Wright provides analytical profiles of computer scientist Edward Fredkin, entomologist and socio-biologist Edward O. Wilson, and chemist-economist Kenneth Boulding. The central theme of the book deals with the information nature of the universe.

He discusses information and communication, randomness, entropy, organization, order, and complexity, genetic and cultural evolution, and evolution of consciousness and awareness. He notes that entropy and growing disorder in the second law of thermodynamics is accompanied by the growth of islands of order, complexities--like human beings.

Meaning in a Universe Driving toward Disorder and Chaos: There seems to be meaning in this very fact, that in a universe driving toward disorder and chaos, and in which complexity is unraveling, there are islands of increasing order and evolving complexity. It goes even further--the growth of this order is possible only at the cost of increased disorder elsewhere, what physicist Hubert Reeves, in *Atoms of Silence*, terms "the great elsewhere," conveniently provided by the expanding universe.

Can't Quantify Meaning: Getting back to Wright, he seeks to relate some of these engaging themes to information and meaning. In the process, though, he notes that *not only do we not know how to "quantify meaning," but we can't even agree on "what meaning means."*

In one of his three profiles Wright recounts his discussion with sociobiologist E.O. Wilson. He refers to Wilson's *On Man, Nature* and *Biophilia*, which discuss the philosophic crisis facing humanity, the unmet need for religious ferver, where scientific discovery intrudes on religious interpretations--such as evolutionary biology explanations departing from what he terms "the creation myth."

Wright and Edward O. Wilson on Religion and Meaning: The Wilson-Wright discussion is apropos here in the context of the role of religion in meaning of life. Wilson, despite his Southern Baptist upbringing, goes on to express his belief, shared by others, that "genetic imperatives" will ultimately explain our "moral codes." These "orphaned religious impulses," says Wilson, should be attached to a "legitimate belief system--science for instance," and we should, he feels, "cultivate a 'scientific human-ism' that taps the energy of our innate religious drives."

But Wright counters that religion generates a sense of purpose--some point in living, "reason to struggle against base impulse and existential despair, to try to live with compassion, restraint, and dignity." Wilson readily acknowledges the reasonableness of the point, the need to understand "the religious impulse as central to human behavior...in a way that not only explains it naturalistically but harnesses it."

Need More Dialogue on the Source of Religious Drive: He then acknowledges that scientists and theologians have need for more dialogue about the *source of religious drive*. He goes on in a constructive vein of "marrying religious thought" with more scientific study of the source of moral behavior, the purpose being "to make moral reasoning more science-like" but without reducing it to materialism.

Wright then gets wistful about knowing *why he is here* and how he is supposed to behave. But, notes Wright, Wilson doesn't "lose any sleep over life after death." Says Wilson, "I don't worry about my own immortality." Well Wright just can't resist relating what happened to Wilson at the ceremony honoring Martin Luther King Sr. in the Harvard Memorial Chapel.

Split Personalities, One Governed by Reason, the Other, Emotion: The Rev-erend King preached fervently and the congregation joined in one of the old-time hymns right out of Wilson's boyhood. Part way through it, says Wright, "E.O. Wilson, scientific materialist, detached empiricist, confirmed Darwinian," started crying.

The name of Wright's chapter on this aspect of his conversation with Wilson? "God." This touching anecdote illustrates something very important. Most of us are dual or split personalities. The person speaking through a pure theoretical reasoning process is not the same person speaking through his emotions. Reason says one thing, emotional intuition, another. In this vein of thought, many people feel there are *no true atheists*.

Why Am I Here? In Part Two of *Real Philosophy, An Anthology of the Universal Search for Meaning*, the authors Jacob Needleman and David Applebaum address the question of "Why am I Here" ?

They note in his *Confessions*, St. Augustine's confrontational encounter with the question of life's meaning, reminding us that, like St. Augustine, many of us plod along in life without stopping to analyze our life predicament, until some attention-arresting event jolts us into sharp confrontation with the question of our existence.

Oedipus Rex: Some may rationalize inattention to the question of our purpose in life on the ground that life appears to be governed largely by forces over which we have little control. They cite Sophocles' *Oedipus Rex.* Fleeing from his home to avoid the oracle-predicted crimes he is to commit, Oedipus, to his dismay, finds that the more he runs, the more is he caught up in the prophesy's fulfilment.

Only by surrendering claims to personal choice or free will does he gain "a measure of freedom." One comment is that suffering may be simply the "inability to understand one's place in unfolding but predestined events." But to put Oedipus Rex into contemporary perspective, modern science, especially quantum uncertainty, now leads many if not most physicists and philosophers to question the early concept of the totally-predestined universe.

Dostoyevsky--The Grand Inquisitor: The authors also cite the Grand Inquisitor scene from Dostoyevsky's *The Brothers Karamazov* in which Alyosha is confronted with a world in which the power to chose is surrendered. One implication is that *one's very drive for meaning is in itself a cause of suffering*, a drive which cannot succeed and from which we cannot escape.

The relentless probing of why we ask questions leads to the frustrating conclusion that the universe, and the confusing state of things, has *no answers*. As Paul Davies suggests, in *The Mind of God, there may be no end to the chain of regression, unless we abandon the reasoning process to acceptance of God as ultimate cause.*

Dostoyevsky describes three distractions from his search for meaning in this life. The third, is the will to unity, which manifests itself in the attraction to conformity, patriotism, and identification of oneself with some organization. The inquisitor speaks of all that human life seeks on earth as narrowing down to *three goals*: some one to worship, some one to keep his conscience, and some means of uniting all in worldwide union or unity, one "unanimous and harmonious antheap, for *the craving for universality is the third and last anguish of men*."

***The Hymn of the Pearl* --Failing to Pursue the Source of Meaning:** The authors observe that when we fail to inquire into the source of meaning, our lives are condemned to remain only a partial fulfillment of potential. If we do initiate a quest for meaning, there is the additional risk of slipping into inactivity, being sidetracked. Cited is the *Hymn of the Pearl*, from Coptic Gnostic sources of the third century B.C. in which a prince travels to a foreign land in search of the "pearl of wisdom."

Distracted, he grows unmindful and the purpose of the mission slips from his awareness. The implication is that we must be alert to prevent diversion from our search, and to strengthen the capacity to question, despite "fear, confusion, and hopelessness." Reflection and introspection have the potential for guidance to answers--which *may change the very course of one's entire life*.

What is it about the human condition that poses and highlights the question of life's meaning? We are promised a role in fulfilling our own destiny. We attain an awareness of ourselves in relation to the world, giving us the opportunity to participate in the unfolding of a life of meaning.

As we acquire more knowledge, things do not become more comprehensible but only more mysterious.--Albert Schweitzer

CHAPTER 48
OTHER THOUGHTS ON MEANING

Some Commonly Held Views of Elements of Meaning: For many, meaning involves the pursuit of pleasure, gratification of some sort, or perhaps the more subtle and abstract pursuit of 'happiness.' For others, like the Greek philosophers, mean-ing is derived from the acquisition of knowledge. Some find meaning from career pursuits, or from society's approbation or approval, recognition of one's efforts or worth.

For still others, purpose and satisfaction in life is fulfilled through children, friends, a nice home, a late model car, and a satisfying church affiliation. Some creative and communicative acts are needed for many. Some see meaning in discovery or inventive pursuits, medical, scientific, or social research. Creative thought is meaningful for writers, philosophers and theologians. Many see meaning in the creation or perform-ance of music or art.

Why Does a Van Gogh Sell for $82.5 Million? What makes a recently purchased painting worth $82.5 million dollars, a sum which will build a college, a hospital, a highway, or feed thousands of starving human beings?

Vincent Van Gogh's 'Portrait of Dr. Gachet,' painted at the age of 37, during his mental illness and shortly before his final despair and suicide, brought that amount on a purchase by Japanese buyer Ryoei Saito at Christie's in New York, for the Kobayashi Gallery of Tokyo. Two days later, Pierre Renoir's 'At The Moulin de la Galette,' sold for $78.1 million at Sotheby's. Another Van Gogh, 'Irises' brought $53.9 million at an earlier sale at Sotheby's, and Pablo Picasso's 'The Marriage of Pierrette,' sold at Drouot-Montaigne, Paris, for $51.38 million.

Meaning in Human Creativity: What draws such huge sums for human perceptions of thought of objects reduced to small brush-driven, oil-painted canvasses? Certainly they must represent something important, instil meaning of some sort. Some feel that their huge cost is simply economics, investment potential, something that will have increasing demand.

But this begs the question. Why is it in demand, why does a graphic representation of an object, person or idea, drawn by a person on a piece of canvas, have such extraordinary 'intrinsic' or investment worth? Some argue that such worth lies in the *value of human creativity,* the power of expressing human perception, graphically portraying the abstractions of human cognition.

A Closeness for Van Gogh's Tragic Life: It has been noted that the Japanese feel a certain closeness for Van Gogh, intrigued with the tragedy of Van Gogh's life, his final melancholy, the irony of his posthumous success, and possibly his interest in Japanese woodblock prints which may have influenced his work. But whatever the explanation, human search for meaning is a driving force in economics, political events, the arts, and every aspect of society.

Meaning Earlier in Human History: Earlier in human history, meaning consisted primarily in getting food by the hunt, finding a cave or building a shelter for housing, and devising a strategy for survival from predators and the elements. In other words, meaning was built around *survival* itself. Psychiatry notes that the more we acquire and the more our basic needs are met, the more abstract becomes our search, the more subtle our concept of meaning.

With the coming of affluence and wealth, the filling of need, the satisfying of want, the removal of deprivation and the disappearance of suffering, comes an absence of goals, a loss of purpose and meaning, sometimes so exacerbated that it may even lead to suicide--the world, life itself, simply ceases to have meaning and therefore value.

Service to Others: But deeper or more subtle meanings sometimes dominate thought. The philosopher and medical missionary Albert Schweitzer found meaning in the universe in his *Reverence for Life.* Meaning for many is found in a life of giving, service, and sacrifice for the good of others. Although the pursuit of acquisition--acquiring things, may turn into an obsession, a form of addiction, it may also have valid meaning and purpose.

The Pursuit of Acquisition: Much has been written on the subject of the role of *acquisition* in life--whether vice or virtue. Psychiatry, philosophy, theology and scripture all suggest that it has no ultimate meaningful reward and its forms and excesses are counterproductive--or even evil.

Yet the idea of *acquiring* and *saving* is somehow built into us, so presumably it must have some *instinctive* meaning. Of course, prudent acquisitions such as saving for "a rainy day," avoiding dependence on others, providing for one's old age, acquiring to help others, these are viewed as worthy 'meaningful' goals.

William James observed that "The great use of life is to spend it for something that will outlast it." Some of the great parks of Boston are the result of the acquisitive instincts of Ben Franklin who donated his wealth for great causes that would survive him. Acquisitions of the noted philanthropist Andrew Carnegie made possible major charitable and educational institutions. He believed in the acquisition of wealth for the public good. The National Gallery of Art, a great national treasure, is the gift to the nation of the financier Andrew Mellon.

So, in the end, it is the *purpose* of acquisition that governs its meaningfulness. Individuals and churches are often at odds on the question of the biblical obligation for tithing, and the Bible has its admonitions on the pursuit of riches. But it also urges saving for one's future needs and for philanthropy. Many a church building fund goal has been reached by bequests from non-tithers, and for that matter, nonmembers.

Gratification from Divesting: It is said that happiness and the deeper satisfactions do not come through pursuit of *acquisition.* It comes, it is thought, from *divesting*--giving, helping others, meeting the needs of others. We are reminded that the only way of investing in the *ultimate future*, is by giving away to serve some worthy purpose. Noted for his philosophic and religious ideas, novelist John Gardner observed that "When people are serving, life is no longer meaningless."

Meaningfulness and Happiness: Meaningfulness is often associated with hap-piness, the idea that human life's goal lies in finding joy, happiness, and that which is uplifting; that if one is *happy,* the universe and life has sufficient meaning. But in a world so often marked by deprivation, illness, imperfection and repeated disappointment, where lies happiness and meaning?

Some feel that the key to happiness lies in constructing the *right outlook or perspective.* Sociologists point to the fact we find many children of low-income families, unspoiled and with little if anything to call their own, who are seen as happily playing, without a care in the world, while children of wealthier families are dissatisfied, and lead troubled lives.

Psychiatrists observe that being satisfied with the things permitted in our lives, rather than unhappy over the absence of things, can be the key to happiness and meaningfulness. There was a time when there were no automobiles, VCR's or TV's, no home appliances, electricity in the home, even the convenience of indoor bathrooms.

In those times, happiness did not relate to such conveniences. Life's blessings or 'satisfactions' lay in much simpler things, a roof over your head, a fire to sit by in the winter, food to satisfy hunger, and above all, caring human relationships. This simpler life represents the lifestyle of the Amish today, recognized as a well-adjusted society, viewing their lives as meaningful without these modern contrivances.

Mother Teresa and Meaning: In a 'Time' interview in Calcutta, India (Time. Dec. 4, 1989), Mother Teresa was asked whether, in her humbleness, she thought of her-self as having special qualities to be a vehicle of God's Grace in the world. She replied that "It is his work..."I am like a little pencil in his hand....He does the thinking...the writing....The pencil has only to be allowed to be used." She perceived herself as somehow an instrumentality of divinity.

God's greatest gift to her? The poor people, for "they are Jesus for me....prayer into action, *the joy of the poor is so clean, so clear."* Asked about materialism--preoccupation with wealth, she said that *poverty is a freedom*, not a mortification or penance; she *found the rich much poorer*, more lonely, never satisfied, always needing something. And about the future? It is "his concern....Yesterday is gone. Tomorrow has not come. We have only today to love Jesus."

The Central Question: For many, at the core of the search for meaning is the question of why *we* exist, whether we are somehow important in the scheme of things. *What is man that thou art mindful of him?* asked King David (Psalms 8:4). The concern over our existence, built into our very awareness, compels us to seek out the meaning of our presence here, the meaning of the universe and of life.

As the Mothers Teresa of the world demonstrate, it is some credit to the human race that the meaning of existence can translate into *concern for others*. The ultimate illustration is the willingness to die for others--a mother's sacrifice to save her child--or a person's willingness to die for some paramount universal cause. Built in the human psyche is some intuitive recognition of the existence of profound values which transcend even the value of individual life.

Who Am I? Is There Anything Which Is My Own? In his essay on " Man's Ever New and Eternal Challenge," published in the book *On The Way To Self Knowledge*, this question of "Why do we exist," sometimes phrased as "Who am I," is addressed by Dr. Michel de Salzman. He points out that each human being is the only one who can experience his own life, a life in which his characteristics and capabilities were given him, inputted into him, one that reflects determining influences of race, heredity, climate, and more *remote cosmic influences*. But if we are 'given' everything, is there nothing that we can truly call your own?

What Is My Own Is My Attention: So what is my very own-- not handed to me? he asks. Something does seem to remain my own, giving me identity, and this is *an intentional act of knowing* which has a unique capacity for freedom. It is my *attention,* he says, my *becoming present to what is my participation in the unifying actuality of being.*

But attention, he says, is inseparable from *interrogation,* the act of *questioning of the meaning of being.* Behind the obscuring screen of all other questions, he says, is the central existential question, Who am I? Whether or not articulated, it is, he says, *human life's first and last question.*

The Mind Surrendering to the Total Consciousness: Summarizing, he says that it is only the dimension of total being that gives knowledge the levels which the mind cannot grasp. It's the passage from the 'mind-centered' questioning to a total questioning--when the mind stops talking--that true experience of oneself becomes possible, the kind of experiencing during systematic relaxation when mind yields its preoccupying attention to the increased awareness or attention of the whole being.

The Endless Multiple of Questions: But, as Salzman points out, these central questions are an endless multiple of other questions. Subsumed in this ultimate existential question--the root question in all of metaphysics, are all other questions--is there a God that caused the universe? Why did he cause it. And, more relevant, *were we the reason?*

Why are the laws of physics, chemistry, biology the way they are? To bring forth Life? Well, O.K., now that we have reasoned our way into Life, why are there all these dubious things like time and death, disappointment, poverty, disease and evil, rather than timelessness and eternal life, perfection and good? Who needs evil anyway? In essence, we ask what is the meaning of everything--of God, the universe, human life--why are things (everything) the way they are?

Who Am I? In *Real Philosophy, An Anthology of the Universal Search for Mean-ing,* philosophy professors Jacob Needleman and David Applebaum undertake a survey on the universal quest for meaning, which they call "Real Philosophy," and "the love of wisdom." Part One addresses the question of Who am I? In moments of extreme physical stress or profound introspection we experience a fundamental uncertainty about our lives and question our very identity, asking, Who am I?

Descartes--*Cogito, ergo sum*: René Descartes, 17th century mathematician and philosopher was one of the founders of modern epistemology. In his *The Meditations*, he rejected the validity of all knowledge that didn't pass his truth verification criteria, with the purpose of building a foundation of irrefutable truths, if such existed.

He started by isolating something so evident that no scepticism could prevail against it's veracity. His process of refinement led him to the conclusion that because he was engaged in a thinking process to pursue this line of inquiry, that he himself therefore must exist, a thought reflected in his now famous classical aphorism: *Cogito ergo sum*, "I think, therefore I am." He went on to accept the existence of God as an innate idea, and like Kant, rejecting the conventional proofs of God, and offering his own proofs.

The Upanishads: *Real Philosophy* cites several sources of meaning relating to the question of Who Am I? One is the *Katha Upanishad*, one of the Thirteen Principle Upanishads, the writings of a sixth-century Indian philosophical movement which, among other things examine the meaning of life and relation of the individual to the cosmos. The Katha Upanishad speaks of the pathway to the soul as a charioteer, using the parable of the soul riding in a chariot which represents the body senses.

Having understanding, being "mindful and pure," represents the chariot (senses) being reined in and controlled, and thus reaching the goal, the highest place of Vishnu, where one's reward is to be "born no more." But his senses uncontrolled, the charioteer fails to reach his goal, continuing on to reincarnation--to run the race again.

The Milindapahna: Another cited work is *The Milindapahna*, the Questions of King Milinda, a text of classical Buddhism, a body of thought and practice which seeks to free the individual from the barriers to full human expression. Through a question and answer series, the study ends with the recognition that there is no self different from the cosmos as a whole, despite experiencing a progression of different impressions.

[Mother Teresa responding to question whether, in her humbleness, she thought of herself as having special qualities to be a vehicle of God's Grace in the world.]It is his work...I am like a little pencil in his hand....He does the thinking...the writing....The pencil has only to be allowed to be used." [She perceived herself as somehow an instrumentality of divinity.] Mother Teresa

CHAPTER 49
THE ENDLESS SEARCH FOR MEANING

Human Search For Meaning: *Man's Search For Meaning*, a best seller throughout the world and the 'book of the year' of many colleges, including colleges of religion, is the work of the eminent Austrian psychiatrist and founder of logotherapy, Dr. Viktor Frankl, whose dehumanizing death-camp experience gives his work an extraordinary intensity of existential analysis.

Meaning Differs from Person to Person and Moment to Moment: Frankl underlines that the meaning of life differs between people, from day to day and even moment to moment. It is not the meaning of life in general--an abstraction--that matters, but the *specific unique meaning or mission of a particular person's life at a specific moment in time.*

Human life asks of others the question of meaning, an exercise in futility, because *life* asks this question *of him.* And how must he reply? Well he replies by being responsible and it is this *responsibleness* that is the essence of human existence. Frankl's responsibleness, interestingly, is close to Immanuel Kant's categorical imperative of duty.

Self-transcendence: Frankl goes on to say that meaning is found in the outside world rather than within the self. *Thus self-actualization is attainable only through self-transcendence.* This view reflects an aspect of pragmatism, that abstractions, meaning, is to be found not in ethereality of idea, but in practical applications in the real world: God's greatest gift to Mother Teresa? The poor people, for "they are Jesus for me....prayer into action."

Scriptural Parallels: It is interesting to observe that these sophisticated obser-vations of a learned human life of science, burned into his consciousness from the depths of despair in the concentration camps, are paralleled by similar scriptural exhortations to transcendence--to love your fellow human being, be compassionate and forgiving, practice charity and humility, dispense justice and eschew selfishness.

Frankl's Three Avenues to the Discovery of Meaning: So "actualization through transcendence" is his key to meaning. There are three avenues to achieving this, to discovering meaning, he suggests: the triad of creation, experience, suffering & transformation:

First, by doing a deed or creating a work.

Second, by experiencing a value, something or someone beyond oneself, such as loving another.

With respect to the *Third*, in *Man's Search For Meaning*, he characterizes this third as 'Suffering.' But in his essay on *Man's Search For Ultimate Meaning*, he particularizes on "suffering," citing the duty of transforming one's *predicament*, an incurable disease, into *achievement*--converting tragedy into triumph.

Meaning Available to the Very Last Moment: Meaning, he insists, is available even to one's last moment-- because then *human life may bear witness to human potential* in its darkest hour. The person who is so 'provided' that he is spared suffering or struggle, no longer has a goal, cannot strive, and faces the abysmal *existential emptiness*, a personal nihilism--existing without meaning.

If human life sees meaning in *something* he is willing to suffer or even die for --and perceive it as ennobling rather than degrading, the ultimate actualization through transcendence. Thus, implied in the grant of life, is the duty to utilize and optimize life to the *last vestige of vitality*.

Do You Invent Your Own Essence--Or Discover It in You? Frankl, considered the father of logotherapy, an adaptation from psychotherapy, asks where our own personal constructs of purpose, meaning--and derived values--come from. Existentialist philosopher Jean-Paul Sartre, author of the monumental *Being and Nothingness*, had suggested that *man, adrift in a meaningless universe with a terrifying freedom to chose,* constructs his own essence.

Frankl asks, does the human *invent* itself, that is, *design its own essence*, as Sartre and other existentialists have suggested, or does the human *discover or detect* from within itself, its essence--the meaning of existence. Does it *find it* in itself? Frankl feels that we don't invent it--that *each of us comes equipped with an essence to be detected and identified, to be discovered by us.*

Understanding the Human Mind to Understand the Universe: It is interesting to note here that science often focuses on the need to understand the human mind first before seeking to understand the nature and meaning of the universe. As University of Mass. Physics and Astronomy Professor Edward Harrison notes, in *Masks Of The Universe,* it is the aggregate of human minds, within society groupings that shapes the many different 'universes' of human perception from time to time and from society to society.

In a delightful display of fine classical scholarship, Harrison describes briefly the sequence of 'universes' representing human life's changing understanding of the Cosmos and related ideas, presenting an especially erudite and exhilarating view of ancient and modern cosmology.

The Platonic System of the Mind: Interestingly, the question of the source from which human life finds meaning, whether detected from within itself, or constructed from perceptions of the surrounding environment, has preoccupied philosophers from early times. Harrison observes that an earlier mythical universe of the gods, gave way to the "Platonic system of the Mind," operating according to a "universal plan *fully known to the human soul before inhabiting the physical body* [and] discoverable by inwardly-directed inquiry."

Anticipating much later philosophical concepts, Plato, student and renowned chronicler of Socrates, proposed that 'reality' consisted of 'forms' or 'ideas' rather than their material examples. He described scientists preoccupied with material applications of ideas as "deaf to the voice of the soul" and "lost...pursuing perishable forms."

But Detecting Our Preset Essence Seems to Impinge on Free Will: But built into the question of whether we construct our own essence, and set of meanings, or whether we detect what is prearranged in us, forces us to reconsider the meaning of free will. If we have existentialist Sartre's "terrifying freedom to chose," then, at least, we have finally optimized free will. If we say, on the other hand, that our essence, meanings and values, are already preset, and we simply find them through introspection, then we have been handed our essence and meanings, and our free will is at least constrained.

Frankl may be correct in his analysis, but for those that favor free will--and a conveniently indeterministic universe--they must surely be disappointed with coming into the world with our views of meaning prefigured. Despite his predetermined theism, St. Augustine, you recall, was able to rationalize the existence of free will and, of course, the unavoidable existence of evil, despite an omnipotent and omniscient God who knew in advance of all human life's coming peccadillos.

Ultimate Meaning: In his *Man's Search For Ultimate Meaning,* psychotherapist Viktor Frankl points out that unlike *values*, which are passed on through tradition, *meaning* can not be 'given' to anyone, it can only be found by one's own 'lonely conscience.' Meaning must be found--and fulfilled--by each individual.

The human search for meaning--Frankl's "will" for meaning--is a "primordial motivating force" in human nature, increasingly frustrated in modern society by the abysmal feeling of *emptiness and futility*, the "*existential vacuum.*" We subconsciously seek to achieve free time to pursue meaningful activity and then can't handle that freedom.

What persons need, says Frankl, is not biological homeostasis, a "tensionless state," but the existence of tension discharged via the "striving and struggling for some worthy goal, "*an awareness of meaning worth living for.*"

Rescuing the Transitory into the Past: He points up an arresting and often overlooked fact, a noteworthy event which occurs to each human being each day. We are, each day, dealing with the transitory, the *possibilities* in life. But once we bring each possibility to fruition--convert it to reality of specific action--we "*rescue it into the storehouse of the past,*" *secured for all eternity through actualization.*

Life never lacks meaning, he says, but it is up to us to discover it. *Human self-transcendence is lived out by giving of ourselves,* "serving a cause or loving a person." Thus Dr Frankl and Mother Teresa share the same beatific vision.

Inventing Values: We have noted Frankl's view that we discover or detect meaning from within. Although many share this view, others look upon the source of meaning with a different perspective. Physicist Steven Weinberg, for example, in his "Confronting O'brien," one of the delightful essays in Clifton Fadiman's *Living Philosophies*, notes that "in our value postulates, we are not responding to some cosmic imperative, but rather *inventing values for ourselves as we go along*."

It is no doubt true that some values surface, emerge so to speak, as a byproduct of new needs and experiences. But whether we *invent* them or *detect* them from our inner self is, perhaps, debatable. One can make the argument that the two sources are the same, i.e., when they 'emerge' into our focus of awareness they constitute our 'invention.'

Why Do These Societal 'Universes' Change? We have noted how our impressions of the universe and events and phenomena in the world around us influence our outlook and temper our perceptions of things. But why do these societal universes constantly change? In *Masks of the Universe*, Edward Harrison asks whether there is some principle of natural selection at work in this process.

Should not cosmologists study the pathology of universes, as psychologists study the pathology of the mind? he asks, to understand why some universes are psychotic, as in the case of the 'witch universe' of the Renaissance in which poor, innocent women were singled out for punishment and death, *citing the Bible as the authority* for such human travesty.

The Human Brain and the "World Pictures" of Individuals: But individuals are not all confined to the cosmic belief systems of their societies, says Harrison; they have their own "world pictures" in which they also live. These world picture impressions are the result of the physiological and neurological processes by which light rays, sounds and other signals are processed by the sense organs and transmitted to the brain via the neurological pathways utilized to govern the flow of information to the brain.

The Brain Studying the Brain Is Society Trying to Understand Itself: Edward Harrison quotes Colin Blakemore's BBC Reith Lectures, *The Mechanics of the Mind,* citing our limited knowledge of how the brain works, asking how, therefore, can we have confidence in the brain's reconstruction of the external world.

In concluding, he says, "*The brain struggling to understand the brain is society trying to understand itself.*" All of this, of course, underlines the essentially fragile mechanism by which we measure our comprehension of the universe. One is reminded of Kant's admonition of the limits on our ability to know the external universe as limited by the frailties of our mental equipment.

Each Brain Map of the External World As a World Picture: The events of the external world lack direct correspondence with the processes of the mind. Each 'mapping' of the external world by the brain forms a unique configuration--world picture, says Harrison, which is peculiar to each person and those shared in common are peculiar to the sharing society. Harrison paraphrases Thomas Huxley's article in the first issue of the 'Nature' journal, that the Universe fashions pictures of itself in the human mind.

Psychotherapy--Accepting Others: Thomas P. Malone, in his paper, "The Christian Sacred Tradition and Psychotherapy," published in *On The Way To Self Knowledge* (Needleman & Davis, Knopf, 1976), says that God, deciding he had made a mistake [in the Garden of Eden scenario], corrected it by *allowing human life to sin--and so liberated him "to become a moral creature."* There was no way for human life to begin as God-- human life had to work toward it, and the 'road to Godliness' was paved with free choice, freedom to sin. Thus, *humanness must embrace capacity for sinning as a precondition to the Ascent.*

We search for God, he says, and pray for grace to continue the search. But what is this grace I seek, he asks? After thirty years in psychotherapy, he decided, he says, that *before you can help anyone to be different, you must accept them as they are.* In this context of psychotherapy, "acceptance of another is what I mean by grace."

Dr Jacob Needleman, in his *The Heart Of Philosophy*, points to the need for human beings to think in new categories and view life within a vast, new frame of reference. He points to the need to awaken the deep impulse in human nature toward that which is higher than ego, that which may place you in authentic relationship with God.

Philosopher Immanuel Kant speaks of the will to good--which he calls the *Sense of Duty*--as nothing less than the power which elevates human life above itself and reveals the sublimity of our highest nature, our "intention toward moral law" which contrasts with our lesser self of nonconforming conduct.

Our Relationship to Time and Unity in the Cosmos: This searching is an expression not only of our concern for the most precious thing to each of us, continued existence, but our need for knowing our relationship to God and the unity of the cosmos--if the two are indeed different. Elsewhere we discuss this matter of the duration of life in terms of the geometry of spacetime and the deep-seated anxiety at the heart of existentialism, exploring the thoughts of Kiekegaard, Heidegger, Sartre and the other existentialists, as well as Einstein's mysterious characterization of distinctions between past, present and future as being an illusion.

In his *Perfect Symmetry*, Heinz Pagels points out that human concern over existence provides the basic drive for relentless pursuit of knowledge of the nature of the universe, the reason for our presence--and our likely ultimate destiny.

> [Paraphrased]God, deciding he had made a mistake [in the Garden of Eden scenario], corrected it by allowing human life to sin--and so liberated it to become a moral creature. So humanness must embrace capacity for sinning as a precondition to the Ascent. --Thomas P. Malone

CHAPTER 50
BAD THINGS HAPPEN TO GOOD PEOPLE, LONE-LINESS AND MEANING--RISKING ONE'S LIFE FOR ANOTHER

WHEN BAD THINGS HAPPEN TO GOOD PEOPLE

When Bad Things Happen to Good People: In the popular book of that title, Rabbi Harold Kushner cites Proverbs 12:21 that "No ills befall the righteous, but the wicked are filled with trouble." He cites Isaiah 3:10-11 "Tell the righteous it shall be well with them, for they shall eat the fruit of their deeds. Woe to the wicked, it shall be ill with him, for what his hands have done shall be done to him."

He goes on to say that the idea that God gives us what we deserve and our misfortunes are the result of our misdeeds, is too simplistic and leads to undesirable consequences in that it creates guilt and blame when there is no cause, it results in people being disappointed in God and themselves.

But most importantly, it is a misdescription of our reality. He cites the world's sufferings on a mass scale--indiscriminate or deliberate killings, derelictions in nursing homes, etc. as demonstrating the unfairness and suffering by the innocent, the death of the righteous unfulfilled, as we wait for the emergence of God's plan. Alas, he says, "the world...is not so neat a place as the prophets would have us believe."

Does God Want Us to Suffer? We tend to assume, he says, that God causes us all our sufferings, deserved or not, for reasons as unexplainable now as they were to Job, whose question of God about his undeserved afflictions met with the most enigmatic response in literary history. Well, Kushner suggests another approach--that *perhaps God is not the source of our suffering*, that perhaps there is some reason "other than the will of God." The implication is that even God has limitations and that certain necessary characteristics of the universe require certain negatives.

Have We Happened on God's Finite Limitations? Well here we are back to Leibniz's "best possible world"; God did his best. Built into this idea, however, is the fact that God does, after all, have some finite limitations; that human presence in the universe can come only with built-in limitations. We know that the 'perfect symmetry' of the primordial nothingness, or the starting cosmic particle in which all the fund amental forces were unified, went through a symmetry-breaking process in which the fields and forces separated to generate a radiation of basic particles which cooled to allow atoms, then molecules, matter and life.

But the question persists of why evil persists, in the face of God's omnipotence and omniscience. And we have explored some suggested explanations. In R. C. Sproul's *Now That's a Good Question*, he sees the answer in the Genesis fall of humanity allowing death and suffering to enter the world. Others, of course, can't accept the notion of the world plunged into human and natural calamity for a single primitive transgression.

Eternal Life As a Redeeming Feature of the Suffering World: If God's omnipotence is, after all, constrained by some innate unavoidable characteristics of a universe which can generate life, then there is some redeeming and compensating aspects of this fall from the original perfect symmetry. This 'fall from perfect symmetry' has been noted by contemporary physicists such as Heinz Pagels and Steven Weinberg.

Immanuel Kant suggested that it is imperfection in our lives that assures us of a rectifying afterlife. Since justice in the world is mandatory, an *a priori* feature, then a justice unfulfilled in our earthly life assures us of a rectifying afterlife. While this may be a redeeming thought, it does not explain, of course, why a little less suffering and depravity would not suffice to satisfy the principle.

LONELINESS AND MEANING

Meaning Requires That You Believe in *Something*: Well, with the various possible explanations which science, philosophy and religion offer for the imperfect state of things, how do we get to the bottom of all of this? What should we believe? We have *got to* believe in *something*, muses Harold Kushner, author of *When Bad Things Happen to Good People* and *When All You Have Ever Wanted Isn't Enough.*

Religious Faith As Antidote To Loneliness: He sees religion as an "outlook on life, a way of seeing and thinking." He sees religious artifacts as symbols of meaning, of commitment to principles; the marriage ring as an expression of loyalty and closeness. Echoing Dr. Frankl's description of a prevailing emptiness, he sees religious faith as playing an essential role in filling the *existential vacuum,* "an antidote to loneliness...the greatest spiritual problem of our time."

Poverty, disease, and grief are ultimately abided if their burden is shared; the blessings of health and prosperity are meaningless without the transcendental vitalism of sharing with human life and, through religion, with God.

We Must Know That Someone Cares and Death Is Not Nullification: Perhaps the most basic human need, Kushner says, is to know that *someone cares*, that we matter to someone, and that our choices mean something in the scheme of things. The human being must know, he says, that "faith in the reality of things unseen," is faith in the nature of human soul as eternal; that death is not the obliteration of what one has stood for, not a nullification of one's existence. In this context, we discuss in the chapters on Relativity, Time, and Death, Einstein's famous and mysterious reference to distinctions between the past, present and future as being illusion.

Meaning needs expression in the thought that we continue to live on, following death, not only in the fleeting memory of human life but in the eternity of the Mind of God. A life without religious faith has failed of its human potential, a potential offered only to our species--to reflect His image.

Meaning and Psychoanalysis: Much has been written on the relationship between meaning and loneliness, sometimes in the context of *existentialist philosophy* and of a fundamental emptiness in life, as discussed earlier. Psychiatrist Dr. Viktor Frankl, in *Man's Search For Meaning*, underlines that it is not the meaning of life in general--an abstraction--that matters most, but the *specific unique meaning or mission of a particular person's life at a specific moment in time.*

Meaning Is Found in the Outside World: Frankl, as we noted earlier, goes on to observe that *meaning is found by involvement in the outside world* rather than within the self, through *self-transcendence*; not in mere ethereality of idea, but in practical applications in the real world, like Mother Teresa's all-consuming work with the poor. So transcendence is a key to meaning. We noted earlier Frankl's three avenues to finding meaning, one of which was *experiencing something beyond your self*, like loving someone.

Meaning through Friendship: Perhaps the finest expression of this "experience beyond oneself" lies in the bonds of *friendship*. The very definition expresses this thought. Emerson observed that "The only way to *have* a friend is to *be* one." He was, of course, echoing Proverbs 18:24, "a human life that hath friends must show himself friendly." And John 15:13, " Greater love hath no human life than this, that a human life lay down his life for his friends."

Friendship may be the greatest single human need for a life of meaning. Said Emerson, "Our chief want in life is somebody who shall make us do what we can; this is the service of a friend." And "The human being who has a thousand friends has not a friend to spare." Harvey Cushing observed that "In the life of a young man the most essential thing for happiness is the gift of friendship." But one need not limit this to the young. Perhaps it is the older in our society who need friendship the most.

Loneliness As the Absence of Friendship: There may be many antidotes to loneliness but none so effective as friendship. Addison mused that "Friendship improves happiness and abates misery, by *doubling our joy, and dividing our grief.*" Plato observed that "true friendship is...infinite and immortal." La Fontaine noted that "Friendship is the shadow of the evening, which strengthens with the setting sun of life."

Existentialism speaks of the prevailing emptiness in many lives. Dr Frankl spoke of religious faith as playing an essential role in filling the *existential vacuum,* "an antidote to loneliness...the greatest spiritual problem of our time." Although religious faith may prove elusive for many, there is another antidote for loneliness-- friendship, perhaps God's greatest gift to us.

The Mid-Life Emptiness: Psychotherapists are constantly called upon to deal with many symptoms that are traceable to loneliness. In *All Alone,* in the April, 1991 Ladies Home Journal, Margaret Rosen discusses *The New Loneliness Of American Women.* She observes that single, divorced, and widowed women are "understandably lonely sometimes." But what's going on, when women who have wonderful marriages, terrific kids, a beautiful home, a job they love, report that they are still lonely, that *something's missing.*

The Modern Pace Moves Old Friendships out of Phase: This mid-life anxiety and sense of isolation, notes University of Washington sociology professor Pepper Schwartz, derives from several factors. For one, the condition of being in the same place and time with friends having mutual problems, like on the popular T.V. shows, no longer exists. Events in the lives of friends don't occur in tandem, conflicting needs alter living patterns so that they are no longer proceeding in unison or parallel and their lives move out of synchrony.

And, she observes, this lost synchrony "erodes the very foundation of friendship." Also, Schwartz notes, we no longer have the more intimate ‘get-together’ neighborhoods of yesteryear which fostered sisterly closeness. The accelerated life-pace forces choosing priorities in which friendships are neglected. And new friendships are hard to cultivate in the work pressures. No matter how much love there is, no one person, she feels, can fill all one's companionship needs and, she thinks, while men may *hear*, women *listen*.

Two Kinds of Loneliness: So loneliness derives from unmet needs for caring relationships, for a confidant. Rosen quotes a pioneer in loneliness research, University of Mass. at Boston professor Robert S. Weiss, as suggesting two kinds of loneliness: 'emotional loneliness' from the absence of an attachment figure like a mate, a condition capable of causing negative physical symptoms, and 'social loneliness,' the absence of friends, a condition which promotes loss of self-esteem and meaningfulness in life.

University of Tennessee psychology professor Warren Jones suggests that some confuse quantity with quality, emphasizing numbers of friends over depth of friendship. He also notes the loneliness-prone personality who is critical, conversation-monopolizing, or just given to uttering insensitivities.

Loneliness Linked to Illnesses: Loneliness, the sense of emptiness, has been linked to heart disease, hypertension, various addictions, depression and fatigue. Divorced, widowed or bachelor middle-aged men are more likely to die than married men, according to a University of California, San Francisco research report which singled out loneliness as the most likely factor.

But one may lack a 'friendship' component even in the marriage state. One psychotherapist case study speaks of the married woman patient who complained of an emptiness, a loneliness in her marriage. She expressed herself as loving her husband but not really liking him; he was her lover but he was not what she most needed him to be, *her friend*.

Combating Loneliness: The starting point to combat social loneliness, says Rosen, is to own up to the problem, stop blaming yourself, and if necessary, seek counseling. Seek new forms of community participation, become a volunteer. In other words, *work* at your friendships. So we're back to the idea that we started with, transcendence, experiencing something beyond yourself. You must *be* a friend to *have* a friend.

Personal Relationships: In *Real Philosophy*, professors Needleman and Applebaum, under the caption "The Spectrum of Love," address pitfalls and pluses in personal relationships. Insecurity or indifference in personal relationships can deprive us of the benefits of association with others and make us more vulnerable to the social trauma.

They observe that our neglect of the value of intimate relationships for promoting our own understanding may produce or enhance pessimism in critical matters of caring, respect, love and communication with others.

Aristotle and Are Relationships a Matter of Expedience? But are these relationships essential? Or is the human being by nature a solitary creature who turns to others only out of weakness or for some gain? To claim that people are not social animals but bond only for protection or reproduction is to argue that all relationships are a matter of self-serving.

In Aristotle's "Friendship," from *Nichomachean Ethics*, he rejects this point of view. In studying the needs of the individual, he says, without friends no one would choose to live, though he were wealthy. For him, living is a matter of following the dictates of the most authoritative inclinations in one's character.

Furthermore, says Aristotle, we are gregarious by nature and it is our nature to live with others, not from fear or drive to dominate, but on terms of sharing, with no conflict between friendship and self-interest. One loves one's friends, he says, for the same reason that one loves oneself--to cultivate what we believe to be most worthy of admiration. Someone who is not guided by good principles himself will be unable to appreciate the good of others.

In Aristotle's outlook, a society is organized according to the same laws which create order within a person. Both consist of intellect, appetite and emotion; both have potential for strife and disharmony, or for concord and harmony; both require careful education and taking differences into account. Plato's view is that people form relationships in accordance with their nature. He speaks of the "interconnectedness of all things," and in the face of this, the shrinking ego.

Enlarging the Ego to Include the Being of Others: To value a relationship more deeply is also to cultivate it. This duality in human life is portrayed in another way by Gabriel Marcel, in *The Mystery of Being*. At one extreme, for him, is the experience of self-consciousness displayed on entering a room in which one knows no one. The situation induces a physical and intellectual tension which prevents the formation of relationships. He suggests that the struggle to find a meeting-place, against the ego's denial of the other's reality, is the struggle for existence--the struggle to enlarge a sense of self to include in it others' claims to being.

Relationships and Pleasure: The authors ask what role pleasure plays in the field of relationships. We seek to establish bonds with those we find pleasant to be with and to avoid those that make us feel uncomfortable. But we forget that a relationship can also evolve from opposition. Jesus said "Love thine enemies," which may suggest that adversaries seek in each other what they share in common.

The Fable of Esarhaddon, King of Assyria: In the Fable of Esarhaddon, King of Assyria (from his *Twenty-Three Tales)*, Tolstoy observes, you can make life better within yourself only by destroying the barriers that divide your life from that of other beings, and by regarding others as yourself and loving them. Bent on a policy of revenge, King Esarhaddon, through the intervention of a stranger and a dream sequence in which he *becomes* his own foe, sees the world through the eyes of his foe, comes to see that executing his enemy would mean destroying himself as well.

We rarely have a sense of oneness with those we find annoying, but there are conditions under which understanding can coexist with negative emotions. This points a way towards bonding in spite of opposition. What is involved is the willingness to explore the perceptions of others in formulating one's own conduct.

Relationships and Attention: a basic ingredient in any relationship, say the authors, is that of 'attention.' Awareness is necessary before a relationship can arise. Conversely, inattention, not taking the other seriously, forms a barrier. We fail, for instance, to see what the other person is really trying to communicate to us. And we react instead in accordance with our own preconceived notions.

Concern for the welfare of another requires us to know a great deal both about that person and about people in general. That we are rarely adequate to the task underlines the extent of the responsibility that caring implies. Could it be, one asks, that our general casualness frequently gets in the way of deeper, more penetrating relationhips?

In *Real Philosophy*, professors Needleman and Applebaum cite the *Sermon on the Mount* where Jesus stresses the need for being receptive to the substance of external relationships, and being tranquil, to avoid hostility.

RISKING ONE'S LIFE FOR OTHERS

Morton Hunt on Extending the Circle of Compassion: Einstein's reference to the need for widening the human circle of compassion, brings to mind Morton Hunt's book, *The Compassionate Beast*--What Science Is Discovering About the Human Side of Mankind. Hunt tells of Samuel Poliner, a Polish Jew who was saved as a child by a Catholic woman when his family was killed by Nazis.

What Makes People Risk Their Lives for Others? Poliner, now an American university professor, studied the gentiles who risked their lives to rescue Jewish people from the Nazis. He found that in general they were not more adventurous, more religious or more negative to Naziism. But he found that they had more empathy and were influenced by parents whose compassionate responses were *translated into caring action*. They were doing just what Einstein spoke of, extending their circle of concern and compassion to others in trouble, the essence of altruism.

The Unexplainability of Sainthood: In a recent television interview with a man similarly rescued from the Nazis by a compassionate villager, he was asked whether he himself would have done the same in reverse circumstances. He replied that he would like to think that he had the courage to do the same thing if only his own life were at stake but that he was uncertain whether he would have risked the lives of his own children--as this woman had done.

Asked to explain the woman's extraordinarily courageous and compassionate action, he replied that he was unable to explain it. After a moment's reflection, he felt somehow that he had the answer. In a voice filled with reverence, *She was a Saint*, he replied.

The Human Being Becoming Part of God in Action: Perhaps this extending of human circle of compassion, is what sainthood is all about. What seems to be involved are persons who somehow *rise above their human status,* and sense their role as an instrumentality or extension of God. *At that exalted moment of transcendental inspiration they become a part of God in action.* Frail little Mother Teresa is seen by many in that light, as figuratively infused with the power of God, transcending human frailties.

Pierre Teilhard de Chardin's Convergence of Life and Matter: But getting back to Heinz Pagels's existentialism, the preoccupying human search for the 'ultimate foundation' of existence, is organic, built into biology. Denying such spiritual impulse, says Pagels, simply "twists our humanness."

Interestingly, Pagels cites the French Jesuit paleontologist-philosopher, Pierre Teilhard de Chardin, who viewed the *universe as a living being* in transformation, all life and matter being intertwined and converging to an *Omega point*. Regarding the end of human life, says Teilhard, the "adventure of the world will have ended in the bosom of a tranquil ocean, of which, however, each drop will still be conscious of being itself."

MATERIALISM AND MEANING

The Tension Between the Material and the Spiritual: There has always been the tension between those who find everything in the universe is ultimately explainable scientifically in terms of 'matter or the material, and those who find 'answers' in the immaterial, in spiritual or theological terms.

But there is growing appreciation that so-called scientific materialism, and religious or philosophical transcendentalism are not mutually exclusive. This idea of the disappearing distinction between the material and immaterial emerges as new scientific phenomena are explored, especially in quantum mechanics, where what is viewed as particles dissolves into uncertain particle-wave states, consisting more of information than matter, where matter, energy, information and mind become associated.

Most discussions on the appearance of life lead inevitably to a discussion of whether all of reality in our universe is explainable in terms of (1) a single material substance which includes mind and spirit as simply part of the physical brain, or whether there exist two independent substances, material (body) and immaterial (mind), as René Descartes and other 'dualist' philosophers asserted.

Physicist Freeman Dyson asserts that "matter is weird enough" so that it does not limit God's freedom to do with it what he pleases; that *scientific materialism and religious transcendentalism are neither incompatible nor mutually exclusive..*

Desmond Bernal and Transitions of Life to Ethereality: Dyson cites crystallographer Desmond Bernal whose vision of humankind soared into the future to explore the *transition of life from material substance into ethereality,* a new life which would "conserve none of the substance and all of the spirit of the old," *consciousness becoming ethereal, and particle masses communicating by radiation, ultimately resolving into light.*

Metaphysical Extensions of the Energy Cycle of the Universe: A number of scientists have sought, as Dyson does, to connect cosmology and biology by speculating on the universe as a progression from radiation energy, to matter energy, to mind energy, from 'materiality to 'ethereality.' From the physics of the three energy forms of radiation, matter, and mind there is a comfortable but metaphysical extrapolation, a picture of the universe as the progression from Idea-God, to energy-radiation, to structured rest-energy-matter, to Life, back to etherealized energy-mind, back to Idea-God.

Distinguishing between Meanings of Materialism: In this context the term 'materialism' conjures up two quite different ideas. One 'materialism' is so-called 'scientific materialism,' that everything in the universe is physical and that subsumed in that physical state are will, mind, spirit and feelings, all aspects of the physical brain. Another refers to the concept that material possession and physical satisfactions are the highest value, human life's goal. Our discussion relates to the former meaning, that of scientific materialism, whether everything in the universe represents some form of materiality.

Increasingly physicists recognize that new concepts about the materiality of energy are surfacing. We noted physicist Freeman Dyson's assertion about the "weirdness"of energy and "God's freedom to do with it what he pleases" and his proposition that the concept of scientific materialism and religious transcendentalism need no longer be viewed as mutually exclusive or even incompatible.

This idea was given depth in the beautiful and soul-searching introspections of Pierre Teilhard de Chardin, about the universe as a living being in transformation, life and matter converging to the Omega point where the "adventure of the world will have ended in the bosom of a tranquil ocean, of which, however, each drop will still be conscious of being itself."

The Interrelationship Between Matter and God: His work in geology and paleontology in China and South Africa influenced his life toward a realization of the unity of God and the universe and the value of human endeavour, "the interrelationship between matter, soul and Christ," as reflected in the cosmic, the universal Christ. Earthly work and aims are not eschewed. "The seeking after, and waiting for, the kingdom of heaven" must not "deflect human activity from its natural tasks" and the "closeness of our union with [God] is in fact determined by the exact fulfillment of the least of our tasks."

Built into his descriptions are scientific and theological concepts which transcend differences between theism and pantheism to emphasize worldly effort and, by implication, to sanction aspects of materialism and humanism. Interestingly, his views encountered difficulties within the ecclesiastical community, to the point that he felt it necessary to explain his philosophic orientation to the Vatican by a letter in which he explained his profound feeling for the organic realness of the World.

The Divine in the Mundane--Sanctification of Human Endeavor: In the translation of his work in *The Divine Milieu,* he unabashedly speaks of "the sanctification of human endeavor." Nine out of ten Christians, he regrets, think of work as a “spiritual encumbrance,” and as a result they "lead a double or crippled life." By virtue of the Creation and the Incarnation, he says, "nothing here below is profane.... Why should there not be men vowed to the task of exemplifying, by their lives, the general sanctification of human endeavor....men...who would devote themselves...to carrying out in the sublime spirit...the basic tasks which form the very bonework of human society?"

The Spiritual Power of Matter: In his discussion he denies the profaneness of matter and speaks of the spiritual power of matter, the tangible setting in which we live. "It attracts, renews, unites and flowers. By matter we are nourished, lifted up,...invaded by life. To be deprived of it is intolerable. Who will give us an immortal body?"

Teilhard de Chardin speaks of the unity of human life and God and the universal Christ. He describes the whole human family, "the human layer of the earth," being under the "organizing influx of the incarnate Christ." He describes the universe in sacramental terms. The Eucharistic transformation is “the sovereign influence upon our human natures; the sacramental species being formed by the totality of the world and the duration of the creation as the time needed for its consecration.”

PART XIV--DEATH AND MEANING

RANDOM THOUGHTS ON DEATH

This world is the land of the dying; the next is the land of the living. --Tyrone Edwards

Life is real! Life is earnest!
And the grave is not its goal;
Dust thou art, to dust returneth,
Was not spoken of the soul. --Henry Wadsworth Longfellow

We should weep for men at their birth, not their death. --Baron de Montesquieu

To die is landing on some distant shore. --John Dryden

The stars are mansions built by nature's hand and, happily, there the spirits of the blest dwell, clothed in radiance, their immortal rest--Wordsworth

Premeditation of death is premeditation of freedom. He who has learned how to die has unlearned how to be a slave....There is nothing evil in life for the man who has thoroughly grasped the fact that to be deprived of life is not an evil...The same passage that you made from death to life, without feeling of fright, make it again from life to death.... --Montaigne

Out of eternity the new day is born; into eternity at night will return. Carlyle

How gaily a man wakes in the morning to watch himself keep on dying. --Henry Haskins

Sleep is lovely, death is better still, not to have been born is of course the miracle. --Heinrich Heine

Man is the only animal that contemplates death, and also the only animal that shows any sign of doubt of its finality. --William Ernest Hocking

Surely God would not have created such a being as man ...to exist only for a day! No, no, man was made for immortality. --Abraham Lincoln

Immortality--a fate worse than death. --Edgar A. Shoaff

One of the proofs of immortality of the soul is that myriads have believed it--they also believed that the world was flat.-- Mark Twain

When I die I shall be content to vanish into nothingness...No show, however good, could conceivably be good forever...I do not believe in immortality, and have no desire for it. --H.L. Mencken

The average man, who does not know what to do with his life, wants another one which will last forever. --Anatole France

I am going to seek a great perhaps; draw the curtain, the farce is played. --Rabelais

Nothing can happen more beautiful than death. --Walt Whitman

First our pleasures die--and then our hopes, and then our fears--and when these are dead, the debt is due, dust claims dust--and we die too. --Shelley

God's finger touched him, and he slept. Tennyson -- In Memoriam

The Gods conceal from men the happiness of death that they may endure life. --Lucan

We sometimes congratulate ourselves at the moment of waking from a troubled dream--it may be so--the moment after death. -- Hawthorne

There is no death! What seems so is transition; this life of mortal breath is but a suburb of the life elysian, whose portal we call death. --Longfellow

It is impossible that anything so natural and so universal as death could ever have been designed by providence as an evil to mankind. --Anon.

This world is a dream within a dream; and as we grow older, each step is an awakening....Death the last sleep? No! it is the last and final awakening! --Sir Walter Scott

The goal of all life is death. --Sigmund Freud

The gift of God is eternal life.--Rom. 6:23

I am the resurrection, and the life; he that believeth in me, though he were dead, yet shall he live.--John 11:25

CHAPTER 51
FINDING MEANING IN DEATH

Human Beings Wonder about God, the Universe, Life and Death: From time immemorial, the human psyche has been filled with speculations about the nature of the universe and God, the meaning of life, why life begins and why it must end, why death. Even more, one wonders what death *means*--whether there are some aspects of life that survive death and continue in some way.

No question is more considered by philosophy and theology, and at critical times at least, more pervasive in the human mind. Interestingly, the meaning and imponderables associated with death have always played a significant role in the social cultures of the times and have had a powerful influence on the way people conduct their lives.

A person's greatest enigma--and sense of desperation--visits when a loved one is taken. The greater the love and caring, the more profound, the more inexplicable and the more devastating is the sudden deprivation by death. How can such a deplorable event be rationalized as the action or inaction of an omnipotent and omniscient God? Where is God's compassion, love, mercy, and in permitting such an event? Yet the event is endlessly repeated and God's thoughts remain shrouded in mystery--as the Book of Job attests.

The Study of Philosophy and Theology Prepares for Death: Some have said that the study of philosophy and theology is truly to prepare us for acceptance of death. The philosopher Schopenhauer observed that *without death neither philosophy or theology would amount to much.* Most so-called philosophic issues are somehow ultimately subsumed in the question of the nature and significance of death.

It is said that those who study the philosophy of life and death are better prepared to accept death as a part of the normal scheme of things, perhaps as another and more mysterious chapter of life. Cicero thought that *a philosopher's entire life was preparation for death.*

Those who make practical preparations, provisions for their loved ones, *extending their love, caring, charity and compassion into the future* by providing for certain events, projects, memorials, and causes of interest to them, are likely to be more prepared for acceptance of death.

The Mystery of Death: Often quoted as a cryptic expression of the mystery of death, the great unknown which faces all human life, were the final dying words of French humanist, Rabelais, "I am going to seek a great perhaps..." Religion, philosophy and science have addressed this mystery without definitive reassurance of its meaning and, in the end, all our analysis is speculation.

Attitudes toward the event, its up-side or down-side, remain subjective according to the circumstances which govern society's culture and perspectives at any particular time. Socrates, having chosen death over exile, in his farewell remarks, noted "The hour of departure has arrived, and we go our ways, I to die, and you to live. Which is better only God knows."

Stages of Dying: Dying is something everyone endures, essentially alone. Nothing, therefore, so engages the mind and the emotions as does the ending of one's life or that of a loved one. Except in the case of sudden death, an individual who is ill centers hope on medical treatment and possible cure.

People often go through a series of stages in accepting the reality of their own mortality. Diagnosis of a terminal illness brings shock, which soon gives way to denial. This denial may take the form of searching for any possible cure for the disease. From denial the patient may go on to anger, anger at himself, at everyone around him, and even at a God who seems remote, disinterested, and deaf to prayers for recovery.

Anger eventually gives way either to hope for a temporary respite or to deep depression over the impending event. This grief may then turn to resignation and acceptance of the inevitable.

How an individual responds depends, of course, on philosophical and theological perspectives one may have acquired, the perceived quality of one's past personal life, and the meaning attached to past experience. For most people it is probably true, as Sir Thomas Browne said, that *the long habit of living indisposes us for dying*. For those to whom life has been an ordeal, death may come, in Hamlet's words, as a consummation devoutly to be wished. For most, the unknowns are a source of apprehension, anxiety and trepidation. On the other hand, people who are reassured by deep religious faith or made tranquil by philosophic reassurances, are often able to face dying with composure and even a sense of serenity.

We Emerge from Nothingness Preoccupied with Duty and Continuance: In no way consulted, each person emerges--from some kind of mysterious pregnant nothingness--to become a new 'awareness' of the universe, something not previously in existence except as potentiality. And with this mysterious emergence comes an uncertain and contingent life-charter, the duration and terms of which are either as yet undetermined, or predetermined but concealed.

And we come fully-equipped with consciousness, self-awareness, and a host of subjectivities: a sense of relationship to God and the unity of the universe, a 'conscience' replete with obligations and, as Immanuel Kant insisted, impressed with the the paramount sense of duty. But we come with something else, an intuitive sense of the value of this life, its fragility, and an abiding concern for its 'continuance.'

Death--Where Quantum Theory Draws Science Closer to Theology: It wasn't so long ago that science and religion were considered at odds in perspectives on the finality of death. Segments of western theology understandably held out the hope of some sort of continuance, a physical or spiritual re-embodied survival, in the case of Christianity, associated with salvation.

Science, on the other hand, while deferring to theology in spiritual matters, also understandably looked upon physical death as essentially that, a medical physical fact. In biology terms, since the physical brain, thought to be the source of mind, died with the body, the mind then, including consciousness, self-awareness, recollections of the past, all associated with the brain, simply ceased to exist.

But strangely, developments in physics have engendered new concepts, derived from quantum theory in which the objective physical universe dematerializes into an informational uncertainty state involving *subjectivity* and raising challenging questions not only about death but about the meaning of life, whether we are part of some computer-driven virtual reality.

The Human Mind As Part of "God's Mental Apparatus": In *Infinite In All Directions,* Princeton physicist Freeman Dyson asserts that the human mind may be considered an extension, an integral part, of *God's mental apparatus.* The contemporary view of many physicists allows for the possibility of the *existence of the mind apart from the physical body*. But what does that mean? So where might the mind exist, then, if not part of the physical body?

Scientists suggest that it may exist *in the quantum*, the *immaterial* reality that underlies and interconnects everything in what appears as our physical universe of atomic and molecular substances: stars, planets and people. Physicists are confounded by quantum theories which imply, to some at least, that all of physical existence is, in a sense, part of a *virtual reality*, subjective and constructed by mind.

And So Profound Questions about the Meaning of Life and Death: This, of course, raises profound questions about the nature and meaning of the life-energy-aggregates of which we consist. But even more does it raise questions about the nature and true meaning of death. In a sense, *both* life and death, like time and space, may be subjective constructs, some sort of virtual reality. Moreover, relativity theory tells us something more, that our perception of time as passing, bringing death, is an illusion. *All time coexists, it's simply there.*

Schopenhauer and Pierre Teilhard de Chardin: Philosopher Schopenhauer asserted that individual consciousness created by separation from cosmic consciousness at birth, rejoins the cosmic consciousness on death. The highly-respected philosophic Jesuit paleontologist Pierre Teilhard de Chardin goes further--upon death, the individual discarnate consciousness returns to its cosmic source, but retaining its worldly individuality, a thought which has given hope to many.

Eternal Existence? Many, influenced both by science and theology, are led to the conclusion that we and our loved ones exist eternally in some dimension; that God, however defined, would not waste or trivialize the beauty and meaning of our worldly experience, growth, and refinement, by an arbitrary and senseless electro-magnetic switch-off. In one way or another, our consciousness, painstakingly nurtured here in a physical environment, must have some meaningful survival. There must be some mind-conservation law similar to energy conservation.

We will explore differing religious concepts of afterlife, of spiritual survival and re-embodiment, including Christianity's continuing religious dialogue over the meaning of the "saved" life. There are those who see their continuance, via salvation, as an immediate spiritual or spiritual-body extension in another realm; those that see an intermediate purgatory; and those that see only a final resurrection with the Second Coming in the end days.

Near Death Experiences: In the course of our discussions we will also be exploring death-related phenomena including the so-called Near Death Experiences. These 'experiences' involve: a sense of being physically but not mentally dead, a serenity and painlessness, a feeling of *release from the body*, a sensation of going through a tunnel, seeing people of light, seeing a 'being of light,' undergoing a review of the person's life, a disinclination to return, and a post-death-experience change in personality. The number and similarity of these reports, and the similarity in the elements of the experience, have given many, including medical doctors, encouragement that our lives don't end in just oblivion.

Questions about Death: Part Four of Needleman and Applebaum's *Real Philosophy* asks questions about death: Why is there death? What is the meaning of death for the living? What, if anything, happens after death? What attitude is appropriate to an awareness of the presence of death in life? Such questions, the authors suggest, call for a "fuller examination of human existence" in view of its inevitable end. The authors seek to convey certain ideas about death.

Defying Death by Intensifying Involvement: Unavoidable events, note the authors, often lead to anxiety, dread, and fear. In the same way, the universality of death grabs our attention providing the occasion to look at our own life more dispassionately. One must avoid being so overwhelmed by the sense of doom that everything is rendered meaningless. But, according to existentialist Heidigger, this presents a challenge to defy death by intensifying our involvement and 'authenticity.'

Using Death to Renew the Search for Understanding: So, constructive anticipation of death then changes death from an end to an agency for a new beginning, to renew the search for understanding. The sense of our personal death, Heidigger argues, derives ultimately from the mystery of conscience, conceived not as a moral faculty but as a source of knowing how to be fully what one is. To attend to this moment of awareness gives one a deep knowledge of how to live with one's own mortality. Death is not to be fought, but rather to be acknowledged and used to advantage, to be worked with.

Satre--Death as the 'Repository of Frozen Possibilities': But isn't death the destroyer? For Nobel-prize-winning French novelist and existentialist Jean-Paul Sartre, in *Being and Nothingness*, death is a 'termination,' a door closing. Satre sees life as a *diminishing package of possibilities for life to turn into actualities*.

In death, he says, life becomes *destiny,* in which everything slips into the past, described as "a repository of frozen possibilities"-- all life-options exhausted. The "realm of death" is the state of nothingness--nothing to be done, finality attaching to all preexisting potential. Death, says Satre, represents the *final arrest of temporality*.

Appreciation of this fact threatens to extinguish the very essence of meaning in life--the very capacity to further question. But whether it makes us aware of more possibilities or restricts our field of action, death is inexorably connected with suffering and the painful side of human existence. As Shakespeare notes in *Hamlet,* though we do not know what awaits us upon death, the release given by death seems preferable to the prolonging of life's final unbearables.

Heidigger's 'Being and Time': Philosopher Martin Heidigger's central focus was existentialism, the question of 'being,' and, in *Being and Time*, he explores the deeper level of being human, the level of receptivity to the call of conscience, in which we have heightened cognizance of death, of our transient existence, and of our awareness of who we are, our humanness or *Dasein*.

The call is silent, because it is a call back from the forsakenness of abandonment into *the reticence of existent potentiality-for-being*. The concept that Dasein is both the caller and the called has not lost its obviousness. Conscience manifests itself as the call of *care*. The one being called, this same Dasein, summoned to its "ownmost potentiality for being." The call of conscience, that is, conscience itself, observes Heidigger, with its ontological possibility in that Dasein, in the very best of its Being, is care.

Tolstoy's "The Death of Ivan Ilich"*:* In *The Death of Ivan Ilich*, Russian novelist and thinker Leo Tolstoy, author of the classics, *War and Peace* and *Anna Karenina,* poignantly describes the difficulty of letting go of life. Ilich is suffering terribly, subconsciously realizing that he is dying. Yet he is unprepared to accept the reality. Uncertain about what he had done with his life, he experiences an agonizing and touching reexamination of his inner life.

He asks, Why hast thou done all this? Why didst thou bring me to this? Why, why didst thou torment me so terribly? And Ilich wept--because there were no answers. "Go on, strike me! But for what? What have I done to Thee?" One of the greatest tensions of the terminally ill is the fear that certain missteps in life are yet to be corrected. Tormented, Ivan Ilich asks: "And what if I go away from life with the consciousness of having ruined everything given me without being able to mend it? What then?"

Achieving the Final Freedom: Tolstoy suggests that Ilich's pain is the result of certain illusions he holds regarding himself and his impressions of what others think of him. In the last stages of that terrible loneliness, spent from anguished self-examination, *reliving his past life without being able to change anything*, he finally relinquishes and lets these troubling impressions fade from awareness.

And as he accepts the *unchangeability of the past* and the inevitable reality of his ending--something previously denied, Ilich finally lapses into the experience of the comfort of a new freedom. The discussion continues on the proposition that, as evidenced by much earlier thinking, an understanding of one's death is a "powerful agent of freedom."

Life As a Series of Phases: The discourse turns to the religious and philosophical notion that a person's existence passes through a number of phases, forever returning to its beginning. In this way of thinking, if death follows life, life must follow death--rebirth, regeneration, and resurrection. Thus death is the doorway to a new stage of life and only a death not followed by rebirth delivers freedom. And what is the best preparation for dying in a proper state? *Living wisely.*

Plato's *The Myth of ER*: "Real Philosophy" cites the idea of preparation for death as conveyed in Plato's "The Myth of Er" from *The Republic.* Er, a soldier, returns from death to explain what he has experienced; that in the next world justice is based on actions here. So Plato suggests that the merit of a person's actions in life determines the kind of his succeeding life and those who search for understanding in this life, will be favored in the next. The story is allegorical, not necessarily suggesting literal rebirth. *We are left,* the authors suggest, *with the problem of trying to understand the soul as bearer of immortality.*

Mind As a Holistic Entity Surviving the Body: The authors of *Real Philosophy* point out that not all thinkers agree that personal identity can be equated with just the person's biochemical composition. In the holistic view--that the whole is more than the sum of the physical components--the mind is separate from the body, leaving open its survival. Quantum theory discussions suggest the existence of mind as a holistic entity existing separately from our time and space, 'in the quantum,' and therefore possibly surviving physical death.

> Without death, neither philosophy or theology would amount to much.--
>
> Schopenhauer

CHAPTER 52
DEATH & THE NATURE OF THE UNIVERSE, CONSCIOUSNESS

DEATH AND THE NATURE OF THE UNIVERSE

Interrelationships between Death and the Nature of the Universe: It is hardly possible to construct a coherent representation of the significance of death without viewing it in the context of the total reality of which it is an inescapable part. This is a reality that embraces life, God, meaning, the origin and nature of the universe, and whether our existence, or for that matter the entire universe, has purpose and, of course, what that purpose might be.

So the real *introduction* and the context in which death must be explored, consists of all the other aspects of existence, cosmology, the origin and development and laws of the universe, the nature of space and time, particle or quantum physics, the emergence of life and consciousness, the existence of and our relationship to God. All these and more are aspects of the same reality of which death is so integral a part. One is led to ask how can death be discussed without also discussing life and its underlying physics, God, and meaning?

Relativity, Spacetime Geometry, and Death: What have physics subjects, like *relativity* and *quantum theory*, got to do with *death*? Well to start, *general relativity* involves that measure of duration of life and death, something called *time*, an integral component of Einstein's relativity theory of the elasticity of time and its interdependence with space, motion and the speed of light.

In a related aspect of special relativity theory, Einstein, in a mysterious expression of consolation to his dear departed friend's sister, insisted that *distinctions between past, present and future are but illusion*, a subject we will be discussing.

Stars, Energy Forms and Death: What, you inquire, do galaxies and stars have to do with death? Well, our body's multiple-structured energy forms derive from elements cooked up in stellar fusion involved in the birth and death of the stars--no stellar explosions, no human life--or death. We are, it is said, incarnated stardust.

And neither can you discern the meaning of existence, of life and free will, and of what persons can know, without discussing the nature of that package of rest-energy phenomena we label 'life,' consisting of body and mind. Are mind and body two aspects of a single substance or, as the famous French mathematician-philosopher René Descartes asserted, two distinct substances, leaving the possibility that the death of one does not necessarily imply the discontinuance of the other?

Cosmic Evolution, Quantum Theory, and Death: And one can't discuss the 'figures of light' seen in Near Death Experiences, or the apparitional aura that appear in grief experiences, without discussing quantum energy phenomena.

Earlier we discuss cosmology, theories of how the universe started and quantum cosmology, the universe as starting from the primordial *quantum vacuum nothingness* in which there occurred a *quantum fluctuation* triggering into existence an incredibly small embryonic universe point, a superforce of unified physics embrac-ing space, time, the fields and forces, dimensionality, and the potential for matter, mind, and life.

In the inflationary universe model, this infinitely small mini-universe seed exponentially inflates, and in that very process, there is created space, time, and the positive radiation which cools to structure today's universe. The inflation subsides into the linear or less than-the-speed-of- light Big Bang sequence of separating forces and our still expanding universe.

Death Requires a Review of the Whole Cosmic Sequence: So you must at least consider the whole sequence: creation, energy, matter, molecular organics, life, the human species, and consciousness, to discuss intelligently the subject of one of the aspects of this process, biological life and death.

And neither can you go through the sequence without exploring primal supersymmetry and unification theory, energy fields, the fundamental forces, the mysteries of particles-wave duality, the indeterminate energy states of probabilistic quantum physics, and the science of the growth of order, organization, and complexity built into a universe supposedly dominated by the thermodynamics of entropy, disorder and chaos.

Death and Cosmic Informational Software: Especially relevant to death are certain aspects of quantum theory. Underlying our physical universe of forces and energy fields is the quantum or subatomic world in which mysterious energy phenomena become nonmaterial, and take on characteristics best described as informational software that invoke concepts of cosmic mind and consciousness.

Fairness, Eternal Life, and Values: Certain values and basic moral concepts are viewed by some in philosophy and theology as being *a priori*, built into life itself. Philosopher-theologian-mathematician Immanuel Kant, in his *Metaphysics of Morals*, asserted a moral foundation for theological assertions such as the existence of God, immortality, and reward or penalty proportionate to our conduct, described as "denying knowledge to make room for faith."

He argued that life continuation after death was assured by the existence of justice as a built-in characteristic of the universe. This is so, he thought, because unhappiness, pain, suffering, and injustice are unfairly and unequally experienced in this life, requiring that such inequities be reconciled beyond the span of this life.

Since justice is built into the fabric of the universe, there must of necessity, he believed, be continuance of life after death to vindicate the principle, to reconcile and square things. So fairness, built into the fabric of a divinely-created universe demands life continuance.

In Eastern religions, heavenly life eternal must be earned by proper conduct through successive reincarnations in this world until the slate of iniquities is fully cleansed. So repetitions of undesirable life experiences are guaranteed until accounts are balanced by moral lessons learned and practiced.

Life, Death, and Moral Values: The philosopher-scientist Immanuel Kant 'disproved' all the conventional proofs of God of Thomas Aquinas but reasoned God back into existence via the concept of justice and the *a priori* need for moral values and an innate sense of duty, moral values built into matter, into our very existence.

> This world is a dream within a dream; and as we grow older, each step is an awakening....Death the last sleep? No! it is the last and final awakening! --
> Sir Walter Scott

CONSCIOUSNESS AS SURVIVING DEATH

Mind, Soul, Consciousness and Spirit: Is continuation of our organic-living framework essential to the survival of our higher consciousness or awareness? This question leads inevitably to consider the nature and source of consciousness; whether there is a 'life-force' somehow infused in our bodies at birth; whether mind and body are the same or independent substances. Here we must further consider the meanings of terms like 'mind,' 'consciousness,' 'soul,' and 'spirit.'

Soul, Spirit, Mind, Life Force: In his chapter on Mind and Soul, in *God and the New Physics*, Davies observes that the concept of the *soul* is "hopelessly vague," the term being used primarily in theological contexts. According to many theological treatments, such as Van Harvey's *A Handbook of Theological Terms*, neither the Old or New Testament contains any 'developed conception' of the soul; but early theologians like Thomas Aquinas have sought to relate the term more to the Greek idea of the soul's immortality.

So theological treatments tend to track the term to the writings of Plato and Aristotle, probably best captured and represented in the thoughts of Aquinas dealing with aspects of survivability of the soul. However, for the most part, the term soul has come to be used generally as a synonym for 'mind.'

The Catholic Encyclopedia speaks of the soul as "the center of thought activity." Other definitions speak of the soul as the 'animating principle' in human life, and theological definitions tend to imply survivability or immortality. The term 'spirit,' has been used, of course, in various theological contexts. In contemporary parlance it is sometimes used to connote the idea of the 'life' or 'animating force' in mind or consciousness. But it is, perhaps, used more frequently in the sense of 'mind' or 'consciousness.'

Interchangeable Terms: For our purposes, then, as Davies does, we'll use the terms mind, consciousness, selfawareness, subjectivity, spirit and soul, as essentially interchangeable, even though, as noted, each of the terms may have somewhat different usages in various contexts.

Davies, incidentally goes on to discuss survivability of the soul or consciousness, noting that these terms all deal with the consciousness as something *nonphysical*, nonmaterial, ethereal. In general terms, he views both life and consciousness as holistic features, phenomena emerging from the growth of order, organization, and complexity, and its informational composition.

Other Dimensions and Universes: We have explored, in several chapters, certain aspects of physics, and our physical existence in our spacetime universe of four dimensions, the possible existence of other dimensions, and even of parallel or adjoining universes which may or may not share our physical characteristics and with which we may or may not be able to interrelate. Physicist Steven Hawking sometimes speaks of these universes as primarily mathematical constructs, while other physicists, like physicist Fred Alan Wolf, view them as real, part of our quantum reality.

Consciousness, Space, and Time: Having noted that mind and consciousness are nonmaterial entities that exist *somewhere,* Davies goes on to discuss the question of *where* mind or consciousness might be located, and whether it exists in time (our spacetime dimensionality).

He discusses the idea of spacetime being divisible into abstract quantum subunits, of the universe extending beyond our presently conceived spacetime, leaving an 'elsewhere' of quantum spacetime bits, and of mind as possibly imbedded in a higher-dimensional space. So, he notes, consciousness may not be in the space of our universe at all.

But what about consciousness and time? He asks whether consciousness is in our time dimensionality. You recall discussions about whether God exists 'in time,' and the problems of how he may be thought of as performing temporal functions, such as planning, if He does not exist in time. Well, Davies feels that both human mental processes that are time-dependent, and our very perception of time, require our consciousness to exist in time.

The Mystery of the Source of Consciousness: Like the source of life, the source of consciousness (mind, spirit, soul, self-awareness, higher consciousness, etc.) has also been a matter of considerable dispute. Is consciousness simply a manifestation of the physical brain or does it have some other explanation? Is life and consciousness implicit in matter or are these emergent properties? Is there a 'mental' substance, as some philosophers, like Descartes and Kant, asserted?

Are individual consciousnesses interconnected, part of a single cosmic mental fabric at the underlying cosmic quantum level? This question has meaning in the context of death, if we are to evaluate survivability and immortality potential in scientific terms, as some physicists are now doing.

Consciousness As Emergent Order: In considering the source of consciousness we noted earlier that many scientists, like physicists Freeman Dyson in *Infinite in All Directions*, and David Layzer, in *Cosmogenesis*, suggest life as the product of increasing order--the amount of information encoded in a system. Something *qualitatively new* appears, Layzer says, at each level--molecules, organelles, cells, tissues, organs, and communities. It is, he says, a *new form of order,* just as much a part of the external world as energy and matter.

The Impression of Design: Similar views are more elaborately expressed in physicist Paul Davies' *The Cosmic Blueprint*, in which he describes the universe in terms of its *creativity* and its "inbuilt facility for making interesting things happen." He cites the *predisposition* toward more organized and complex states where 'higher-level laws' and 'new principles' are free to come into play. The universe, he says, summing up, "has organized its own self-awareness," evidence that there is "something going on behind it all." *The impression of design is overwhelming*, he declares.

Something Mysterious Emerges: But the nature of *consciousness* and subjectivity is more abstract and uncertain than that of life in general. We know all kinds of lower life forms that do not appear to possess the faculty of selfawareness, consciousness. In his *Cosmogenesis*, Harvard astrophysicist David Layzer describes consciousness as a *metaphysical emergent property* of living organisms, a part of the cosmic evolutionary process in which elements *that did not previously exist,* emerge in a *hierarchy* which includes all previous elements. This is an expression of the so-called *holistic* view.

Consciousness As Radically Contingent: Layzer goes on to suggest that emergence of consciousness could not have been predicted; that it is *radically contingent*; that, *scientifically*, it has 'no rhyme or reason'; and consequently that it is 'emergent' in the *metaphysical* but not scientific sense. The event of consciousness is viewed more as a holistic phenomenon, the emergence of a higher level in the hierarchy of developing organization and complexity. Similar views are expressed by physicist Paul Davies and Gregory Bateson.

Consciousness As Derivative, Implicit in Matter: But other physicists suggest the reductionist point of view, that life, consciousness and mentalness are indeed, physical, implicit in matter, in particles. Physicists Freeman Dyson and David Bohm, view particles themselves as possessing an elementary form of intelligence and consciousness, suggestive of mind as being implicit in matter, part of the hierarchy of intelligence represented by the universe.

Consciousness As Holistic Rather Than an Inserted Substance: Paul Davies, in *God and the New Physics*, says that mind and body are not dual components (Descartes' dualism of physical-body substance and mental or mind substance) but "different concepts drawn from different levels in a hierarchy of description," and without any mysterious extra substances (a reference to an inserted 'life force' of vitalism, or the mental or soul substance of dualism).

Survivability of the Consciousness: There are different opinions on whether the individual human consciousness and self-awareness can survive the dissolution of its organic framework--death. Both reductionism and scientific materialism, for some at least, suggest that consciousness is dependant on its originating organic format.

The concepts of holism, quantum theory, spiritualism and religious transcendentalism, on the other hand allow for disembodiment of the consciousness, an independent existence and a reassimilation of individual consciousness with the cosmic consciousness.

Consciousness As Surviving Death: French Jesuit paleontologist Pierre Teilhard de Chardin specifically posits *individual consciousness not only as surviving death*, but as retaining its uniqueness. Whether this is an intuitional insight, born somehow of faith or insight derived from scientific rationale, is not clear. It is probably the result of both these powerful influences in his life.

Physicist Paul Davies, as we've noted, speaks of the survivability of the consciousness since it is not a physical part of our spacetime universe. Certain versions of quantum theory suggest that the consciousness is part of an underlying cosmic mental fabric; it consists of nonphysical information and software and so has potential for surviving death of the body with which it is associated.

How Much of Our Self Relates to Our Bodies? But physicists who speak of survivability of consciousness also ask how aspects of our consciousness might survive. The question is highly relevant because so much of what we *are* is directly related to our body. We know that physical deficiency and illness alters our disposition, outlook, philosophy and our personality.

We also know that strong bodies generate sports-mindedness and promote ideals of Olympic competition, while body weakness or infirmity predisposes toward intellectual pursuits. Pain and suffering, aside from their related physical impairment, alter the mindset. Love, hate, disgust, and all the other emotional qualities are induced or affected by physical causes or events.

Personalities are Altered: Personalities are altered by endocrinological changes. In one clinical case, for example, a person's deeply religious outlook and preoccupation disappeared altogether when hormone imbalances were corrected. Drugs can produce or alter mental states, take you from depression to euphoria and back.

Brain surgery can totally alter personality. Many of our thoughts, hopes, prayers, and plans for the future are related to our physical needs and condition. We also know that much of our personality, disposition, and sets of inclinations and proclivities, are physically inherited or influenced characteristics, differing from body to body, family to family.

So What's Left of Us When We Die? So Davies asks, what, then, is left of us when the physical part of us, around which our personalities have been built, is gone? If all physically-dictated or influenced emotional qualities and personality characteristics are wiped out at death, what remains to describe us? Davies asks whether we would retain an acquired sense of humor, of reverence, of boredom. It is debatable, he says. But what of memory? We are a product of experience which gives us our sense of identity.

And What of Memory? Scientists disagree on what and where is memory, some associating it with various locations in the brain, others suggesting that it exists as an unknowable somewhere else, like consciousness and self-awareness. "It is not clear," says Davies, "that someone with no memory can really be thought of as a person at all."

Making Scientific Sense of Immortality: Davies goes on to discuss a number of principles and concepts relating to individuality and says that some of these ideas "hold out the hope that we can *make scientific sense of immortality*, for they emphasize that the essential ingredient of mind is *information*. He cites the analogy that Beethoven's Fifth Symphony does not cease to exist when an orchestra finishes playing it. So, he concludes, "the mind may endure by transfer of the information elsewhere."

While research efforts leave little room for the traditional idea of the 'soul,' he observes, "they leave open the possibility of survival of the personality." It is interesting to note here the oft-repeated religious proposition that our physical life is given us to refine and hone the spirit given us at birth through the wealth of mellowing and enlarging experience that our physical lives produce.

Pre-existence and Post-existence: The question of where our consciousness comes from, sometimes becomes the question of where did our 'existence' come from and is that existence eternal? If we had some kind of 'pre-existence' in some inchoate form, as Plato asserts, then this will influence our view of its continued eternal character, the post-death continuation.

It was St. Augustine's seemingly prescient view that all time exists, having been laid out at the creation as one of our four dimensions, strangely enough, a notion supported by Einstein's relativity theory of the geometry of the spacetime continuum. In this context, at least, it might be asserted that, in a sense, we, as does Christ in the New Testament, exist from the beginning and throughout eternity.

In his discussion as to whether the soul, consciousness, transcends time, Paul Davies has difficulties with "timeless souls." He cites the Catholic Encyclopedia: "The notion that God has a supply of souls that are not any body's in particular until he infuses them into human embryos is entirely unwarranted by the evidence....The soul is created by God at the time it is infused into matter."

He notes that this same "temporal dilemma runs through all discussions of immortality." He notes that Jesus spoke of "life everlasting," implying unending time but he refers to the difficulties of reconciling this with God's laws of physics, suggesting that "there may be no 'everlasting' anyway."

Part of a Unified Cosmic Consciousness? Do we originate as part of a primordial and infinite, unified consciousness? This is sometimes expressed as whether we originate as an undifferentiated part of the universal mind, or a part of the unity of God. As we discuss elsewhere, we may be viewed, in a sense, as being as old (and young) as the universe, composed of 'rest-energy' forms fused in the stars.

Despite our cultural and personal diversity, many think that quantum theory implies that we all share a common source, the unity of the primordial cosmic mind. As the Eastern religions highlight, even in our diversity and conflict, our enmity and wars, our cultural tensions, we are One.

Related Theological and Philosophic Ideas: Some theological and philosophic ideas related to cosmic or divine unity are the immanence of God in human life; human life as an extension of God; the human mind as an extension of the mind of God--as described by physicist Freeman Dyson, as part of Spinoza's single universal substance, and the totality of the universe as the living universal being, as described by Teilhard de Chardin.

The Universe as Mind and Life: We have discussed the universe as consisting of a single substance, energy, and the universe as a single entity, mind. And 'mind' has been equated by some, to 'spirit,' 'soul,' 'higher consciousness,' and so on--all terms implying theological immateriality and ethereality rather than materiality.

Here, terms 'World Mind' and 'World Soul' surface, sometimes with essentially pantheistic implication, the idea of the universe equating to God, or alternatively, as an extension of God, or as a living 'organism,' just as the *Gaia Hypothesis* deals with the Earth as a single organized living organism.

The Ultimate Perfect Symmetry of Unification: The concept that underlying the physical universe is something mental, cosmic mind, has a certain appeal for those that seek foundation for reassurance that life is not transitory. In this 'scientific transcendentalism,' cosmic mind becomes a central theme in the search for the unified harmony of the primordial perfect symmetry.

Cosmic Mind, Scientific Materialism, and Death: One dividend of the concept of cosmos as mind is a new outlook on scientific materialism. As physicist Freeman Dyson declares in *Infinite in All Directions*, scientific materialism and religious transcendentalism lose their mutual exclusivity and dualism, monism, and pluralism are subsumed in one--mind.

Parapsychological happenings, for example, are invariably associated with energy manifestations such as poltergeist manifestations of physical impact of some sort. Death experiences, discussed later here, invariably describe places and figures of intense light or manifested in some energy-filled 'aura.' And so spirituality and ethereality on the one hand and energy and matter on the other, may simply represent different descriptions of the same reality.

Bodies as Simply Mind Containers: Bodies, in this view, are simply nature's structures to house aggregates of mind and, as physicist Freeman Dyson points out, may take different forms in the future as intelligence spreads throughout the far reaches of the universe.

Thus the implications in these theories of the interrelated nature of matter, mind, life and death are profound. In its ultimate expression, *science and theology may be seen as simply two expressions of the same reality.* This is a point of view Einstein repeatedly expressed when he spoke of astronomy and cosmology as expressions of theology in the search for understanding the Mind of God.

Hegel--Universe as Mind; Berkeley--Universe as Mental Substance: Interest-ingly, this description of the mental nature of the universe is also suggestive of philosopher Georg Hegel's universe of pervasive mind and unfolding thought and Bishop Berkeley's single mental substance. One may or may not make the final extrapolation to Spinoza's single substance, God, but may leave God in the exclusively transcendent role of theologian Karl Barth, or theologian Paul Tillich's God as above and outside of the universe, or as being itself.

Schopenhauer: Life As *Individualization* of Unified Consciousness: If our life vitality as individuals originates as part of that cosmic unified soul and consciousness, then our birth, thought philosopher Schopenhauer, is a form of *individualization* into our temporary, physical, more limited and finite form. On death, our spirit--our "determined infinite" form or spirit, to cite philosopher Feuerbach, is reassimilated. But here, death is real in the sense that it represents the effective *end of our 'separateness.'*

The Physical Universe Versus the Quantum Nonphysical Universe: As discussed earlier, quantum theory introduces the concept that our physical world is built on a micro-universe not of particle building blocks but of uncertain quantum energy states influenced by human observation.

Physicist David Bohm, in his *Wholeness And The Implicate* Order, suggests that another more fundamental and nonmaterial reality underlies our seemingly physical universe, his "explacate" order, of three dimensions and the dimension of time derived from the spacetime geometry of general relativity.

Our perceived physical universe, in this view, is but a holographic projection of a deeper, multidimensional and unifying level of nonphysical reality of mind, consciousness and information--his implicate order--in which all 10^{89} particles in the universe, are an interconnected whole into which all time is enfolded.

Separation of Mind and Body As an Abstraction: He views this unity of life and nonlife, consciousness and matter, as a living organism, characterized as possessing 'purposefulness' and 'intentionality.' He speaks of the separation of mind and body, mind and spirit, as an abstraction and of this subquantum level as "idealism, spirit, consciousness."

So when we look at the universe, all physical things like galaxies, planets, mountains, oceans, people and plants, we have here the remarkable phenomenon of one aggregation of particles observing "other" particle aggregates and speculating on their identity and significance as though they were something or someone else when in fact we are, at the subquantum level, as some physicists have put it, "observing ourselves."

Life and Death and the Firefly: Using Bohm's subquantum implicate underlying unified (undifferentiated) order of the unified whole, one analogizes Schopenhauer's "individualization" into our physical reality with Bohm's physical reality as a holographic representation of an underlying interconnectedness. The implication is that we emerge from the cosmic darkness, Bohm's fabric of the underlying cosmic mind, to light up for a moment like a firefly, as Schopenhauer's individualization, and then, as our light goes out, we reassimilate with the unified mind from which we originated. Do we thereafter nonexist as an individual consciousness?

Survival of Individual Consciousness: Whether the "individual consciousness" survives this individual discontinuance we call death, is discussed by French Jesuit philosopher and paleontologist, Pierre Teilhard de Chardin. Perhaps in a flash of intuitive insight, he concludes that *the individual consciousness survives* our re-assimilation with cosmic consciousness.

Life as Energy *Individualizations*: Early philosophic theories constructed universal life force models, the universe as mental substance--God or world soul. If our life vitality as individuals originates as part of the primordial energy center that became the structured universe--and us--then many, including the Eastern Religions, thought in terms of a *cosmic unity* of everything--universal soul and consciousness.

Our birth, thought philosopher Schopenhauer, is then a form of *individualization* into our temporary, physical, more limited and finite form, of a portion of the cosmic unity which is the universe. On death, our individualization, our 'determined infinite' form or spirit--to use philosopher Feuerbach's concept--is reassimilated, reunified.

Death, Mathematics and Dimensionality: Death, here, while *real* in the sense that our individualized state is real, represents not total nullification but merely the *dis-continuance of our mental separateness* and its reassimilation with cosmic mind, perhaps describable as a quantum or dimensional change.

Information as the Ultimate Components of Matter: The ultimate fundamental components of matter, say some physicists, like computer scientist Edward Fredkin and physicist John Wheeler, are bits or units of information, part of the underlying informational mental subquantum reality, and part of physicist John Wheeler's "meaning" universe consisting of "meaning" software. For those with an abiding interest in the meaning of life, and with the portents of death, the implications of these relationships must be significant.

Movement and Travel Via Information Transfer: Physicist David Bohm speaks of motion, the movement of particles, trajectory, not as a physical movement of an object but as a series of foldings and unfoldings from the underlying quantum level of information into our reality. In this way, so-called psychokinetic manifestations--mentally-induced movement of objects--are essentially transfers of the information representing an object from one place to another.

Death and Rebirth As Information Transfer and Rematerialization: Possible implications surface on the meaning of these relationships and death. The time limits of our physical life, a mere cosmic flash in the time of our spacetime geometry world, might be bypassed by a dematerialization at one time and place and a rematerialization elsewhere in time and space. Perhaps this is another description of the process of death and rebirth or reincarnation, in which the information of an earlier life dematerializes, to be recreated or rematerialized in a later life--via information transfer.

Death As 'Delivery' from Individual Existence: To philosopher Schopenhauer, death was the most powerful phenomenon that could lead human life to awareness of his authentic nature and the outcome and purpose of his life. With his pessimistic outlook toward life, he saw death as the dissolution of individual will--the source of evil, and the moment of redress and chance for atonement.

For him, individuality, the personal physical life, rather than being a source of self-actualization, was the enslaver of the will. Redemption could come only with the freedom of death and the achieving of the ensuing final existence. The fear of death, he felt, came from the false notion that the entire 'self' perished when in his reality, death freed his real self from the captivity of his physical being.

Not Born at The Moment of Birth or Die at the Moment of Death: Life, as seen by some, is a succession of deaths and rebirths as we die from what we were yesterday and are reborn to what we are today, to die again for tomorrow's rebirth. In this view, life and death overlap even in the state dominated temporarily by life, and death grows coincidentally with the development of life, until life's vitality diminishes and coalesces with death at the moment of loss of physical vitality.

Georg Simmel: Only the Individualized Self Dies: In his essay on The Problem of Death, in *Understanding and Believing*, Joachim Wach cites philosopher Georg Simmel that *death is immanent in life and a formative agent of life.* Simmel considered that, in the fashion of the philosopher Hegel, life (thesis) requires death (antithesis) as the "other" toward which it moves until the thesis and antithesis merge (Hegel's synthesis) in something higher, the universal "will." Here, one's awareness is enhanced by death and the *transcendental self survives the death of one's 'individualized' self.*

Mortalism Versus Life Continuation in Earlier Societies: Opposite concepts of death as final--mortalism, and the idea of various alternative forms of life continuance, go back to the earliest cultures. The idea of some form of life continuation after death, while not universal, appears to have dominated various societies and is traceable to the earliest civilizations.

The presumed form of this 'life-continuance' has varied, of course, between different cultures, some involving re-embodiment or transmigration into new bodies. Burial practices of some early civilizations made elaborate provisions for a physical afterlife. Others supposed a nonphysical or spiritual survival.

> Premeditation of death is premeditation of freedom. He who has learned how to die has unlearned how to be a slave....There is nothing evil in life for the man who has thoroughly grasped the fact that to be deprived of life is not an evil...The same passage that you made from death to life, without feeling of fright, make it again from life to death.... --Montaigne

CHAPTER 53
DEATH & IMMORTALITY, SALVATION, WHAT KIND OF AFTERLIFE

What Do People Believe about Afterlife? Newsweek of March 27, 1989 reported results of a 1989 Gallup poll showing that ninety-four percent of Americans believe that God exists, and seventy-seven percent that there is a heaven. More look for a reunion with family than being with God, since, apparently, they never got that acquainted with God.

Most look toward a life in heaven that continues an improved earth life. Some fear a lonely heaven; some are even concerned that there might be an absence of conflict to which they have become accustomed. Seventy-four percent look forward to finding *humor* in heaven. Fifty-eight percent believe in Hell, a mere six-percent expected assignment there, and fifteen percent encountered a so-called near-death experience.

As Kenneth Woodward points out, Notre Dame University theologian James Burtchaell notes that current thinking about heaven and hell is about "character, not context," about divergent states of "intimacy with and alienation from God." One creates his own alienation from God by what one chooses or fails to do.

In the concluding sentences of Woodward's article, he makes a meaningful philosophic final observation. Hell, as Burtchaell points out, is *the alienation we chose for ourselves*; heaven, suggests Woodward, is for lovers--of others and of God; and "The hell of thinking about heaven is that we cannot imagine--or trust a love that surpasses our own understanding."

Mortalism and Immortality--Plato and the Forms or 'Universals': Perhaps it was the dualism of Greek philosopher Plato that provided the most profound inspiration and influence on Western religious thought on immortality of the soul as separate from the body.

In *The Republic*, Plato, student and chronicler of Socrates, recounts the arguments of Socrates, in his *Myth of Er*, for the immortality of the soul. In *Phaedo*, Plato argues for the two-fold proposition of both the soul's eternal pre-existence and immortality--eternal post-existence. Interestingly, even contemporary theology generally avoids discussion of the pre-existence of the soul.

In *Timaeus*, Plato describes the world soul and individual souls as the work of the *Demiurge*, the individual soul being all-knowing and eternal before entering a body which then obscures its earlier knowledge. Life's goal is emancipation of the soul from the body. Plato's immortality of the soul arguments are largely attributed to his Theory of Forms or eternal and immutable ideas--"universals," of which corporeal or physical things are imperfect and perishable applications.

Plato sought to prove the immortality of the soul and the intellect which represented the higher realm of eternal ideas--the body to dust, the soul to contemplate the eternal realities or universals. St Augustine was influenced by these eternal forms of Plato in his views on eternal beings. Plato's indestructibility of the soul was accepted by Thomas Aquinas, one of the great philosophical thinkers in early Christianity, to become, eventually, a cornerstone of Catholic and Christian theological thought.

The Soul Imperishable Because It Is Indivisible: One of the arguments for imperishability of the soul is that compound substances, like the human body, are perishable by reason of their 'divisibility.' Thus material bodies being divisible, are destructible conglomerations, whereas the soul is basic, not further 'divisible,' and therefore permanently existing. The analogy in contemporary physics is to the divisibility of molecular aggregates into atoms and atoms into the basic particles--quarks, electrons and the gluon force particles.

Aristotle and Soul as Perishable: Aristotle, pupil of Plato, on the other hand, did not accept the soul as a universal, treating the soul as mind and substance, inseparable from the actuality or vitality of the body and so perishable with the body. His view is closer to modern medical mind-body physiology but departs from contemporary theological thought of a separate spiritual component.

The original biblical promise of human immortality, based on obedience, was, of course, in some views of biblical terms at least, lost with Adam's original sin of disobedience. And so death, and salvation for some, and some possible ultimate resurrection, is the human lot. Life, says the Old Testament, is like the grass that withers, the cut flower, the smoke, the disappearing shadow (Isaiah, Psalms, Job).

Thomas Aquinas Adopts Plato's View--Immanuel Kant Questions: Strongly influenced by Greek philosophy, Thomas Aquinas, perhaps the greatest of the early Christian philosophers, adopted Plato's view of immortality of the soul. It survives to this day in Catholic and other Christian theology, although other philosophers, like Immanuel Kant, pointed out that while basic substances may not 'divide,' they can diminish to nonexistence.

While some quantum theorists see mind as part of the cosmic interconnected quantum and therefore imperishable, for the most part, modern physiology and neurobiology of course, view the mind as an organic adjunct, the functioning of a physiological-neural complex, part of the brain and its electronic and glandular interrelationships with the entire body.

French Philosopher Descartes' Separate Soul: The proof-demanding, French mathematician and rationalist philosopher, René Descartes, famous for his declaration *cogito ergo sum,* "I think, therefore I am," asserted that the validity of all knowledge required mathematical reasoning for proof, except for certain basic logical truths that were self-evident. And the mind-body distinction, known as 'Cartesian dualism,' was one of these truths.

In Descartes' view, God was the absolute substance, and mind and body, the two relative substances--dualism. Human life involves a nonphysical mind lodged in a physical body. Locke was of the same view and said all we can know is ideas and so the soul's existence is a product of faith.

Berkeley, Hobbes, Hume, Kant, Spinoza: The view of dualism, material body and distinct and separate immaterial soul or mind, has had its critics. Forms of 'monism' suggest that the body, its mind, spirit or soul, are parts or extensions of one material substance. Hobbes was of that view. On the other hand are the views that there is but a single substance and this is mental, the physical being a mental construct.

To the Anglican Irish philosopher, Bishop George Berkeley, all was mind and so-called physical 'things' represent something the nature of which has no meaning except as defined by the mind, characteristics which cannot exist without the mind, unless it be the mind of an eternal spirit. In this view there is the one substance, mental. He also asserted that nothing can 'exist' without the knower, an idea also associated with quantum theory.

Hume felt the nature of the universe, and the existence of God, was unknowable and our knowledge simply a succession of impressions. Spinoza asserted the existence of only a single substance, God, of which the soul was part. This pantheistic view of everything as part of God does simplify things but does not represent the mainstream of scientific and theological thought although 'New Age' thinking points in this direction.

Renowned German mathematician, philosopher and theologian Immanuel Kant insisted that we can't know what exists out there, the 'thing-in-itself'; that we can "know" only what our mental construct delivers to our perception. But since the notion of soul has merit, we can, he thought, 'assume' its existence. While he rejected the conventional proofs of God, Kant asserted that *God's existence was a moral imperative.*

Human Life's Primary Anxiety: The primal human anxiety is existential, a conscious and subconscious preoccupation with the possible discontinuance of his existence. Psychiatry and related sciences, deal constantly with mental states related to this underlying concern. The strongest human fundamental drive is survival, yet nothing is more certain in his physical life than its discontinuance through a process called death. And thus the critical importance of dealing with the subject and meaning of death --and its keeper, time.

Survive to 'Continue' in Another Life Form? Despite the contemporary treatment of mind, consciousness, spirit, and soul as interchangeable terms, by religious indoctrination and, to some extent perhaps, by intuition, we tend to think of our 'minds,' in the sense of mental apparatus, as associated with and part of our bodies, but our "souls" or spiritual selves as somehow separate entities. Whether there is some part of us that 'continues,' is the subject of differing beliefs and conjecture, all falling short of objective certainty.

What does philosophy, theology, and science have to say of the finality of death? Many are swayed by religious views on immortality and salvation. But others are comfortable with constructing their own concept of death apart from religious interpretations.

In our modern society, religious or denominational affiliation has as much or more to do with social influences and geographical convenience as with subtle doctrinal concerns. However, it is therapeutic and constructive in dealing with the imponderables of death to explore differences in religious approaches to this most intense of human concerns.

Victory over Death in Christianity: As we have seen, religious views on the finality of death differ widely. Through faith, belief in and union with Christ, Christian theological material speaks of immortality through salvation, a powerful force in contemporary religion. But New Testament salvation is not necessarily available to all, and many will not qualify. Whether New Testament salvation goes to the predestined elected, as Calvin, Luther and others have argued, or to those whose faith, belief, and dedication qualify them, we discuss in the chapter on Religion, Denominations and Meaning.

St. Paul's Spiritual Afterlife: The most prevalent view of continued life in Christianity is probably St. Paul's account of the afterlife of spiritual re-embodiment in heaven for the saved. However, there is also a wide acceptance of continuance, not just as a "spiritual re-embodiment," but as an actual new physical body form. There is also the view of renewal as the delayed final bodily resurrection on Christ's Second Coming, a view closest to the Old Testament final physical resurrection.

'Willing' Faith: But faith and belief do not spring into being simply through expression or determination of individual desire, will, or purpose. Faith and belief may elude the most intense and prolonged search. Seemingly, faith and belief are not a matter of will but of grace--sometimes bursting into reality in an unexpected flash of intuition.

For certain, faith and belief are not a fringe benefit of 'signing up' for denominational affiliation, nor, it is said, do they follow as the reward for good works, although a lifelong pattern of good works is thought to reflect the grace of faith and belief.

John Calvin and Salvation for the Elect: We are reminded of the views of John Calvin, one of the great leaders of the Reformation, who revitalized St. Augustine's predestination theism that salvation is only for those 'elected' from the reprobate, predestined at the creation to be absolved of original sin and saved. In this view, human life can do nothing, by way of 'good works' to 'earn' salvation. Views of human life's nature vary from human life as corrupted by original sin or inherently unworthy, to the Arminian view of human life as being inherently honorable but having the option of falling from grace through debasement in the exercise of free choice.

Human Life's Intrinsic Nature: Views of the intrinsic nature of human life have fluctuated from time to time and between differing cultures and religious orientations. But on the whole Western thought tends toward belief in free choice, assigning blame for breaches of society's standards and inflicting disciplinary consequences for violations of laws reflecting those standards. Yet our prisons are filled with individuals whose fate was seemingly sealed by their inherited predispositions and imposed environmental negatives.

Parallels in Theology and Physics: It is interesting to observe that Calvin's theological views on predestined salvation espouse St. Augustine's deterministic uni-erse. The 18-19th century French mathematician-philosopher Marquis Pierre Simon de Laplace, reflected the view of classical physics of his time, the Newton era. This

was the then classical physics view of the universe as having been predetermined at creation, a confirmation of the earlier theological views of St. Augustine. But with the advent of the 'uncertainty principle' in quantum physics, the view of the universe as totally predetermined, came into question, although there are those that rely on the relativity theory coexistence of all time to hold on to the idea of the predestined world.

"O Death Where Is Thy Sting?" And so St. Paul's "O death where is thy sting," denying the power of death over the saved Christian, may be the goal sought by the many but achieved by the few. But there are differing interpretations of Scripture which are not so confining and absolute.

No Old Testament Counterpart: There is no precise or comparable Old Testament counterpart to the New Testament conditional offer of salvation and immortality, although there are references to eluding the finality of death by avoidance of sin, and, of course, the final resurrection and judgment of the dead.

Although there are, of course, prayers in practicing Judaism which refer to a spiritual continuation, to the rejoining of loved ones, we find no unequivocal prospect of immortality as held out by salvation in New Testament Scripture. Immortality of the soul, it is suggested, found its way into Jewish theological thought from Greek philosophy rather than from the Old Testament.

Apart from the final resurrection, the Old Testament clearly treats death as death, as in Ecclesiastes 9:5 "...but the dead know nothing." In Psalm 146:4 "...he returns to the earth; in that very day his thoughts perish." In Genesis 3:19 "For dust you are, and to dust you shall return." In some interpretations, the dead are "asleep" until the final resurrection.

Old Testament Expressions of Immortality: As noted, both the Psalms and Ecclesiastes speak of the finality of death. But there are expressions of immortality in Deuteronomy, Isaiah, and Ezekiel. Ecclesiastes 12:7 speaks of the spirit being reunited with God. Spirit of human life is also referred to in Job 32:8. This spirit is not the same as soul but the person's 'essence,' his character and personality spoken of in

Proverbs 20:27 (and in New Testament Hebrews 12:23). Both the good and the bad departed are dead, asleep (Deuteronomy, Chronicles, Kings, Psalms, Job); and they rest together until the final Judgment (Ecclesiastes).

Christianity sometimes refers to immortality of the soul going back to Thomas Aquinas, one of the most respected of the early Christian philosophers, who was strongly influenced by Plato's view, described earlier, on immortality of the soul.

But the prevailing view is that, like the Old Testament, death is death, except resurrection in the end time, or unless saved by salvation through grace. In John 5:28-29, Jesus speaks of "...all who are in the graves will hear His voice and come forth--those who have done good, to the resurrection of life, and those...evil, to the resurrection of condemnation."

Talmudic Writings and Contemporary Views: The Babylonian Talmud, it is said, has several hundred arguments in favor of belief in the final resurrection. It is interesting to note that religious teachings contained in early rabbinic interpretive writings reflect a more diversified view of life after death than the Old Testament itself, influenced by contemporary Greek philosophy.

In the delightful March 27, 1989 Newsweek article, further discussed below, writer Kenneth Woodward quotes Jacob Neusner, Brown University Judaism scholar, as expressing the conviction of modern Jewish Orthodoxy and Conservatism that in the next world, individuals will enjoy a communal life with God; that sin is atoned for in this life through repentance, and through the ultimate atonement, the death experience itself.

Moslem Views of Death: In Islam, God, Allah, is the same one and only God as in Judaism, and the same as the God of Christianity, except, of course, the absence of the impress of the Trinity. While the prophets, from Abraham to Jesus are honored in Islam, the greatest of the prophets is Mohammed, who was enabled, through the angel Gabriel, to recite the divinely revealed verses of the Koran, the Holy Book of Islam.

Upon death, the souls of the righteous, and those killed in battle, go to Heaven; the soul of the sinner to Hell, both to await the Day of Judgment. At Judgment Day, the souls of sinners are destroyed, unbelievers assigned for eternal torment, and the pious assigned to heaven or paradise.

Paradise preserves the same role of women as the Islamic fundamentalist view of that role in this life. The trend away from this primarily supportive role for women in the hereafter is said to have been arrested by resurgent Moslem fundamentalism.

Variations in Islamic Theological Thought: However, there are many Moslems whose view of heaven may be closer to those of the other Western religions. Perhaps because in many parts of the world, Islam is practiced close to large populations practicing the Eastern religions of Hinduism and Buddhism, there are many Moslems who believe in reincarnation, but not necessarily as captive to the inadequacies of their prior life conduct. In the same way, many Moslems in close contact with Christian communities, have a somewhat less differentiating view of the role of women in the hereafter.

Seeking Escape from Rebirth in Hinduism: As in most religions, it is possible to find differing points of view on the nature of events after death. But generally, in Hinduism, the death of the body is an event in a sequence of deaths and rebirths (transmigrations) of the Self. Rebirth is not a reward but an imposed discipline based on prior conduct. Death is a stop between successive lives; and rebirth is the result of one's prior-life conduct, the quality of the new life depending on the earlier-life conduct. Also important, is the frame of mind one achieves at the moment of death because this will influence the new life state.

When desires are finally overcome, the cosmic-self is discovered, one becomes Brahman, and the true inner human-self is released from the bondage of the cycle of rebirth and death to become immortal. In a land of pain, poverty and deprivation, the religious concept of the desire to escape the hardships of life is understandable.

Buddhist Thought about Death--Captivity of the Wheel: In Buddhism, the human mind is, to use William James' term, a "stream of consciousness," a series of continuous perceptions. Upon death the life force continues functioning. Life, like the universe, is an unending rapid succession of births and deaths and there is no absolute beginning or end. The body is a mixture of aggregates of sensations and perceptions (reminiscent of David Hume's "bundle of perceptions").

Death, in Buddhism as in Hinduism, is followed by rebirth. The quality of death and that of rebirth will vary as may have been earned by conduct. But the ultimate goal is to be freed from the captivity of the wheel of rebirth and death; and to achieve the eternal bliss, the peace and serenity of Nirvana. Death is escaped by avoiding death in rebirth, through the dissolution--over a period of life terms--of desires, thus achieving the state of 'desire-lessness,'and earning avoidance of a new rebirth with its pain and suffering.

Different Meanings of Heaven and Immortality: We see, between the great religions, highly divergent views on the meanings of life and death and afterlife. Perceptions of the values of rebirths are conspicuously different. There is a substantial diversity of meanings when we deal with terms like 'immortality,' 'rebirth,' 'resurrection,'' spiritual re-embodiment,' 'salvation.'

One perceived difficulty with the concept of the inevitability of rebirth in the Eastern religions is that there is no continuing consciousness of prior lives. You are simply left to infer that a poor present life is the result of error in prior-life performance and, conversely, a good new life, the reward of earlier-life good conduct.

Even within Christianity, the meaning of the 'saved' life, the form of life continuance, has been the subject of religious and theological debate. There are those who see their continuance as an immediate spiritual or spiritual-body extension in a spiritual realm; those that see an intermediate purgatory state; and those that see only the final resurrection with the Second Coming in the end days. But Christian views even of 'spiritual re-embodiment' differ.

The conventional view emphasizes the spiritual nature of such life. But others view the spiritual re-embodiment as reinvesting human spirit in a new physical body, as expressed for example, by Baptist minister, the reverend Jerry Falwell, in an April, 1989 TV talk show.

Early Judeo-Christian Thought Contemplated Reincarnation: Although presently popular Judeo-Christian thought does not directly address reincarnation (with some biblical exceptions such as Ezekiel 37:1-28, and Revelation 20:12 and 13-15), searches of historical materials show that concepts of reincarnation were in fact part of the early Judeo-Christian culture. Some of the early religious writings reflect the presence of these ideas that salvation brought actual physical resurrection of the body. This is not unexpected, however, since Eastern reincarnation culture, like Greek philosophy, had strong influences on the development of Judeo-Christian thought.

The Fifth Ecumenical Council: It is speculated however, that the influence of the Emperor Justinian prevailed over the sessions of the Constantinople Fifth Ecumenical Council, to eliminate these materials contained in the earlier texts that competed for adoption in the New Testament. But not every reference was excised. For example,

Jesus' reference to Elijah's return as John The Baptist. Recall that in Malachi 4:5, Elijah didn't die but, was whisked away in the whirlwind, with the prophecy of his return.

Jesus' Eulogy for John the Baptist as the Returned Elijah: And in Matthew 11:14-15, we have the eulogy of Jesus for John the Baptist in which Jesus refers to John the Baptist as the returned Elijah: "And if you wish to accept it, that he is Elijah who was to come. He who has ears to hear, let him hear." It is interesting to note that the modesty of John the Baptist made him uncomfortable about his characterization by Jesus as the returned Elijah. Some may see a distinction between Elijah's 'reincarnation' as John the Baptist, and his 'reappearance' (the same life) as John the Baptist.

The importance of Malachi's prophecy of return, in the practice of the Hebrew religion, was significant not only in Jesus' time but even to the present. To this day, there is the practice, in the Passover Service, of setting a place for the special guest, Elijah, who, it is anticipated, will appear at the Service, since his promised return is momentarily expected.

Immortality and Reincarnation: Many books have been written dealing with immortality and the continuity of life. While contemporary theological views on immortality deal more with survival or revival of the soul, there was a time when, apart from the eastern religions which, as we have noted, embrace reincarnation, early Greek philosophy, as exemplified by Plato, thought of human immortality in terms of a cycle of births, deaths and rebirths--reincarnations.

Is There Life before Life? In the July, 1990 Plain Truth, a publication of the Worldwide Church of God, David Albert writes on *Reincarnation--Is There Life Before Life*. When a person dies does his existence cease or is there some surviving life form? Have you lived before--will you live again, he asks. He goes on to describe growing belief, by many in the Christian community, in reincarnation, transmigration of souls, reappearance from earlier personages, even different animal life forms. Not only being born again but, as he says, of being "born again and *again* and *again*."

Recalling Prior Lives: He speaks of surprising "evidence" of so-called past lives elicited during heightened suggestibility through hypnosis, perhaps a product of "manipulation," intended or otherwise, involving information lying dormant in the brain, perhaps brought into the memory bank from previously read books or movies.

But he goes on to cite instances of persons recalling in precise detail a previous life and being able suddenly, in a hypnotic state, to speak fluently in *another language* of which they had no conscious knowledge whatever.

The Concept of an Immortal Soul Does Not Derive from the Bible: Asking where can we learn the truth about these things, Albert goes on to identify God, through the Judeo-Christian Bible, as the source of such knowledge. He cites various passages in the Bible not only to negate multiple reincarnations but to disprove the commonly held belief in a continuously existing immortal soul.

He speaks of the idea of an immortal soul as traceable to ancient religions before Judaism and Christianity, like that of the ancient Egyptians, wrought into Western thought via Plato's imprint on Greek philosophy, and early Christian theology through Plato's influence on Thomas Aquinas.

The Source of the 'Great Deception' about Immortality: Albert pinpoints the Biblical source of the 'great deception,' the idea of immortality falsely implanted, as Genesis 3:4, where Satan tells Eve that "You will not surely die." Albert cites St. John's Book of Revelation which speaks of the casting out of the great dragon, the serpent of old "called the Devil and Satan, who deceives the whole world."

So details of former lives that cannot be known by ordinary mortals must come, he suggests, from *unnatural sources.* And what are these sources? They are, he notes, nonphysical sources, the spirit world of Satan in which live the still powerful and rebellious fallen angels who know the past and live on through the centuries.

But There Are Other Views on the Source of Knowledge of Past Lives: But there are other explanations for evidence of past lives. The Eastern religions, which as we have seen, happen to have concepts of cosmic unity which are very close to contemporary scientific views of cosmic evolution and quantum physics, hold that deaths and rebirths are simply part of our physical reality in the cosmos.

Another view is that information of past lives comes from the spirit world of God's angels. Still another view, as described by the Jesuit paleontologist and philosopher Pierre Teilhard de Chardin, is that the individual human awareness or consciousness and its memory aspects, survives physical bodily death when it merges with cosmic mind.

The New Physics Suggests Another Explanation for Reincarnation: Still another explanation lies in the new physics, discussed earlier, which posits an underlying nonmaterial reality of cosmic mind, an interconnected fabric of mind, consciousness and information. In that quantum reality reincarnation might simply represent a re-emergence of 'information' from an earlier state.

In our discussion of quantum reality we discuss the material world, including life, as a holographic projection of information existing at some subquantum level inter-connecting the universe. In this view, we are holographic information projections into our physical reality and so death might affect the projection but the underlying us--information--continues and can thus be reprojected into another physical reality or new life. And so, in this context, rebirth may simply be nature--quantum physics at work.

DEATH AND SALVATION

Salvation Concepts: This chapter addresses the subject of New Testament sal-vation. Van Harvey's *A Handbook of Theological Terms* speaks of salvation in the Catholic view essentially as "the deification of man by participation in supernatural grace," thus also embracing the role of the sacraments and the Church in "mediating immortality."

In Protestantism, says Van Harvey, salvation is closer to "restoration of a broken personal relationship," involving "forgiveness of sins" and reestablishing "personal communion with God," where resulting grace is akin to "divine favor." In practice, it may be seen as redemption or rescuing of the soul, through Christ's sacrifice and belief in Christ as Savior, from the eternal alienation.

This original alienation is believed by many to have derived from 'original sin,' the disobedience in Eden. Troubling to many its potential for the exclusion of most of humankind, either through the limiting doctrine of the primordial 'election,' or through inability to achieve faith and belief, despite upright living.

In Christian Thought, Are Non-Christians Denied Salvation En Masse? It would seem that one must ask whether God, the creator of all human life, for whom He has the greatest love, would deny salvation to most of the brotherhood of human life. Such a denial would extend not only to those who are non-Christians but to those Christians, of whom there are many, who are unable to achieve the requisite qualifying level of faith and conviction. This would condemn by far the majority of the human race to extinction and damnation as the reprobate.

The Quandary of Some Scriptural Material: This, unfortunately, is the compelled view of many in the Christian Church at large. Thus many pastors are confronted with this quandary and paradox of Christian belief on entitlement to salvation. The life of Jesus is, after all, considered by much of the world, as the most sublime expression of love, compassion, and forgiveness. Given that faith and belief are not simply willed into existence but are a gift, something that mysteriously happens, the exclusionary doctrine of salvation of the elect troubles many in the church at large.

What could be further from the nature of Jesus than to interpret scripture as withholding salvation by reason of unachievable faith, despite intense striving, and despite a dedicated lifetime of good works, charity, compassion and unselfish love?

And So Christianity Looks Compassionately for Explanations: But Christianity has shown an inventive reservoir of love, compassion, and forgiveness of human frailties. And so segments of Christian theology have developed theories of exception from the harshness of the election doctrine. For example, there is the concept of eligibility for salvation through *the universality of the spirit of Christ* in all humankind. And there is the doctrinal exception in Catholicism based on the principle of *invincible ignorance* (1963-1965 Second Vatican Council).

The Universalist Church is said not to accept the mass denial of salvation. Evangelicals, according to a recent study by University of Virginia sociologist James Hunter, currently have difficulty viewing worthy nonbelievers as condemned. The idea, for instance, of one of the world's greatest men of all ages--and, to many, God's special gift to human life--the revered Mahatma Gandhi, residing in Hell because of his particular amalgam of enlightened beliefs, is at least disconcerting.

But, continues the study, many fundamentalists do appear doctrinally-troubled with a heaven limited to those few with proclaimed scriptural prerequisites, especially in view of the many parts of the Old and New Testament that express the Fatherhood of God, the Brotherhood of all men, and God's infinite love for the entire human family.

A Newer View of Some Church Groups: But there is a rather broad spectrum of religious thought that sees something deeper, more reasonable and understandable than the simplistic division of all eternity into the heaven and hell of ancient times. Harvard Divinity School theologian Gordon Kaufman describes the *decline in the heaven-hell concept* and concludes in a whimsical note that there is little 'future' for heaven and hell.

But Kaufman's real, more serious meaning, is perhaps nicely captured in Notre Dame University theologian James Burtchaell's description of the current thinking of heaven and hell as dwelling on the divergent states of "intimacy with and alienation from God." The conceptual essence of the thought so expressed, has universal appeal and seemingly transcends religious boundaries that divide the great religions and denominations.

The Doctrine Commission of the Church of England: In *The Mystery of Salvation*, the Doctrine Commission of the Church of England, rejecting earlier views on hell and salvation, noted that "Christians have professed appalling theologies which made God into a sadistic monster and left searing psychological scars on many."

In its January 1996 report, the Commission pointed out that belief in a literal hell of fire, agony and everlasting punishment has, except for the very conservative, given way to the view of hell as symbolizing a state of annihilation or nonexistence for those rejecting the love of God. But a March 1995 Gallop poll suggests that in Great Britain some one-fourth still believed in a literal hell of fire and eternal damnation.

Other churches have also given expression to the idea of hell as a metaphor for alienation from God. The U.S. Episcopal Church catechism speaks of punishment as "eternal death in our rejection of God" and the Roman Catholic Church catechism, of the punishment of hell as "eternal separation from God."

Observations on Developments in Catholic Theology: Of great interest is the growing awareness of trends in Catholic theology which suggest that the oldest Christian church, with the longest tradition of conservatism and exclusivity, seems to have the greatest institutional reservoir of strength for producing theologically innovative material with ecumenical potential. An example of this is in the work of the Catholic Biblical Association which is essentially academic and nonsectarian in its membership and thrust.

While there are repeated expressions of concern from 'on-high' over elements of a Catholic theology sometimes perceived as straying from orthodoxy, there is, nevertheless, a profound influence, both within and outside of the Roman Catholic Church, generated by this trend. Those who dream of a 'reconciled' Universal Church, may again think instinctively of the potential of this great institution as a source of the kind of transcendental thinking required for ultimate universal religious harmony. The revered Mother Teresa, for example, devout Catholic, ministers to the Hindu poor.

Unbelievers in Islam: Although the fate of 'unbelievers' in Islam is ominous, the practical outlook of individual Moslems is indeed tolerant of other faiths. Despite more recent examples of Moslem fundamentalist impatience with dissent, as for example, against followers of the Baha'i faith in Iran, historically, at least, Islam has practiced far more tolerance for other faiths than it does for divisions within its own faith. Judaism is biblically enjoined to practice tolerance for other faiths. Hinduism and Buddhism is similarly tolerant of other beliefs.

WHAT KIND OF AFTERLIFE?

Eternal Life and Leibniz's "Best Possible World": There are the many who not only cannot see "heaven" or eternal life without a physical body but cannot see it without the good, the bad and the indifferent inherent in our kind of world--the "best possible world" of the great German theologian, mathematician and philosopher, Gottfried Leibniz.

The Spiritual Afterlife Difficult to Grasp: The meaning of an afterlife of 'spirituality' or 'spiritual re-embodiment' is not easily grasped. Even if 'understood,' however, it may have uncertain appeal for many whose desire for heaven is simply to resume their lives with their loved ones who have gone on--to hold their hand, to go to and come home from work and be greeted by them.

What they really yearn for is to have their loved ones return to *a heavenly reconstruct of this world*--or at least this kind of world. Theologian R. C. Sproul speaks of Christianity as believing in a resurrection of the body, so, as he says, "we're not looking for a ultimately pure soullike existence without a body."

Condemned to an Eternal Idyllic Life? There are many who would resent an imposed eternal idyllic life. If one didn't like it, he could not even be bored to death--or bow out for the eternal sleep. The right to change one's life style, change one's values, to chose pain and sacrifice--if the goal warrants, the right to an end--to the oblivion of death, if such exists, all these 'privileges' are precious, too dear to trade for an unchangeable, idyllic, no-problem, 'eternal' spiritual idyllic life of some sort. They want a 'pinch-yourself-to-know-you're-there' heaven, with a real, honest-to-God, sometimes painful, body--free of major impairments, naturally.

Meaning Comes from Suffering and the Need to Strive: Down through the ages human life has come to understand the real 'meaning' that comes from the ups and downs in life, from deprivation and pain. Many of the 'have-everything, don't-have-to-work,' people in our world, finally commit suicide to get away from that same 'no-problem,' idyllic and to them, meaningless world.

An 'eternal' idyllic life, from which there is no escape to a life of challenge, of striving, of deprivation, could become the world of the eternally-damned, the ultimate Hell. Which is why some feel that our life on earth, somehow, *is heaven*.

Besides, so many of our innocent pleasures are body-related. A world without body-cooling breezes, the touch of a loved one, the fatigue of a good day's work, hot dogs and beer, ice cream cones, and without chocolate-covered nuts or raisons, such a world, to many, would hardly be 'heaven,' hardly worth it. Neither would a world without good hot days and a handy lake or pool to plunge into, a shaded path to walk down, holding hands.

No, say many, our 'eternal life' will have to be restructured before it gets our vote--or our confidence that it exists. Give us instead, the painlessness and serenity of eternal sleep--until the new dawn, the renewed universe from which our sleeping consciousness can reemerge, to enjoy again, in some unknown place, God's new life forms for us, with some pain and imperfection, some goals to achieve, adversities to overcome, victories to celebrate, and tears to shed.

Preservation of Life in Some Abstract Form Not Enough: The built-in longing for continuation of life after death is not, as theologian Karl Barth acknowledges, simply for some abstract preservation but for continuation--for life's completion. We long to see and be with our departed loved ones, to continue life as we know it. To many, the oblivion of death is preferable to an eternal life, pain-free but devoid of the challenges and pitfalls of the life we know and have come to accept--Leibniz's imperfect but "best possible world," the best God could contrive for us.

Does Each Human Being Construct His Own Heaven--His Gift from God: And so it would seem that each human being constructs his own heaven, the kind of life he would have for himself--and his loved ones. Can it not be that this is one of the rights granted by God to each human being, each of God's children--to live in a heaven of his own choosing, even if this be the repeat of the imperfect life on earth?

There are many who believe that there are an infinity of worlds; that each of us lives in his own world, a world different from that of any other. In his *Masks of The*

Universe, Edward Harrison describes the various "masks" or societal universes constructed in every age by each society and the infinity of "World pictures" constructed by each human being to reflect his own perceptions of the world.

The Soul of the Departed: In his *Man's Search For Meaning*, Dr. Viktor Frankl poignantly and convincingly describes the essentiality of suffering challenges to a universe of meaning. Intuitively we seek to communicate somehow with the soul of the departed--to find reassurance of their continuance. The story is told of the person who sent a letter each Mother's Day to his mother at some discontinued address, wondering if somehow there was some sort of divine forwarding system that took over from the U.S. Postal Service.

Those who have parapsychological experiences of communications from the departed have a conviction, sometimes born of the experience, of communing with the inner self or soul of the departed. Some forty percent of Americans, according to a poll in 1988 claimed some kind of contact with the departed.

We have discussed theories which explain this as part of the interconnected sub-quantum reality, the underlying quantum connection. We can receive such communication not as a result of some paranormal power, but because we and the departed are all part of a single interconnected universal consciousness, each part available in some way to the other, part of physicist David Bohm's so-called 'implicate' order.

The Argument That a Meaningful Universe *Requires* That Life Continue: We see the argument frequently made, born of intuition, that God is not arbitrary; that God's handiwork is intelligent and meaningful; that it is unthinkable to consider for one moment that God would bring souls into being, perfect and hone them to precision, greatness, beauty, wisdom and discernment, and then whimsically destroy or abandon them as though they meant nothing.

Buttressing this argument of a law of conservation of life is the analogy of the law of conservation of energy. Following that analogy, our lives would change in form but not cease to exist. Further, as manifestations of God, extensions of the mind of God, as Freeman Dyson puts it, we cannot perish, we co-exist with God. We might speculate on the form of our continuation but not on the fact of continuation--or so the argument goes.

> The goal of all life is death. --Sigmund Freud
>
> The gift of God is eternal life.--Rom. 6:23

CHAPTER 54
NEAR-DEATH AND OUT-OF-BODY EXPERIENCES

Clinical Death and Revival Experiences: There are perhaps eight million Americans who have had NDE or near-death-experiences, according to the Journal Of Near Death Studies, and as reported recently in the quarterly publication, *Special Report*. Dr. Elisabeth Kubler-Ross (*On Death and Dying* and *Death-The Final Stage of Growth*), Dr. Raymond Moody (*Life After Life*), Dr. Melvin Morse (*Closer To The Light--Learning From Near-Death Experiences Of Children*), and others, write of those with so-called 'near-death experiences,' where they were brain-dead, technically dead, for a short period, perhaps in a hospital bed, only to return to life and regain consciousness.

When Does Death Occur? Time of death has always been the subject of debate. Actually, death, like life, is a process spread over time. Death is not really instantaneous and certain signs of life may continue for some time after the heart ceases to beat and breathing therefore stops, unless heart-lung equipment is employed, such as in certain major surgery.

But the ability to use artificial means to continue certain body functions triggers the question of when a person dies. Determining cessation of heart-lung functions is not difficult but 'brain death,' including questions of the existence of measured electrical activity is another matter and such confusion raises issues of morality, religion and even law.

And even if brain death is said to ensue, machines can continue heart and lung act-ivity and there may be instances in which some brain activity might theoretically restart. In any event, there are the myriad of questions about authorizing discontinuance of machine-induced life signs, leading to state 'living-will' laws permitting persons to authorize discontinuance of artificial life-prolongation techniques.

The Sensation of Being Discarnate: There are many reports of persons, in some life-threatening emergency, perhaps on a hospital operating table when the heart may have ceased to beat, who have the sensation of becoming or being discarnate, that is, detached from their bodies, viewing their bodies lying below, not responding to emergency resuscitation efforts, for example.

Crossing over--Being Received by Figures of Light: They report feeling lifted or transported; going through a tunnel of darkness and emerging into an opening of great light, having crossed to the 'other side.' They report seeing figures of great light; being in the presence of what they describe as a guardian-angel figure, having a sense of euphoria, serenity, wellness; sometimes, seeing long-departed family members. They recall being told that it is not yet their time, that they must, for now, return and finish their stay.

The Similarity of Experiences: The relative similarity of these documented experiences give both Dr. Kubler-Ross and Dr. Moody the intuitional feeling that there is more than coincidence involved. Both medical doctors, they share deep-seated convictions that these NDE events provide evidence of life-continuation after 'death.' As medical doctors, their sophisticated insights are entitled to more significant weight than might be attached to the insights of the more impressionable, the religious, the unsophisticated, or the superstitious.

Experiences of the Very Young: Dr. Melvin Morse's reports of such NDE events are noteworthy because the subjects are children, some very young and consequently less likely to be subject to acquired or preconceived impressions of heaven or the suggestive influences of religious ideas in general, and without knowledge even of the subject of near-death experience.

Involved were persons who were not simply unconscious but in a state in which there were no brain, neurological, or other physiological life signs; they were, clinically speaking, brain-dead, even though the brain cells were still viable and restoration of faculties therefore still theoretically possible.

Some T.V. Accounts: A number of T.V. programs have been devoted to this subject. ABC's Dr. Timothy Johnson, who happens also to be an ordained minister, noted, in an ABC 20/20 show, that although these experiences are consistent with, or may be associated with oxygen-deprived, brain-chemistry hallucinatory phenomena, that after taking into account other recognized medical or scientific phenomena, there were still residual aspects of these experiences beyond known medical explanations.

The Experiences of Children: Dr. Melvin Morse, who catalogued, analyzed, and recounted the near-death experiences of young children, persons too young to be influenced by auto-suggestion or religious indoctrination, notes the remarkable similarity of the experiences of persons whose tender age precludes any earlier cultural absorption of such concepts.

Reluctant to Return to Their Bodily Life States: Appearing on the Larry King Live show, Dr. Morse's responses to questions were unequivocal that medical experience cannot account for these experiences. One particularly striking feature of these experiences is the fact that individuals who have had these experiences no longer have the normal fear of death although they may have a fear of undergoing the traumatic sequences of the dying process itself. Many reported that once embraced in the serenity of this NDE 'after-life' discarnate state, they recall themselves as being reluctant to 'return' to their bodily life states.

Near Death Experiences and Higher Consciousness: Elsewhere here we discuss the increasingly complex organization of the brain, the emerging subjectivity in both biological and cultural evolution. In a similar vein, professor Kenneth Ring of the University of Connecticut at Storrs, exploring near-death experiences in his *Life at Death* and *Heading toward Omega*, asks whether near-death experiences may be part of human life's "evolutionary thrust toward higher consciousness for all humanity."

Ring cites the transcendental outlook resulting from these near-death experiences, the sense of enhanced values and loss of fear of death. He expresses an interest in using testing devices like EEG, PET and CAT scans to explore these as energy experiences.

William Wilson's Summary of Near Death Experiences: In *Seeing The Light*, in "Special Report," freelance writer William Wilson provides an engaging review of near-death experiences. As do Moody and Kubler-Ross, he enumerates the elements common to most of these experiences as: a sense of being dead, a serenity and painlessness, a feeling of *release from the body*, the sensation of going through a tunnel, seeing people of light, seeing a "being of light," experiencing a review of the person's life, a disinclination to return, and a post-NDE change in personality.

The Similarity of Experiences Has Meaning: The number and similarity of these reports, and the similarity in the elements of the experience, have given many encouragement that there is something more to our lives than oblivion and life nullification. On the other hand, there are, of course, various theoretical medical explanations offered for these sensations or impressions.

One of these explanations is that this trip through a tunnel and emergence in light is a final 'reliving' of the birth experience; another that these may be brain-chemistry-induced hallucinatory phenomena of a body which is clinically dead but still not physiologically dead; that is, brain chemistry is still at work despite indetectibility of brain activity. The difficulty here is that at this stage of science, so far as we know, mortification, the totally dead body, will not permit revival to tell of the experience.

In the New Physics, Mind Is a Separate Nonmaterial Something: Another explanation for NDE relates to the notion of individual consciousness as an individualized part of cosmic mind, consciousness, and information inherent in the nonphysical quantum reality. In that theory the mind, associated with the brain at birth, is something separate from the brain, dissassociating itself at the time of bodily death.

Although we can't be certain what NDE experiences mean, we are free to speculate on their meaning and accept or reject their relevancy according to our own insights. In the final analysis, each member of the human race constructs his/her own 'reality' and view of the meaning of death. That these reports are reviewed and evaluated by persons with medical credentials tends to confer some appearance of credibility.

Consciousness Independent of the Body : These near-death experiences, involving release of the consciousness and perception powers from the physical body, are also the subject of discussion as 'out-of-body' experiences not necessarily associated with the occasion of a near-death experience. We discussed earlier the concept of mind or consciousness as not dependent upon the body but as a part of the underlying quantum informational and mental-connectedness of the universe.

In their *Mind-Reach*, Harold Puthoff and Russell Targ of the Stanford Research Institute discuss experiments in *remote viewing* by people of remote sites and events, suggesting that the underlying informational level of the universe is available to some as a medium of their thoughts.

More Than One Level of Consciousness? One theory is that a human being operates at more than one level of consciousness; that there is a form of vitalism or superconsciousness invested in each human life at birth for the duration of the body's mortal life, which, at death, becomes *discarnate*, departing from the body. Some may refer to this as the spirit or soul. Most people intuitively feel what they might describe as a *soul* presence, in themselves and in others.

The Discarnate Consciousness: And so this superconsciousness, which becomes discarnate in near-death situations, may be viewed as part of the underlying cosmic mind, consciousness, and quantum information system--the *quantum connection*, between everything, all particles and energy forms in the universe. This discarnate spirit has been perceived by many as hovering over the burial site in the presence of beloved survivors.

Remote Viewing and Crossing the Time Barrier: The Puthoff and Targ Stanford Research Institute experiments in 'remote viewing' by people, of remote sites and events, even crossed the present-future time barrier to a view of events in the future.

Human Perceptive Powers: The remote viewing phenomenon suggests that the mind is able to focus on and view events beyond the reach of the ordinary immediate body perception ability; that there are human perceptive powers able to draw on an informational level underlying that of the ordinary physical world. These cases seemingly confirm notions of quantum physics that our energy-matter universe is an extension of this underlying universe of information, with all the implications relating to the phenomenon of death.

Hypnosis and Reincarnation--Hypnotic Recollection of Past Lives: In addition to the clinical near-death experiences, there are other quasi-scientific phenomena worth exploring. Patients under hypnosis have 'remembered' what they believe to be past lives, in considerable detail. One explanation offered is that they represent a form of 'genetic memory' passed on from earlier members of the ancestral line. We see parallels in trained mice genetically passing on some of their training. But increased interest is being displayed in studies of what are regarded as para-psychological events.

Memory of the past is also explainable, as we have noted above, by reference to the underlying quantum cosmic mind or consciousness and the ability of some to tune in to that cosmic *resonance* or fabric of information underlying our physical world.

Regression to Recall Prior Lives: Even though hypnosis is now an accepted medical practice, there is much skepticism about the efficacy of memory regressions in hypnotherapy to recall events in a prior life. There is much written on the subject and some rather striking 'examined cases.' In *Past Lives--Future Lives*, Dr. Bruce Goldberg, a dentist, hypnotherapist and member of the American Society of Clinical Hypnosis, describes the idea of reincarnation and the karmic cycle.

Justice and the Golden Rule Don't Seem to Work in One Life Span: The thrust of these materials is the concept that the ideas such as justice, compassion, reward, the golden rule, have little relevancy in a single-life existence but are fully functional in the karmic cycle of multiple lives which provide for the reconciliation of the 'balance sheet' of conduct between lives. Injustice in this life is rectified in another life.

Some views of the karmic concept assert the existence of the over- soul; that each soul is part of a larger world soul. Death simply terminates the temporary body part of us. But our real eternal soul-self continues in another plane or dimension, eventually to enter a new body, and a new chapter in the life of the soul. This idea of life as an individualization of the larger cosmic life, is implicit in the work of the philosopher Schopenhauer. Based on the hypnotic case histories, the death experience seems to have little, if any, relationship to the individual's religion.

The Post-Death Experience: In the hypnotic trance induced to relive an earlier death experience, the 'cross-over' after the prior 'death,' produces a floating sensation, feelings of inner tranquility and a recollection and review of the events of the 'prior' life. The soul's new post-death-experience life-form is not encumbered with our physical laws and moves about without regard to time, space or earth dimensionality. It should be noted here that the arrow of time has no application at the quantum level.

How Does the Karmic Cycle End? In this frame of karmic reference, God does not punish; we self-inflict our punishment, just as our earth body's conscience punishes us. When does the karmic cycle end? It ends, as in the Eastern religions, when you meet all your karmic obligations set in motion by your conduct and your thinking; when you have set things right, restored the balance in your karmic account, when kindness to your fellow soulmates is your way, and selfishness has been conquered. You achieve this end the old-fashioned way, you earn it!

> Surely God would not have created such a being as man ...to exist only for a day! No, no, man was made for immortality. --Abraham Lincoln

CHAPTER 55
PARAPSYCHIC ASPECTS OF DEATH

Introduction: Many find difficult barriers to rational belief in continuation of life in some form following physical death of the body. Some of these barriers may be lowered by exploring the esoterica of psychic phenomena. This chapter discusses psychic and paranormal phenomena and their underlying concepts as they relate to or bear on the phenomenon of death.

Human Divinely-Derived Mind as Providing Parapsychological Powers: As we have discussed earlier, some physicists, like Princeton physicist Freeman Dyson, speak of the human mind as a derivative from the mind of God, sounding more like that of a theologian than of a physicist. It is not such a stretch then, to attribute to humans, psychic or parapsychological powers. From time to time, mystics, personages with seeming psychic vision step into the pages of human history.

Psychic Power and Memory of the Future: The extraordinary predictions of Nostradamus, the powers of deeply religious or contemplative men of the past, the biblical prophets, all exhibit indications of special psychic power, such as prescience or knowledge of the future, sometimes referred to by physicists as *memory of the future*. But knowledge intuitively or inspirationally arrived at may, after all, be as valid as so-called conventionally-derived knowledge. Einstein, Newton, Hawking, and others, all had conceptual insights, 'hunches' or premonitions, in advance of corroborative mathematical or other proof.

Speculations on the Nature of Communication at the Psychic Level: Given the evidence of the existence of psychic or paranormal skills in certain persons, science has explored possible explanations, looking for possible new laws or new applications of known laws. Mind-reading and mental telepathy, seemingly inconsistent with normal electromagnetic communication, have been explored in terms of possible as yet undiscovered communication media.

Psychic Power and Biblical Prophecy: Another psychic manifestation, with many believing adherents, is that of 'automatic writing' by individuals psychically attuned to receive communications from higher levels of consciousness. The biblical prophets, considered to have been men of unusual psychic skill, are thought to have had access to such *higher levels of consciousness*.

And, if human mind is an extension of the mind of God, as physicist Freeman Dyson suggests in *Infinite in All Directions*, then divinely-inspired scripture becomes more scientifically-oriented. Mohammed was said to have written the Koran at the direction of the angel Gabriel, a possible analogy to such higher levels of consciousness.

Projecting Mental Influences over Great Distances: Other phenomena, such as so-called subliminal hypnotic effects, assertedly used in advertising, and perhaps politically, or even militarily, have been tested both by the U.S. and other countries. Similarly, tests have been conducted of the power of the mind to project its presence and influence at great distances.

Communication Between Twins--Higher Levels of Consciousness? We're used to thinking of communications in terms of familiar energy transactions, ordinary electromagnetic transmissions like radio and telephone. But there are communication forms that obviously involve other principles with which we are not yet familiar or comfortable. This applies to communication between identical twins, sometimes thought of as a single life in two bodies (a single sperm impregnating a single egg which then subdivides, with identical genetic composition).

Pain and headaches in one identical twin is often felt, registered, in both, and young twins in many cases communicate (unconscious criptophasis) in a language known only to both. They register the same brain waves and often communicate or understand each other without vocalizing or other conventional communication.

Twin Communication Analogies in Physics: Analogies are sometimes introduced to suggest a basis in physics for what is sometimes referred to as psychic communication power, as thought to exist, for example, in identical twins. One example is the nature of all particles or energy units in the universe as basic units of 'information' or 'intelligence,' part of an underlying universal information system. All information and knowledge, in this 'information system' theory, is reflected in the quantum reality of the new physics suggesting an underlying unified cosmic consciousness and fabric of information.

The Quantum Connection:: Quantum theory posits an instantaneous 'informational' link, between all particles in the universe, representing the primordial unity of the universe, sometimes referred to as the *quantum connection*. This phenomenon is thought possibly to explain the simultaneous registering of pain by distant identical twins. In this analogy, the information is felt simultaneously, that is, without involving the normal lapse of time involved in electromagnetic transmission governed by special relativity speed-of-light limitations.

While this may sound imaginative, remember that we continue to learn more about the universe. Many of the things we now know--and can do--like how to go into space, to other worlds, once fanciful science fiction, are now part of our reality. Human-made satellites have now left our solar system to journey into the interstellar space of the Galaxy, the science fiction of just a generation ago.

Sensing the Presence of the Departed: Many have had the experience of being at a funeral and sensing the 'presence' of the departed. There are many who return to the grave site regularly, and meaningfully 'converse' with the departed. Some go to present a problem and come away with a sense of having been counseled. While this seems strange, even morbid to some, to others it is perfectly natural and the grave-side session is looked to with anticipation. George Burns frequently visited the grave of his beloved Gracie, with whom he is said to have carried on conversations.

Parapsychological Phenomena: There has long existed a group in society, termed by some, the 'spiritualists,' who assert that afterlife is demonstrated by authenticated instances of phenomena that have no basis in the known physical laws. Examples include telepathic communication between close but widely separated individuals, or communication received from a person no longer living.

Other examples include: clairvoyant perception by one individual of events occurring elsewhere; psychokinesis--the production of motion or levitation in objects through the application of psychic powers; and premonitions and prescience--the knowledge of events before their occurrence.

Such occurrences involve powers not explainable in terms of commonly discussed laws of physics. According to special relativity, electromagnetic transmissions would be limited by the speed of light. Yet these communications appear to be free from the application of that law of physics, a phenomenon, as we have noted above, explainable through resort to the quantum-connection theory of an underlying inter-connected fabric of cosmic information and mind.

A Nonphysical Cosmos of Transcendent Experience: In *Mindsteps To The Cosmos*, astronomer Gerald S. Hawkings organizes the evolutionary development and experience of human life into *mindsteps*, the stages by which human thought evolved. Mindstep 6 involved discovery of "a new field," a non-physical field related to the nonphysical cosmos, perhaps involving a spiritual domain, the existence of a vital force, a soul, continuation of life.

He cites Nobel prize-recipient Sir John Eccles, renowned for his work on human brain structure, who mused that the cosmos was not running down (entropy) "for no meaning." He wondered whether this present life is "all to finish in death," or whether there was to be an unknowable "transcendent experience."

Does God Speak to Us, on Occasion, through the Departed: But some will suggest other explanations. One, for example, is that God is universally present and consequently present within us; that when we carry on conversations with the departed we are engaged in soliloquy, or perhaps dialogue, with a part of the God within us. Do perceived discussions with the departed perhaps represent communication through, or perhaps even from, God? Or are they perhaps one and the same? Many feel their departed loved ones are somehow 'rejoined' with God.

Subconscious Awareness of God? Many of us tend quite naturally to think of our departed parents, for instance, as being 'there,' somewhere, with the power to see and tell hidden truths, to commune with our souls and fathom our thoughts. Perhaps this is a metaphor for our subliminal awareness of God and cosmic unity in which the departed now share.

The Logic of Immortality: We sometimes think of ourselves as human manifestations of God, our minds being extensions of His mind, as physicist Freeman Dyson suggests. As such, perhaps we ultimately share in his immortality, omniscience and ubiquity. What more natural sharing of activity than communication from the departed.

We have spoken elsewhere of continuance of the spirit or soul as a much more logical and meaningful than a wasteful total extinguishment of each individual that God brought here and nurtured to high levels of intelligence, experience, and accomplishment. Why would God allow these accomplishments, refined souls, to be wasted? It makes no sense. As we have discussed, conservation principles must have some spiritual analogy for refined souls.

More on the Eternal Balancing of Accounts: As previously described, in some theories, the discarnate spirit is exposed to a 'Karma' review of the record of performance in various lives. In this concept, one's rewards and punishments transcend any one life. Wrongs of one life may be redressed in another and acts of greatness and generosity correspondingly rewarded.

The Convictions of Some; the Doubts of Others: The point of this discussion is to reinforce the impression that, like the biblical account of Job's demand of God to explain his unearned predicament, the human mind will not lightly surrender its perceived right to rational explanations of the mysterious universe. While there is a high degree of conviction of many intelligent individuals who subscribe to the authenticity of these parapsychological phenomena, they remain a minority in a cynical world.

It is interesting to note that actual extrasensory perception testing, by Duke University experiments with telepathic communication, has repeatedly shown that the mathematical odds against these psychic communications being simply chance is huge indeed. So random chance does not account for these psychic phenomena.

Science and Life Continuation: Built into disciplines of science are its demands for certainty and provability which will always cloud acceptance of intuitional metaphysical concepts, like survival of the soul or spirit, or the concept of eternal life, near-death experiences, and so forth. But we have also noted new physical explanations for what was earlier thought to be metaphysical.

But even science now speaks of additional dimensions in the universe, as yet undiscovered laws of physics, the 'quantum connection,' 'memory of the future,' the existence of parallel universes, reality as subjective, human mind as an extension of God's mind, and many other once-fanciful developments. So we can no longer casually dismiss psychic phenomena and models of mind survival following physical death.

> The average man, who does not know what to do with his life, wants another one which will last forever. --Anatole France

CHAPTER 56
TIME, QUANTUM THEORY AND DEATH

Relativity, Quantum Theory, Availability of Past and Future: Any discussion of death must also deal with the nature of time. Relativity and quantum theories offer new insights into the nature of time, possible time travel, and the *availability of both the past and the future in the present.*

Relativity, Quantum Theory and Time: Special relativity theory tells us that time is a geometric dimension of the universe; that all time, past, present, and future, coexist; that time does not 'pass,' its just there and we encounter time as *we move* along the 'world line.' As physicist Fred Allen Wolf observes, in his *Parallel Universes*, we "march on in time....through an imaginary dimension of space called time, moving at the speed of light."

Mysterious Relationship between Life, Death, and Time: There is a *mysterious relationship between life, death, and time* which surfaces in the context of special relativity theory where time is not an absolute but relative and all time, past, present, and future, coexists, part of our dimensionality, a geometric fourth dimension of our universe.

Einstein's Enigmatic Letter on Time as an Illusion: We recall Einstein's enigmatic letter on the illusion of the passage of time. Einstein had carried on long correspondence with his dear friend and mathematician Michele Besso on the nature of time. Besso, preoccupied with the meaning of time, had written Einstein again and again about time; what is time and what is meant by irreversibility of time.

On Besso's passing, Einstein wrote to Besso's sister a letter of consolation which has mystified physicists and philosophers alike: "Michele has preceded me a little in leaving this strange world. This is not important. For us who are convinced physicists, the distinction between past, present, and future is only an illusion, however persistent."

Just what did Einstein mean? How would the coexistence of all time somehow diminish the tragedy of the death of a loved one or somehow bear on the meaning of that death? Since the letter was a consolation letter, it is assumed that he meant his words as consoling; that somehow, the coexistence of all time should have special meaning and relevance to one's philosophic outlook on death.

In developing his theory of special relativity, Einstein had spent a year after the Michelson-Morley experiments which had failed to prove the existence of an 'ether,' in trying to resolve the problem of the *contradiction* between the invariance of the velocity of light and the theory in classical mechanics of added velocities. Einstein notes that he finally struck upon the answer.

"Time Cannot Be Absolutely Defined": Meeting his friend Michele Besso, on whom he had been trying his theories, he announced to Besso that "an analysis of the concept of time" was the answer to this apparent contradiction. "Time cannot be absolutely defined," he said, and there is "an inseparable relation between time and signal velocity." *Time is dimensionally interwoven with space.*

Fascinated with Einstein's new approach to time, Besso sought further clarifications from Einstein. What were the implications of this theory of time? Einstein's incremental responses to Besso's queries further intrigued Besso, as they have so many others in the years that have since gone by.

A Fundamental Misconception of the Nature of the Universe: Einstein was a serious and contemplative person, with a definite philosophic bent. This arresting observation which implied that Besso's earlier passing meant something different than appeared, that difference related to the illusion of time, was as enigmatic as it was fraught with potential for redefining the meaning of death. These thoughts implied that our concerned preoccupation with the consequences of the passage of time and its relationship to death, was, in some strange way, based on a fundamental misconception of the very nature of the universe.

Philosophic or Theological Implications beyond Pure Physics? The implications for many are that Einstein had grasped not only the 'light' and velocity invariance concepts of relativity but that his discovery of certain aspects of the nature of time somehow had implications beyond pure physics--they had crossed over into metaphysics; that he had grasped some deeper meaning in our reality; something, to him at least, apparently reassuring; something about time that influenced Einstein's concept of death in our universe as being somehow less menacing than it appears in our accustomed outlook.

Why Did His Discovery Somehow Give Him More Equanimity? He was pre-pared to accept his friend's death with more equanimity and serenity than might have been expected, given the deep friendship that had existed between the two. Somehow, this resulted from his newer conceptions of the true meaning of time in the universe.

Sadly, Besso's passing ended these dialogues with Besso and a more elaborate explanation of Einstein's thoughts on why the nature of time might reassure us in concerns over death, eludes us.

Time Doesn't Pass--Its Just There: According to relativity theory, time doesn't pass, it is just there, a dimension of the universe, part of the space-time geometry. Your consciousness of the present, moves along your world line, experiencing aspects of your life as you reach them. In this view, the events in the portion of your world line that is beyond the now point are already there and you simply 'happen' on them when you reach those points.

Under this concept, things don't 'happen' to you, but you happen to these events by encountering them, as in Dobson's "Alas, Time stays, we go." There are divergent views on this idea, of course, and to our subjective perception, it is time that seems to be passing. And this brings us to quantum subjectivity.

Time, Space, Life and Death--Part of Some Simulated Virtual Reality? The fact of having used this concept of the coexistence of all time in his consolation letter is an intriguing insight into his concepts of relativity and has been discussed by scientists as carrying the mystifying implication that death itself is an intrinsic part of this illusion of time. Scientists speak of time and space as mental constructs, part of the virtual reality constructed by our mental processes.

Somehow implied is that time, space, life, death and time are all an aspect of some virtual reality, a subjective mental construct, perhaps part of some computer simulation in which we share, as described by mathematician Frank Tipler in his *The Physics of Immortality*.

Quantum Theory Introduces Even More Startling Concepts of Time: Certainly Einstein's thoughts would bear on the central theme of existential philosophy, to say nothing of their theological implications. It is interesting to note that similar 'reassurances,' something 'less menacing,' also surface in the context of quantum theories introducing subjectivity and suggesting the existence of a nonmaterial interconnected universe of information-mind, of which our consciousness is part.

But quantum theory introduces startling concepts of the nature of time and what we perceive as a concrete physical universe, our so-called 'physical reality.' This physical reality, it turns out, involves subjective observer-dependent quantum uncertainty states, in which not only is the universe not so *physical* as we thought but it is perhaps *not alone* but existing along-side other, so-called *parallel universes*.

Parallel Universes and Time: Physicist Fred Allen Wolf, noted quantum physicist and author of such books as *Taking The Quantum Leap*, in his *Parallel Universes*, demonstrates how science fact is stranger than science fiction. The very existence of our universe may be may be *part of a complex of parallel universes*, even existing in our own space in some spooky way, he says, "perhaps only atomic dimensions away, but in a higher dimension of space" an extension into what physicists term *superspace*.

Parallel Universes Required By Absence Of Big Bang Observers: Elsewhere here we note that according to quantum theory, the universe we experience is one which resulted from the presence of observers; that everything in and about the universe was *resolved* into its present state by the acts of observation and that act of resolution then has the effect of ruling out all the theoretically possible alternative states.

Well, at the time of the Big Bang, quantum theory says that because there were then no observers, the universe was in some 'superposition of quantum possibilities' and so, says physicist Wolf, there *must have been parallel universes* then because until there occurred a quantum state selection or resolution by the act of observation, quantum uncertainty rules required the occurrence of *all possible scenarios*.

But, asks Wolf, what was the universe's initial radius, who measured it, and when was it measured? Well, he asserts, "It is our observations *now* that are determining the past," *including the Big Bang*!

New Notions of Time and the Human Mind: And, it is suggested, these concepts are producing *new notions of time and the functioning of the human mind*. Physicist Wolf goes on to discuss our decision-making process, suggesting that the mathematics of quantum physics demonstrates that *the future as well as the past enters our present time*, that time is not fixed, time travel is possible, such as through spacetime warps created by black-holes; that if parallel universes exist, as he believes, that time travel must be theoretically possible.

Time, Quantum Forces, and Death: Do time machines exist now, asks Wolf? According to Einstein's relativity they exist within black holes and possibly within electrons, says Wolf, so that our bodies, through electronic feedback, in effect, become time machines with simultaneous access to past, present, and future. So not only does relativity theory imply the coexistence of all time but quantum theory, as he says, also "implies the existence of both the past and the future 'simultaneously' with our own time."

So relativity theory and quantum theory converge to arrive at the same treatment of time. But this is not so surprising perhaps when current unification theory suggests that relativistic and gravitational effects are themselves traceable to an original quantum state when all the forces including gravity were unified in a single superforce.

The implications of these ideas, of course, impinge on our notions of the meaning of death, not only on the question of whether we really die, in the sense of a *total discontinuance*, but whether, *to start with*, we really *physically* live, at least in the 'cast-in-concrete' context in which we are accustomed to think of our so-called physical world.

Quantum Theory Raises Questions on the Nature of Death: Quantum theory seems to be moving us in the direction of wondering about the nature of our reality and whether we may be taking our 'physical' state too literally. We may be simply holographic constructs, projections of something which is essentially informational and mental, in which case, of course, not only the 'fact' of death but the 'fact' of life must be reevaluated.

In Paul Davies *The Mind of God* and Frank Tipler's *The Physics of Immortality*, the point is made that we do not know whether we are just part of some Computer Operator's simulated reality. Of course, that Computer Operator might be some natural manifestation of super intelligence--or it might be the supernatural God, assuming, of course that the God we think of as functioning outside the physical laws, supernatural, is indeed doing so. As we have seen, many miracles now have natural explanations, even those of Jesus, as astronomer-minister Hugh Ross points out..

Self-Awareness States in Quantum Terms: We tend to forget that we are, essentially, an aggregate of mysterious energy phenomena characterized by uncertainty and comprised of fields and fundamental forces so configured by nature as to produce the holistic states of consciousness and self-awareness we describe as 'lives.'

But in quantum terms, these self-awareness states may be viewed as simply minute individualized pieces of an interconnected cosmic energy mental state, temporarily detached and, through holistic biological processes infused with this self-awareness effect.

If, as some suggest, God can be considered as Cosmic Mind, then we may simply be fleeting, temporary self-awareness portions of Cosmic Mind, as Freeman Dyson suggests, extensions of the mind of God, destined to be rejoined to the parental cosmic mental source from which we originated. Interestingly, this is essentially, philosopher Schopenhauer's description of death, the reassimilation with cosmic unity.

More on Time and Quantum Theory: In *Beyond the Quantum*, Michael Talbot reviews aspects of the "new physics" and comments on implications of physicist David Bohm's observations of the universe as a "stupendous multidimensional hologram," in which time and space are simply projections of the deeper order in which past, present and future are enfolded.

As we recall in the discussion of relativity and time, Einstein's geometric spacetime universe of relativity theory envisions divisions of time unfolded as segments forged by the spacetime geometry, implying that all time, past, present and future, exists simultaneously. And so the concurrency of all time was part of Einstein's geometry of relativity, even before the new physics highlighted a reality of quantum connectedness.

If Time Is an Illusion, Life and Death Are Part of That Illusion: One sees several paths leading from this idea. One, the idea that time is just there and you simply experience events as you happen on them along the time line, is suggestive. *It implies not only that the passage of time is an illusion but that death, the consequence of the passage of time, must also be an illusion.* In fact, our very advent, birth, and all so-called 'subsequent' events in our development, including even the aging process itself, and finally death, seemingly exist as a function of our subjective perceptions.

Illusions as Part of Our Reality: And if our entire existence is a holographic quantum mental construct, then our death or that of a loved one is also a virtual reality. But if these events are not 'real' in some sense, then at least the illusion or our perceived reality is still something presumably arranged by God. And if our divinely-arranged perceptions create that 'reality' for us then, at least in those terms, death is perceived as real and 'virtually' devastating.

So Here We Are, Back to Dealing with Our Reality, Real or Illusory: This would seem to bring us back to coping with our actuality, real, virtual or illusory. So it would seem that we have little to fall back on, having followed the path of these illusions, except, for many, at least, what they started with, faith--faith in God's existence and in the meaning which our universe must have to warrant its real or illusory existence.

Other Thoughts on the Theology and Physics of 'Afterlife': Deep in the human psyche is a yearning to 'continue' life after our sojourn in this world. To humans, 'acceptance' of an afterlife of a 'spiritual body' may, however, be as elusive as belief in post-death family reunion. There are many who, through faith and religion alone, cannot find their way in these matters.

Insights on the Nature of the Universe: But perhaps there are not unreasonable explanations of the nature of the universe that give some insight, some support for belief-acceptance of afterlife. Einstein's special relativity theory equation, described as defining the universe, $E = MC^2$, tells us that everything in the universe is a single constituent, energy--everything that exists is some form of energy.

We have learned that the universe transitioned from radiation to matter to mind, to asking questions about death. We have seen how the original Big Bang particle radiation energy cooled to became elements of matter like hydrogen and helium, and through stellar fusion, to became carbon, oxygen, silicon, iron, etc.

And we have seen how these elements gravitated together to produce planets, then complicated molecular aggregates leading to life; and, in turn, how the process continued, generating mind and consciousness; and how these developments lead to abstractions and the search for answers to questions like why death?

Mind as Underlying the Universe: These phenomena of the creation suggest that underlying the physical universe and structured energy forms is mind, cosmic mind. Since mind may be said to represent life one may view the universe as a single energy-life force. Under the laws of energy conservation, energy, may change form but not be destroyed, a fact which may suggest approaches to belief-acceptance of afterlife.

Individual Consciousness as Part of the Underlying Cosmic Unity: In physicist David Bohm's view referred to above, of the implicate order of quantum interconnectedness, in a multidimensional universe in which "everything interpenetrates everything," *the brain particles are interconnected with the other particles in the universe* in which the apparent physical reality is a holographic projection of a deeper interconnected or unified order. And so the human consciousness is a part of the quantum cosmic consciousness.

This quantum particle interconnection is more fully treated in such works as Nick Herbert's *Faster Than Light, Superluminal Loopholes In Physics*, and Michael Talbot's *Beyond the Quantum*. As we have noted elsewhere, the lines between natural and supernatural are always being redrawn with the unfolding knowledge of the laws of physics and in particular, quantum theory.

Universal Consciousness Inherent in the Subquantum Order: Physicist David Bohm sees the universal consciousness inherent in the subquantum undivided interconnectedness, his implicate order, as providing a link to the individual consciousness which represents the holographic representation of the underlying interconnected cosmic consciousness.

Our individual consciousness derives from and continue as a part of the imperishable cosmic consciousness. In this context, the only thing that changes on death, is the discontinuance of the individual holographic projection represented by the physical body of the deceased. Earlier we spoke of consciousness as pure information and as an individualized projection from an underlying subquantum fabric of intelligence and information, something non-local but part of the cosmic interconnectedness.

The point is that we tend to judge the likelihood of life continuing by the limits imposed by the laws of physics that are presently known to us. But here we see evidence of other phenomena, not governed by those laws. Some take these phenomena to represent applications of other laws and to support the existence of a spiritual dimension.

Human Mind As Part of Mind of God; The Weirdness of Energy: Dominant western metaphysics posits the sacrosanctity of human life as a manifestation of God. Physicist Freeman Dyson, in *Infinite in All Directions*, perceives the human mind, which as we have seen, derives from energy, as an extension and actual part of the Mind of God. Dyson notes that energy is 'weird' enough for God to do with it anything he desires. God's laws of physics, conservation, tell us that we and our mentality are, in effect, a manifestation of growing order in the universe, mandated by the laws of physics, something eternal, and that can transition but not be destroyed.

Death As the Merging of Individual Life Force: These extrapolations have led many to assume that death represents simply the merging of our temporary life-force with the eternal energy-life force, cosmic mind. So even one who questions existence of soul or spirit, may accept the existence of something else, the mind as an individualized expression of a universal life-force.

Philosophers of the past, like Schopenhauer, without knowledge of physics, came to similar conclusions. So upon death, individual energy-life force reconverges with cosmic life-force-mind, the universal intelligence. It is this very thought that was so strikingly postulated by the Jesuit paleontologist Pierre Teilhard de Chardin.

So there are aspects of quantum physics which somehow bear on the metaphysics of theology. From the tiniest subatomic particle to the largest galaxy cluster formations, in every element or grouping of elements of matter, whether biological or other, there is something in common, an energy-mind component, part of a cosmic, inter-connected universal intelligence. It is in this context that the universe is sometimes referred to as a living organism.

This universal intelligence is manifested in different forms and stages of order and complexity: galactic, stellar, human, animal, plant, sometimes even human-created, such as computer-artificial-intelligence. In a sense, all are energy quantum states in various stages of unfolding.

> Life is real! Life is earnest!
> And the grave is not its goal;
> Dust thou art, to dust returneth,
> Was not spoken of the soul. --Henry Wadsworth Longfellow

CHAPTER 57
GRIEF AND THE ACCEPTANCE OF DEATH

The Diversity of Thought on Death: Down through the ages, views on the significance and meaning of death have varied widely. Some abhor and others welcome the event. Some even find humor in it. Mark Twain observed that "One of the proofs of immortality of the soul is that myriads have believed it--they also believed that the world was flat." "Immortality--a fate worse than death."-- said Edgar A. Shoaff.

"...No show, however good, could conceivably be good forever...I do not believe in immortality, and have no desire for it." Credited to H.L. Mencken." Anatole France: "The average human, who does not know what to do with his life, wants another one which will last forever. Rabelais: "I am going to seek a great perhaps; draw the curtain, the farce is played."

Some Classical Thoughts on Afterlife: But a universe of meaning has almost universally triggered thoughts of an afterlife. Said Tyrone Edwards: "This world is the land of the dying; the next is he land of the living." Abraham Lincoln urged: "Surely God would not have created such a being as human life...to exist only for a day! No, no, human life was made for immortality."

In classical literature, some accept death's final inevitability. Shelly wrote: "First our pleasures die--and then our hopes, and then our fears--and when these are dead, the debt is due, dust claims dust--and we die too."

Wrote Tennyson: "Most of us do not accept death as the end of everything. "God's finger touched him, and he slept."

Lucan wrote: "The Gods conceal from men the happiness of death that they may endure life." Hawthorne: "We sometimes congratulate ourselves at the moment of waking from a troubled dream--it may be so--the moment after death."

Sir Walter Scott: "This world is a dream within a dream; and as we grow older, each step is an awakening....Death the last sleep? No! it is the last and final awakening!" Schopenhauer: "Death is the extinguishment of the individualization representing each life, to be rejoined in the cosmic soul." Rom. 6:23: "The gift of God is eternal life."

The Study of Philosophy and Theology Prepares for Death: Some have said that the study of philosophy and theology is truly to prepare us for acceptance of death. The philosopher Schopenhauer asserted that *without death neither philosophy or theology would amount to much.* Most so-called philosophic issues are somehow ultimately subsumed in the question of the nature, meaning, and reason for death. Cicero stated that *a philosopher's entire life was preparation for death.*

Those who make practical provisions for their loved ones are likely to be more prepared for acceptance of death. By so doing, they *extend their love, caring, charity and compassion into the future*. Such provision includes arranging for such necessities as insurance, bequests, and memorials.

Attitudes toward Acceptance of One's Own Death: There are wide differences in human attitudes toward death in general, toward one's own death and toward the death of others. Death of a loved one brings mourning, grief, and confusion about meaning, matters we will be discussing. One's own death is another matter. The young tend to think of death as something that happens to others, to the aged, and their own old age is vague and remote.

But by middle age, most persons have encountered the realism of death and the pangs of grief in the death of a parent, or perhaps, a friend, associate, or fellow church member. As retirement looms and old age comes into sharper focus, however, notions of death emerge more frequently and thoughts occur about providing for future eventualities.

As Old Age Approaches: As old age approaches, death becomes more real and relevant and at the same time, most people become more philosophically accepting. But aging symptoms remind one of one's mortality. Retirement permits more time for introspection, about what it means to die, whether one exists no more or whether there is a 'continuation,' and what that might be.

Depending on one's religious orientation, one thinks of concepts of immortality, resurrection, salvation, reincarnation, heaven, spiritual re-embodiment. For the most part our thoughts are in the direction of avoiding death and minimizing our anxieties. We see around us those with philosophic or religious acceptance and those beset by anxieties and unwilling to come to terms with their mortality--and that of their loved ones.

Transcendence through Accepting Anxiety: Theologian Paul Tillich, like Soren Kierkegaard, renowned for his Christian existential insights, points out that human life, beset by anxiety over the inevitability of death, has the duty of accepting that anxiety. By doing so, one achieves tranquility and transcendance; and further, that human life's striving for the ultimate is, by definition, *religious activity*.

Interestingly, Einstein, whose theories of general and special relativity have transformed the notions of spacetime geometry, asserted that the astronomers, dedicating their lives to exploring the mysteries of the universe, were also, in a sense, engaged in religious activity, searching for God

Accepting or Even Welcoming Death: But there are other factors which remind us all of death and even make death less unwelcome. Intense and persistent pain and suffering, an inherent pessimistic nature, subjection to deep adversity and malaise, becoming lost in existential meaninglessness, undergoing of psychiatric difficulties, all these conditions can generate deep anxieties and predispose toward death as a release and relief from a life perceived as too burdensome, purposeless, or plagued with suffering or deprivations.

Some will even seek death for an ennobling cause--the death of heroes. Socrates, given the choice of exile or death, chose death. In death, he thought, either he would sleep the great sleep or commune with other philosophers who preceded him. Given the option of saving his life by leaving his beloved Athens--and his daily pursuit of philosophy, or death by the Hemlock drink, he chose death.

The Human Dies Alone--The Final Search for Meaning--and God: Regardless of one's personal outlook toward death, there is the almost universal last moment of uncertainty, the desperate last craving to grasp the meaning of everything before the thinking process ceases. And in that last moment of awareness, before surrender, one senses the utter aloneness of that final journey. This is so, even if blessed with the comforting presence of a loved one. In that lonely moment of intuition of the end, human life reaches out--for the hand of its Maker.

How Do We *Focus* on the Meaning of Death in Our universe: It is hardly possible for us to construct a coherent representation of the significance of death without viewing it in the context of *our total reality*--the universe, God, life and some conceptualization of the meaning of it all. As we become more preoccupied with mortality we recognize that the subject needs be considered in the broader framework of the nature of the universe; human life's place in the cosmos; the nature of God; and the very nature of mind and matter.

Coping with Grief--Acceptance of the Death of Loved Ones: But our concerns about death of our loved ones, our parents and mates, probably preoccupy us far more than thoughts of our own mortality. Perhaps the greatest pain we will suffer in life is the devastating trauma introduced in our lives by the death of our beloved. The need to find a way to cope with this experience is perhaps the most critical need we face in life and much has been written on the subject, as we will be considering.

In *When will I Stop Hurting*, Jean C. Kolf points out that although each death situation is unique, there are "common denominators" and those who grieve should know that they are not alone in either their feelings or experiences.

Grief Because We Were Fortunate Enough to Have Loved Someone: Death of a loved one is a psychological and physiological wound, sometimes more damaging than a physical wound. It is both difficult to deal with and 'impossible to understand.' Why does grief happen? Because, says Kolf, *we were fortunate enough to have loved someone.*

Would we exchange our grief for never having known and loved our lost loved one? Were the love and happy memories of the deceased worth the present pain? Would we have turned away from the love and caring relationship to have avoided the pain of the loss? Few would wish to have paid that price.

Stages of Grief--Multiple Emotional Reactions to Grief: Many speak of the various 'stages of grief,' the process of coming to terms with the change in circumstances following the loss of a loved one. Kolf speaks of the sequence as shock, sighing, crying, anger, and depression, and the varying degrees and kinds of pain one suffers.

Stages of Dying: People often go through a series of stages in accepting the reality of their own mortality. The death of a loved one, or diagnosis of one's own condition as terminal, first introduces shock, then denial, then anger at the world, at friends and family, and at a God who is seen as uncaring and without compassion. Anger may give way to depression, grief and resignation. The long habit of living indisposes us for death, noted Sir Thomas Browne. But a hard life and suffering may predispose for death which, as Shakespeare notes in Hamlet, may be a consummation devoutly to be wished.

How long does grief last? In an article, *Guidelines for Aiding the Bereaved Through Death*, Eloise Cole speaks of "acute grief" as lasting from one to three years. This period can be diminished, and healing promoted, by encouraging the bereaved to give expression to pent up feelings, which if repressed can lead to emotional breakdown, or worse.

Anger at God, the Sense of Abandonment: God promises, Psalms 34:19, that he will never leave us during our times of trouble. Sometimes grief-caused anger is directed at God. God has forsaken us when we needed him most, is the reaction of many upon the passing of a loved one. Anger at God is a common response to the death of a loved one. If God is omniscient and omnipotent, as religion asserts, He need not have permitted this tragedy. Where were His love and compassion, is a common question, especially if the death comes early in life or from unexpected causes.

Loss Often Generates Guilt: There are few indeed who have not experienced guilt of some sort following the death of a loved one. Why didn't I take better care? Or make him/her stop smoking? Or overeating? Why was I not more attentive to his/her concerns? Why was I not more tolerant of his/ her frailties? The torment goes on--self-punishment and mental torture over that which can't be altered.

Unforgiveness: The passing of a loved one often focusses on some act or failure by someone close to the deceased whose conduct somehow diminished the meaningfulness of the life of the deceased--a son or daughter, or brother or sister, too preoccupied to call. Grief over your loved one's passing is often translated into unforgiveness of derelict conduct by someone who was or should have been close to the deceased and somehow failed to provide the critical support. Unforgiveness damages our lives and the grudge-bearer often "hurts the bearer more than the recipient," says Kolf.

Resentment at the Deceased: Grief leads, at times, not only to anger at God but even to "negative feelings" about the deceased. Why did he leave me, unprotected, to fend against life's adversities all alone? What happened to the departed mate's marriage vows to love and protect "until the sun shall cease to shine?" Just *eating in a restaurant alone* is often a dreary reminder of the terrible loss, the realization that a vital necessity, *companionship,* has been lost. One's very identity comes into question.

The sense of aloneness may be overpowering. So much so that ideas of escape through suicide emerge. Psychiatric or psychological counseling, at this point, may become critical to survival. Friends, who serve a crucial role at such times, should be involved, their availability a reassuring element in the life of the bereaved.

Grief, Memories, and Coping: The first year after the passing is the most difficult, of course, although grief can last a lifetime. Often the persistence of grief is tied to the treatment of memories, reminders of the loved one who is gone.

There are coping mechanisms, though, which are often helpful if not allowed to become an obsession: Retaining selected pictures or articles dear to the departed; the giving of flowers for an alter; memorial plantings that will flourish, flower, or bear fruit; the playing of favorite tunes; the retelling of favorite anecdotes. Even companionship with pets are consoling and a demonstrated therapy.

Sometimes a husband or wife survivor loses those caring friend relationships which were, it turns out, dependent on the couple status. It's a "world of twos," we sometimes learn the hard way. Often, even genuine, longstanding friendships between couples fail to survive the loss of one of the foursome. Survivor relationships may decline, eventually to slip away.

We Can Influence What Kind of People We Become: In *When Bad Things Happen To Good People*, Harold Kushner discusses how the pain of loss can be made "meaningful" rather than "pointless empty suffering."

While we may not understand why we must suffer, and can't control the events which produce it, we can have some influence on "what sort of people we become" as a result. Pain can, for example, make us "sensitive and compassionate," rather than "bitter and envious." We can turn grief into something "meaningful" and constructive, rather than "empty and destructive."

Holistic Effects of Mood Elevation--Laughter: Holistic medicine points out that directing thought to modify moods and outlook can produce therapeutic, health-promoting effects on the body. Even laughing has a place in grief. People feel guilty if made to laugh during the grief period, for instance, upon recounting of some humorous experiences in life with the deceased.

But laughing and mood-elevating behavior produces hormonal reactions which improve the body's immune defenses. Laughter releases health-promoting endorphins into the bloodstream which vitalize the immune system. Grief and depression have the reverse effect.

Aids in the Grief Process: Writers on death and grief describe various aids to dealing with grief. In *When will I Stop Hurting*, Jean C. Kolf cites four: Facing the reality of the loss (letting go); experiencing the pain of grief (crying, expressing anger, frustration, resentment, and talking it out); adjusting to the altered environment (acceptance); and reinvesting emotional energy elsewhere (reaching out and getting involved).

Staying busy, traveling, learning new skills, furthering one's education, joining clubs, church activities, pursuing social outlets, all contribute to getting through the grief process. Don't make shrines of a favorite location or object of the departed unless it has the therapeutic potential, such as fond memories.

Unprepared for the Final Parting: *On Being Alone, AARP Guide For Widowed Persons* cites the "Til death do us part," marriage vow. But, it asks, "who among us is ever prepared for that final parting?" Most of life's challenges, it notes, allow time to prepare: Parenthood has a nine-month pregnancy, marriage has its courtship and engagement; divorce has its buildup and separation. While death may be slow in some cases, often it is sudden and overpowering; widowhood is a role almost never anticipated.

Couples just don't go around making 'dry runs' and rehearsing 'what if' scenarios. Death, for most couples, is an eventuality hardly prepared for. Nevertheless, for most, the grief and mourning process which follows does have an end, incomplete though it may be.

Grief as Healthy and Appropriate: For most, marriage is their biggest emotional investment. Lose your spouse or close family member and you lose something of yourself. The grief, pain, and suffering you go through is unavoidable. Grief and pain have even been described as "healthy and appropriate," not a very sensitive description of things but nevertheless true in functional terms.

On Being Alone points out that while your spouse has passed on, "the relationship lives." Some weep and publicly grieve, others hold it in, giving the appearance of composure. But holding back, stifling the grief, prolongs and makes more debilitating the cleansing process. Venting grief through emotional release, on the other hand, shortens the process.

Imagining Your Loved One Is in Your Presence: Because men tend to hold back displays of emotion, they sometimes suffer more prolonged grief. Expressing emotions is part of nature's healing process. Stages of grief may not follow any prescribed order. But shock, numbness, and disbelief are normally the first reactions. These give way to feelings of separation, emptiness, depression, yearning, pining, longing for 'restoration,' even dreaming or imagining the presence of the loved one.

John Erskine's Experience: Writer John Erskine felt, by his own personal experience, that "the dead do not leave us, but in some fashion continue here as faithful companions, sustaining and inspiring us. We find them in familiar places, in the home, in the garden, on the street. This constant resurrection of the dead is for me a simple fact, part of my human acquaintance with the daily mystery and beauty of life."

Despair: For most, when reality sets in and you realize that your mate's passing is not a dream from which you will awake, a devastating and demoralizing desolation, despair, and fatalistic resignation may set in, leading to genuine physical or psychosomatic illnesses, and in extreme cases, even death. We all know of cases in which survivors have lasted only a few months, or even weeks, after loss of their mates. But for most there is a natural, if gradual healing process.

Grief Diminishes but Doesn't End: While grief has no real ending it does diminish to a point of acceptability or at least *endurability*. Part of you will mourn forever, your dedication to the loved one. But intensity gradually diminishes and "acceptance does mean you can cherish your memories and at the same time, understand one has to go on alone and try to make the best of it."

Protecting Yourself from Yourself: Grief impinges on rational and objective thinking. Impulses to initiate various actions that might mitigate aloneness, often surface. Some may be constructive, others dubious. The AARP *On Being Alone* lists some *temptations*: Quitting your job; moving in with family; moving your residence.

Any or all of these steps may ultimately be desirable but the caveat is that such urges for immediate actions be resisted in favor of *steps that are not irreversible*, such as: taking a leave of absence; renting your residence temporarily; visiting with family for a short stay before deciding to move in; taking a long vacation away before pulling up stakes. Hasty actions are often regretted, especially if irreversible.

Grief and Friendship: The role of family and friends is critical in dealing with grief. Religion and philosophy extol the virtues of friendship as the Gift of God. Sometimes the value of life and the meaning of existence narrows down to friendship, both within and outside the family. Perhaps the greatest loss of a mate is the loss of friendship at the heart of the marriage relationship. More often than not, the mate is the confidant, the listener, the soother and comforter, the refuge in troubled times. Having all that taken away, in one fell swoop, represents the disaster of a lifetime.

For the bereaved, friends are looked to for comfort and reassurance. It is the listening and caring friend that permits the expression of grief, the cleansing and catharsis of pent up emotions, the communication of the agonies which aloneness keeps in repression.

Your Friend Has Lost a Loved One--What Do You Say and Do? Your friend is confused and grief-stricken over loss of a mate and you just don't know what to do or say. In Vogue of March, 1996, a young lady, Crystal Gromer, writes of her own experience on loss of her husband. The truth is, she says, that there is nothing anyone can do, to "make the deceased stop being dead."

An Etiquette of Grief: Even religious traditions, she notes, do not help with the social questions-- whether to call, what to say, or avoid saying. But she feels, there *is* an "etiquette of grief" we can acquire, more "a way of being" than a set of instructions. And this requires allowing a place for sadness. We tend, she notes, to treat grief as an illness, to be treated away in a doctor's office, with some medicated balm. The quick-cure urge is hard to suppress, she says.

American Optimism About a Quick Cure is Bad Manners: None of us wants to see friends suffer, and we want to help. But, she comments, American optimism about a quick palliative in the grief situation is simply bad manners. And so are the clichés that seek to diminish grief by citing a possible worse scenario, like "At least he didn't suffer," adding, "Time heals all wounds." These truisms and platitudes, she feels, don't make the real and current dilemma any better, they just trivialize it.

Because she was young, some grief ameliorating observations were offered her: "At least you're young," and "At least you have your life ahead of you." This language of attempted comfort may be appropriate if confronted with a bad haircut or lost job but grief from death of a loved one is something apart. We need to acknowledge the importance of that and accept its presence-- "not try to make it disappear."

Grief Behavior Involves Doing As Well As Not Doing: Tone is also important. We need to avoid the disrespect of being over-cheerful on the one hand, or unduly "reverential," softening the voice to show sympathy and understanding.

An etiquette for grief, she observes, should tell us more than what not to do. It should tell us what to do as well. We can extend ourselves. We should attend the funeral or memorial service, regardless of our reluctance, in order to swell the ranks of the mourning community. The church or other place of respect, she says, should be full of human faces and human sounds. Though an individual presence may not be noticed, all in attendance 'make' the ceremony. Your presence is a part of a life and contributes to the memorial.

Write Down Your Recollections of the Deceased: Never, simply sign a greeting card, Gromer suggests. She feels that use of phrases like "Words cannot express" do not contribute anything useful since words are what we have to work with in giving written expression to our bereavement thoughts. She suggests writing down one's recollection of the deceased, as clearly as you can. This is what you can give, she says, and it is a gift, another picture.

Never Too Late for Expressing Thoughts in Grief Situations: Don't seek excuses for not expressing yourself on grief situations. It is never too late to do this, she insists, so don't allow yourself the excuse of the passage of a month or two. Often, she thinks, we refrain from saying something truly meaningful for fear it may hurt. But, pain is not the worst thing in the world. Neither is embarrassment, and that should be the lesson for those of us who would like to help. Both evasion and forgetting are worse.

Harder to Deal With Grief of an Acquaintance: So we can lend our presence at the funeral, we can lend our memories afterward , and we can continue lending ourselves and our services. Despite the difficulty in expressing thoughts to those close to us it may be harder, she says, to know what to say to an acquaintance who is grieving. Sometimes the fear of erroneous action immobilizes us. But, she says, don't scoot down nearby aisle of the drug store or walk out another door to avoid someone. You've no doubt been spotted, she says, and the memory of that "failed moment," however fleeting and unintentional, endures--on both sides.

Vulnerability and Embarrassment in Giving Solace--Deeply-Human Activity: You don't have to begin a conversation with "I'm so sorry." She suggests: "I' m so glad to see you" or I'm so glad to see you out" or "I'm so glad to see you back." The giving and receiving of solace, the very vulnerability of opening ourselves to it, despite embarrassment, despite potential rejection, seems to her the most deeply human of activities. She feels that these lessons of behavior in the face of grief situations are learned, rather than instinctive, and difficult rather than easy. An etiquette for grief, she says, should "help us to live with what is hard as well as to share what is heartfelt."

Aloofness from Friendship with God: Aloofness from God during one's lifetime has a terrible price. Perhaps the ultimate in friendship comes to one who, through life has been close to God and came to see God as Friend. People who find comfort in a deep soliloquy with themselves sometimes have the feeling of talking with God, a vital channel of communication. But others don't find God within themselves and seek God or substitutes for God, through their friendships. They may not be able to be in touch with God except through the God perceived as manifested in others.

Planning for the Worst: How can the trauma of death be diminished through the preparation offered by anticipation? One way is to face the inevitability of death and plan for it. Too morbid you say? Perhaps, but it is evidence of the depth of love between mates when they face this inevitable reality and act to minimize or diminish trauma through some reasonable planning. Get up wills, living wills, powers of attorney, and durable powers of attorney. Make sure property is held in the form that will best ease the transition from the married status to widowhood.

What If? Facing up to this 'what if' ultimate eventuality requires exploring the likely scenarios. Consider: where the survivor will live and whether a move will be necessary or desirable; what the survivor income will permit; the proximity to children or other family members; what support group will be available; what church or other organizational affiliations should be initiated or nurtured.

The Downside to Planning for the Eventuality of Death: Unfortunately, this "planning" process may have its down side. It may subconsciously influence the outlook and pattern of living; it may influence behavior in unpredictable ways. Questions about possible remarriage of a mate, for example, may be highly sensitizing and many will wish to shy away from it. Most people find it difficult or impossible to consider remarriage of their mates or, for that matter, their own remarriage. Human frailties creep into these scenarios to generate tensions.

Concern About Dying Rather Than Death: How do we come to terms with the reality of death and to accept its inevitability? The highest concern about death, for many, is that of the death process itself rather than the actual event.

Experience tends to show that the preoccupying fear of death is more often for the loved ones than for self. As for one's own life, when life is in its youth or prime--beautiful, fulfilling, thrilling and vibrant, the very thought of death--discontinuance, will be troubling. But a life dominated by pain, suffering and deprivation leads to acceptance of, or even welcoming of death as the ultimate serenity.

The Composition of Grief: Understanding the nature and composition of grief often helps one to find one's way along the difficult road back. Grief differs in intensity for each person according to his makeup. The sense of *aloneness and loss* is the most immediate and the most profound. For many, whose own lives have been tightly interwoven with the vitality of their beloved departed, the loss may overwhelm them and introduce them to the existential abyss of meaninglessness. The value of life--their own--may be lost or diminished to desperation.

But a related concept may be a source of some comfort, that the *life of the departed is now somehow merged in their own* and from this may eventually emerge a new sense of purpose--that their own life must now do for both of them and a sense of duty to optimize that melded life.

Grief Forces the Search for Meaning: Grief immediately challenges one's assumption of a universe with meaning and *launches one on a search for the meaning* of life, often with an impending sense of the futility of the endeavor. The restoration of faith in a universe of meaning becomes crucial to the survivor.

The Uncaring God? As we noted earlier, the loss may quickly translate into anger and resentment, especially of a meaningless universe and an irrelevant or uncaring God. He left us to mourn. He left us to be alone in an empty world. By what twisted logic, one asks, does God rationalize the pain and suffering, grief and despair, thrust upon us? Infuriating, is God's inexplicable answer to Job, out of the whirlwind, that it is not our place to question the wisdom of God's actions.

Where is the compassionate God? Where is the God of love and justice? We can not accept God's explanation to Job--and we repeat and repeat the question--why? Why did God lead me down the primrose path into this great love only to tear it from me?

Grief As Measure of Selfishness: But in psychotherapy, we learn, sometimes to our surprise and perhaps dismay, that much of our sorrow and resentment on the death of a loved one comes in no small measure from self-centered causes; that they don't entirely represent concern for the *departed* as much as they represent anger for loss to the survivor.

The departed, after all, no longer suffer, they are free from pain and illness, protected from harm and accident, no longer bear the burden and indignities of illness or old age, or of earthly responsibilities and problems. They are 'home,' with the beauty of their souls and spirits forever enshrined and sheltered.

Death, Like Life, Must Have Its Purpose: But if we believe in God deep down, and believe that life has a purpose, then we must believe, as difficult as that may seem, that death too has some unfathomable but salutary purpose. Here faith meets its most difficult test--when our demand for an explanation gets Job's frustrating admonition from God not to question God's wisdom.

Yet we know--have always known in our depths--that the price of life is death--having to leave this life; that the price of having love, is having to lose it, in one's own death or by the death of the beloved. It is said that reconciliation with your own mortality helps, in subtle ways, to accept that of your loved ones. Often, however, death of a loved one intensifies the existential anxiety of the survivor.

Love, However Reduced in Time Is a Fact for All Eternity: Death involves the mysteries of the dimension of time and one needs to grasp the profound significance of the time that is shared in love. The love between two people, of even a single day, or a single moment, is an important fact, for all eternity. A lifetime of a hundred years is of little more significance in cosmic time than one of a hundred days or minutes. The fact that two people were born, lived and loved, is a cosmic event for all eternity, whether it be for a human lifetime or for an hour.

Would You Have Your Loved One Grieve for You? It has been said that you have not fully loved another unless the other dies first so that you may have the experience of loving the departed through all the agony of mourning--and across the great void; that this is the ultimate love that can be experienced. And would you wish that the departed had survived--to mourn your own passing? Or would you, in your love for the departed, accept the passing before you so that the departed would not have experienced the torture of grief for you?

Kant's Assurance That Moral Law Requires Immortality of the Soul: The German philosopher Immanuel Kant, considered by many to be the greatest philosophic thinker of all time, and author of the famous "categorical imperative," said, in his *Critique of Pure Reason* that "It is plain that the hope of a future life arises from the feeling, which exists in the breast of every man, that the temporal is inadequate to meet and satisfy the demands of his nature."

Kant also expressed the thought that it is Reason that gives humans the idea of soul as the culmination of all mental processes. Humans cannot become absolutely good during their mortal life. Thus the immortality of the soul is essential so that demands of moral law may be met. In eternity the human soul goes on to perfection and a fulfillment of the demands of moral law. As we have observed earlier, these are the central elements of the philosophy inherent in the rebirths of the Eastern religions.

Socrates Classic Philosophy on Why Not to Fear Death: In Plato's *Apology*, Socrates says that there is a migration of the soul from this world to another. That if death be in the nature of sleep that to die is gain, for eternity is then only a single night. But if death is the journey to another place where all of the dead abide, what can be of greater good than this? What would not a human being give to converse with the great philosophers? For this would continue the search for true knowledge. For besides being happier, the human would then be immortal. Therefore, be of good cheer about death and know of a certainty, that no evil can happen to a good person either in life or after death.

Not Afraid of Coming to an End: Seneca, in his Letters to Lucilius, said: "What is death?...either a transition or an end. I am not afraid of coming to an end. This being the same as never having begun, nor of transition, for I shall never be in confinement quite so cramped anywhere else as I am here."

The Fundamental Problem of Life Is Coping with Death: The Spanish poet, novelist and philosopher Miguel De Unamuno points out that the fundamental problem of life is the necessity of coping with the idea of death; that each human life lives in the agony of conflict between the Will's need for a life after death, and the Reason's denial of life after death. Therefore, if our existence is to be meaningful, we must accept this frustration and, in spite of our awareness of death, drive for action from the inner core of our existence in order to provide balance for our dependence on impersonal and unpromising reason.

The Value of the Thoughts of Great Minds of the Past: These are a few random thoughts of some of the great minds of the past who have dwelt upon the meaning of our beginning--and our end. They do not give us all the answers and perhaps they raise more questions than they answer. They rationalize, they reason, and of course, they speculate. Perhaps there are no answers--or perhaps in the final accounting, the answers are different for every human life.

But these thoughts may reassure us that our own uncertainties have been matched by the uncertainties--and insights--of wise men who have gone before us. The collective wisdom of all of the great philosophers and theologians must add up to something, to some worthwhile appraisal of the meaning of life in the universe. At least we know that others have trod the paths of sadness and grief, speculation and wonderment, and of faith and belief.

Listen to the Counsel of Your Deceased Loved One: Had roles been reversed, you, the survivor, would have wanted your deceased to be counseled; to be strong in the anxiety and grief over your passing; to have faith that all was not futility; to know that there must be some grand purpose--and some "continuation."

Accept this counseling as it were your loved one's to you, and be strong to make your life as meaningful as you would have insisted of the departed. There must be--there is--some great purpose in life, in our beginning and in our end. Pursue it to the end, with dignity, and determination, to make your life meaningful as you seek answers, for yourself, to these age-old questions.

Out of the Night of Suffering: Out of the night of suffering comes the dawn of serenity and out of the crucible of despair comes meaning--and finally acceptance. Nothing is safe in life until rescued into the past, and so the beauty of the life of the departed, intertwined with the survivor's, is now eternal.

Death may be but a transition to another dimension. We learn in physics that nothing is destroyed, it only changes form. This law of conservation prevails throughout the universe. It would seem unlikely that God, having created and nurtured a beautiful soul, would save bits of energy but fail to apply his law of conservation to something so precious as souls.

Accept the Nostalgic Love: Perhaps this vale of joy and tears, into which we have been existentially thrust, is simply our preparation for the eternal, our gateway to eternal life. The loved one who passes before us is spared the grief of our own passing. It is said that no love can be complete until nourished by the tender love of mourning and the nostalgic love transmitted across the great void.

In psychotherapy, we learn, sometimes to our surprise and perhaps dismay, that much of our sorrow and resentment on the death of a loved one comes in no small measure from self-centered causes; that our resentments do not entirely represent concern for the *departed* as much as they represent anger for loss to the survivor. --Author

PART XV--PREDETERMINATION AND FREE WILL

RANDOM THOUGHTS ON DETERMINISM AND FREE WILL

Everything that happens in the world is part of a great plan of God running through all time. --H.W. Beecher

Either all is chance, and being but chance is of no consequence, or God rules the world, and all is well.--Andrew Lang

There is no such thing as chance; and what seems to us the merest accident springs from the deepest source of destiny. --Schiller

Chance is but the pseudonym of God for those particular cases which he does not choose to subscribe openly with his own sign. --Coleridge

Life is like a game of cards. The hand that is dealt you represents determinism; the way you play it is free will. --Jawaharlal Nehru

Chance is a word void of sense; nothing can exist without a cause. --Voltaire

All chance [is] direction which thou canst not see.-- Pope

That God has foreordained everything is self-evident.-- René Descartes

Creative human activity...one aspect of human freedom...is unpredictable in the same way [as] biological evolution. We are...as Socrates and Democritus believed...free because we are, [essentially] the authors of our own lives [and] the openness of the future and of our own ability to help shape it reflects a deep property of objective reality.--David Layzer

CHAPTER 58
THE PROBLEM OF FREE WILL

Meaning and Fundamental Questions: Down through the centuries, philosophers, whether of primarily scientific or of theological background, have sought answers to what they perceive as the most 'fundamental' questions in the effort to find meaning in human existence.

While there are some differences in point of view, so-called fundamental questions can be said to relate to the following subjects: *God*--the existence and nature of God; *Creation*--the source and reason for the creation of the universe, the existence of something rather than nothing; *Death*--whether life continues in some form after death; *Free will*--Does human life have free will, true choices--raising the question of whether the universe is governed by chance or is predestined from the creation.

Free Will and the Advent of Quantum Theory: Some suggest that all the ques-tions are subsumed in the first--the existence and nature of God. But the age-old question of free will, so important to meaning, took on spirited renewal following the advent of 20th-century quantum or subatomic theory with its underlying uncertainty principle. This was seen as demonstrating the underlying role of chance in the workings of the universe, thus altering the earlier perception of a predetermined, cause-and-effect universe of Newtonian classical physics.

The Great Debate over Free Will--All Time Laid out at the Creation: From earliest times, philosophers have argued the question of the existence of human free will, whether it is real or illusion. Is the universe we inhabit deterministic, predetermined at the creation, or is it indeterministic--governed by chance? Does the uncertainty principle of quantum mechanics signify a chance universe?

Fifth-century St. Augustine concluded that time came into being with the Creation which laid out all events in time, thus predetermining all future events. This remarkable conclusion about time and the creation happens to comport with classical mechanics, including even relativity theory in which all time is said to *coexist*. In the context of his treatment of time as the fourth dimension, Einstein observed that distinction between past, present and future was an illusion.

So St. Augustine's conclusion that everything was predetermined at the Creation was seemingly confirmed by Newton's cause-and-effect classical mechanics, and was the view held by many, including Einstein, despite the advent of quantum uncertainty. As we shall see, there is disagreement even to this day about predestination and the existence of free will.

Written in the Stars? Despite quantum uncertainty which was thought to introduce chance into the course of events, many individuals still believe that everything is essentially predetermined, 'written in the stars,' that free will is an illusion. Many if not most persons, however, likely feel that apart from certain limits imposed by heredity or society they are actually free to make choices, possess free will, and are therefore generally accountable for their actions.

This view serves the salutory purpose of helping to reinforce dedication to morality, principle, and observance of biblical injunctions. But is this view of possessing free will scientifically supportable? What do physicists say about this? Despite the advances in physics and their philosophic implications, we still find wide variations in the conclusions drawn by scientists and others, which we will explore.

Hard and Soft Determinism: Pragmatist William James speaks of hard determinists--everything caused by external forces, and soft determinists-- human life not being compelled but nevertheless governed and conditioned by environment and heredity. Psychology points out that powerful subconscious forces dictate much of behavior. A recent report in Nature Genetics, for example, identifies a gene affecting human personality traits, accounting for behavior called 'novelty-seeking,' that is, behavior that is extroverted, impulsive, quick-tempered, excitable, and adventurous.

Endocrinologists point out behavior changes when glandular irregularities are corrected. In one case, a deeply religious person, preoccupied with God, religion, and experiencing religious visions, lost interest in religion when a gland deficiency was corrected. So, it is thought, free will has its built-in influences.

Between Determinism and Free Will: The strict 'determinist' view holds that everything is causally accounted for; the 'indeterminist,' that events are not cause-and-effect but happen as the result of chance. Physicists point out, however, that chance and chaos are themselves causal instruments producing *directionality* toward certain (preferred) results.

But between the opposing extremes is the view that while heredity and environment are predetermining or causal, that the self in each individual possesses a power to decide against the influences of this preconditioning.

Is The Question of Free Will a Pseudo Issue? British philosopher Gilbert Ryle suggests the question of free will as being a *pseudo* question, that there is no such thing as 'will' to be subject to being 'free'. He suggests that there is no way of validating the existence of some "occult volitional instrument" which can only be inferred from observed behavior. Our actions are, after all, the consequence of the concatenation of a complex amalgam of interrelated forces, emotions, physical consequences of biological mechanisms of the body, all of which influence behavior.

Free Will As a Limitless Multitude of Conditions: One's action may be dictated by any or all of a limitless multitude of conditions: one's state of digestion, lack of sleep, stress, anxiety, glandular imbalance, daily successes or failures, love life, addictive behavior, attitude toward Deity, or some deeper primordial cosmic influences. To equate this vast array of influences to the 'exercise of will,' the argument goes, is too simplistic a step and therefore inherently inaccurate.

Meaning and Free Will--Why Is the Question of Free Will Important? Why is the question of the genuineness of free will considered by so many to be of such importance? Because free will addresses much of what we consider as related to the *meaning* of our existence. Concepts of morality, principle, normalcy, ethics, for example, have little meaning if we do not have true choices. If we're essentially automatons, preprogrammed at the creation, why should we go through the elaborate decision-making processes we do, and hold ourselves responsible for our moral decisions, if we can change nothing--the outcomes having been predetermined.

As noted, the predeterminists, like St. Augustine, argue that all time was laid out at the creation and all events then predetermined; that the starting universe bore the imprint of *free will choices destined to have been made* by men throughout their stay here. Many find difficulty in accepting the proposition that primordial-predetermined choices represent contemporary free will.

Free Will and Society's System of Justice: Our system of justice assumes that we do have free will; many crimes requires actual 'criminal intent.' Accidental homicide, not having been 'intended'--not freely chosen--is not the heinous crime of premeditated murder, and accordingly, is less punished. Evidence of a person's 'givens'--his inherited nature and perhaps dehumanizing environment-- lack of education and moral training, poverty, starvation, drug infestation-- all these may be considered in arriving at punishment, though not in the decision of guilt or innocence.

Rehabilitation Rather Than Punishment: The outlook that criminal sentencing should have rehabilitation as its purpose, rather than punishment, is often traceable to feelings that offenders may not have all that much really free will. Determinism suggests that rehabilitation rather than punishment should be society's goal; that punishment should be left to divine justice and society concern itself more with behavior-influencing aspects which fit under the general heading of 'rehabilitation.'

Society Does Not Rely on the Divine Justice System: But the social contract envisioned by French philosopher Jean-Jacques Rousseau does not rely on divine punishment. Society's need for motivation in crime deterrence, and a practical system of bringing miscreants to justice, requires a separate punishment by society here in the physical world. The early church was often protective of fleeing criminals, placing more credence on divine rather than society's judgments.

Kant's Three Metaphysical Problems: As noted earlier, in his deliberations on meaning and the nature of the universe, 18th century philosopher-mathematician Immanuel Kant, posed three 'fundamental metaphysical problems': Does God exist? Is there life after death? And, *do we have free will?*

Because the answers to these questions were viewed as 'metaphysical,' unverifiable, Kant concluded that we can only speculate about such things. Actual reality--what is actually 'out there,' declared Kant, is forever constrained by the limitations and structure of our thinking mechanisms. Einstein, on the other hand, thought our mental processes up to the task of determining reality.

Frank Tipler Addresses Kant's Questions: In *The Physics of Immortality,* Tulane University mathematician Frank J. Tipler uses his *Omega Point Theory* to provide specially-defined 'yes' answers to all of Kant's three questions, offering 'mathematical proof of the physics involved.

Because of advances in physics, Tipler regards Kant's questions as now amenable to *scientific* conclusions rather than just metaphysical speculation. He distinguishes, for example, between relativistic, quantum and cosmological theory versions of free will, which we review later. Tipler points out that the history of science has typically turned insoluble metaphysical problems into problems of physics and "then solved them."

Quantum Theory and the Philosophy of Determinism: In *A Brief History Of Time*, Stephen Hawking wonders why quantum theory with its uncertainty principle has not had a greater influence on mainstream philosophy in the direction favoring indeterminism and the existence of free will. Actually, quantum theory has had a strong influence and the notion of total predetermination has now been largely abandoned.

One recalls the determinism of French mathematician and astronomer, the Marquis de Laplace. Like Isaac Newton, he concluded that all, including human choice, was cause and effect, predetermined by the laws of physics, leaving no room whatever for chance. Consequently, he thought, everything was also theoretically (though not practically) predictable.

Since the advent of quantum uncertainty theory, however, many, if not most, philosophers and physicists have renewed their belief in the existence of significant indeterminism, and consequently the existence of meaningful free will.

Quantum Theory, Relativity And Determinism: Of course, there have been and perhaps still are, physicists who question the role of quantum theory in rescuing us from a predetermined universe. Before quantum theory, Einstein had presented his deterministic theories of relativity and time which say that all time was laid out at the creation by the geometry of spacetime. For many, the fact of the *coexistence of all time* does imply that all events are predetermined.

Einstein's early view of the then-developing quantum theory was that the uncertainty principle applying at the subatomic or quantum level is an 'enabling mechanism' of some sort that concealed an underlying determinism. We know, for example, that there is statistical predictability in quantum effects, and the macro universe of galaxies and people do indeed follow classical cause-and-effect principles despite the working of the underlying quantum mechanics.

In his debates with Danish quantum physicist Niels Bohr, Einstein never totally gave up on the causally-determined universe. He was unable fully to accept quantum uncertainty--that God would, as he termed it, "play dice" with the universe. As it turned out, Einstein's intuition, so right about relativity, spacetime geometry, and mass-energy equivalency, appears to have been misdirected in questioning God's resort to chance as inherent in quantum uncertainty. Bohr, of course, finally urged Einstein to stop telling God what to do.

Do Relativity and Quantum Theory Conflict on Determinism? So classical physics, including relativity theory, suggesting a deterministic universe, and quantum theory, involving probabilities inherent in the uncertainty principle, give a first impression of being in fundamental conflict. In *God and the New Physics*, physicist Paul Davies takes note of the argument that classical relativity conflicts with quantum theory on determinism and free will.

But Davies concludes that any perceived conflict is illusory. Although relativity tells us that the universe is extended in time as well as space, it does not tell us whether there are "rigid links of cause and effect." Relativity, he says, simply forbids us to divide spacetime into "universal instants of time."

Determinism As Not Fatalism: Davies explores degrees of determinism and freedom and in the process, addresses the question of how can you punish someone for spontaneous acts without a prior cause? He says we must not confuse 'determinism' with 'fatalism' which asserts that future events are entirely beyond our control--what will be will be.

Science or Theology--Reality or Illusion: It is asserted by many that what is the nature and origin of the universe is ultimately a theological question; that energy, matter, and the laws of physics represent simply the physical mechanism for divine idea.

From the beginning, philosophy asks about our 'reality,' the nature of our universe, the character of physical phenomena as being 'real' or just the product of our mental constructs. Einstein, master of dimensionality and spacetime geometry, mysteriously referred to the distinctions we perceive between past, present and future as illusion. In a somewhat different vein, physicists speak of our perceptions of time and space as mental constructs, a *virtual reality*.

Free Will and Classic Causal Physics: The Newton-Laplace measurable universe of classical physics, buttressed by the spacetime geometry of Einstein's theories of relativity, with all time laid out, seemingly validated earlier assumptions of an essentially predetermined universe. Then, as we have noted, came quantum mechanics, Heisenberg's uncertainty principle and Bohr's 'Copenhagen interpretation.' These supposedly posited a universe of quantum uncertainty, in some measure implying an indeterministic (chance) universe, with the resulting philosophical and theological implications.

Early Philosophy and Theology Assumed a Predetermined Universe: As far back as the Greek philosophers and early Christianity the idea was expressed that the *creation* of the universe established everything, including all time and the entire causal chain of happenings inherent in time. Even before St. Augustine, the Greek Pythagoras and Heraclitus also viewed the universe as laying out human destiny and fate.

The Sophists, Socrates, Plato, Aristotle, Zeno and the Stoics: This causal chain was the view of early Greek philosophy except for the sophists who saw in human life possibilities beyond an inescapable process of fate. Human life, they thought, could influence and be the measure of its destiny. And this, according to Plato, was also the view of Socrates in his relentless pursuit of knowledge as influencing destiny.

Plato, the great chronicler of Socrates, shared this view of human freedom, unfortunately, a freedom in which the physical senses might dominate reason. For Aristotle, human life had total free choice. Both virtue and evil “lie in our power.” So Socrates, Plato and Aristotle rejected a determined universe in which the impression of free will must be an illusion.

For Zeno and the Stoics everything was predetermined, an unbroken chain of causation--God's will by fate ordained--but with an exception, the freedom to obey or violate moral law. Here they held with Plato and Aristotle. Some say that this ‘exception’ represented an early form of modern pragmatism--accept the world as it seems to work, never mind the esoterics.

Early Christianity Deterministic in Outlook: Augustine and Jerome observed that time started with creation and, as we have noted, all events were predestined by God's act of creation. In his *Confessions,* St. Augustine emphasized the logic of predestination in a universe designed and created by God. A ‘creation’ that left things to chance, uncontrolled by God, just didn't go well with the church in those days. Interestingly, Einstein had the same misgivings about God using chance, ‘playing dice’ with the universe.

Calvin's Fatalism: Although at variance from much of modern Christian theology, early Christianity generally subscribed to Augustine's laid-out-for-all-time universe. Marcus Aurelius, Roman emperor and a Stoic, writes in his *Meditations*, that "Whatever may happen to you was prepared for you from all eternity."

But views on determinism and free will in the early Reformation were not monolithic. Arminianism, based on the teachings of 16th-17th century Dutch theologian James Harmens (Jacob Arminius), as contrasted with Calvinism, challenged complete predestination in favor of 'conditional' predestination. The Arminian view was that Christ's death atoned for all, and all who believe will be saved, except that believers could so act and think as to fall from grace.

St. Augustine and Determinism: Unanimity, however, escaped early Christianity--as it does to this day. Despite everything being laid out by creation, predestined, human life was originally created with eternal life and free will, thought St. Augustine. But he believed that the corruption of Adam and Eve, through original sin, devolved on all descendants, condemning them, the reprobate, to mortality unless 'elected' for salvation at the Creation.

In this view, we are brought into the world corrupted and doomed to eternal damnation unless predetermined for salvation by election. Understandably predestination lost some of its original vitality as the years went by but was reinvigorated during the Reformation.

The Pelagian Heresy: Welsh Monk Pelagius, a contemporary of St. Augustine, was troubled by this idea of eternal torment through inherited corruption and found the idea so abhorrent that he gave explicit expression to his objections. Pelagius asserted that human life's choice of sin was not attributable to inherited corruption, but simply human life's own choosing, that God gave human life honest-to-God free will.

And for these compassionate views--which elevated human life out of inherent corruption, and in the process revealed God in a better light--the Pelagian view was, at St. Augustine's urging, declared heresy. But many Christians today are quite willing to associate themselves with the Pelagian Heresy.

Ecclesiastical Determinism Followed by Scientific Determinism: It is interesting to observe that when early science broke from ecclesiastic constraints, the world encountered something even more formidable--the repressive constraints of 'scientific' determinism, implicit in the Newton-Laplace classical mechanical, causal physics.

Science, freed from ecclesiastical determinism by a mellowing society, seemingly substituted its own shackles of scientific concept, thus reintroducing determinism, including the causally-induced chain of human events.

Space-Time Geometry and Determinism: "Alas, time stays, we go," said Austin Dobson, *The Paradox Of Time.* Time doesn't pass, it is just there. With the advent of Einstein's relativity theories, the universe was found to be the geometric expression, not of space and time but of spacetime. Time is seen as an interacting, in-place dimension of the geometric structure of the universe.

Spacetime geometry in relativity suggests time as laid out from the beginning, irreversible--an integral part of the geometry of the universe. Your consciousness of the present, moves along your world line, encountering and experiencing events already laid out. Thus events don't happen to you, you happen on or encounter them.

Genuineness of Free Will in a Determined Universe: As we have seen, before quantum theory, philosophy and science were generally disposed toward the predetermined universe. Classical relativity theory seemed to reinforce this. Elaborate philosophic and theological arguments had been built up hypothesizing the determined world and, with appropriate rationalizations, God's role in human life's predetermined exercises of free will in a universe into which imperfection and evil are structurally ingrained.

Theological arguments, of course, have gone both ways. One is that this exercise of will is genuinely optional even though the selected choice is predetermined and known to God all along; that one is free to decide and choose, even if it is one's 'destiny' to have made that choice. The other argument, is that the predestined choice reflects only pseudo-free will, that free will is illusion. So human life here is captive to decisions which are the fiat of creation. *Ergo,* no free choice.

Philosophy and Theology Wrestle with Free Will and Evil: More recent Western theology, even before quantum theory, trended toward the acceptance of free will as genuine--a determinism diluted, so-to-speak, to permit genuine freedom of choice in which our moral successes and failures would affect our future--in this life and in the hereafter.

Abelard and Aquinas both asserted the existence of free will, despite St. Augustine's causal chain of creation, although Aquinas accepted original sin of Adam as devolving on all men, thus requiring divine intervention for salvation. Humans, in the view of John Scotus, had completely free will.

Both St. Augustine and St. Jerome concluded that the universe was determined at creation. And so they wrestled with the presence of evil in God's predetermined universe; whether human life could therefore really be blameworthy. St. Augustine conscientiously wrestled with various logic configurations that might absolve God of permitting evil.

Redefine Evil? One approach to reconciling evil with God's omnipotence and failure to prevent evil was the idea of redefining evil in a way that diminishes God's culpability, and in which evil is, well, not so evil. Evil, in this definition, became the 'absence of good,' simply a negative. Moreover, this 'absence of good' was, in this view, something essential to the workings of the world--there could be no good without a correlative bad or evil. So evil is essential to the reality of good.

Rationalizations of Free Will Are Obscure: But most see rationalizations revolving around the concept of free will in a determined universe as obscure and labored and see evil as just that, evil, and whichever the universe is--predestined or chance--well, that is what defines the true extent of human free will.

In the predetermined universe, God shares in the responsibility of human choices. He knew of them (omniscient) from the beginning and, though all powerful (omnipotent), refrained from changing them. And so, it is felt by many that God must judge, temper punishment, and arrange salvation, accordingly because even God is inherently bound to apply moral law and dispense justice. Indeed, Kant observed that proof of God lies ultimately in the existence of morality.

If, on the other hand, we have a truly chance universe, then free will is essentially genuine, and human life, though compromised by the handicaps of genetic or environmental constraints, is responsible, to some degree at least, for the consequences of its choices--and may properly be so judged and punished, not only by God but by society.

'Gut Feelings' vs. Philosophy: In the work-a-day-world of everyday life, most people are not given to philosophic introspection and analysis about whether the universe is or not predetermined, or whether there is real or illusory free will. One makes decisions and choices by intuition or 'gut feel.' The average person does not stop to examine and consciously deliberate and evaluate choices in philosophical or theological terms, though such activities may be implicit in the cognitive processes themselves.

If pressed to describe the process by which one makes choices, one is likely to attribute such choices to a combination of inherited 'nature' and chance--the 'breaks' (good or bad). Of course, one's nature, in the evolutionary theory at least, is molded from the earlier causal and chance events of the evolutionary process. "Life is like a game of cards," said Jawaharlal Nehru, "The hand that is dealt you represents determinism; the way you play it is free will."

Divergent Views in Theology: Theology too, has divergent views on determinism. The theological implications are there on a grand scale. Calvinism's outlook toward predestination, original sin and arbitrary election for salvation, as theologians point out, were influenced by St. Augustine's predeterministic theism. So human life may successfully eschew evil, perhaps, because of the disciplined conscience wrought into him, or perhaps to give credence to his having been (s)elected for salvation.

Whatever the case, 'free will,' choices in a determined universe, were written into creation. But prevailing Protestant theology asserts human free will--at the same time that it asserts access to salvation; and so it tends toward a universe of indeterminism--interestingly, as does science, influenced by probabilistic quantum theory.

Laplace's Deterministic Universe and Quantum Theory: Based on the progress in the absolute measurements of classical physics, especially the work of Newton, as we have noted, the 19th-century French mathematician and astronomer, Marquis de Laplace, asserted the deterministic character of the universe, allowing the theoretical prediction of future in the universe--even including human behavior.

Although the extrapolation of Laplace's views on absolute measurability in classic physics to the nature of the universe as deterministic had seemingly only modest influence on the theology of the time, it did influence the then contemporary philosophy and at least raised the same theological questions as troubled St. Augustine.

Then Came Quantum Theory: But this was before the discovery of quantum theory which emerged primarily from the contributions of a half dozen or so physicists. Max Planck had discovered the quantization of energy exchange, his black-body radiation law, introducing a constant, the measure of discrete packages or quanta in energy exchange. Einstein went on to extend the quantum principle to light quanta--photons.

Heisenberg and Schrödinger--The Uncertainty Principle: Two other scientists, independently arrived at the existence of the 'uncertainty principle' introducing an array of numbers--probabilities, Werner Heisenberg by way of matrix mechanics and Erwin Schrödinger by way of probability waves, wave mechanics. Niels Bohr and other scientists like Born, and Pauli, contributed to this work leading to the development of quantum theory, the probabilistic nature of the micro particle world.

From a distance, for example, the behavior of a particle like the electron appears predictable. But the more closely one looks at it, the greater its unpredictability, the more violent its 'vibrations,' so that its behavior can only be expressed statistically, as a matrix of numbers, an 'array of probabilities.'

Thus, unlike the predictability from specific measurements in classical mechanics, quantum mechanics posits the impossibility of accurately, simultaneously determining both position and momentum of subatomic particles like electrons, relying instead on probabilities. So Laplace's absolute measurements of classic mechanics in the laws of physics was replaced by Heisenberg's uncertainty principle.

Quantum Theory Diminishes the Level of Determined Events: Particles exist *only on a contingency basis*, says Heinz Pagels in *The Cosmic Code*, quoting Max Born. Unlike classical mechanics predicting specific measurements, the wave function specifies that an atom will have the probability of 'a specific energy level.' Pagels sums up the effect of quantum probability theory on earlier classical theory. Quantum theory, he says, *causally determines* only the *probability distribution* of events, not the outcome of specific events.

It is interesting to note how some have interpreted quantum theory as relegating determinism to the museum of antiquities while others see it as a further partial confirmation of determinism.

Mortimer Adler's Ten Philosophical Mistakes: American philosopher Mortimer Adler, Director of the Institute for Philosophic Research in Chicago, has been a distinguished editor for the *Encyclopedia Britannica*, the *Synopticon*, and *Great Treasury of Western Thought.* In *Ten Philosophical Mistakes*, Adler singles out "ten subjects about which mistakes are made," one being "Freedom of Choice."

Adler's Four Freedoms: To better define issues, he chooses to divide 'freedom of choice' into four distinct freedoms. The *first* is the freedom or ability to *do as we wish* or please. Within practical limitations, everyone has this, he says, but subject to obstacles of coercion, duress, and the necessary wherewithal--you can't eat high on the hog with an empty wallet. Adler suggests a *second,* freedom of suffrage, "political liberty."

His *third* freedom is *moral freedom*, the *freedom to will what one ought* to according to the code of morality--to do or not to do what one *should*. This is a *conditional* freedom, available only if reason controls emotion. Unlike the first two freedoms, possessed by all to some extent, the third is one, he says, you either have totally, or you don't have at all.

The Fourth Freedom, Freedom of Choice--The Issues Not Joined: The *fourth* freedom, which he describes as the *freedom to will as one chooses*, generally known as freedom of choice, he recognizes as being beset with philosophical and scientific dispute. If we don't really have this fourth, freedom of choice, then how, he asks, can we account for the first freedom to do what we want, or the second to exercise our rights of suffrage.

Because of the way the issues are presented in the expressed views of the determinists, he concludes, it is not possible to demonstrate the correctness of the determinist or the free-choice point of view. He suggests the issue confusion as being traceable to the determinist presentation in such manner that the *issues are not joined.*

Freedom of Choice As Outside Quantum Uncertainty: Adler cites the *suggestion by some philosophers and scientists that quantum uncertainty provides room for freedom of choice.* But, as he evaluates the views of the various physicists, "more sober minds," he says, dismiss this argument of free choice derivable from quantum theory.

Quantum indeterminacy is just not the kind of indeterminacy involved with freedom of choice, he concludes. As we will see from the varied physicists' views described below, physicists simply do not agree on the consequences of probabilistic quantum theory on the uncertainties in our daily lives, as distinguished from particle uncertainty.

Freedom of Choice As Being Outside Quantum Uncertainty: The action of the will, says Adler, is outside the physics of quantum indeterminacy. The will, he says, is not a physical phenomenon like the functioning of the neural-driven senses but an immaterial, intellectual property of human consciousness not governed by causal laws of physics. So there are, in his view, two distinct indeterminacies, one of matter, quantum physics, and one of the nonmaterial will. But notwithstanding Adler's view, we do know that there are quantum effects in the biological thinking processes of the human brain.

What Happened to the Predetermined Universe? And so, getting back to quantum theory, if we cannot measure the present state of the universe, without altering it and introducing uncertainties, we cannot, so some suggest, predict the future. Thus, laid out at creation was a universe into which quantum indeterminacy was an integral part, and so, it would seem, one not so predetermined, at least at the quantum level.

Chance and Quantum Theory As Essential for Evolution: There are those who suggest that the quantum theory is essential to the universe as we know it; that randomness-chance itself is a necessary mechanism for an unfolding universe. Hubert Reeves, in *Atoms Of Silence*, comments on the role of uncertainty and chance in quantum theory. Each particle, he says had the ability of combining or transforming.

But a 'fortuitous event,' chance, was required for a heavier nucleus to form, for molecules to meet in the primitive ocean, for the cell to form, for a cosmic ray to produce a mutation. And so, he says, *the organization of the universe demands that matter abandon itself to the games of chance*.

The Cosmic Computer Program Not a Deterministic Program: In *Perfect Symmetry*, Heinz Pagels engages in metaphysical "Reflections." His first is "The Cosmic Computer," in which he describes the metaphor of the universe as a cosmic computer for which the quantum particles are the 'hardware,' the laws of physics the 'software,' and the evolution of the universe as the execution of the 'program.'

The evolving universe can be seen as executing a program specified in the laws of nature, although, says Pagels, it is *not a deterministic program like those in digital computers.* The ultimate 'output' of this cosmic computer remains uncertain but we already know, he points out, that its program has provided complex 'subroutines' identified with life.

But what kind of universe gets created? Pagels cites Einstein's questioning whether God had any choice, whether the laws would leave any physical constants unspecified. If all the constants are specified, then God had no choice; but if some are not specified, than other types of universes are possible and other universe programs can be run by the cosmic computer, says Pagels.

Trying to Establish Determinism As Producing Indeterminism: In his *Cosmic Code*, Heinz Pagels asks whether there exists beyond quantum theory some sort of 'subquantum theory' used by the *All-knowing Mind* (scientists are frequently more comfortable using these generic descriptors, rather than the term 'God') to determine the universe.

But quantum theory, he says, makes this impossible; that even an 'all-knowing mind' must use experience, in the course of which "the rest of the deck of nature gets randomly shuffled again." In the ultimate extension of the uncertainty principle, Pagels says that "the very act of attempting to establish determinism produces indeterminism."

Astrophysics and Quantum Theory: British physicist Stephen Hawking has studied the application to astrophysics of phenomena of quantum mechanics. We are forced, says Hawking, to turn our search for an understanding of the universe "from our theory of the extraordinarily vast to the theory of the extraordinarily tiny," the theory of quantum mechanics.

And so, he says, quantum gravity, the ultimate enigma of unification theory, lies, it would seem, not so much in the exhilarating and spectacular nature of astrophysical phenomena but in the subatomic reaches of quantum mechanics. *Quantum theory, he says is an essential characteristic of our universe.*

The Role of Mind in Quantum Theory: We earlier discussed the 'anthropic principle' which implies a relationship between our presence in the universe and the nature of the physical laws of the universe--that the universe is as it is, because we're here. Quantum theory, according to many scientists, says that the universe is governed significantly by chance ingrained in the laws of quantum physics.

But it is also suggested that quantum theory gives an extraordinary role to human mind in the universe; that the observer, the human mind, is somehow an essential ingredient of quantum mechanics; that mind somehow enters into the evolution of an inchoate universe. As physicist John Wheeler asserts, the existence of the universe requires "observer-participancy."

So here we have quantum theory suggesting, to some at least, an indeterministic universe of chance, and at the same time, the anthropic principle that the universe is the way it is because we're here, implying to some at least, our necessary presence, an aspect of determinism and design.

Quantum Mechanics--Connects the Human Observer to the Atom: Quantum theory highlighted the dominant role of mathematics in theoretical physics, and quantum theory, says Pagels, became the most powerful mathematical tool to explain physical phenomena that *ever fell into human hands*. Its conceptual impact was unparalleled in the history of science and technology and put us in touch with *new* *components of the cosmic code.* The hallmark of achievement in this century, he says, will be this contact with the quantum universe, *an apparatus between the human observer and the atom.*

Other Views on Causality and Chance: In *Atoms Of Silence*, physicist Hubert Reeves refers to "effect and cause" as a "fruitful' pair," the "handmaidens of philosophy and science," since ancient Greece--and even before. We see events and seek their cause, an idea rich and useful. It teaches us that everything is "not left to chance," reality is not total entropy or chaos. There is, he says *a certain 'determinism' to things.* He points out that science is constantly revealing new causes for effects and asks whether the universe will someday be found to be *completely determined,* despite quantum uncertainty.

Freedom of Will and Quantum Mechanics: Edward Harrison, in his *Masks of the Universe*, after discussing the various 'universes' constructed by society at different stages in history, goes on to discuss determinism and free will and the ' many-worlds' interpretation of quantum theory in which all possible alternatives do in fact occur, in different universes.

Stapledon's 'Exfoliating Universe': Harrison cites Olaf Stapledon's *The Star Maker* in which *all alternative courses of action are taken*, creating many "distinct temporal dimensions and distinct histories of the cosmos." The multiple alternative courses of the many creations and the uncountableness of their combinations leads to "an infinity of distinct universes exfoliated (separated) from every moment of every temporal sequence in this cosmos."

Freedom of Choice As an Illusion in the Rational Universe: *Freedom of choice is an illusion in a rational universe*, Harrison says, for all is determined by the inviolable laws of the universe. So liberty to do or go is intolerable, for it contradicts the rational determinism of the universe. Hence, you take both paths. Says Harrison, this is *the many-worlds interpretation of the freedom of will. Free choice is an illusion, and all that can happen, must happen.*

Branching Universes Expend Free Will: The point in this explanation of the many-worlds quantum-uncertainty interpretation is that *these branching universes account for and exhaust the alternative choices of which free will consists*. Our free will is expended by exercising *all choices* and so we're left with an illusion of free choice. Of course it has been said that our universe is an illusion, a virtual reality of which our illusory free will is a part.

The Growth of Order in the Degenerating Universe: In his book, *Cosmogenesis, The Growth of Order in the Universe,* Harvard University astrophysicist David Layzer takes the reader on a grand tour of cosmic evolution, on to biological evolution and the development of the human brain. He notes the paradox of the buildup of chemical, structural and biological order in a universe whose general trend is one of energy-degradation or decreasing order, sometimes expressed as increasing disorder or entropy.

Evolution Is Blind but Humans Can Make the Future: Layzer explains that evolution is blind but human beings can foresee the future, and by inventing and pursuing goals, *make* the future. To introduce his views on determinism, quantum uncertainty and the extent of human freedom, he asks whether this view of human development is consistent with science or at least with the (novel) scientific view he has constructed in his book.

Reconciling Human Freedom with Causal Physics: Layzer points to the difficulty of *reconciling intuitive conviction of human freedom with an objective view of the universe as a causally-connected sequence of events.* Value, ethical, and legal systems assume freedom of choice. Creativity is an expression of freedom, as philosopher Henri Bergson suggests, but *the unpredictability of creative acts, says Layzer is not the same unpredictability as random outcomes because while they are unforeseeable, there is a "coherence with what has gone on before."*

Freedom Comes from Choice Rather Than Randomness: A free act, he suggests, is the outcome of a deliberative or creative process with generally predictable consequences. While some have suggested quantum indeterminacy as allowing freedom in an otherwise deterministic universe, *free choice comes not from randomness but from choice.* He cites Plato and Aristotle as injecting chance and necessity as forces opposed to predesign in the universe.

Choices Create Order That Was Not Prefigured: Common experience shows outcomes not predetermined but resulting from the deliberative process--"a process unfolding in time." Layzer argues that the neural processes which connote the subjective experience of freedom are processes which create *order that did not exist before* and were not "prefigured in earlier physical states."

He notes that "We are free because we are, to a considerable extent, the product of our own lives," as Democritus and Socrates believed. Our ability to shape our future, he concludes, is *a property of objective reality.*

Do All Law-abiding Processes Have Predetermined Outcomes? "Do all law-abiding processes have predetermined outcomes?" He answers by saying that "Outcomes are determined by laws plus initial conditions" and the *outcomes are undetermined to the extent of the initial unspecified conditions.* Thus predetermined outcomes are not required. The neural processes which produce "novel and unforeseen products," such as art, music, inventions, and which imply freedom, "are indeed creative (non-predetermined) processes," he concludes.

Freeman Dyson and Determinism: In the concluding chapter of his *Infinity In All Directions,* an updating of his *Gifford Lectures* given in Aberdeen Scotland, physicist Freeman Dyson refers to *five points* in the no-man's land between faith and reason: the origin of life, the human experience of free will, the prohibition of teleological or purpose explanations in science, the argument from design proof of God, and 'final aims,' ultimate purpose of the universe.

On free will, he credit's Schrödinger in his *What Is Life*, as most clearly formulating the problem--to reconcile the human experience of free will with "a belief in scientific causality" (determined universe). He says that both *narrow-minded science* and *narrow-minded theology* stand opposed to free will. He cites Monod's *Chance and Necessity* and the "orthodox theology" of an omniscient and omnipotent God as denying free will.

For those who want both God and free will, he suggests the God of the Socinian heresy, the God who is growing with us in the evolving universe and therefore free from the blame of perfection--of being omniscient, omnipotent, and still permitting evil and imperfection.

He describes the human brain's perception and cognitive functions as governed by random processes--not predetermined, and *God or universal mind as also operating on random processes*. However one may read this interrelationship between classical physics and quantum theory, Dyson's bottom line is unmistakably free will.

Viktor Frankl and Free Will: Viktor Frankl, well known in psychiatry circles as the father of logotherapy, and author of best-seller, *Man's Search For Meaning*, suggests, in his sequel, *The Unheard Cry For Meaning*, that even the extension of scientific research into psychophysics did not bring us closer to solution of the mind-body and determinism-nondeterminism (free choice) problems--the "unity in diversity which could be the definition of human life."

Perhaps the contradictions in the pluralism of science, he muses, can only be reconciled beyond all of them, in the next higher dimension, in the unity of reality--available only in the human dimension, the *unitas multiplex*' as Thomas Aquinas defined human life.

However, perhaps enough has now been written in quantum theory and neuro-biology, as illustrated above, to suggest that perhaps we have broken through into that next higher dimension of Frankl's reconciliation process.

Determinism in the Context of the Necessary God and Contingent Universe: In *The Mind of God*, physicist Paul Davies also discusses determinism and free will in the context of the time-honored problem of the necessary God who creates a contingent universe. He notes that volumes have been written in an attempt to break out of the contradiction between the timeless immutable necessary God and the contingent, changing, time-affected universe. He discusses one escape route, the Dipolar God and Wheeler's Cloud.

Plato, says Davies, confronted the paradox of necessity versus contingency by proposing two Gods, the eternal and necessary Good, and the contingent Demiurge, responsible for the changing material world. Davies proposes to describe this situation as two related aspects of a single "dipolar" God, as adopted by advocates of process theology.

Process theology views the universe as a process with 'directionality,' rather than a collection of physical objects. The 'flux of time' plays a key role in process philosophy, which, notes Davies, asserts the "primacy of *becoming* over being," stressing the indeterminism of nature, rather than Newton's cause and effect universe.

Indeterminism with Purpose: Here, Davies, noting that the future is not ingrained in the present, suggests that "there is a choice of alternatives, by forsaking reductionism in favor of the holistic view of the universe as the whole of its parts." A human being is not a collection of atoms but something represented by higher levels of organization, introducing teleology or behavior suggesting purpose.

The advent of quantum physics replaced the notion of the universe as a deterministic mechanism, he notes, one with open systems capable of reaching ordered behavior despite underlying randomness. He cites the existence of self-organizing principles, not traceable to conventional physics, governing behavior of systems in their movement toward higher organizational levels.

Quantum indeterminism, says Davies, permits God to remain transcendent while His basic character and purposes remain unchanging and eternal. Thus, Davies says, "timelessness and temporality are folded in a single entity."

Here again is one more approach reconciling the timeless, unchanging, necessary God with the contingent and temporal world.

DETERMINISM AND FREE WILL IN RELATIVITY AND QUANTUM THEORIES

Determinism and Free Will As Metaphysics or Physics: In *The Physics of Immortality* Frank Tipler cites Kant's fundamental problems, one being whether free will existed, and Kant's characterization of this question as 'metaphysical' because the human mental apparatus does not permit knowing true reality of the conceptual. Tipler declares that the question is no longer metaphysics but is a matter of physics, explainable in terms of relativity, quantum theory, quantum cosmology and his Omega Point Boundary Condition.

Determinism and Free Will in Relativity and Quantum Theories: Because of the quantum uncertainty principle, many are used to thinking of quantum theory as standing for indeterminism. But Tipler states that basic quantum theory is deterministic, *despite* the uncertainty principle. Tipler goes on to explain the existence or non-existence of free will in terms of "omniscience" and the "coding of information of the whole in part of the whole."

The Failure to Reconcile God's Omniscience and Human Free Will: Tipler notes that freedom [inherent in indeterminism] is lost *when total information about the whole is coded in a part of the whole.* Similarly, he observes, "if God is pictured as separate from spacetime, and if total information concerning the structure of spacetime is coded both in God and in spacetime, then spacetime has no freedom; it is completely determined by God's omniscience."

Tipler explains that when in general relativity, there is no "background spacetime," and information on everything past or future is "contained implicitly in the initial data," then there is "nothing new... in a deterministic theory like general relativity."

Nonrelativistic Quantum Mechanics as Deterministic: He goes on to discuss (noncosmological) quantum mechanics as being entirely deterministic, somewhat surprising, as we noted, since quantum mechanics involves the chance aspects inherent in the 'uncertainty principle.'

He then describes the 'Copenhagen' and 'Many-Worlds' interpretations of quantum mechanics and the famous Schrödinger's cat experiment. In 'Copenhagen,' we encounter the wave function of the cat as dead or alive. But in 'Many-Worlds' theory, the wave function of the cat is neither that of a dead cat or a live cat but the "the *sum* of the dead cat and live cat wave functions."

Omega Point Theory Adopts the Many-Worlds Interpretation: Since the Omega Point Theory is fundamentally a *quantum cosmology*, he says that he is compelled to adopt the Many-Worlds quantum interpretation. Since the Copenhagen Interpretation assumes that 'wave function reduction' eliminated quantum effects on cosmological scales an exceedingly short time after the Big Bang, the universe, in that theory, is no longer significantly quantum.

No Reduction in the Wave Function in the Many-Worlds Interpretation: But in the Many-Worlds Interpretation, Tipler points out, there is no reduction of the wave function, so physical reality is completely described by the wave function of the universe, and, says Tipler, the universe is *just as quantum now as it was in the beginning.*

Human Beings As Quantum Objects--Permits Resurrection--and Free Will: Tipler's exposition then turns to the assumption that human beings are quantum mechanical objects, like everything else, from which he goes on to prove that it is possible for every one of us to be *resurrected* one day--and live forever. Furthermore, by applying the Many-Worlds Interpretation to the ontology of quantum cosmology, he says, he will be able to prove that we probably have free will.

The Future As Deciding What Will Emerge from the Present: Tipler continues with a discussion of the future as deciding what will emerge from the present, with interesting material from theologian-physicist Wolfram Pannenberg. He discusses determinism working in both time directions, both past and future influencing the present. He also discusses how 'ontological indeterminism' does not follow from 'epistemological indeterminism' and how 'quantum-gravity indeterminism' differs from the nonrelativistic quantum version.

Roger Penrose on Determinism and Quantum Theory: In *The Emperor's New Mind, Concerning Computers, Minds, and the Laws of Physics*, theoretical physicist Roger Penrose, associated with Stephen Hawking on the mathematics of singularities, discusses quantum theory and determinism. He goes on to discuss the alternative quantum theories, the uncertainty principle and the many-worlds and parallel universes theory.

He notes that while what he calls the **R** aspect of quantum mechanics, involving the "state vector reduction" (collapse of the wave function), introduces uncertainties and probabilities into quantum theory, the other aspec-- which he describes as the evolution process **U**, is governed by the Schrödinger equation. This aspect, he notes, gives us a *completely deterministic evolution of the wave function*, telling us how the wave function evolves in time.

Where Does All This Leave Us? However, he observes that there is no clear rule as to when the probabilistic rule **R** should be invoked, in place of the deterministic aspect **U**. He asks: What constitutes 'making a 'measurement'? Why and when do squared moduli of amplitude "become probabilities?" Questions like these, he acknowledges, are deep and puzzling.

Where, he asks, does all this leave us? He suggests that we must strongly consider that quantum mechanics is simply *wrong* when applied to macroscopic bodies; that the laws **U** and **R** provide only approximations to some more complete theory; one in which quantum theory will undergo *fundamental change*.

Quantum Mechanics in Brain Activity: Penrose goes on to discuss quantum mechanics in brain activity. Penrose asks whether our present rules of classical and quantum theory are sufficient for an adequate description of our brains and minds. He asks: Is the brain to be regarded as 'observing itself' whenever a thought emerges into conscious awareness?

Present theory, he says provides no clear rule as to how quantum mechanics could take this into account. He notes also that digital computer actions themselves depend critically on quantum effects and we must inquire how we might make an entirely classical object behave like a digital computer.

It seems to him that we need some further fundamental changes which would make the **R**, the wave function collapse aspect of quantum theory, into an *actual* process before we can explain the way in which we think. A plausible case can be made, he says, that there is an *essential non-algorithmic ingredient* to conscious thought processes. So, he asks, what actual physical effects might constitute a consciousness influencing the action of the brain?

> Creative human activity...one aspect of human freedom...is unpredictable in the same way [as] biological evolution. We are...as Socrates and Democritus believed...free because we are, [essentially] the authors of our own lives [and] the openness of the future and of our own ability to help shape it reflects a deep property of objective reality.--David Layzer

PART XVI--CONCLUDING THOUGHTS

CHAPTER 59
RECAPPING

Review and Comment: Instructors are taught to describe what they are going to present, then present, and finally recap or explain what they have presented. Well, in that tradition, this chapter briefly describes what was presented in the book. This Recapping Chapter reviews selected thoughts from the book, with a few new twists.

Meaning--Developing Confidence in a Rational Universe: We wonder about existence, God and the universe, life and death, and seek reassurance that we have the *right picture*--that it all makes sense and generates confidence; that everything is explainable, meaningful, and consistent with our everyday observations of life. We recognize that the nature of our *reality* cannot be fathomed without an ample foundation of essential knowledge from which to construct a coherent description.

We noted earlier that this search for understanding must address certain fundamentals such as: the existence, nature and meaning of *God*; how the universe was created; why the human mind is attuned to the mathematical underpinnings of the laws of physics; the mysterious fact of the strangely felicitous constants of physics which assure the emergence of life and mind; and the capabilities and limitations of human thought.

Challenging Questions: Questions continue to occur to us. Is the universe *designed*? Does it have purpose? And if so, does that purpose embrace our presence here? What if there is *no reason* for existence, if the universe, and life, are all simply the inexplicable outcome of random processes? Does this foreclose meaning? Or, as several physicists suggest, could it be that we may simply be 'virtual beings' in a virtual reality derived from a computer program, run by "someone or something' outside the universe? What of meaning then?

The Blurring of Science and Theology: Questions about the meaning of the universe and existence of human life, in the past century were addressed primarily by philosophy-theology. Increasingly, though, it is addressed by science, as the separation between metaphysics and science blurs.

Scientific explanations are emerging for phenomena earlier considered matters of philosophy, theology, even miracles. Relevant is the emergence of 'quantum genesis' theories, derived from quantum and relativity mechanics, suggesting self-generation of the universe from a primordial quantum void that 'pre-existed' the Big Bang. Questions about the Beginning, once primarily the province of theology-philosophy, as theologian John Polkinghorne notes, are now fraught with science, especially quantum theory.

Science Tells Us Much about the Universe: Science now tells us much about the universe, such as its *size*, a sphere about 15 to 20 billion light years across (with 10^{89} particles)--which represented time period or age, as physicist Frank Tipler notes, in *The Physics of Immortality*, is "only a tiny fraction of reality." Time, in any meaningful sense, is said to have begun at 10^{-43} second, and at 10^{-34} second, the universe commenced an exponential expansion, at faster than the speed of light, in fact some 10^{50} faster than the present linear rate of expansion.

The universe 'began' in a Big Bang some 15 to 20 billion (earth-years) ago and, predicts Tipler, is likely to exist *another hundred billion years*, so that, as Tipler emphasizes, *the future comprises almost all of time and space*. The universe cooled enough for the initial or primordial matter (about 80% hydrogen and 20% helium) to form within four minutes from creation. Science also tells us that since *chaos* occurs on all size scales, intelligent beings would be able to use the universe's 'instabilities' to manipulate matter motion, so that the "actions of intelligent life in manipulating physics cannot be ignored in the evolution of the far future."

Time for Scientists To Rethink the God Hypothesis? Tipler notes, in connection with his Omega Point God theory, that physics has reached the point where it is time for scientists to *reconsider the God hypothesis*. And, he declares, "The time has come to absorb theology into physics, to *make Heaven as real as an electron*."

Astronomer-minister Hugh Ross's *The Creator and the Cosmos* discusses *divine* origins, the personal theist, supernatural Creator God, and recent scientific discoveries about the Big Bang. With considerable persuasion, he interprets these as a validation of divine creation and corroboration of certain biblical passages, although he discredits Tipler's Omega Point God theory, presumably on the basis that in that theory it is physics which creates Tipler's God, rather than God creating physics.

Physicist Paul Davies, in the *The Mind of God*, on the other hand, discusses the *natural* sources of creation, with emphasis on the *regression* of universe-creation causation to *logical necessity*, the *logic of mathematics*, and *ultimate good*, all of which one may, for want of a better explanation, call *God*, but still part of some 'natural process."

Hawking's Quantum Self-Generated Universe: Noted, is theoretical physicist Stephen Hawking's *A Brief History of Time*, in which Hawking offers a quantum or 'quantum-gravity' universe-self-generation cosmology approach to the creation with his spatial or 'imaginary time,' and his 'no boundary' proposal in which there is no true beginning because quantum uncertainty fuzzes out any true beginning point.

Hawking's quantum-generated universe is the product of the primordial void quantum uncertainty. In his *Creation Revisited*, Oxford chemistry professor Peter W. Atkins goes one step further, describing everything that exists as having its source in *mathematics*, his main theme being that this creation is a natural process, *without outside intervention.*

An Omega Point God? Also cited is mathematician Frank Tipler's *The Physics of Immortality,* in which Tipler demonstrates, through physics and mathematics, that human intelligence will spread through the expanding universe to become the *Omega Point God.*

To avoid the end of all life that would accompany the 'heat death' of an 'open' forever-expanding universe, Tipler hypothesizes a *closed* universe, one of sufficient mass to stop expansion and start contracting. But contraction models assume a collapse of the universe into a 'big crunch' singularity, which also ends all life. Assuming that life is a requirement for the universe, Tipler gives us the physics and mathematics to assure that life survives the contraction.

Resurrection of All Life: In Tipler's model, the 'big crunch' singularity of the contracting universe is avoided by the action of an evolving collective intelligence to achieve a *Taub Universe* of sequenced collapse, allowing survival in which the evolved collective intelligence, the Omega Point God, by computer information management, resurrects all prior life, in computer-improved form. Does his theory mean we have no God existing *now*? No. Physics assures the Omega Point God's existence *now*, says Tipler, because in both relativity and quantum theory, *all time co-exists*.

Redefining God in Terms of Physics--The Omega Point Evolving God: Tipler points out that it will be necessary to "redefine God in terms of physics, which won't be easy." But he proceeds to do just that, define God, omnipotence, omniscience, personal, and transcendent, a*ll in terms of physics*. Tipler, like Freeman Dyson and Rupert Sheldrake, believes that God evolves with cosmic evolution and this will permit us to determine "how God then generates the universe." Tipler's Omega Point is or becomes God, representing the point of infinite knowledge, omniscience, and occupation and control of the universe--omnipotence.

Stranded without Meaning? But the Omega Point God aside, the quantum-genesis cosmology confronts us with the dilemma of the quantum-self-generated universe. As a random or chance event, it would be a "purposeless" universe, as P.W. Atkins asserts in *Creation Revisited,* or "pointless" as Steven Weinberg suggests in *The First Three Minutes*.

As we have discussed, the meaning of the universe and life is naturally influenced by one's view of the origin of the universe. We normally view *meaning* in the context of the existence of a personally-involved Creator who provides meaning through all the vast implications of the divine Creation, and through revelation reflected in the various Holy Scriptures.

But if, as quantum cosmology suggests, the universe was self-generated by primordial, pre-existing quantum uncertainty, then, as astronomer-minister and cosmologist Hugh Ross declares, in his *The Creator and the Cosmos*, there was no purpose behind the creation. Humankind would be on a trip to nowhere, stranded in an abyss of meaninglessness.

A Comprehensive Description: But apart from who or what triggered the creation, what can one say about the meaning of the existence of an incredibly complex universe of radiant energy which metamorphoses into biological organisms housing mind and selfawareness, able to construct or identify exquisite abstractions--and wonder about the meaning of everything?

What is the role of the human mind in constructing reality? Is physical reality a computer-generated *virtual* reality? Why does the human mind somehow resonate with physical and quantum structures of the universe and its mathematical substrate?

Meaning and Morality: Meaning and morality are often associated. In his discussion of morality Immanuel Kant speaks of the generally-prevailing 'predisposition for good,' respect for 'action conforming to duty.' One thing in our soul about which we must wonder, which is exalting, he declares, is the 'original moral predisposition' in us by which we judge ourselves unworthy if we cater to 'unlawful satisfactions.'

Meaning and God-- Human Moral Predisposition: Kant refers to this mysterious built-in predisposition for moral conduct as being so "incomprehensible," as to suggest divine origin, acting upon the human spirit "to the point of exaltation," exciting the sense of "sublimity of his moral destiny." So Kant, having decimated conventional arguments for God, goes on to argue *God's existence as derived from the existence of the innate human predisposition to morality.*

Meaning and Religious Concepts: Thousands have died for and, for that matter, killed for, deep-seated religious beliefs, as illustrated by the various religious wars like the Crusades, inquisitions, and trials for heresy. Galileo risked his life and suffered house arrest for his views endorsing Copernicanism, that the earth, like the other planets, evolved around the sun, challenging the anthropocentric view of earth as the center of the universe.

Meaningfulness in Martin Luther's perceptions were linked to religious concept. He risked his life in his reformation crusade with his Ninety-five Theses posted on the door of Castle Church in Wittenberg, and in particular, his determined stand on justification or salvation by faith alone, the basic doctrine of the Reformation.

"Here I Stand, I Can Do No Other": Called by the ecclesiastical authorities to recant at the Diet of Worms, his very life at stake, he declared, "Here I stand, I can do no other. God help me. Amen." There are almost innumerable Protestant denominations stemming from the Protestant Reformation, divided in some aspects of church doctrine and/or organization. Baptist and Congregational churches, for instance, place ultimate authority in the congregation rather than church hierarchy, reflecting the Reformation concept of the priesthood of all believers.

The Matter of Grace --The Most Crucial Concept in Christian Theology: One of the most meaningful concepts has to do with *grace,* representing perhaps the most crucial concept in Christian theology, because it refers to the 'free and unmerited act' through which God restores to himself his creatures who were 'estranged' (the alienation from Garden-of-Eden disobedience--original sin).

The Diversity in Outlook on Meaningfulness: What is the meaning of and in life? In the everyday world we see various expressions purporting to convey meaning. "Life is a Beach" says one bumpersticker. "Love is all there is," says another. Human life is God's noblest creation," says one essay. One scientist expresses the ultimate put-down of human life, describing it as "a disease of matter."

We see meaning reflected in the dedication to some cause. This is illustrated, for example, by Mother Teresa's dedication to helping the poor, Van Gogh's pursuit of artistic creations, Albert Schweitzer's medical missionary effort and *his reverence for life*, Martin Luther's mission to reform the church, Einstein's drive to redefine the universe, George Washington's drive to free his country.

An Endless Set of Meanings: There is a virtual endless set of expressions of meaning, values, and purposes in life, some sublime, some seemingly absurd. Where do we start in the search for meaning? How are we to decide what really is and is not important?

Each person's perception of the universe, and of meaning, is influenced by genetics, the intellectual and practical influences of his environment, the society in which he lives, and even the state of his health. Paleontologist and essayist Stephan Jay Gould notes that "We cannot read the meaning of life in nature," we must construct it ourselves "from our own wisdom and ethical sense."

We discuss the term meaning, first in the sense either of meaning of the universe and in the sense of the meaning of existence, with the emphasis on meaning of life. In *Three scientists and Their Gods*, science writer Robert Wright, engaged in the constant search for "evidence that life is not devoid of meaning," notes the views of scientists about relationships of information to meaning in a universe which consists ultimately of information.

Meaning of Life in General: Wright cites the *Encyclopedia of Philosophy* as speaking of two contexts of meaning. The *first* sense is related to divine purpose, whether a supernatural intelligence created us to serve some end, a reference to *purpose of life in general.*

The *second* meaning of life relates to meaning or meaningfulness from the standpoint of individual lives, reflected in dedication or commitment to some important cause or mission: Van Gogh's creative pursuits, Albert Schweitzer's medical missionary effort and reverence for life, Dr Viktor Frankl's dedication to defining meaning through selftranscendance--doing for others or permanently changing something for the better, *implying an element of immortality.* It is in individual lives that we find the deep existential concerns, the human search for meaning.

Can't Quantify Meaning: In *Three Scientists And Their Gods*, Robert Wright provides analytical profiles of three scientists whose themes relate in some way to *information and meaning.* In the process, Wright notes that *not only do we not know how to quantify meaning, but we can't even agree on what meaning means.*

Why Am I Here? In Part Two of *Real Philosophy, An Anthology of the Universal Search for Meaning*, the authors Jacob Needleman and David Applebaum address the question of: “Why am I Here?” They note in his *Confessions*, St. Augustine’s confrontational encounter with the question of life's meaning, reminding us that, like St. Augustine, many of us plod along in life until some attention-arresting event jolts us into sharp confrontation with the question of our existence.

What is it about the human condition that poses and highlights the question of life's meaning? We are promised a role in fulfilling our own destiny. We attain an awareness of ourselves in relation to the world, giving us the opportunity to participate in the unfolding of a life of meaning.

Some Commonly Held Views of Elements of Meaning: For many, meaning involves the pursuit of pleasure, gratification of some sort, or perhaps the more subtle and abstract 'pursuit of happiness.' For others, like the Greek philosophers, meaning is derived from the acquisition of knowledge. Some find meaning from career pursuits, or from society's approbation or approval, 'recognition' of one's efforts or worth.

For still others, purpose and satisfaction in life is fulfilled through children, friends, a nice home, a late model car, and a satisfying church affiliation. Some creative and communicative acts are needed for many. Some see meaning in discovery or inventive pursuits, medical or scientific or social research. Creative thought is meaningful for writers, philosophers and theologians. Some see meaning in the creation or performance of music, or art.

Meaning in Human Creativity: A Van Gogh Sells for $82.5 Million. Why? What makes a recently purchased painting worth $82.5 million dollars, a sum which will build a college, a hospital, a highway, or feed thousands of starving human beings? Why does a graphic representation of an object, person or idea, drawn by a human being on a piece of canvas, have such extraordinary 'intrinsic' or investment worth? Some argue that such worth lies in the value of human creativity, the power of graphically portraying the abstractions of human cognition.

Service to Others: But deeper or more subtle meanings sometimes dominate thought. The philosopher and medical missionary Albert Schweitzer found meaning in the universe in his *Reverence for Life,* life being perhaps the universe's ultimate purpose. Meaning for many is found in a life of giving, service, and sacrifice for the good of others.

William James observed that "The great use of life is to spend it for something that will outlast it." The great parks of Boston are the result of the acquisitive instincts of Ben Franklin, and the National Gallery of Art from the acquisitive habits of Andrew Mellon. So, in the end, it is the *purpose* of acquisition that governs its meaningfulness.

The Bible has its admonitions against the pursuit of riches at the same time that it urges saving for one's future needs and for tithing and philanthropy.

It is said that happiness and the deeper satisfactions come not through pursuit of *acquisition* but from *divesting*--giving, meeting the needs of others. We are reminded that the only way of investing in the *ultimate future*, is by giving away to serve some worthy purpose.

Meaningfulness and Happiness: Meaningfulness is often associated with *happiness*, the idea that the human goal lies in finding joy, happiness, and that which is uplifting; that if one is happy, the universe and life has sufficient meaning. But in our imperfect world some feel that the key to happiness lies in constructing the *right outlook or perspective.*

In early times, happiness did not relate to modern conveniences. Life's blessings or 'satisfactions' lay in simpler things, a roof over your head, a fire to sit by in the winter, food to satisfy hunger, and caring human relationships. This simpler life represents the lifestyle of the Amish today, recognized as a well-adjusted society which views life as meaningful without modern contrivances.

Mother Teresa was asked whether she thought of herself as a vehicle of God's Grace in the world. She replied that she was a 'pencil in the hand of God.' God's greatest gift to her? The poor people, for "they are Jesus for me....prayer into action." About wealth? *Poverty is a freedom, not a mortification.* She found the rich much poorer--always needing something.

The Central Existential Question: For many, at the core of the search for meaning is the question of why *we* exist, whether we are somehow important in the scheme of things. *What is man that thou art mindful of him?*, asked King David (Psalms 8:4). The concern over our existence compels us to seek out the meaning of our pres-ence here. It is some credit to the human race that the meaning of existence can translate into concern for the existence of others. Built into the human psyche is some intuitive recognition of *the existence of profound values which transcend even the value of individual life.*

In *Man's Search for Meaning* Austrian psychiatrist and death-camp survivor Dr. Viktor Frankl, notes that *meaning differs from person to person and moment to moment*. What matters is the specific unique meaning in *a particular person's life at a specific moment in time.*

We ask of others the question of meaning, an exercise in futility, he says, because *Life* asks this question *of us.* And how must we reply? Well, says Frankl, by being responsible and it is this *responsibleness* that is the essence of human existence. Frankl's 'responsibleness,' interestingly, is close to Immanuel Kant's concept of an innate, divinely-inspired sense of duty-morality.

Meaning Through Self-transcendence--Three Routes to Self-Transcendence: Frankl goes on to say that meaning is found in the outside world rather than within the self. *Thus self-actualization is attainable only through self-transcendance,* to be found not in notion or idea, but in practical applications in the real world.

For Frankl, there are three avenues to "actualization through transcendance" as the key to meaning, namely, the triad of creation, experience, suffering & transform-ation: *First*, by doing a deed or creating a work. *Second*, by experiencing a value, something or someone beyond oneself, such as loving another. *Third*, through experiencing 'suffering.' In his essay on *Man's Search For Ultimate Meaning*, he particularizes on 'suffering,' citing the duty of transforming the *predicament* of an incurable disease into *achievement*--converting tragedy into triumph.

Meaning and Death: Critical to meaning is to understand and confront the phenomenon of death; the *reality framework* within which death occurs; what death *means*, and how to anticipate and deal with its devastating effects. Deep in the human psyche we all seek some foundation to justify our hopes and beliefs. We need this foundation to undergird our persistence, to rationalize and accept the present, and to have the confidence to project into the future.

Search for Meaning As the Essence of Philosophy: We have noted the search for meaning is the essence of philosophy and have explored what the great philosophers through the ages have identified as the fundamental philosophic problems.

What are these problems? As we noted, Immanuel Kant's listed three: Does God Exist? Is there life after death? and Do we have free will? We also noted the work of philosophy historians who have listed other groupings of philosophic questions which we have discussed.

Exploring Several Other Questions: Recent developments lead us to explore several other commonly-suggested questions: Does language totally communicate ideas of the mind? Why does the universe (as Oxford Professor P. W. Atkins suggests in *Creation Revisited)* reduce to mathematics? Does all metaphysical or non-empirical knowledge (not derived from senses-or physically verifiable) consist of 'nonsense' or valueless observations, and serve only to clarify language, as suggested by some philosophers (the 'logical positivists')?

On the Shoulders of the Great Philosophers--Challenging the Future: The history of philosophy, as S.E.Frost, Jr. observes in *Basic Teachings of the Great Philosophers*, is the story of how different philosophers have created different patterns, proposed different solutions to the "puzzle that is human experience." He suggests that we are at the pinnacle of centuries of struggle with the problems of human life, *standing on the shoulders of the great philosophers* of the past, who admonish that we pick up the torch and carry on from them.

This, of course, is exactly what is now transpiring as we explore concepts of the co-existence of all time, the unification of all forces and fields, the existence of parallel universes, the extension of biological intelligence through forms of artificial intelligence, and the devising of methods of spreading intelligence throughout the universe, as physicists like Freeman Dyson and Frank Tipler portray.

The Universal Quest for Meaning as the Search for Guidance: In *Real Philosophy, An Anthology of the Universal Search for Meaning*, philosophy professors Jacob Needleman and David Applebaum describe their survey on the 'universal search for meaning,' which they call "real philosophy," and "the love of wisdom." They describe the goal of philosophy--the search for meaning, as the exploration of human thought for *guidance.*

The underlying fundamental drive 'within the human psyche,' they conclude, has its origin not so much from biological or social sources, as from the *inborn thirst for meaning.* They cite death-camp-survivor-psychiatrist Viktor Frankl's conclusion in *Man's Search for Meaning*, that the deepest driving force within every human being is the *existential imperative--the need for meaning,* without which no one can live, and that which 'defines our humanity.'

Foundations of All Human Activity Being Reevaluated: We see a trend in the physical sciences which sometimes blend Eastern thought on the unity of all things with the 'new physics,' invoking 'holistic' concepts' where the whole is more than the sum of its parts, involving new concepts of consciousness and quantum reality. And so, they suggest, the *intellectual foundation of all human activity is being re-evaluated.* All of which lends new vitality in the search for meaning.

The Mystery at the End of the Universe--The Chain of Causality: In the search for meaning, philosophers, theologists and physicists have always gone back to the question of what is the source of *existence,* in order to find some foundation on which to structure purpose and meaning.

As minister-astronomer Hugh Ross points out in *The Creator and the Cosmos*, the existence of God, as the source of the universe and of human life, is essential to establish that our existence is divinely purposeful and therefore our moral and ethical guidelines, and our destiny, is pre-planned --not just some incredibly fortuitous coincidence of chance.

In his final chapter of *The Mind of God*, physicist Paul Davies returns to his starting question of where the chain of causality ends. He reminds the reader of the earlier discussion of the tracing of causes to God and God's "necessary existence" or, the alternatives discussed, that is, the necessary existence of the universe itself, or of the laws of physics or mathematics or the unified superlaw, or of a necessity related to ethics--called the Good.

Does Existence Defy Explanation? And after reviewing some of the options, Davies asks whether it is possible that the reason for existence has no real explanation in the normal sense? He refers to the mystical insights (mostly mathematical) experienced by various scientists, and physicist Hoyle's reference to a superintelligence guiding evolution through quantum processes.

A Superintelligence Acting at the Quantum Level: Hoyle's God is a 'teleological' God, says Davies, guiding the world toward an ultimate state in the endless future. And by acting at the quantum level this superintelligence can inject these ideas from the future into the human mind. He refers to Einstein's own experiences with cosmic religious feeling and we note science writer David Peat's experience with "a remarkable feeling of intensity that seems to flood the whole world around us with meaning...."

Meant to Be Here: Davies feels that we are barred from ultimate knowledge by the same rules of reasoning that trigger our pursuit of explanation. So what does he believe? He concludes that he cannot believe that our existence is just a "quirk of fate." Our involvement, he notes, is too intimate and the presence of mind in the universe is certainly of fundamental importance.

Finally, he declares that via conscious beings the universe has generated self-awareness. A Fact which is not trivial, not a minor byproduct of *purposeless* forces. " We are truly meant to be here," he concludes.

Will There Be a Tomorrow? In *Real Philosophy, An Anthology of the Universal Search for Meaning*, philosophy professors Jacob Needleman and David Applebaum note that any inquiry into the human condition inevitably confronts the dimension of time. What has been our past? What is our future--and do we have a future? The phenomenon of self-sustaining consciousness makes it difficult to think of the future in terms of an end, of finality, but rather do we think of the future in terms of some process of change, further unfolding.

We know that in the distant future, when its fuel is exhausted, the sun, our home star, will become a red star, to expand and evaporate its planets. By then, life will long have been gone--disappeared as though it never existed. Indeed the universe itself, depending on its mass, will one day end, either in a catastrophic contraction of a closed universe--the Big Crunch, or in the chaos of the 'heat death,' of eternal expansion of an open universe.

But much sooner is the potential for self-destruction by nuclear winter or humanity-induced irrevocably-damaged ecosystem. Unless we have learned to modify our physical human structure and travel to the stars, as physicists Freeman Dyson and Frank Tipler suggest, our civilization--or its progeny--may be doomed to extinction. This would represent a catastrophe of indescribable proportions--unless wiser and more fortunate extraterrestrial life forms exist somewhere out there to pick up the torch.

Freeman Dyson and Frank Tipler on the Future: Few physicists have addressed, in so fascinating a way, the question of what the future holds for survival of the human species in the eons to come, as does physicist Freeman Dyson in *Infinite in All Directions.* Dyson gives expression to his thoughts on the survival of intelligence, its spread throughout the galaxy and the universe at large, and the physics of the processes involved.

In his Omega Point Theory, outlined in *The Physics of Immortality*, mathematician Frank Tipler explores the spread of intelligence through the universe to become the Omega Point God and assure us of resurrection, in more trouble-free form, through computer simulation.

Tipler notes our preoccupation with the past but suggests we should be *more concerned with the future since there is where most of reality lies*, the future, being some 30,000 times larger than the past. He asserts that you won't necessarily understand the future by reference to the past.

Is God's Existence Provable? Tipler's book, with its supporting mathematical explanations in the appendices, raises many challenging questions. In the October, 1994 issue of *Omni* magazine on "Science and Religion," Tipler, under the caption "Interview," answers some questions which have been raised.

One important question is whether God's existence is provable. Looking at the total reality, says Tipler, God "just falls out" in "an essential and intrinsic way" from a consistent physical analysis. He's not just artificially injected to fill an information gap. God truly represents the Omega Point, the "ultimate future."

Mind Has Taken Control of Matter: In his *Infinite in All Directions*, Freeman Dyson crosses over into metaphysics and theology, giving expression to the vision of future mind over matter. *Mind*, he says, has taken control of matter on this planet, "fulfilling a law of nature that requires this and makes the process unstoppable." And, if we waver, he adds, others will--or may have already done so. He observes that mind has taken three eons to control this planet and three more billion years may pass before it covers the galaxy.

Will Life and Intelligence Survive Forever? If, as Dyson hypothesizes, the universe is open and expanding forever, is it possible, he asks, for life and intelligence to survive forever and transmit information across the constantly expanding galactic distances? Life and intelligence, he feels are potentially immortal, with available information growing as the universe's temperature declines and free energy reserves diminish. He thinks that intelligence forms throughout the universe can forever preserve a communication network, exchanging ideas.

Life and Intelligence Molding the Universe to Its Own Purposes: The energy requirments for survival and communication, he calculates, are "surprisingly modest." New things will be happening, he says, and no matter how far we look into the future, there will be new information, new worlds to explore, a constantly expanding domain of life, consciousness, and memory. He sees a universe of life, growing in richness and complexity, surviving forever and communicating with its neighbors "across unimaginable gulfs of space and time." There is a substantial scientific basis, he notes, for hypothesizing that life and intelligence can succeed in "molding this universe of ours to their own purposes."

The Point at Which Science Ends and Theology Begins: What will mind choose to do when it controls the univese? That is a question, he notes, we cannot hope to answer. When mind has expanded its physical reach and its biological organizations "by human powers of ten beyond the human scale," we can no more expect to understand its thoughts and dreams than a Monarch butterfly can understand ours.

Mind can answer our questions only as God enigmatically answered Job out of the whirlwind. In contemplating the future of mind in the universe, he muses, "we have exhausted the resources of our puny human science. *This is the point at which science ends and theology begins."*

Speculations on the Future Must End in Fantasy: In his last chapter, "Butterflies Again," of *Infinite in all Directions,* Dyson observes that every speculation concerning humanity's long-range future must end in fantasy. Nobody can foresee the "quirks of history," and the "accidents of human wisdom and folly," that will determine ultimate outcomes.

Learning to Think in A New Way: But getting back to the concluding chapter of *Real Philosophy* on 'Tomorrow,' the authors ask whether humanity is to make its own future or whether it is already 'there' as his destiny. They cite Bergson's view, in *Creative Evolution*, that human consciousness is synonymous with invention and freedom, opening humankind to an evolution in which human life intervenes or even determines that process.

We often desire to 'know the unknowable' and allow imagination to roam as it extrapolates the present to conjure a picture of the future. And it may be that the future will announce itsef to us in a dream, suggest the authors, but we need to be able to extract the message from our own "wishful interpretations." Speaking of dreams and access to the future, it is pertinent to cite the work of physicist Fred Alan Wolf.

The Superspace of All Possibilities: This notion of our mind's possible access to the future is highlighted by physicist Fred Alan Wolf in *Parallel Universes, The Search for Other Worlds,* in which "reality" must be defined in terms of quantum terms which consists, he suggests, of a gigantic superspace, representing "the mathematical space of all possibilities." Wolf says we might think of this mathematical superspace as "the mind of God" and in which our minds float in "superspace."

Minds--Sensing Possibility Waves From Past & Future To Create Existence: This freely-associating mind, he says, is able to cross time barriers, "sensing the future and reappraising the past." As "time machines," our minds sense the flow of possibility waves from both the past and the future. And, Wolf says, in his view you can't have "anything like existence without this 'higher form of quantum reality.' "

The Most Significant Insight Since Advent of Quantum Theory: He believes that this insight into the workings of quantum physics is *the most significant insight into 'this strange landscape' since the advent of quantum physics*. If it is a stable hypothesis, he says, it will revolutionize our view of the world. It will say in effect that time is not a barrier. *The future exists now, and so does the past*. Finally, from this perspective, we see that we are both the "choosers of subsets of reality and the constructors of existence."

Any statement which attempted to express a consensus of scientists about religious and philosophical questions would miss the main point. There is no consensus among us.--Freeman Dyson

CHAPTER 60
CONCLUDING THOUGHTS

Biblical and Big Bang Creations in Conflict? Eastern theology speaks of the unity of the universe as cosmic mind. In conventional cosmology, Western theology speaks of God as spiritual and personal, Creator of the universe and human life, all powerful, and all knowing.

In quantum cosmology, physicists like Stephen Hawking, Heinz Pagels and Paul Davies speak of the universe as self-generated from the quantum-uncertainty-dominated, pre-big-bang, quantum mega-universe void, essentially a mathematical, pregeometric, quantum energy state of the potential of the laws of physics. The two approaches, Creation by God, and quantum self-generation, have been characterized by many as being in irreconcilable conflict, a view which contributes to promoting an atmosphere of hostility between science and religion.

Some Reconciliation: As we have noted, some religiously-oriented scientists, like astronomer-minister Hugh Ross, sponsor of 'Reasons To Believe,' an organization of religious scientists and their supporters, speak of the latest proofs of the Big Bang as evidencing the existence of the God of the Bible, citing in particular certain passages of the New Testament.

They question the quantum cosmology of quantum void self-generation of the universe as essentially metaphysical speculation while, for the most part, accepting Big Bang cosmology as compatible with and corroborative of the Biblical Creation scenario.

The *Creationists*, of course, see the Big Bang physics as incompatible with the biblical creation scenario, based on inerrancy of the bible, including the interpretation of the days of creation as 24-hour earth-rotation days rather than vast time epochs based on astronomical or cosmic-time 'days.' (The earth's rotation day was longer in the past and will be shorter in the future, the result of the moon's gravitational pull on earth.)

Reconcilability of God and Quantum Void Creation: In considering Big Bang cosmology, many are willing to accept the notion that the biblical account is not intended as a handbook on physics but allegory, an account of divine involvement in the creation in terms understandable to primitive society.

But there are suggestions that so-called self-generation from the primeval mega-universe quantum nothingness state, devoid of time, matter, forces, and dimensionality, might be viewed as a metaphor for the biblical creation *ex nihilo* by a God who exists outside of time and space. Obviously that primeval quantum void represents a powerful *creative vitalism*, which generates a physical particle or mini-universe of unified force, space and time, that unfolds into our dimensional universe of space, time, and matter, further unfolding into highly personal human life and mind.

Cosmology and God: Physicist Freeman Dyson characterizes this highly personal human mind as *an extension of God's mental apparatus* so that *human personal* nature is a derivative of God's personal nature. The individual human mind is growing in cognitive capacity as a part of group mind, humankind and society as a whole, in its power of comprehension and God-derived creative power.

Mathematician Frank Tipler asserts that this growth of societal mind will spread throughout the universe to become the *Omega Point* omnipotent and omniscient God--who exists *now* too because of the relativistic and quantum coexistence of all time.

As we have seen, some religious scientists cite the relativistic Big Bang cosmology as confirming evidence of the Biblical Creation. Others suggest quantum genesis creation from the primordial void as a metaphor for Divine Creation *ex nihilo,* God using quantum mechanics to create the universe.

Use of the Term 'God': In the chapter, *What About God?,* in his *Dreams of a Final Theory*, Nobel Laureate Steven Weinberg notes that if there were anything we could discover in nature that would give us *some special insight into the handiwork of God*, it must be the final laws of nature. So, he observes, it was understandable that Stephen Hawking should discuss the laws of nature in the context of "the Mind of God."

Weinberg notes theologian Paul Tillich's observation that among scientists only physicists seem capable of using the word 'God' without embarrassment. Some people, he notes, have views of God that are so *elastic* that it is inevitable that they will *find God* wherever they look for Him. The word 'God,' he notes, can be given any meaning we like.

Weinberg's Definition of God: But if the word 'God' is to be of any use, he suggests, it should be taken to mean "an interested God, a creator and lawgiver who has established not only the laws of nature and the universe but also standards of good and evil, some personality that is concerned with our actions, something, in short, that is appropriate for us to worship." Our chapter on *Efforts to Define God* chronicles many of the attempts at describing God. This performance specification for God by Weinberg is not a bad start as a functional approach to defining God.

Using God As an Abstraction: He is reminded that scientists and others sometimes use the word 'God' to mean something *so abstract and unengaged* that such a God is hardly distinguishable from "the laws of nature." Weinberg cites Einstein as once saying that he believed in "Spinoza's God who reveals Himself in the orderly harmony of what exists, not in a God who concerns himself with fates and actions of human beings."

God As 'Order' or 'Harmony': What difference does it make to anyone, asks Weinberg, if we use the word 'God' in place of 'order' or 'harmony,' except perhaps to avoid the accusation of being an atheist--having no God. Though we are free to use the term 'God' in that way, he observes, it seems to him that it makes the concept of God not so much wrong as *unimportant*.

God and Sensibilities: To find meaning, this book addresses many subjects and, in the process, ventures deeply into physics which governs the behavior of the universe and into both theology and philosophy, especially philosophic thought about religion. As philosophers have observed, believers in aspects of God, whether theists, deists, pantheists, nontheists or agnostics, <u>all philosophize about religion and God</u>, arriving at a vast array of differing conclusions. And these differences involve profound and fragile sensibilities. People and nations have fought and died for their religious convictions.

Without God, the Universe Lacks Authentic Meaning: For myself, as any reader of this book soon realizes, my concerns for defining God run deep. This book addresses the human search for meaning--of the universe and of life. In *The Creator and the Cosmos*, astronomer-minister Hugh Ross notes his view that *without God* any meaning of the universe, and life, is *artificial* and contrived, even though it may serve some pragmatic end. There is likely a broad consensus that without God the universe has no *authentic* meaning because there would be *no authenticating source*, no *author*.

Reader Reactions: One early reviewer of this book read it as biased in favor of theism, that is, suggesting it goes too far in favoring a scientific foundation or underpinning for, and by implication, my own personal belief in, the existence of a theistic personal God. But another reviewer perceived an *ambivalence* on my part, the absence of an unequivocally-expressed conviction favoring a theistic personal God. A third questioned my seeming use of the *metaphor* to equate or at least reconcile the physics of relativity and a quantum-generated <u>c</u>reation universe with the biblical Creation. So what *is* my personal view of God?

Mother and God: Earlier in my life, I had rather firm convictions of God's existence because Mother *knew* God existed-- and that was enough for me. Who was I to question the authority figure in our family. And besides, her mother had assured her that God was watching her every move.

But as time passed, my unquestioned acceptance and certainty diminished as there unfolded the panorama of an often miserable world both of human-induced mindless slaughter, many killings in the very name of God, and nature-sponsored human disasters of even greater magnitude, such as by floods, famines and plagues that regularly visited Earth, to say nothing of periodic global destruction from periodic comet or asteroid strikes.

Imperfection on a Vast Scale: Certainly, imperfection on such a vast and cruel scale could not be the product of an omnipotent, omniscient, caring and loving God. Or could they? The picture was not made any brighter by the untimely, inexplicable, and senseless deaths of loved ones. It is virtually impossible to question the fact of the existence of a higher power, something we can call 'God.' But as theologists point out, there is a conspicuous gap between what religion promises and reality delivers, suggesting that our knowledge of God is limited at best. Several thousand years later, we are as confused as was Job after God's enigmatic response to Job's desperate request for explanation.

Sunday School and God: Many have been the discussions over God's existence and the definition or meaning of God in the King's Men's Class of Alexandria, Virginia's Baptist Temple. Class leader and prominent attorney Cullen Jones, with patriarchal demeanor--and voice, repeatedly assures us of God's existence, even though, at times, he may find difficulty reconciling this with the actions of the miscreants he regularly defends. His instruction style is to start the lesson with a general discussion of some aspects of daily life and then casually slip into the subject of the lesson.

Master Mason and decorated veteran Bob Bullock is always there to defend God's existence and accept His at times seeming indifference to human or natural disasters, when some, myself included, wonder about it all. Down through the years, Pastors Wayne Yawn, Mike Allen and scholarly Nancy Foil's sermons about God's existence, availability to humans, and the power of faith have, of course, been faithfully addressed, and reassuring-- at the time.

Then there is business executive and community leader Sam Dickens, who didn't let many meetings go by without questioning just what is meant by the term 'God' in particular contexts. The class, of course forgets how the question was previously answered and each meeting brings a somewhat different approach to the question. But then, this class is not the only group in the world with this problem of defining God and explaining his enigmatic role in human life. Sam's repeated question about explaining God has not only intrigued me but, I found, preoccupies many of the world's gifted theologians, and, in a way, influenced the content of this book.

God as Undefinable? There are many, theologists included, who deny we can define God--because He is beyond defining. As theologian A.J.Heschel puts it, *God is ineffable*, beyond comprehension or description. Whether we will, as some physicists suggest, some day find or identify a spiritual force, and make Divine contact, is not altogether unthinkable but in the meantime each human being inevitably draws his own portrayal of God.

"Cannot Agree on...Who God Is": You recall our earlier reference, in the chapter on the Changing God, to the World-Wide Church of God publication, *Groping In The Light.* Author John Halford traces the "new way of looking at the world," brought about by the startling findings of science in this century. After neatly summarizing some of the remarkable scientific advances, such as quantum and relativity theories, he gives plaintive expression to frustrations over *lack of any corresponding progress in revealing information of God.*

"It is a sad commentary, he observes, that, after 2,000 years of Christianity and some 3,500 years of biblical tradition, philosophers and theologians still *cannot agree on even the most basic explanations of who God is, and his purpose for creating man.*" Halford goes on to note that what is known of God derives from Biblical revelation, a view of course, shared by Western theology.

The Ultimate Question: The ultimate question--about God, says theologian Abraham Heschel, in *Man Is Not Alone*, begins with the ineffable--the inexpressible. The question about God inquires what all things stand for, the transcendental; the mystery shared by all men and a dimension of all existence. In this view, all meaning tracks back to the existence of God.

My Intention to Influence Belief? Nothing would be more gratifying than by this book somehow to have provided a base of knowledge which would promote faith--and the reward of belief. However, my *intention* in undertaking the book was not evangelical but simply to explore *reality* as derivable from the expressions of the world's great wisdom, without preconceived notions, and hopefully *without intentionally leading in any preferred direction*.

For the convinced theists with unwavering faith, nothing here will have separated them from that faith--indeed, prior convictions may have been reinforced. For those uncertain of God's existence or meaning, perhaps they will have experienced a broadened framework for further reflections. Convinced *nontheists* may or may not have found in these materials further grounds for their point of view but some uncertainty may have crept in about foundations of nontheism.

My Own Experience: In my own experience, at many times and places, have I seen powerful manifestations which *overwhelmingly demonstrate the existence of God*, or at least the work of some higher power. But sadly, on too many occasions, have I observed His manifest absence, inattention or indifference--where only God could have make the critical difference.

Physicists and God: As we have seen, there is no monolithic physicists' view of God. While there are many physicists who are self-declared nontheists, there are others who are devout believers. But, interestingly, even the nontheists address the existence of *something* to which the term 'God' is associated, though the references may not qualify as the personal theistic God.

Physicist Paul Davies, for example, in *God and the New Physics* and *The Mind of God,* speaks of the ultimate cause or source of everything--ultimate regression, the logic of mathematics, or the 'good,' as something one may wish to call God. Physicists sometimes cite the principle of Occam's Razor as possibly supporting God as the ultimate cause, since that is by far the simplest of the various increasingly more complicated alternative explanations for the creative source.

In Infinite in All Directions, Physicist Freeman Dyson speaks of the Socinian God who is growing with the universe. And in *The Physics of Immortality*, mathematician Frank Tipler presents his theory of the Omega Point God, the final point of growing universal intelligence, who exists *now,* says Tipler, because of the coexistence of all time.

Theoretical physicist Stephen Hawking, in *A Brief History of Time*, asks what it is that brings the abstract mathematical equations to the physical reality of our universe and accounts for the master law of physics in which all physics is subsumed. He suggests that, knowing the answer, we would *know the Mind of God.*

Coping with These Different Portrayals of God: As we noted earlier in the book, one of the contributing problems in defining God is the definition of "universe,"which raises questions of God's transcendence. If our universe is one of quantum cosmology's large or even infinite number which derive from the primeval quantum void and from the "many worlds" quantum uncertainty interpretation, then a 'transcendent' God must transcend this totality of universes, not just ours.

But Dyson's Socinian (imperfect but) growing God is a God of this (our) universe. The same is true of Tipler's Omega Point growing God, even though Tipler defines the universe as everything there is, so that "universe" includes God and therefore raises 'transcendence' questions. Hawking's Mind of God also deals with the God of this universe even though he speaks of the quantum void generating other universes and "baby" universes. In this context many universes must have come and gone through eternity (whatever that means since there is no "time" in the pre-Big- Bang quantum void).

Was God "growing" in each of these universes? As evident, the picture of God growing in perfection--as in Dyson's Socinian God, and Tipler's Omega Point God-- in quantum cosmology's infinite number of universes, presents conspicuous problems of transcendence, omnipotence, omniscience, and varying levels of finite limitations.

Questioning an All-Caring Personal God: But getting back to our universe, considering the manifest prevalence of evil in the world, the devastating natural calamities, Steven Weinberg , Nobel Laureate for his work in unification theory, does not see the universe as really good evidence of the existence of God, or at least an all-powerful, all-caring personal God whose demonstrated preoccupation is to cherish and protect us at all times. On the other hand, neither does he question Steven Hawking's use of the metaphor of the Mind of God in the finding of a master law of physics--unification theory.

But Weinberg isn't attracted to any analogy between God and the laws of physics, or the forces of nature. Yet scientists like astronomer-minister Hugh Ross do a creditable job of showing just such a relationship. While I have difficulty in seeing the absolute unquestionable existence of God as derivable from the Big Bang and the anthropic principle, I find myself closer to Ross's view than Weinberg's, despite Weinberg's superb credentials and despite P.W. Atkins charming account of the creation as "without outside intervention."

In *Creation Revisited*, Oxford chemistry professor and nontheist P.W. Atkins declares that mathematics, existing by necessity, as some assert for God's existence, generates the physical universe, *without any outside intervention whatever.*

But even in his denial of outside intervention one might still 'see' mathematics, or the logic of mathematics (in his view the ultimate regression or primal cause) as a metaphor for God. Einstein and other physicists have given expression to their view of God in terms of mathematics or as a mathematician.

Interestingly, while Atkins denies "outside intervention," he also provides a metaphor for the laws of physics. He says that the so-called laws of physics is a name we use to describe the natural "behavior" of the universe--how the universe conducts itself. But "behavior" begins to sound more like what we attribute to God. The physics laws-equals-behavior metaphor seems to reintroduce God.

God As Existence: Theologian Paul Tillich, who noted the freedom with which physicists bandy about the term 'God' in their references, rejected the reference to God as a *being*, supernatural or otherwise, rejecting the term as demeaning. For him, God was *'being' or 'existence' itself.*

In one fell swoop this all-encompassing definition of God as "being itself," renders *nonacceptance of God's existence virtually impossible*. How can one challenge the existence of 'existence'? Tillich asserted that God must be invoked even in the atheist's very denial of God.

But not every theologian is comfortable with Tillich's God as 'being" or 'existence' itself, because of the pantheistic implications--reminiscent of Spinoza's God as the totality of everything, including the universe and human life. Another objection is that, in a sense, that sweeping definition may be thought of as embracing philosopher and mathematician Leibniz's "bestpossible world" concept, criticized as an acknowledgment of the finiteness of God, a limitation on His omnipotence.

Existence and Personhood: But Tillich's definition of God as 'existence' might better explain the limitations of the world without denigrating God or trying to rationalize His failure to prevent evil or restrain the life-destructive natural forces. The totality of existence would understandably include everything, the good and the bad, the primitive and the advanced, the mundane and the sublime. Perfection is understandable only in the context of imperfection. Of course, God as 'existence' puts in jeopardy the working of miracles except as seeming 'miracles' have under-lying *natural* explanations.

The Natural vs. the Supernatural: On the question of the supernatural vs. the natural, in *The Creator and the Cosmos* astronomer-cosmologist and minister Hugh Ross points out that some of the New Testament acts of Jesus, thought of as miracles, are explainable in the *extra dimensions,* the existence of which we now recognize and, of course, available to God, even to a natural God.

It has been suggested that in the light of scientific progress, and the knowledge about the universe it has brought, the term 'supernatural' refers not so much to acts *in defiance of nature* but to those aspects of nature beyond our present comprehension, involving knowledge with which we are presently unfamiliar.

Everything Needed Built into Nature? As discussed earlier in the book, physicists sometimes speak of the 'natural' rather than 'supernatural' God, implying that nature contains within itself everything needed for affirming God's existence, including the conduct of a working universe, with self-corrective mechanisms.

In this context, God may be seen *as* nature rather than something 'above' or 'outside' of nature, some-thing compatible with Tillich's God as 'existence.' In this context, Jesuit astronomer William Stoeger notes that we need to be able to *see how God could be working with the natural processes revealed by contemporary science.*

A Distinction Without a Difference? One may well ask what is the distinction, if any, between a personal theistic God and a quantum phenomenon, a power of mathematics, or unified physics, or logic, which go on to do exactly what we are used to thinking of God as doing, producing human minds which dwell on abstract concepts like reverence and holiness? From acorns great oaks grow. The end product, something *personal*, namely human life, suggests a *personal* creative vital source--whether we call that source God, mathematics, logic, or any other descriptor.

The Ultimate Regression: In this context, then, the causal regression ends with something which for all intents and purposes we have described under different names. In *The Mind of God*, physicist Davies follows this line of reasoning to regress back to physics or the mathematics of logic, which one might call God. But to astronomer-minister Hugh Ross's disappointment, he stops just short of the God of theism.

In such a setting, of God functioning through or as the laws of physics, and physics meaning the behavior of the universe, terms like theism, deism, pantheism or even atheism begin to lose some of their distinguishing significance. As noted earlier, physicist Steven Weinberg, in *Dreams of a Final Theory*, criticized the use by scientists of the word 'God' to mean something closer to "the laws of nature." Some, myself included, may have difficulty with his peremptory disavowal of that analogy.

Theologian Paul Tillich's God as 'being" or existence also comes close to equating God with the natural state of things-- nature. Tillich's God as existence does not foreclose either the personhood or spirituality of God and derivative personhood of humankind.

The ultimate composition or ingredient of 'existence,' as contrasted with sheer nothingness-- if there is such a thing-- is energy. We are reminded that Einstein's equation for the universe (energy equals mass times the speed of light squared) means that the universe consists of forms of energy, the forces, and space and time. We ourselves, in quantum terms are quantum system forms of configured energy.

Cosmic mind, human mind-- a derivative from the mind of God as Freeman Dyson suggests--are energy phenomena. Quantum theory tells us that underlying energy is *information-mind*, suggesting that the minds of God and derivative human mind are related to each other through energy relationships.

Demonstrating God's Existence? Speaking of scientists and God, we are reminded of the arguments of astronomer-minister Hugh Ross, in *The Creator and the Cosmos* and other publications, who makes a persuasive case for God's existence as demonstrated by the extraordinary coincidences in the life-promoting constants of physics-- the anthropic principle.

As we have seen, Ross also sees the work of God in the Big Bang, recently confirmed by several space probes such as the NASA COBE satellite finding of primordial temperature ripples in the fabric of spacetime. While he opts for the supernatural God, he makes a good case for the proposition that miracles have natural explanations, mostly associated with multiple dimensionality, a natural phenomenon.

Something Profound Emerging: We have noted the increasing interest of physicists in bringing God into discussions on cosmological physics and mathematics. It appears that something profound and compelling is emerging at this stage of human progress, something that requires us to rethink our evaluations of the past. These evaluations tended to *compartmentalize* physics from theology--and from God, something which Hugh Ross declares as a *scientific chauvinism*--equaled only by similar divisive counter-chauvinisms. Chauvinisms on the part of those theologians who reject the existence of demonstrated science-theology interrelationships.

Evolving, is the appearance of significant and meaningful relationships between the laws of physics, God, and human life. As theologian-physicist John Polkinghorne puts it, *quantum theory is shot full with God-human life implications*. Scientific and theological explanations about God and the universe are becoming rationally less distinguishable. Physicist Paul Davies, in *God and the New Physics and in the Mind of God,* notes the urgency of reevaluating our ideas on the God-physics relationship.

Emerging Scientific and Religious Convergence? Emerging, is awareness that *something* which, for want of a better description, is called 'natural processes,' is doing just about everything which has been attributed to the supernatural God. Personhood and exquisite abstractions--like love, holiness and religious sensibilities-- emerge through the growth of order. As physicist Richard Feynman so aptly noted, the atom and the abstractions inherent in human thinking are *equi-distant from God.*

The primeval mega-universe quantum nothingness state, which generates the universe may, arguably at least, be analogized with the biblical creation *ex nihilo* by a God who is both immanent and transcendent, existing outside of the time and space of our universe--just as does the mysterious quantum void.

Obviously that primeval void preexisting the Creation, represents a powerful *creative vitalism,* which generates a physical particle or mini-universe of unified force, space and time, which unfolds into our dimensional universe with space, time, and matter. That universe further unfolds into highly personal cognitive life and mind capable of exquisite and highly personal abstractions, something we associate with God.

Quantum Theory, Abstractions, Personhood, and God: Thus, emerging from the quantum, comes the physical universe, life, consciousness, self-awareness, and mind, equipped to comprehend abstractions that signify personality and personhood, elements that we regularly associate with God and the Divine Creation.

So, despite the resistance of some, one may well see this creative process, arising from a vitalized quantum source, as a metaphor for God. Abstractions and ennobling spiritual concepts, whether we name the source God or something else, represent ideas which, to a large extent, define the human dimension in its relationship with God.

> After close to two centuries of passionate struggle, neither science nor faith has succeeded in discrediting its adversary. On the contrary, it has become obvious that neither can develop normally without the other. And the reason is simple. The same life animates both. Neither in its impetus nor its achievements can science go to its limits without becoming entangled with mysticism and charged with faith. --Pierre Teilhard de Chardin

EPILOGUE

BELIEF IN GOD, FAITH, AND PRAYER

Prayer to the Undefinable God Whose Existence Is Not Provable:

Readers of this book, and of other books that deal with God, the Universe, and human life, seem inclined to ask:

> What is the logic behind prayer to a mystical God that defies human understanding and whose very existence theologians acknowledge as not capable of proof? And as well, the related question, if faith eludes one, is prayer self-deception?

This thought-question occurs to most churchgoers at some time in their lives, such as upon the death of a loved one or perhaps even in the midst of some prayer. It is difficult to find "authoritative" answers to such questions. But in *The Quest for God, A Personal Pilgrimage,* Paul Johnson directly addresses this question in his chapter: "Talking to the God we do not know and cannot prove exists."

As a well-known British historian of the Catholic faith, a journalist and frequent contributor to leading British and American newspapers, he has seen fit to venture into writing in the field of religion and wishes to share his thoughts with others.

The Crucialness of Prayer: We may not be able to understand God, he observes, or even to believe in him, confused by "contradictions and improbabilities of faith.... But we can all pray, the one resource that can never be taken away," though we may be penniless, imprisoned, stricken, even bound and gagged. "It is the last weapon of the weak, the starving, the helpless, the puzzled, the unsure.... *the most powerful weapon of all.*" He suggests that it is not only the Pope that has a private line to God, each of us has.

God's Existence Independent of Your Belief: You may not believe in God, but that *does not prevent God believing in you.* God's existence is independent of your belief in him, he notes. So it makes good sense for the unbeliever to pray for faith, just as it makes sense for the incomplete believer to pray for more faith, *the greatest possession* any of us can have.

So to pray for faith is the most ambitious of all prayers, "you are asking God for the key to everything else" It makes sense, then, that "those who should pray hardest are precisely the agnostics and atheists and doubters."

The Nature of Prayer: He goes on to discuss the nature of prayer, noting the particular beauty and importance of the 150 psalms, some written as long as 3,000 years ago, "resonant," "powerful," and used by Jesus in his discourses. "There are countless prayers, to be said, accumulated over thousands of years "during which the Judeo-Christian tradition has shaped itself." One should also make use of spontaneous prayer, "doing privately what the Quakers do so impressively in public."

In the End: And in the end, that is what the love and worship of God is all about, ''to turn our minds and if possible our bodies too, away from earthliness to perfection, from doubt to certitude, from the self to goodness, and from the flesh to the spirit."

Prayer is always an answer even it is not invariably the complete answer. There are countless prayers to be said, accumulated over thousands of years during which the Judeo-Christian tradition has shaped itself. They exist for all occasions, all temperaments....A prayer written by the dictates of one's own heart must be especially dear to God.--Paul Johnson

ACKNOWLEDGMENTS

Introduction: Novelist, poet and scientist Johann Goethe observed that *Everything has been thought of before, but the problem is to think of it again.* Much if not most of the information and ideas expressed in this material are traceable to previously expressed scientific, theological and philosophical views, The element of the author's originality lies primarily in the composition and juxtaposition of these materials and resulting perspectives.

Special Tribute: is due for use of concepts of the many authors whose thoughts have been called upon to present the mosaic represented by this book. Particular credit is due certain authors for their contribution to the field of knowledge from which ideas have been drawn. Their books are must reading for achieving perspective on our subjects:

To chemistry professor P.W. Atkins, *Creation Revisited*, W. H. Freeman & Company, for his neatly written, fascinating ideas on the role of mathematics in the creation.

To physicist Paul Davies, *The Mind of God*, Touchstone, Simon & Schuster, for his erudite and analytical treatment of theological concepts in scientific contexts and innovative thoughts on the interface between science and theology.

To psychiatrist Viktor Frankl, *Man's Search for Meaning*, Simon & Schuster, for contributions to the understanding of the source of the deep driving force in the search for meaning.

To physicist Harald Fritzsch, *The Creation of Matter, The Universe from Beginning to End*, Basic Books, for his thoughts on science and religion and the relationships of God and meaning.

To physicist George Greenstein, *The Symbiotic Universe*, for his challenging and stimulating treatment of the anthropic principle.

To theoretical physicist Stephen Hawking, *A Brief History of Time*, Bantam, Books for his unique contributions to quantum cosmology and his concept of the association of a final theory with the Mind of God.

To scientist Nick Herbert, *Quantum Reality*, Doubleday, for his fascinating treatment of the significance of quantum theory to our notions of reality.

To theologist Abraham Joshua Heschel, *Man Is Not Alone*, *A Philosophy of Religion*, Farrar, Straus & Giroux, Inc., for his sublime contributions to understanding God-human relationships.

To John H. Hick, *Philosophy Of Religion*, Prentice-Hall, for his exceptional materials relating to the philosophy of religion and the so-called arguments for God.

To Physicist David Layzer, *Cosmogenesis*, Oxford University Press, for his emphasis on biological aspects of the evolving universe and his erudite treatment of physics and free will.

To Jacob Needleman and David Applebaum, *Real Philosophy--an Anthology of the Universal Search for Meaning,* Penguin, for their analytical approach to the pursuit of meaning.

To physicist-philosopher Heinz Pagels, *Perfect Symmetry*, Bantam Books, for his beautiful and philosophical analysis of progress in unification theory and intellectual rewards in the pursuit of scientific truths.

To Hubert Reeves, *Atoms of Science*, MIT Press, for his fascinating thoughts on beginnings and evolution of life.

To astronomer-minister Hugh Ross, *The Creator and the Cosmos*, How the Greatest Scientific Discoveries of the Century Reveal God, NavPress; and *The Fingerprint of God*, for his moving and persuasive arguments for science as demonstrating the existence of God.

To astronomer and space scientist Carl Sagan, *Cosmos*, *The Cosmic Connection*, and *Dragons of Eden*, *Speculations On The Origin Of Human Intelligence,* Random House, for his fascinating sweep through cosmology and the development of the human mind.

To scientist Michael Talbot, *Beyond the Quantum*, Bantam Books and *The Holographic Universe*, Harper Collins, for simplifying quantum theory for the lay public and exploring quantum reality.

To mathematician-physicist Frank J. Tipler, *The Physics of Immortality,* Doubleday, for his extraordinary efforts to provide physics explanations for theological concepts.

To physicist Steven Weinberg, *The First Three Minutes*, Bantam Books, and *Dreams of a Final Theory,* Pantheon Books, Random House, for his candid insights into relationships between unification theory, and cosmic purpose.

To physicist Fred Alan Wolf, *Taking The Quantum Leap*, Harper & Rowand *Parallel Universes*, Simon & Schuster, for his profound insights on how quantum effects and parallel universes affect the human mind.

BIBLIOGRAPHY

The publications listed below were valuable sources in the search for scientific and metaphysical conceptual materials relevant to the ideas the author wished to explore. Many of the publication sources of ideas called upon here also represent examples of superb classic literature, quite apart from technical content. While similar or identical ideas recur as one goes from one book and author to another, somehow each reflects some novelty and provides reinforcing experience.

Mortimer J. Adler, *Six Great Ideas*, MacMillan
Karen Armstrong, *A History of God*, Ballentine Books (Alfred A. Knopf, Inc.)
P.W. Atkins, *Creation Revisited*, W. H. Freeman & Company
A.J. Ayer, *Philosophy In The Twentieth Century*, Vintage Books, Random House
John D. Barrow & Joseph Silk, *The Left Hand of Creation*, Basic Books, Inc. (Oxford University Press)
Richard Bergland, *The Fabric Of Mind*, Penguin Books
John Boslough, *Stephen Hawking's Universe*, Avon Books
Nigel Calder, *Einstein's Universe*, Penguin
Rudolph Carnap, *Philosophic Foundations Of Physics*, Basic Books
Eric Chaisson, *Cosmic Dawn*, Penguin Books
Paul Davies, *The Mind of God*, Touchstone, Simon & Schuster
Paul Davies, *God & The New Physics*, Simon & Schuster
Paul Davies, *Superforce*, Simon & Schuster
Paul Davies, *The Cosmic Blueprint*, Touchstone, Simon & Schuster
Paul Davies and John Gribbin, *The Matter Myth*, Dramatic Discoveries that Challenge Our Understanding of Physical Reality, Touchstone, Simon & Schuster
Paul Edwards and Arthur Pap, *A modern Introduction To Philosophy*, The Free Press
Albert Einstein, *Out of My Later Years*, *Ideas And Opinions*, Bonanza Books
Clifton Fadiman, *Living Philosophies*, Doubleday
Timothy Ferris, *Coming Of Age In The Milky Way*, Anchor Books, Doubleday
Richard Feynman, *Surely You're Joking, Mr. Feynman*, Bantam Books
Anthony Flew, *An Introduction To Western Philosophy*, Thames And Hudson

Viktor Frankl, *Man's Search for Meaning*, Simon & Schuster
Viktor Frankl, *Man's Cry For Meaning,* and *The Unconscious God,* Simon & Schuster
Harold Fritzsch, *The Creation of Matter, The Universe from Beginning to End*, Basic Books
S.E. Frost, *Basic Teachings of the Great Philosophers*, Doubleday
John Gribbin, *In Search of the Big Bang*, Bantam Books
John Gribbin and Martin Rees, *Cosmic Coincidences,* Bantam Books
Editor, Nahum N. Glatzer, *The Dimensions of Job*, Schocken Books
James Gleich, *Chaos*, Penguin Books
George Greenstein, New York; William Morrow, *The Symbiotic Universe.*
John Gribbon and Martin Rees, *Cosmic Coincidences*, Bantam Books
Edward Harrison, *Masks Of The Universe*
Gerald Hawkings, *Mindsteps To The Cosmos*
Stephen Hawking, *A Brief History of Time*, Bantam Books
Stephen Hawking, *Black Holes And Baby Universes*, and Other Essays, Bantam Doubleday Dell Publishing Group, Inc.
Robert Hazen and James Trefil, *Science Matters*, Doubleday
Nick Herbert, *Quantum Reality*, Doubleday
Abraham Joshua Heschel, *Man Is Not Alone*, *A Philosophy of Religion*, Farrar, Straus & Giroux, Inc.
John H. Hick, *Philosophy Of Religion*, Prentice-Hall
Paul Johnson, *The Quest for God*, Harper Collins
Thomas E. Katen, *Doing Philosophy*, Prentice-Hall
David Layzer, *Cosmogenesis*, Oxford University Press
Michael Lemonick, *The Light At The End Of The* Universe, Princeton University Press
C.S. Lewis, *Mere Christianity*, Macmillan Publishing
Alexandria Luria, *Lectures On Language And Cognition*, Wiley
John Macquarie, *Existentialism*, Penguin Books
Frank S. Mead--Samuel S. Hill, *Handbook of Denominations*, Abigdon Press
Jack Miles, *God: A Biography*, Knopf
Jacob Needleman and David Applebaum, *Real Philosophy--an Anthology of the Universal Search for Meaning,* Penguin
Jacob Needleman, *The Heart Of Philosophy*, Bantam Books
Needleman & Davis, *On The Way To Self Knowledge*, Knopf
Heinz Pagels, *Perfect Symmetry*, Bantam Books,
Heinz Pagesls, *The Cosmic Code,* Simon & Schuster
F. David Peat, *The Philosopher's Stone*, Bantam Books
F. David Peat, *Superstrings*, Contemporary Books, Inc.

"Reasons to Believe" publication--Hugh Ross
Roger Penrose, *The Emperor's New Mind, Concerning Computers, Minds, and The Laws of Physics*, Oxford University Press
Hubert Reeves, *Atoms of Silence*, MIT Press
Hugh Ross, *The Creator and the Cosmos, How the Greatest Scientific Discoveries of the Century Reveal God*, NavPress; and *The Fingerprint of God*
John K. Roth, *Problems Of The Philosophy Of Religion*, Chandler Publishing
Bertrand Russell, *A History Of Western Philosophy*, Simon & Schuster
Gilbert Ryle, *The Concept of Mind*
Carl Sagan, *Cosmos*, Random House, and *The Cosmic Connection* and *Dragons of Eden, Speculations On The Origin of Human Intelligence,* Random House
Roger Scruton, *From Descartes to Wittgenstein*, Harper & Row
R. C. Sproul, *Now That's a Good Question*, Tynedale House Publishers, Inc.
Michael Talbot, *Beyond the Quantum*, Bantam Books
Michael Talbot, *The Holographic Universe*, Harper Collins.
Paul Tillich, *Systematic Theology, Dynamics Of Faith,* Harper
Frank J. Tipler, *The Physics of Immortality,* Doubleday
James Trefil, *The Dark Side Of The Universe,* Scribners
Mary Warnock, *Existentialism*, Oxford University Press
Jefferson Haynes Weaver, Editor, *The World of Physics* (3 vols.)
Steven Weinberg, *The First Three Minutes*, Bantam books and *Dreams of a Final Theory,* Pantheon Books, Random House
Michael White and John Gribbin, *Stephen Hawking, A Life in Science*, Penguin Books USA, Inc.
Alfred North Whitehead, *Science And The Modern World*
Ludwig Wittgenstein, *Philosophical Investigations,*
Fred Alan Wolf, *Taking The Quantum Leap*, Harper & Row
Fred Alan Wolf, *Parallel Universes*, Simon & Schuster Inc
Worldwide Church of God publications such as *God Is....*
Robert Wright, *Three Scientists and Their Gods*, Harper & Rowe
Omni Magazine: Omni Publications, International Ltd., 1965 Broadway, New York, N.Y. 10023-5969. *The Enigma of Distance*, George Zebrowski, Nov. 1992; Science and Religion, October 1994; Margaret Wertheim; *Interview with Frank J. Tipler*; October 1994, *Is Science Rational*, George Zebrowski, June, 1994; *The Consciousness Wars*, Robert J. Killheffer, October 1993
Discover Magazine, various articles: Walt Disney Magazine Publishing Group Inc., 114 Fifth Ave, New York, New York 10011-5690

THE AUTHOR

Science: The author's early interest in science was heightened by his work as the Grants Administrator of the National Science Foundation, further intensified by exposure to extensive space science materials while involved in the 'acquisition process' for NASA Apollo and various space science programs and projects.

An abiding interest in cosmology, and his science environment in the National Science Foundation and NASA, led the author to wonder about different views of the origin of the universe. How and why did primordial quantum phenomena transition into fields, forces, dimensionality, and, ultimately, into human mind which somehow resonates with the mathematical substructure of the universe.

Philosophy and Theology: An interest in philosophy was encouraged by his study, as a Law School student, of jurisprudence, of the philosophic foundations of abstractions like justice and truth, which later matured into full-scale study of the history and foundations of philosophy. A fascination with theology, religious thought, and God-related concepts had its foundation in exposure to a succession of religious experience in various denominations.

Writing: An inclination toward research and writing derived from legal research and disciplined writing for various university law publications, writing for and editing the Federal Bar and other professional Journals, preparing proposed Federal legislation, and regulatory material for the Federal Register.

Lecturer: Having had many 'lecturer' assignments in agencies such as the U.S. Army, GSA, National Science Foundation and NASA, the contents of the book are set up to promote its use as an interesting tool of instruction in cosmology, physics, and theology.

Government Service: Included U.S. Army and Army Reserve Judge Advocate, various responsible positions in the Treasury Department, GSA, National Science Foundation, NASA, and a Congressional study commission. Among his recognition-for-merit awards was the Arthur Flemming Civil Service Award awarded annually to the "Ten Outstanding Young Men in Federal Service." The author holds a Juris Doctor degree from Boston University and is a member of the Bar of the United States Supreme Court.